W9-CUK-404

THE AQUEOUS PHASE BEHAVIOR
OF SURFACTANTS

COLLOID SCIENCE

Editors
R. H. Ottewill and R. L. Rowell

In recent years colloid science has developed rapidly and now frequently involves both sophisticated mathematical theories and advanced experimental techniques. However, many of the applications in this field require simple ideas and simple measurements. The breadth and the interdisciplinary nature of the subject have made it virtually impossible for a single individual to distil the subject for all to understand. The need for understanding suggests that the approach to an interdisciplinary subject should be through the perspectives gained by individuals.

The series consists of separate monographs, each written by a single author or by collaborating authors. It is the aim that each book will be written at a research level but will be readable by the average graduate student in chemistry, chemical engineering or physics. Theory, experiment and methodology, where necessary, are arranged to stress clarity so that the reader may gain in understanding, insight and predictive capability. It is hoped that this approach will also make the volumes useful for non-specialists and advanced undergraduates.

The author's role is regarded as paramount, and emphasis is placed on individual interpretation rather than on collecting together specialist articles.

The editors simply regard themselves as initiators and catalysts.

1. J. Mahanty and B. W. Ninham: Dispersion Forces.
2. R. J. Hunter: Zeta Potential in Colloid Science.
3. D. H. Napper: Polymeric Stabilization of Colloidal Dispersions.
4. T. G. M. van de Ven: Colloidal Hydrodynamics.
5. S. P. Stoylov: Colloid Electro-optics.
6. R. G. Laughlin: The Aqueous Phase Behavior of Surfactants.

THE AQUEOUS PHASE BEHAVIOR OF SURFACTANTS

ROBERT G. LAUGHLIN

The Procter and Gamble Co.
Miami Valley Laboratories
PO Box 398707
Cincinnati
OH 45239-8707
USA

ACADEMIC PRESS

Harcourt Brace & Company, Publishers

London   San Diego   New York   Boston

Sydney   Tokyo   Toronto

₀649769x
CHEMISTRY

ACADEMIC PRESS LIMITED
24–28 Oval Road
LONDON NW1 7DX

U.S. Edition Published by
ACADEMIC PRESS INC.
San Diego, CA 92101

This book is printed on acid-free paper

Copyright © 1994 ACADEMIC PRESS LIMITED

All rights reserved
No part of this book may be reproduced or transmitted in any form
or by any means, electronic or mechanical including photocopying, recording,
or any information storage and retrieval system
without permission in writing from the publisher

A catalogue record for this book is available from the British Library

ISBN 0-12-437745-9

Typeset by Alden Press Ltd, Oxford and Northampton, Great Britain
Printed in Great Britain by the University Press, Cambridge

TP994
L258
1994
CHEM

Professor Friedrich Krafft, Heidelberg University, *c.* 1890 (Section 14.2). Professor Krafft's work on aqueous soap solutions was instrumental in spurring the interest of chemists in the physical science of soaps. The crystal solubility boundary of soaps and other surfactants has come to bear his name. This picture was provided by Professor William B. Jensen, Oesper Professor of History of Chemistry, from the Oesper Collection in the History of Chemistry, University of Cincinnati.

Dedication

This book is dedicated to my father, Frank Laughlin,
who instilled in me (by word and by example)
a philosophy and the perseverance
necessary to complete large tasks.

Contents

Part III STRUCTURAL CHEMISTRY

Part IV MOLECULAR CORRELATIONS

Foreword

One very important aspect of physical chemistry is that of equilibrium between different phases. This subject had its quantitative foundations in the 19th Century with the thermodynamic analysis of J. Willard Gibbs in 1875, and the recognition of its application to Physical Chemistry by H. W. B. Roozeboom in 1884. The recognition of its applications in surfactant–water systems and, in particular, to the complex phase equilibria which exist in soap–water systems, owes much to one of my predecessors at Bristol, J. W. McBain, the First Leverhulme Professor of Physical Chemistry in the University of Bristol. In more recent times the synthesis of many hundreds of new surfactants and the important part played by these in modern life, both in the home and in technology, has increased the need to understand at a basic level the behaviour of these fascinating molecules. These systems challenge the scientist to provide both an understanding of the states in which these molecules exist, and the thermodynamic changes which underlie them.

It is perhaps surprising, in view of its considerable importance, that scant attention is presently paid in undergraduate courses to teaching students about the Phase Rule and the nature and usefulness of phase diagrams. This is even more surprising in view of the fact that the preparation of most chemical products in common use involves the application of phase science. Moreover, as a research area a study of phase equilibria still offers enormous potential for new and exciting discoveries. This is particularly so with surfactants in water, where small changes in molecular structure can produce quite dramatic changes in phase behaviour and in their industrial performance.

Dr Robert Laughlin has spent many years of research probing in detail the relationship between the molecular structure of long chain molecules and their phase behaviour, and more recently in developing significantly improved methods for the acquisition of phase information. His skill, enthusiasm, and insight have produced quite a unique understanding of these subjects, and I am sure that many will have cause to be thankful that he has been persuaded to put this knowledge into book form. I can recall with pleasure many hours spent with Bob discussing the behaviour of long chain molecules, and have benefitted greatly from his wisdom. I am sure that many readers will also benefit from the distillation of this wisdom into a book which gives in a readable and precise manner the basis of phase science, and provides a survey of a subject which has not hitherto been treated in a comprehensive manner. This will help to guide and to stimulate much new research, as there are assuredly many areas of surfactant phase science still to be explored.

Ronald H. Ottewill
Bristol, 1993

Preface

This book is intended to serve as a comprehensive reference to the aqueous phase behavior of surfactants. The focus is on those aspects of surfactant phase science likely to be of value to those who seek to exploit these materials in developing surfactant-based technologies. The approach taken is phenomenological and nontheoretical, so the contents should be understandable to anyone with a bachelor's degree (or higher) in chemistry or chemical engineering. A course in undergraduate physical chemistry will be helpful, but is not essential.

While the book was not written primarily for use in an academic course, the principles put forth are intended to be scientifically rigorous and the data included have been critically reevaluated. This book could therefore also serve as a reference source for those who are teaching or studying the general subject of phase science. Anyone who studies this text will become very familiar with the phase behavior of the sodium chloride–water system, as well as of surfactants. While salt and water do not form many of the various kinds of phase structures that exist in surfactant systems, mixtures of these two familiar compounds display most of the various kinds of phase chemistry that are encountered in surfactant–water systems. (The only exception is the absence of liquid/liquid equilibria). The salt–water system constitutes a superb teaching diagram for illustrating the principles (and also the utility) of phase science.

There presently exists within academic physical chemistry an intense preoccupation with its theoretical and quantum mechanical aspects, and a comparatively low level of interest in experimental phase science and thermodynamics. Viewed from the perspective of an industrial research chemist, this situation is difficult to understand and impossible to justify. It has been not only the author's experience, but that of many industrial colleagues, that knowledge of phase behavior and the ability to utilize this knowledge is broadly useful during industrial research and development. The subject can be fascinating scientifically, as well. This information is particularly relevant when physical processes that involve changes of state are kinetically slow, mechanistically complex, or produce time-stable colloidally structured mixtures. In such cases the equilibrium state is unobvious and may rarely be attained, so that an independent determination of the equilibrium phase behavior is essential if one is to diagnose correctly the kinetic or colloidal phenomena that exist. Doing so is of considerable value industrially, because diagnosing problems correctly is the key to addressing them efficiently.

The aqueous phase behavior of surfactants is not a subject of interest to industrial chemists alone; it can also be of considerable interest in a purely academic context. Phase structure is the highest level of structure that is fully defined simply by selecting a system and stating the conditions under which a particular mixture exists. The development of improved fundamental concepts that are capable of encompassing the considerable complexity found in surfactant phase behavior poses an immense intellectual challenge. The difficulties are exacerbated by the fact that the magnitude of the energy differences between states is often extremely low during many surfactant phase transitions.

Further, it is important to recognize that our present knowledge of surfactant phase science is far from complete. Substantial gaps in the experimental facts exist, probably because the methodologies used in the past to determine phase diagrams have been incapable of handling some systems. Many such gaps are explicitly noted in the text.

Finally, it was deemed necessary in writing this book to review critically the basic concepts of "phase" and of "interface". Much of the historic confusion in surfactant phase science stems from an inadequate comprehension of these basic concepts. This is particularly evident in connection with the biologically important surfactants (the polar lipids). The lack of any satisfactory binary polar lipid–water phase diagram probably represents a barrier to progress in this area (Section 11.3.2.7).

For these reasons, the opportunity to collect in a cohesive and relatively permanent form the ideas and information regarding the phase behavior of surfactants (which have been garnered in the span of 35 years of research on both the basic and the applied aspects of surfactant science) became strongly appealing once the idea was seriously proposed. This project was therefore undertaken with a degree of enthusiasm, notwithstanding the considerable and sustained effort that would obviously be required.

Acknowledgements

The author is deeply indebted to many colleagues who have contributed significantly to this book. Foremost among these is Professor R. H. Ottewill, FRS (Bristol University) who first proposed the project, contributed helpful comments as to the general form that the book should assume, and wrote the foreword. Professor Ottewill also expedited the search for information regarding the historic role that McBain played in surfactant phase science, since McBain's pioneering work was conducted at the Woodland Road Laboratories at Bristol University a short walk from the present Department of Chemistry building.

Professor Krister Fontell (University of Lund, now retired) was a graduate student with Ekwall at Åbo Akademy in Turku, Finland. He was not only helpful in providing authoritative information regarding Ekwall's historic role in this field, but on numerous other occasions. Dr R. Strey (Max Planck Institut, Göttingen) provided useful biographical data on Krafft, whose work at Heidelberg University at the turn of the century may be regarded as the beginning of the modern era of surfactant science. Professor W. B. Jensen (University of Cincinnati) provided the picture of Krafft that serves as the frontispiece.

The principal referees for the manuscript were Drs Y.-C. Fu, D. N. Rubingh, and J. R. Hansen (all of Procter and Gamble). Dr W. L. Courchene (retired) critically evaluated the chapter on thermodynamics. Dr B. T. Ingram (Procter and Gamble, Newcastle) provided useful counsel in the early stages, and Ms M. E. Crews assistance in searching the literature. Dr F. C. Wireko provided critique of the X-ray aspects, and Mr M. Mootz and Ms S. M. Thoman supplied figures related to the X-ray crystallography of surfactants. The author is indebted to his management for their genuine support of this project. This included Drs C. D. Broaddus (Vice President of Research and Development, retired), D. J. Peterson (Director, Corporate Research Division), and R. P. Oertel (Associate Director).

Professor R. S. Hansen (Iowa State University, retired) contributed Fig. 4.1 on the free energy of water and has been a consultant in thermodynamics for more than two decades. Professor B. Widom (Cornell University) provided consultation regarding the Griffiths and Wheeler concepts of field and density variables. Numerous discussions of surfactant phase science with Dr P. Kekicheff (Australian National University, Canberra), Professor M. C. Caffrey (The Ohio State University, Columbus, Ohio), Professor B. Lindman, and Professor H. Wennerstrom (both at the University of Lund) have been useful. Professor Wennerstrom also provided input regarding the treatment of theory in the book. The author has benefitted immensely over a period of many years from numerous personal contacts with chemists at this University.

Professor M. L. Klein (University of Pennsylvania) supplied the computer drawings depicting the various geometric structures that exist in surfactant phases, and

Dr D. M. Anderson (State University of New York, Buffalo) made helpful suggestions regarding the figures for phase structures. Dr R. M. Hill (Dow Corning) provided authoritative information regarding the structures of silicone surfactants, and Dr J. R. Minter (Kodak) partially reviewed the manuscript. Other noteworthy contributors included Professors R. Aveyard (University of Hull), Y. Talmon (The Technion, Haifa), F. Meeks (University of Cincinnati), and Dr Th. F. Tadros (ICI, Bracknell).

Dr Galina G. Chernik (St Petersburg University, St Petersburg, Russia), deserves special mention for her strong interest in, and support for, the writing of this book. She contributed highly useful critique, especially as regards the thermodynamics of phase transitions and calorimetry, and brought to my attention Russian contributions to this field. Her assistance is all-the-more appreciated since it occurred mostly during the recent turbulent and difficult years in Russia.

Much earlier, Dr J. C. Lang (Alcon Laboratories, Ft Worth, Texas) made many indirect (but very significant) contributions to the writing of this book. John was a colleague during his research on phase science within Procter and Gamble, and served as the author's mentor insofar as learning the fundamentals of phase science are concerned.

Finally, the author appreciates the encouragement to undertake this project from his wife, Joyce, and assistance in secretarial matters from Ms Cheryl Hippe.

Robert G. Laughlin

Part I

BACKGROUND

Chapter 1

Introduction

1.1 The subject material of phase science

Science is defined in Webster's *Third New International Dictionary* as "accumulated and accepted knowledge that has been systematized and formulated with reference to the discovery of general truths or the operation of general laws". This book is about "phase science" – but what is phase science? The subject material of this discipline can be illustrated by reference to the behavior of mixtures of common salt and water [1].

If a few sodium chloride crystals are mixed with liquid water, the crystals dissolve and a liquid phase containing both compounds is formed. If more salt is added it continues to dissolve, but only up to a finite limit. Within this limit, which is the "solubility", salt and water are said to be "miscible". Beyond this limit, adding salt does *not* further increase the salt concentration in the liquid phase; it is said to be "saturated" with salt. The added crystals persist in contact with the liquid (coexist), seemingly unchanged.

If salt crystals at −5°C and ice crystals at −5°C are mixed using mechanical agitation (as in a home ice cream maker), the two crystals readily interact to form a liquid phase. This liquid too will dissolve more salt crystals up to its solubility limit, but the crystals that coexist with this saturated liquid are *not* the same crystals that were added. They contain two molecules of water per sodium chloride ion pair, and thus differ in both composition and crystal structure from dry sodium chloride crystals.

All the above information, and much more, is displayed in the plot of sodium chloride composition vs. temperature shown in Fig. 1.1, which is the sodium chloride–water phase diagram. The qualitative and quantitative description of phenomena such as the above, formulated in a manner consistent with the general scientific laws governing phase equilibria (the thermodynamics of mixtures, the Gibbs Phase Rule, and conservation of mass), constitutes the *phase science* of the salt–water system. The sum of the various aspects of the physical interactions that occur between salt and water constitutes the *phase behavior* of salt and water (Section 5.1). Phase science is very important to industry and is highly relevant to biology, as well as being academically interesting. It is an important branch of the broader subject of *physical chemistry*.

Phase science is concerned, first, with the influence of temperature, pressure, and composition on the number of phases that exist. Next, it treats the compositions and structures of each phase. Finally, it is concerned with the rates and mechanisms of the processes by means of which phase equilibrium is attained. (These are also the focus of that branch of physical chemistry that deals with the thermodynamics of irreversible processes [2].) These experimental dimensions of phase science provide a

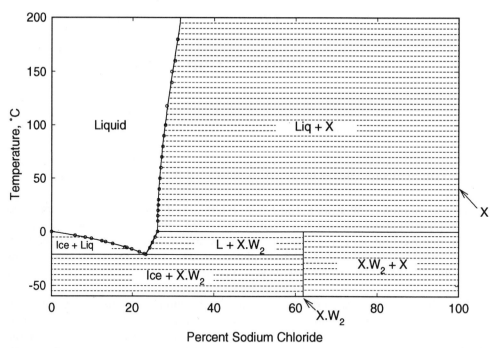

Fig. 1.1. The phase diagram of the sodium chloride–water system between −60 and 200°C. The dashed lines are the tie-lines, which indicate the compositions of coexisting phases (Section 4.8.2). The symbol "X" designates the dry crystal (molecular formula NaCl), and $X \cdot W_2$ the dihydrate crystal having the formula $NaCl \cdot (H_2O)_2$. The phases which coexist within two-phase regions are so indicated; for example, within the area marked "Liq + X" both the liquid and the dry crystal phase are found [1].

basis for the development and testing of theories, which in turn provide a mathematical framework for the quantitative description of experimental results, and serve to correlate, rationalize, and extrapolate them.

Phase science focuses to a large degree on mixtures that are heterogeneous (contain more than one phase), on abrupt changes in the number or the nature of the phases present that occur as conditions are varied, and on the limits of composition, temperature, or pressure beyond which the components are immiscible. For these reasons it is also frequently termed *the science of heterogeneous equilibria*.

1.2   The distinctive characteristics of surfactant phase science

This book is concerned primarily with the phase science of surfactants and water, and to a lesser extent with other aspects of their physical science and with more complex systems.* Like salt, surfactants and water form liquid, crystal, and gas phases. Unlike salt, surfactants and water may also interact to form other kinds of phases that are

---

*The term "water" is used in two distinctive ways. Often it refers to "water substance", that is, the chemical compound having the formula H_2O. The word is also often used with reference to the liquid phase of water substance, however. Both will be used herein; which one is intended should be apparent from the context.

neither liquids nor crystals. These are usually termed "liquid crystal phases", or alternatively "mesomorphic phases" or "mesophases" [3,4]. Some may be appropriately described as "plastic crystals" [5]. The gas phase may coexist in equilibrium with an aqueous surfactant mixture, but since surfactants are typically nonvolatile this phase is ordinarily pure water.

Phases are best distinguished from one another solely on the basis of their structure (Section 2.4.3). Gas phases at modest pressures are distinguished from all other states by their exceptionally low densities, and by their disordered (but statistically well-defined) structures. At the other extreme lie crystal phases, which are the most dense and highly ordered states of matter. Crystal structures are comparatively well defined and relatively easily determined. Liquids are typically less dense than crystals, and have complex and disordered structures that are extraordinarily difficult to determine and describe. The liquid state has nevertheless received considerable attention [6,7].

Liquid crystal phases are discrete states of matter that are less highly ordered than crystals, but more highly ordered than are liquids. Some of the more common liquid crystals include the lamellar phase (which is constructed by stacking together bilayers), the various hexagonal phases (which result from assembling indefinitely long cylindrical units in a hexagonally close-packed array), and a variety of isotropic cubic phases. The broad features of liquid crystal structure are comparatively easily determined. Each liquid crystal state differs qualitatively from every other liquid crystal state in one or more characteristic structural features, and also from gas, liquid, and crystal phases. Chapter 8 is concerned with the structures and properties of the phase states that surfactants form.

Liquid crystal phases invariably span a finite range of compositions and temperatures. The physical properties of these phases vary widely with these parameters, and are also strongly influenced by the molecular structure of the surfactant. For such reasons, it is far better to define and classify phases on the basis of their structures than to do so on the basis of their properties.

1.3 The nature of the complexity of surfactant phase behavior

The phase behavior of aqueous surfactant systems may at first seem to be far more complex than is that of salt and water. While a high level of complexity does exist, it is in one sense deceptive. That is because it arises solely from diversity in the *structures* of the phases that exist. All the phase states found in other systems (gas, liquid, crystal) also exist in surfactant systems. All the phase transformations encountered in other systems (melting, azeotropic, eutectic, peritectic, and polytectic phenomena) are also encountered in surfactant systems. No phase transformation has been found in surfactant systems that is unique to these systems. Thus, no unique phase chemistry will be encountered in surfactant systems – only unique phase structures.

While considerable information has been developed regarding the phase science of surfactants, it is worth recognizing that this branch of phase science is dwarfed by other branches. Only about 1% of the publications on phase science are concerned with surfactants, but this is still a large number and this fact in no way demeans the importance of the subject.

1.4 The influence of system variables and of molecular structure

The parameters temperature, pressure, and composition provide fundamental information regarding all mixtures, and are collectively termed the "system variables". The aqueous phase behavior of a surfactant is influenced both by its molecular structure and by these variables, and both areas will be treated. Once a system has been defined, *stating the system variables fully defines all aspects of its phase behavior at equilibrium.* In this approach molecular structure and the system variables are the independently controllable parameters, and once these are defined the phase behavior, optical properties, rheology, and many other aspects of the physical science of the system are also defined. Colloidal structure and phenomena related to its formation and collapse are *not* defined (Section 2.4.4).

Knowledge of the influence that system variables have on phase behavior is of particular value to process development engineers. Typically these personnel are required to design a process to manufacture a product formulation, using raw materials that have been selected on the basis of performance data (see Chapter 13, and especially Section 13.2). Information regarding the influence of system variables on the physical behavior of their formulation is therefore of considerable value. To illustrate, temperature often strongly influences phase ratios in mixtures containing more than one phase, and phase ratios in turn exert a profound influence on the rheology of such mixtures. Understanding the relationships between rheology, phase behavior, and temperature may suggest ways of resolving problems that would not otherwise be evident. On the other hand, knowledge of the relationship between phase behavior and molecular structure is largely inconsequential to personnel engaged in process development.

At an earlier stage of technology development, an application may have been defined (in terms of performance criteria) but the raw materials for the formulation have not yet been selected. For those engaged in the raw material selection process, knowledge of the relationship between molecular structure and the physical science of surfactants can be of significant value, and phase behavior is an important dimension of this physical science. Properly measuring the utility of the formulation is the uppermost consideration during raw material selection (Section 13.3.1), but knowledge of the phase behavior of prospective raw materials (and its relationship to the structure of the components in raw materials) can measurably assist the selection process.

1.5 Sources of data

Finding data can be a genuine barrier to the use of phase science. For numerous nonsurfactant inorganic and organic systems, all the information required to create phase diagrams (but not the actual diagrams) exists in the various compilations of "solubility" data [8,9,10]. Typically, these references contain only data that have been critically evaluated before it has been accepted. Assistance in producing diagrams from such data is provided in Appendix 3.

For data published since 1967 direct searching via electronic databases (such as *Chemical Abstracts*) is possible, but the investigator must critically evaluate the data found during the search. An extremely useful source of references to phase data is the Wisniak survey [11], which provides references to phase studies on all manner of systems (including surfactants) indexed by chemical name. Compilations of actual diagrams have been assembled for ceramic [12] and metallurgical [13] systems, and phase information (but not diagrams) has been compiled for the biological polar lipids [14].

The most extensive compilations of surfactant phase information that exist are the first broad survey by Ekwall [15], the author's chapters on hydrophilicity vs. molecular structure [16], the more recent survey of nonionic surfactants [17], and a review of the phase science of cationic surfactants [18]. A brief review of the general phase behavior of surfactants has also been published [19].

1.6 The book's organization

Part I of this book provides a broad perspective regarding surfactant phase science. Following this introductory chapter, the phase science of surfactants is placed in context (relative to other aspects of their physical science) in Chapter 2. Terms and concepts that will be utilized throughout the book are also discussed in this chapter.

Part II is concerned broadly with the physical chemistry of surfactant phase behavior. Chapter 3 provides a brief review of those elements of thermodynamics that are particularly relevant to aqueous surfactant phase behavior, and Chapter 4 is concerned with the Phase Rule itself and fundamentals of phase science. The distinctive features of surfactant phase behavior are discussed in Chapter 5, the rates and mechanisms of the phase reactions that surfactants undergo in Chapter 6, and the relationship between water activity (relative humidity) and surfactant phase behavior in Chapter 7.

Part III (Chapter 8) is concerned with the structural aspect of surfactant phase science, and with the optical and rheological properties of surfactant phases.

Part IV is focused on correlations between molecular structure and surfactant phase behavior. The polar functional group is treated in Chapters 9 and 10, and the nonpolar structural fragment in Chapter 11. Both the fundamentals of ternary phase science and the influence of added salts and oils are reviewed in Chapter 12.

Part V provides miscellaneous information that lies beyond the subject of surfactant phase science *per se*. In Chapter 2 (Part I) the relationship of the phase behavior of surfactants to their physical science is presented, while in Chapter 13 of Part V the broader relationship of this physical science to the utility of surfactants is treated. Finally, the history of surfactant phase science is outlined in Chapter 14.

Appendices. Appendix 1 contains a glossary of phase science terms, plus definitions of acronyms for molecular structure and of thermodynamic symbols used in the text. Literature references to binary and ternary surfactant phase data are contained in Appendix 2, which includes all the diagrams found in this book plus literature references for systems whose diagrams are not included in the text. The compounds are systematically arranged on the basis of the hydrophilic group structural classes defined in Section 9.7.

Appendix 3 describes computer methods for the extraction, graphing, and quantitative use of the information contained in published phase diagrams. Finally, Appendix 4 is concerned with experimental methods for the determination of phase diagrams, the purity issue (how pure is "pure enough"), a description and analysis of the methods presently used to determine surfactant phase diagrams, and a process by means of which the quality of phase diagrams can be evaluated.

1.7 The treatment of fundamentals

This book is intended to serve as a source of information on the aqueous phase science of surfactants, and on selected representative phase diagrams. It is written primarily for both industrial and academic chemists or engineers who are working with surfactants, but who are not expert phase chemists. Inevitably, some of the problems encountered will have their origin in phase behavior. A solid grasp of the principles of phase science will greatly enhance the likelihood that these problems will be correctly diagnosed, and thus efficiently and effectively resolved.

The minimal academic background required to comprehend this material is undergraduate education in chemistry or chemical engineering. In the US this typically requires 16 years of schooling, which ideally will have included an undergraduate course in physical chemistry. The reader's comprehension will be both enhanced and facilitated if a firm grasp of basic thermodynamic principles has been acquired. Just in case, a review of those aspects of thermodynamics that are particularly relevant to surfactant phase science is provided in Chapter 3.

The book should also be useful to people having graduate education in fields other than physical chemistry who are confronted with problems related to phase behavior. Like it or not, everyone encounters phase behavior at some point – even if only in preparing a solution of a test material for an experiment. If the material is soluble at the selected composition this process can be straightforward, although sometimes slow dissolution rates must be dealt with. If the material is insoluble at the selected composition (as often happens), then an understanding of phase (and colloid) science is extremely useful. It is hoped that this book will enable interested people to acquire a useful level of knowledge of phase science with a reasonable expenditure of time and study. Because of the intimate relationships that frequently exist between phase and colloid science, some attention has also been paid to the latter discipline.

Tactics. To accomplish these objectives, the pattern typically followed in treating phase science in academic physical chemistry texts has been significantly altered. The derivation of the Phase Rule, for example, is instructional but straightforward. Its derivation has not been repeated, but considerable attention is paid to its full and proper utilization. A firm grasp of the Phase Rule and its proper use is mandatory for those engaged in the determination of phase diagrams, but is also extremely useful to those who utilize this information.

The point of departure in treating fundamentals is analysis of the phase behavior of binary (two-component) systems, rather than of unary (one-component) systems (as is often done). Binary systems are the simplest systems in which composition is a variable. Once binary system phase behavior is thoroughly understood, it is comparatively easy either to return to unary systems or to move on to ternary systems. The sodium chloride–water system, which is a familiar compound whose aqueous phase diagram is well established, has been found to be extremely useful as a basis for discussing the fundamentals of phase science.

The subject is approached from a phenomenological perspective, coupled with a descriptive graphical analysis of the thermodynamic basis for partial miscibility. Phase diagrams constitute a dense form of physical and thermodynamic information, and either considerable thought and analysis or instruction is required to extract this information. Also, the implications of phase diagrams as to the physical behavior of a mixture are far more comprehensible if the general form of free energies of mixing curves for partially miscible systems is known and the thermodynamic origin of partial miscibility is understood.

The book is focused on the thermodynamics and phase behavior of macroscopic systems. The thermodynamics of aggregates of molecules within macroscopically homogeneous phases, which has been approached via the "thermodynamics of small systems" [20,21], is a separate subject that will not be considered.

1.8 The information content of phase diagrams

The phase diagram of the salt–water system (Fig. 1.1) indicates that a 50% mixture of salt and water at 25°C consists of two phases: a liquid salt solution and dry salt crystals. This is the obvious information provided by the diagram, but there is much more.

For example, it can be seen at a glance that water is a poor solvent for recrystallizing salt (although it is possible). This follows from the temperature coefficient of solubility, which is the slope of the solubility boundary. A feeling for the slope can be obtained simply by inspection; if desired, quantitative data regarding this coefficient could be extracted from the diagram. If one did elect to purify salt by recrystallization, the optimal composition and temperature for a particular process, and the maximum theoretical yield for the process, could be predicted.

It is also apparent that the evaporation of water from saturated salt solutions will precipitate dry salt crystals, but that this will only happen if the relative humidity is below a specific value (which has been found to be about 75%). Salt crystals absorb water and are both hygroscopic and deliquescent at relative humidities greater than 75%, but are neither below 75%. The concentration of the salt solution that exists (at equilibrium) at relative humidities >75% can be extracted from the phase diagram, from the dependence of water activity on composition within the liquid phase (Chapter 7).

These considerations allow one to anticipate the state of droplets of sea water that are thrown into the air during wave action, and this information is relevant to the corrosion of metals by salt water spray. Very concentrated salt solutions may exist on metal surfaces exposed to ocean spray even at high relative humidities, because the concentration of these solutions is dictated solely by relative humidity and temperature; it is not influenced by the concentration of salt in sea water.

Analogous phenomena are encountered in surfactant-containing mixtures, and may be approached using the same principles. The evaporation of water from a liquid detergent (at the rim of a bottle, for example) may produce either a liquid, a liquid crystal, or a crystal phase. Which is formed depends on the intrinsic phase behavior of the product and the relative humidity.

Phase behavior and relative humidity influence the extent and consequences of the uptake of water by surfactant powders, and as a result their stickiness and pourability. The sequence of phases that develop at the surface when a surfactant particle is immersed in water (as a result of its swelling by water) can be anticipated in detail from the phase diagram (Appendix 4.4.4). If a lamellar liquid crystal phase is present in mixtures of two liquid phases, it tends to exist at interfaces and may dramatically stabilize emulsions [22]. Such phenomena also occur during the baking of bread, during which the lamellar phase of the surfactants in flour contribute to stabilization of the "foam" produced during the baking of bread [23]. Conversely, the precipitation of an oily liquid phase (Section 13.3.6.4) may kill foams. This is relevant to the use of fatty alcohols as defoaming agents for surfactant solutions [24], and also explains how a fatty alcohol may, under different circumstances, be a foam-stabilizer.

The above examples serve to illustrate two important points:

(1) phase science has a profound and diverse influence on a great many aspects of physical behavior that are of considerable practical significance, and
(2) many aspects of this influence are not intuitively obvious.

1.9 The philosophy

The author's philosophy of research has been eloquently and succinctly stated by Sienko and Plane [25], as follows:

> The acquisition of knowledge is simplified by seeking the principles that underly the field of study. In chemistry, we can consider the principles as being composed of the generalizing statements that summarize observed phenomena and of the theories, or explanations, proposed to account for the observations. Such principles of chemistry serve both as a convenient framework for remembering a large body of information and also as a firm basis from which to make predictions about the unknown.

The approach taken in writing this book was suggested by the author's impression that phase science is not presently in vogue, either as an element of academic curriculae or as the object of academic research. While every physical chemistry text has an obligatory chapter on the Phase Rule and heterogeneous equilibria, the treatment of the subject varies widely. Further, the present level of instruction in phase science (in both undergraduate and graduate courses) is typically meager. As a result, one only rarely encounters chemists who are highly skilled in this discipline. This situation contrasts strikingly with the level of attention commanded by this subject in textbooks and courses during the first half of the 20th century [26].

The present lack of emphasis on phase science belies its considerable importance. Being essentially thermodynamic in nature, it is a cornerstone of physical science. Nevertheless, the evident truth of this statement has not kept the discipline from declining in recent years. Because of this situation, very little will be taken for granted in treating fundamentals.

The thrust of a book such as this is inevitably influenced by the philosophy and prejudices of its author. This book is written from the perspective of a basic industrial research chemist – one who is interested in knowing more about performance data than merely the end result. The author's interest in phase science (and practically all of his training in this discipline) stems from a long and illustrious tradition of scientific and practical interest in the subject within the Procter and Gamble Company. This training was considerably enhanced by the presence of several colleagues who were physical chemists of considerable scientific stature. The most important of these, as regards phase science, were J. M. Corkill (Oxford), J. F. Goodman (Cambridge), W. L. Courchene (Cornell), and J. C. Lang (Cornell).

Experience has shown, time and again, that a solid grasp of phase behavior coupled with an understanding of how to anticipate, manipulate, and exploit it, can be of immense practical value. An in-depth understanding of the theoretical basis for phase phenomena may be of considerable academic interest, but is of secondary importance in a practical sense. While significant progress in the theoretical area has been made, the theory of surfactant phase equilibria will not be addressed in this book. A brief outline of the approach that is currently being taken in this area is presented in Section 8.8.2.

References

1. (1965). In *Solubilities*, Vol. II (W. F. Linke ed.), pp. 958–959, American Chemical Society, Washington, DC.
2. Prigogine, I. (1968). *Introduction to Thermodynamics of Irreversible Processes*, John Wiley, New York.
3. Luzzati, V. (1968). In *Biological Membranes, Physical Fact and Function*, (D. Chapman and D. F. H. Wallach eds), pp. 71–124, Academic Press, New York.
4. Danielsson, I. (1976). In *Lyotropic Liquid Crystals*, pp. 13–27, American Chemical Society, Washington, DC.
5. Winsor, P. A. (1974). In *Liquid Crystals and Plastic Crystals*, Vol. 1 (G. W. Gray and P. A. Winsor eds), pp. 48–59, Ellis Horwood, Chichester, England.
6. Croxton, C. A. (1974). *Liquid State Physics—A Statistical Mechanical Introduction*, Cambridge University Press, Cambridge.
7. Hansen, J.-P. and McDonald, I. R. (1955). *Theory of Simple Liquids*, 2nd ed., Academic Press, London.
8. (1965). *Solubilities*, Vols I and II (W. F. Linke (formerly A. Seidell) ed.), 4th ed., American Chemical Society, Washington, DC.
9. (1979). *Solubilities of Inorganic and Organic Compounds*, Vols 1–3 (H. L. Silcock ed.), Pergamon Press, Oxford.
10. (1963). *Solubilities of Inorganic and Organic Compounds* (Translated from the Russian), Vols 1 and 2 (H. Stephen and T. Stephen eds), Pergamon Press, Oxford.
11. Wisniak, J. (1981). *Physical Sciences Data. Phase Diagrams, Part A*, Vol 10, Elsevier, New York; *Physical Sciences Data, Phase Diagrams, Part B*, Vol. 10, Elsevier, New York; Wisniak, J. (1986). *Physical Sciences Data, Phase Diagrams, Supplement 1*, Vol. 27, Elsevier, New York.
12. NIST Standard Reference Database 31, *Phase Diagrams for Ceramists*, available from The American Ceramic Society, 65 Ceramic Drive, Columbus, Ohio 43214. Phone 614/268-8645.
13. *Binary Alloy Phase Diagrams*, 2nd ed. Available from ASM International, Metals Park, Ohio. Phone 216/338-5151.
14. NIST Standard Reference Database 34, *Lipid Thermotropic Phase Transition Database*, available from the National Institute of Standards and Technology, 221/A320, Gaithersburg, MD 20896. Phone 301/975-2208. Caffrey, M. (1993). *A Database of Thermodynamic Data and Associated Information on Lipid Mesomorphic and Polymorphic Transitions*, CRC Press, Boca Raton, FL 33431.
15. Ekwall, P. (1975). In *Advances in Liquid Crystals*, Vol. 1 (G. H. Brown ed.), pp. 1–142, Academic Press, New York.
16. Laughlin, R. G. (1978). *Advances in Liquid Crystals*, Vol. 3 (G. H. Brown ed.), pp. 41–148, Academic Press, New York.
17. Sjoblom, J., Stenius, P. and Danielsson, I. (1987). In *Nonionic Surfactants, Physical Chemistry*, Vol. 23 (M. J. Schick ed.), pp. 369–434, Marcel Dekker, New York.
18. Laughlin, R. G. (1990). In *Cationic Surfactants Physical Chemistry*, Vol. 37 (D. N. Rubingh and P. M. Holland eds), pp. 1–40, Marcel Dekker, New York.
19. Laughlin, R. G. (1984). In *Surfactants*, (Th. F. Tadros ed.), pp. 53–82, Academic Press, New York.
20. Hill, T. L. (1963). *Thermodynamics of Small Systems*, Vol. 1, Benjamin, New York.
21. Hill, T. L. (1964). *Thermodynamics of Small Systems*, Vol. 2, Benjamin, New York.
22. Friberg, S. (1979). *J. Soc. Cos. Chemists* **30**, 309–319.

23. Larsson, K. (1986). In *Chemistry and Physics of Baking, Special Publication No. 56*, (J. M. V. Blanshard, P. J. Frazier, and T. Galliard eds), pp. 62–74, The Royal Society of London, Burlington House, London.
24. Friberg, S. (1978). In *Advances in Liquid Crystals*, Vol. 3 (G. H. Brown ed.), pp. 149–165, Academic Press, New York.
25. Sienko, M. J. and Plane, R. A. (1957). *Chemistry*, 1st ed., McGraw-Hill, New York.
26. Glasstone, S. (1946). *Textbook of Physical Chemistry*, pp. 693–814, Van Nostrand, New York.

Chapter 2

The Role of Phase Science Within Physical Science

2.1 A general view of physical science

It can be difficult to perceive the relationship of phase science to the more general science of surfactants, and thus its significance and value. It is the aim of this chapter to clarify this relationship. In Chapter 13, the broader relationship of the physical science of surfactants to their utility will be considered.

Physics may be regarded as the most fundamental of the scientific disciplines. In physics the various kinds of forces of nature (including electromagnetic, gravitational, and the other forces) and their action on matter are treated from the perspective of universal laws [1]. Physics is concerned with elements of matter which range in size from subatomic particles to the entire universe, and the laws of physics encompass a wide range of phenomena. In some applications of these laws details of the molecular structure of matter are inconsequential, while in others (chemical physics) the focus is clearly on molecules [2].

Chemistry is that branch of science which is concerned with explicit materials and their properties. The focus on molecules and molecular structure within chemistry is extremely strong. In fact, within some chemical subdisciplines (such as synthetic chemistry) the concern with the molecular level of structure may be so extreme as to nearly exclude consideration of other levels. The reason for this in this instance is readily apparent; it rests on the fact that synthetic goals may only be defined in terms of molecular structure.

To illustrate, suppose that one desires to prepare a material which has chemical or physical properties not found among existing materials. One cannot synthesize "properties"; one can only synthesize molecules that possess properties. Hence, two events must occur during such a program. In the first, a hypothesis as to the relationship between molecular structure and properties is stated and the molecular structures of the materials to be synthesized are selected. In the second, the selected materials are synthesized and their properties are determined. Both events are equally important, and it is for this reason that Part IV of this book exists. Part IV is concerned with the relationship between the molecular structure of surfactants and their phase properties.

Physical chemistry has been described as "the quantitative branch of chemistry" [3]. It is concerned with the measurement of properties, the determination of structure at various levels (often using the methods of physics), the description of the rates and equilibria of both chemical and physical processes, the equations of state which relate state to environmental conditions, and related subjects. Chemical change occurs during some of the processes with which physical chemistry is concerned, but not in others.

In physics a pervasive tendency to develop concepts and theories that are universally applicable exists, and this tendency has carried over into physical chemistry. In seeking universality one must blur the differences that exist between chemical and physical processes. Particularly within the phase science branch of physical chemistry, ignoring these differences introduces serious problems. Chemical processes, when they occur, significantly influence the physical behavior of systems. They introduce new components and, in so doing, increase the number of combinations of degrees of freedom plus number of phases that are allowed (Section 4.5).

The chemical and physical aspects of physical chemistry may be readily distinguished from one another, as follows:

The chemical aspect of physical chemistry is concerned with molecular structure *per se*, and with the characterization of those processes during which changes occur in molecular structure.

The physical aspect of physical chemistry includes the characterization of all aspects of structure except molecular structure, and of all processes during which changes do *not* occur in molecular structure.

2.2 The chemical and physical aspects of physical chemistry

Determining experimentally whether or not a given process is a chemical or a physical process is usually straightforward. Except for ion exchange reactions (see below) covalent bonds are made or broken during chemical reactions, and numerous methods exist for recognizing this event. In surfactant chemistry the proton transfer (acid–base) reaction is especially important. If an uncharged surfactant (AH) reacts as an acid with water (as a base) to form a hydronium salt, H_3O^+,A^-, this is clearly a chemical process. An example of such a reaction is the hydrolysis reaction of an *N*-alkylammonioacetate [4]:

If a nonionic surfactant (B) reacts as a base with water (as an acid) to form a hydroxide salt, BH^+,OH^-, this too is clearly a chemical process. A well-documented example of this process occurs in amine oxide surfactants [5].

Other chemical reactions of surfactants, which are not proton transfer reactions, include hydrolytic cleavage of the ester link of a polar lipid (to form a fatty acid and a lysopolar lipid), of the ester link of sodium alkyl sulfates (to form a fatty alcohol and sodium bisulfate), and of an imidazoline ring (to form an acyclic aminoamide compound).

Association processes that occur within a phase during which molecular structure remains intact are here defined as physical processes – no matter how strong the interaction. Familiar examples include the association of ions in concentrated salt solutions, and of surfactant molecules in micellar solutions of both nonionic and ionic surfactants. The association of water molecules with ions in solution, or with hydrophilic groups, is also a physical association.

In concentrated sodium chloride solutions, the water bound directly to sodium ions is distinctive and has been characterized using vibrational spectroscopy [6]. While it is energetically and structurally distorted from the water molecules in pure water (whose energies and structures, incidentally, vary depending on the numerical values of the system variables and the particular state the water is in), it is nevertheless recognizable as water. The same is true of water molecules in the crystal dihydrate of sodium chloride (Section 1.1). The crystal structure of this dihydrate is known, and intact water molecules may be recognized to exist within this crystal [7]. They have been characterized using both single crystal structure determination and spectroscopy [8] and, just as in the salt solution, they too are severely distorted from those found in pure water. Nevertheless, they exist.

If the pressure on either the salt solution or the crystal hydrate is reduced below a particular value, pure water is liberated and the dry salt crystal is left behind.

Ion exchange chemistry. The dissociation and reversible exchange of ion molecules in solution does *not* constitute a chemical reaction by the above definitions, but an exchange of ions which results in the precipitation of a new salt *is* to be regarded as a chemical reaction. A classical example of the ion exchange chemical reaction is the precipitation of silver chloride from solutions of sodium chloride and silver nitrate:

$$Na^+,Cl^- + Ag^+,NO_3^- \rightleftarrows \underline{Ag^+, Cl^-} + Na^+,NO_3^- \tag{2.1}$$

Ion exchange reactions differ from those chemical processes described above in that covalent bonds are not formed or broken. An intimate relationship has always existed between ion exchange chemistry and phase science, since ion exchange processes are grossly evident only when the ion exchange product precipitates as a separate phase.

The basis for distinguishing chemical from physical processes. A physical process can be distinguished from a chemical process by ascertaining whether or not the composition of the product that is formed can be varied independently of the compositions of the reactants (eq. 2.1). If this is possible a chemical process has occurred, an additional component exists, and the value of C in the Phase Rule (Section 4.6) is increased by one. A change of this nature profoundly alters the physical science of the system (Section 10.7). If the stoichiometric relationships cannot be independently varied, then the value of C is not altered by the process. When this is true the process may be regarded as a physical process, and ignored insofar as the value of C in Phase Rule is concerned.

2.2.1 The chemical aspect

These ideas may be illustrated by comparing the sodium chloride–water with the sodium acetate–water system. Three molecules and two compounds exist in the first system. The molecules are the sodium cation and chloride anion (both of which may be regarded as monatomic, formally charged molecules), and water (which is a polyatomic, electrically neutral molecule). Since the anion and the cation must exist in precisely equal numbers (because of the constraint that salts are electrically neutral), they together constitute the compound sodium chloride. The structures and chemistries of sodium chloride and water are well characterized, their mixtures are known to be chemically stable over a wide range of conditions, and we need not be concerned with chemical processes in treating the physical science of this system.

In the sodium acetate–water system, the chemical science is more complex because acetate anion reacts perceptibly with water to form acetic acid and hydroxide anion. This reaction produces two new compounds (sodium hydroxide and acetic acid):

$$Na^+, CH_3CO_2^- + H_2O \rightleftharpoons CH_3CO_2H + Na^+, OH^- \tag{2.2}$$

The sodium cation is not directly involved, but is paired with hydroxide anion in the product instead of with acetate anion.

Whether or not this reaction influences the phase science of the system hinges critically on the boundary conditions of the experiment. Because the process is well defined stoichiometrically, it does *not* introduce another component if the mixture being investigated is strictly closed (as, for example, in a sealed container). However, if acetic acid is added or removed (e.g. by volatilization), or if the pH is adjusted away from the natural equilibrium value by adding sodium hydroxide or acetic acid, then the system has three components instead of two. (If a different acid or base is added, then yet another component is introduced.) Having three components instead of two introduces one more degree of freedom in applying the Phase Rule and this fact *must* be considered during analysis of the phase science.

A parallel analysis applies to ammonium chloride solutions, except that the ammonium cation reacts as an acid with water (which reacts as a base), while the acetate anion (in sodium acetate solutions) reacts as a base with water (which reacts as an acid).

Chemical reactions closely related to the hydrolysis of sodium acetate occur when soaps (such as sodium palmitate) are mixed with water [9]. The hydrolysis reaction of soaps produces, initially, long chain (fatty) carboxylic acids. These compounds may be regarded as nonsurfactant amphiphilic oils. Such compounds severely modify the solution, surface, and phase physical chemistry of aqueous surfactant mixtures (Section 12.4). In addition, fatty acids interact with soaps to form a number of crystalline "phase compounds"; these contain both molecules, and are commonly termed "acid soaps" (Section 4.15). The equation for formation of the 1 : 1 acid soap of sodium stearate is shown in the equation

Acid soaps have been prepared in which the soap : acid mole ratio is 1 : 2, 1 : 1, and 2 : 1 [10]. Considerable information regarding acid soaps is presented in the book by Small [11].

The counterpart of soap hydrolysis in cationic surfactant salts (such as dodecyl-ammonium chloride) is reaction of the substituted ammonium ion with water according to the equation

$$\text{/\/\/\/\/\/\/\/}NH_3^+, Cl^- + H_2O \rightleftharpoons$$

$$\text{/\/\/\/\/\/\/\/}NH_2 + H_3O^+, Cl^-$$

If the reactants and products remain in solution (as in solutions of ammonium chloride itself) equilibrium is attained extremely fast, and new components are not introduced. If precipitation occurs (as with soaps), then the mixture must be treated as if another component had been added (Section 10.5.2). In both the soap and the alkylammonium chloride systems, the attainment of *both* chemical and physical equilibrium can be kinetically slow and mechanistically complex when such chemical reactions occur. Such processes also introduce complexity in the colloid science of the mixture.

2.2.2   The physical aspect

While the sodium chloride–water system is chemically stable and its chemistry simple, its physical science is relatively complex. This may be illustrated by considering the physical behavior of various salt–water mixtures at 25°C and one atmosphere.

By reference to the phase diagram (Fig. 1.1), one sees that a 20% mixture is a liquid phase under these conditions. This phase has a particular structure and set of properties (density, vapor pressure, compressibility, free energy, enthalpy, etc.). If suddenly disturbed, for example by a small but very fast temperature- or pressure-jump process, the resulting state attains equilibrium too fast for the process to be detected by casual observation.

If water is removed from the 20% mixture so as to form a 50% mixture, the liquid phase is no longer stable. Part of the sodium and chloride ions precipitate as a crystal phase. During this process both ions lose all their water of hydration. Precipitation proceeds until the concentration of the liquid phase is reduced to 26.43%, then seemingly stops (Section 4.17.2). Hours (or days) may be required for equilibrium of state to be attained during the precipitation of crystals. Hence, the rate of equilibration of state is dramatically altered in comparison to the rates that pertain to the 20% mixture. This is primarily a consequence of the fact that the 50% mixture contains at equilibrium two phases.

If a 25% mixture is cooled, the crystal dihydrate $X \cdot W_2$ precipitates instead of the dry crystal X (Section 1.1). $X \cdot W_2$ is a "phase compound" formed by a "phase reaction", which (on the basis of the above criteria) is to be regarded as a physical process. During phase reactions structure at the phase level changes, while structure at the molecular level remains intact. In the formula for crystal hydrates such as $X \cdot W_2$, "X" both indicates that the phase is a crystal, and is a symbol for the molecular formula of the compound (NaCl). "W" represents water, and the center dot "$\cdot$" indicates that the water is physically bound to sodium chloride within this crystal structure. Stoichiometrically well-defined phase compounds exist only as crystal phases (Section 8.3.4). The phase reaction mode of expressing physical processes may also be used to describe reactions having non-integral stoichiometries, however (Section 3.21).

2.3 The dimensions of the physical and colloid science of surfactants

To simplify the language, the term "physical science" will henceforth be used rather than "the physical aspect of physical chemistry". It is recognized that this term may possibly be misconstrued as implying consideration of the physics of surfactants as well as their physical chemistry, but this is not intended. The subject material of this book falls almost entirely within the province of physical chemistry.

The physical science of surfactant–water systems (and of many other systems) may be regarded as having four principal experimental aspects. These include:

(1) a structural aspect,
(2) a thermodynamic aspect,
(3) a kinetic and mechanistic aspect, and
(4) a colloidal aspect.

There exists in addition the non-experimental theoretical aspect, by means of which experimental results may be correlated, understood, and possibly extrapolated. Each of these will be considered in the following sections. The structural aspect will be considered first, because structural information is utilized during elucidation of most other aspects of physical science.

2.4 The structural aspects of physical science

The structure of molecules is extremely important in all aspects of science (excepting some aspects of basic physics, Section 2.1), but there exist other dimensions of structure that also are of considerable importance. It is mandatory in using the term to define what is meant by "structure", because this is perhaps the most ambiguous term in all of science. It has been used to describe the arrangement of subatomic particles within nuclei, of electrons and of atoms within molecules, of organs within an individual, and of the bodies within the universe. We must therefore specify just which among the innumerable "structures" (or more precisely, which kinds of structural information) are relevant to the physical science of surfactants.

There are four that are of principal interest:

(1) molecular structure,
(2) conformational structure,
(3) phase structure, and
(4) colloidal structure.

Each of these kinds of structural information differs perceptibly from the others in an attribute that may be termed "complexity". Complexity is a relative concept, and is determined by the number of variables that must be stated in order to define fully a particular kind of structural information. In the above list, the different kinds of structural information are arranged in order of increasing relative complexity as one descends the list.

2.4.1   Molecular structure

For many purposes, molecular structure is the principal (and sometimes the only) level of structure with which one must be concerned. Molecular structure is defined by stating "atom connectivity", that is, which atom is connected to which. In monatomic molecules (such as neon, sodium, or chloride atoms) there is no connectivity (Section 2.2.1). In the polyatomic water molecule, the connectivity is H−O−H (not H−H−O). In nitrous oxide it is N−N−O (not N−O−N), in methane it is C(−H)$_4$ (not H−H−C(−H)$_2$, and so on. Simply stating connectivity provides information about a molecule such as its electron distribution, bond angles, bond distances, energy states, vibrational constants, chemical reactivity, etc. An immense amount of effort has gone into determining and understanding molecular structure.

2.4.2   Conformational structure

While stating molecular structure conveys a vast amount of information, it does not (with certain exceptions) fully define the structure even of an isolated molecule. Many molecules contain covalent bonds about which rotation can and does readily occur, and bond rotation produces "conformational isomers" [12]. (Exceptions include molecules such as benzene, carbon tetrachloride, and rigid bicyclic molecules such as bicyclooctane. In these molecules major conformational changes require the input of sufficient energy to break chemical bonds.) In surfactants, rotation about carbon−carbon bonds within aliphatic lipophilic chains is facile (except within the crystal phase), and results in the well-known "*trans*" and "*gauche*" chain conformations (Fig. 2.1). There are one *trans* and two *gauche* conformations; one of the *gauche* structures is right- and the other left-handed. Within fluid states a modest energy barrier obstructs the transformation of each of these conformations to another; at room temperature conformational equilibrium is rapidly established.

When conformational isomerism exists it has several important effects. The enthalpy of the *gauche* conformation of butane is about 3.8 kJ mol^{-1}

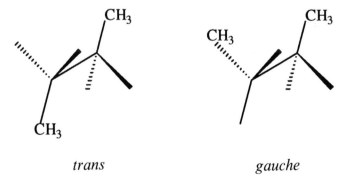

trans   gauche

Fig. 2.1.   The idealized *trans* and *gauche* conformations of butane. It is currently thought that the dihedral angle between the two planes of three bonded carbons in the *gauche* forms in long chains is 67°, rather than 60° as shown in this figure [14].

$(0.9 \, \text{kcal mol}^{-1})$ higher than that of the *trans* conformation. The free energy difference $(1.7 \, \text{kJ mol}^{-1}$ or $0.4 \, \text{kcal mol}^{-1})$ is smaller because there are two *gauche* but only one *trans* form. This energetic difference between *trans* and *gauche* forms is believed to persist in more complex molecules as well. The idealized dihedral angle between the two three-carbon planes in *gauche* butane is 60°, but in recent calculations of the conformational energies of long chain molecules this angle has been taken to be 67° [13].

The presence of many different conformational structures increases the entropy of a phase state. The introduction of *gauche* bonds *increases* the molecular volume, but *shortens* the mean length of the chain. Rotation about the C–C bond in C–CH$_3$ groups does not produce energetically different conformations, but does increase the entropy of the state. Often extremely low temperatures are required to freeze the rotation of methyl groups – even in the crystal state [13].

By counting the number of bonds (*n*) which produce structurally distinguishable conformations, the number of theoretically possible distinct chain conformations may be calculated using the equation:

$$\text{Number of conformations} = 3^n$$

For surfactant lipophilic groups, the number of conformations that can exist is enormous. For example, the C$_{17}$H$_{35}$ chain of sodium stearate could theoretically exist in 3^{15} (14 348 907) distinctive conformations. If one assumes (conservatively) that half of these are excluded from consideration due to their excessive energy (for example, because of atomic overlap), there would still be seven million conformations to consider. It is clear that conformational structure in surfactant molecules must be described on an average or statistical basis, rather than in terms of discrete conformations. The number of conformational states in the tether of zwitterionic surfactants (Section 10.5.3.2) is, by comparison, usually very small.

Stating conformational structure requires defining the dihedral angles between adjacent three-atom planes, as well as defining molecular structure. Because this parameter must be stated in addition to those required to define molecular structure, conformational structure is evidently a more complex level of structural information than is molecular structure itself.

Conformational structure is a structural dimension of both isolated molecules and molecules within aggregates. Information regarding this level of structure in crystals results directly from single crystal structure determination. In fluid states, information regarding conformational structure can be obtained by use of NMR [15] and the vibrational (infrared and Raman) spectroscopies [14].

The activation energy for rotational isomerism varies widely, depending on the electronic structure of the bond and the state within which the molecule exists. Conformational change in crystals is typically difficult, except for processes such as the rotation of methyl groups. In fluid states, rotation about single C–C bonds is facile, while rotation about an olefinic C=C double bond is difficult. As a result, *cis* and *trans* olefins are usually regarded as different molecules rather than as conformational isomers. If a catalyst or some other means of accelerating rotational isomerism were to exist, or at extremely high temperatures, these two isomers should be regarded as a single component. Under such circumstances the distinction between conformational and molecular structure is fuzzy.

2.4.3 Phase structure

The next higher level of structure with which we must be concerned is phase structure. Phases are macroscopic in nature; the concepts of "phase" and "phase transition" do not apply to isolated molecules.

Phase structure may be regarded as "*the manner in which molecules are arranged in space within a phase*". The definition of "phase" is critical to the definition of "phase structure"; this very important issue is addressed in Section 4.18. The coordinates and orientation of many molecules in space, relative to each other, must be stated to define phase structure. Within a phase the conformational structure of an individual molecule may also be defined, but we cannot describe the structure of a phase itself without considering a representative fraction of the entire ensemble. Thus, the information required to define phase structure is still more complex than is that required to define conformational structure.

The structures of the phases that surfactant form will be treated in Chapter 8. For now it is necessary only to recognize that each phase has a characteristic structure, that this structure involves large ensembles of molecules, and that the phase level of structure is more complex than is conformational structure.

2.4.3.1 Defining conformational and phase structure

Both conformational and phase structure are fully defined once the system variables for a particular mixture are specified (Section 4.8). A system is defined by stating the number of components and their molecular structures. Its state is defined by specifying the system variables, and defining state dictates all the levels of structure up to the phase level. Defining state does *not*, however, dictate levels of structure beyond phase structure. Phase structure is the highest level of structure that is fully defined by thermodynamics, and is in this sense a fundamental property of state. The same cannot be said of any of the higher levels of structure, such as colloidal structure.

Except in crystalline phases, the rate at which conformational structure changes is typically extremely fast compared to the rate at which phase structure changes. Perhaps this is the reason why conformational structure can never be varied independently of phase structure. That is, one cannot create a phase of a selected composition at a particular temperature and pressure, and then (by any means that is presently known) alter the conformational structure of the molecules without simultaneously altering the phase state itself. Once phase structure is defined, conformational structure is also fixed.

These comments presume that equilibrium of state exists, of course. As a metastable state gravitates towards equilibrium, both conformational and phase structure will change during the process. The molecules within metastable phases typically display conformational structures which differ from those of equilibrium phases (Sections 5.5.4, 8.3.3). Infrared data can be used to characterize conformational structure within both metastable and equilibrium phases, but these data have not been widely used as yet.

2.4.4 Colloidal structure

Colloidal structure may exist within both one-phase and multi-phase mixtures. Domain structure in one-phase mixtures may be regarded as a dimension of colloidal structure, since it represents a higher level of structure than does phase structure [16]. Domain structure exists in practically all crystals; special care and painstaking techniques are usually required to produce even small single crystals that do not display significant domain structure (such as twinning). Control over the domain structure in structural materials (such as metals and ceramics) strongly influences their mechanical properties. Materials having small domain size have recently been popularized in materials science as "nanostructured materials" [17].

Domain structure also exists within liquid crystal phases, including those formed from pure compounds (Section 8.4.3). Such structure in fluids persists for long times only within phases that are both highly structured and very viscous. Defects within fluid phases of low viscosity rapidly disappear. The steady state concentration of defects that ordinarily exists within liquid crystal phases gives even uniform samples a characteristic turbid appearance (Section 8.4.3).

The term "colloidal structure" is more commonly used to describe multiphase mixtures, and in this context may be defined as "*the manner in which coexisting phases are arranged in space*". Both phase and conformational structure will have already been defined in specifying the system and stating the system variables, but still more information is required to define colloidal structure. From this fact, it is apparent that colloidal structure is relatively more complex than is phase structure.

The above definition resembles superficially that used in Section 2.4.3 to define "phase structure" itself, which is "the manner in which molecules are arranged in space within a phase". The differences between the two definitions are extremely important. In defining phase structure, one is concerned with structure within mixtures that are thermodynamically homogeneous (Sections 4.18, 4.19). In defining colloidal structure, one is concerned with structure within mixtures that are thermodynamically *heterogeneous* since they consist of two (or more) phases. It is how these *phases* – not molecules – are arranged in space that defines colloidal structure.

It is exceedingly important not to confuse structure at the phase level with structure at the colloidal level, but sometimes they are hard to distinguish experimentally. Important examples of this problem include the "coagel" and "gel" states, which (in the case of dioctadecyldimethylammonium chloride–water mixtures) have been shown to be colloidally structured biphasic mixtures of different crystal hydrates and a liquid phase [18]. If a biphasic mixture having colloidal structure is described as a single phase, then the assumed value for P is wrong. Applying the Phase Rule then leads to an incorrect conclusion either with respect to the variance, F, or the number of components, C.

2.4.4.1   The classes of colloidal structures

It has long been recognized that two broad categories of colloidal structure exist; these have been termed "reversible" [19] and "irreversible" [20]. Reversible colloidal structure exists in micellar liquid solutions of surfactants. These solutions have a variance of three, are thermodynamically homogeneous, and are a single phase (Section 4.8.1).

Micellar structure is indeed reversibly formed from discrete molecules, although the rates of exchange of molecules between monomeric and micellar species are a strong function of chain length [21,22,23]. If a nonequilibrium state is rapidly created within a micellar solution, as during T- or P-jump experiments, it rapidly and spontaneously relaxes to a new equilibrium structure [24]. Thus, micellar solutions meet both of the criteria of reversibly formed states [25]:

(1) they are indefinitely stable with the passage of time, and
(2) under specified conditions, the same state is again reached when these conditions are approached via different paths.

The second category of colloidal structures is termed "irreversible colloids". Examples include emulsions of a liquid or a liquid crystal within a liquid, or dispersions of a crystal phase within a fluid phase. Irreversible colloidal structure meets neither of the above criteria for reversible equilibria. It typically changes continuously with time, and is dependent on the history of the mixture. If disturbed or altered, it does not spontaneously revert to the initial structure.

2.4.4.2 The subclasses of irreversible colloidal structures

Two subclasses of irreversible colloidal structure exist, which have been termed "sols" and "gels" [20]. "Sols" are irreversible colloids in which one phase is dispersed within another as discrete, unconnected particles. (Unconnected does *not* mean non-interacting.) "Gels", on the other hand, are irreversible colloids in which the dispersed phase forms a network which occludes the other phase. A molecule in the dispersed phase within a gel may pass throughout the space occupied by the mixture within this phase without ever entering the other phase. This is impossible in sols; for a molecule to pass from one particle to another, it must enter the continuous phase. Molecules in the continuous phase of sols may, of course, pass throughout the space occupied without entering the dispersed phase. One encounters both classes of colloidal structure (reversible and irreversible), and both subclasses of irreversible structure (sols and gels), in surfactant systems.

Regardless of how long-lived a colloidal structure may be, the fundamental differences between reversible and irreversible colloidal structures remain. Faraday's gold sols have remained essentially unchanged for a century and a half, but it is still preferable to regard these mixtures as metastable dispersions of gold in an aqueous liquid rather than as a single phase. If this position is not taken, then two different equilibrium states would exist for the gold–water system and that is an unsavory situation. If one takes as an axiom that there can exist but one equilibrium state for all systems under specified conditions, then gold sols must be regarded as metastable colloidal dispersions that are incredibly long-lived. How best to describe them mathematically, or to model their properties, is an altogether different matter.

Many higher levels of structure exist that are far more complex than is colloidal structure, as for example within living cells. Also, many "lower" levels of structure exist that are less "complex" than is molecular structure. Examples include the electronic structures of molecules and subatomic structure. The above kinds of structural information have been selected for consideration only because they are particularly relevant to the physical science of surfactants.

2.5 The thermodynamic aspect of physical science

Because of its independence of time and path, thermodynamics is a cornerstone of physical science. The thermodynamic aspect of physical science defines the equilibrium state under specified conditions, and phase science clearly falls within the province of chemical thermodynamics. The phase behavior of a system is most clearly understood from consideration of free energies of mixing, and phase separation can be viewed as the consequence of anomalies in the free energy of mixing function (Chapter 3).

Phase diagrams describe the limits of miscibility of the components of a system. They express the range of temperatures, pressures, and compositions within which the components are miscible as equilibrium phases. At the same time, they define the regions within which the components are immiscible – the miscibility gaps. Finally, they are concerned with the discontinuities that separate regions which display qualitatively different kinds of phase behavior as temperature or pressure is varied.

Phase diagrams do *not* provide numerical values for free energies, enthalpies, chemical potentials, etc.; separate numerical thermodynamic studies (e.g. calorimetry) are necessary in order to obtain this information. Nevertheless, it is clear that phase information is to be regarded as being inherently thermodynamic in character.

The thermodynamic aspect of physical chemistry is qualitatively different from the structural aspect, and the determination of structure is approached using entirely different methods than are used to determine the thermodynamics. While in principle the thermodynamics of a system are dictated by the structures of the molecules present, in practice it is presently not easily possible to extract meaningful information regarding the phase science of surfactants solely by consideration of molecular structure. Direct measurements of thermodynamics and structure are necessary to describe the state of a mixture, and would in any case be necessary to test theoretical predictions as to its state.

2.6 The kinetic and mechanistic aspects of physical science

Thermodynamics defines the equilibrium state, and structural information defines the qualitative nature of these states. However, neither kind of information addresses the rate at which equilibrium is approached during a reaction process.

The thermodynamic, kinetic, and mechanistic aspects of phase transformations are all most easily comprehended and analyzed when expressed as "phase reactions" (Section 4.10). The term "mechanistic" is here used in the context of "reaction mechanisms". The study of reaction mechanisms is concerned with the rate of a process, the rate-determining steps, the intermediates that appear during the process, etc. The form of phase reaction equations is analogous to that of chemical reaction equations (for reactions between molecules), but the reactants and products are phases instead of molecules. As noted above, phase structure changes during phase reactions but molecular structure does not.

If a phase reaction occurs by a circuitous route, or via a sequence of steps having different rates, then one must also be concerned with its mechanism. "Mechanisms", after all, are simply the sequence of events during the course of time by means of which processes actually occur. Because of intervention of the time dimension, the study of mechanisms is intimately connected with the study of rates and kinetics. However, a proper mechanistic investigation will invoke other information besides rate data, such as isotopic studies and the correlation of rates with molecular structure.

Just as with molecular reactions, the mechanisms of phase reactions are often unobvious from consideration of the reactants and the products. An example of this principle from molecular chemistry is the mechanism of the acid-catalyzed halogenation of acetone [26]. In the classic study by Lapworth in 1904 [27], it was found that the rate of this reaction is not only independent of the concentration of iodine, but that chlorine, bromine, and iodine all react at the same rate. The rates of these reactions are governed by the concentration of acetone and of hydrogen ion, a result that was very surprising at the time but has since been observed in many analogous processes.

Kinetic studies played a key role in establishing the presently accepted mechanism of this reaction, in which the overall reaction rate is governed by the acid-catalyzed enolization of acetone. Formation of the enol occurs by loss of a proton from a rapidly-formed O-protonated acetone cation [28]. The acidity of the medium influences the extent to which this fast initial proton transfer reaction occurs, and thereby the rate of the slower enol-forming reaction. The concentration or nature of the halogen does not influence the overall rate because the reaction of halogens with the enol to form the halogenated ketone product is very fast, relative to the rate of formation of the enol itself.

$$\underset{CH_3\overset{\displaystyle O}{\overset{\|}{C}}CH_3}{} + I_2 \xrightarrow{H^+} \underset{CH_3\overset{\displaystyle O}{\overset{\|}{C}}CH_2I}{} + HI$$

$$Rate = k * [CH_3\overset{\displaystyle O}{\overset{\|}{C}}CH_3] * [H^+]$$

$$CH_3\overset{\displaystyle O}{\overset{\|}{C}}CH_3 + H^+ \rightarrow CH_3\overset{\displaystyle \overset{+}{O}-H}{\overset{\|}{C}}CH_3 \rightarrow CH_3\overset{\displaystyle O-H}{\overset{|}{C}}=CH_2 \xrightarrow{I_2} CH_3\overset{\displaystyle O}{\overset{\|}{C}}CH_2I$$

In the phase chemistry of surfactants, the peritectic reaction of crystal hydrates (Section 4.10) has, in two instances, been shown to occur in two discrete steps. The overall phase reaction (neglecting stoichiometries) is expressed using the equation:

$$X \cdot W_n \rightarrow Lam^{eq} + X \cdot W_{(n-1)} \tag{2.3}$$

This process does not actually occur, however [29]. Instead, these reactions occur via fast simple (congruent) melting during which a metastable liquid crystal phase is formed, followed by slower nucleation and crystal growth processes during which the dry crystal and the equilibrium liquid crystal are produced. These two phase reactions describe the mechanism of this process, and may be expressed as follows

$$\begin{aligned} X \cdot W_2 &\rightarrow Lam \cdot W_2^m & \text{fast} \\ Lam \cdot W_2^m &\rightarrow X \cdot W + Lam^{eq} & \text{slow} \end{aligned} \tag{2.4}$$

where the superscript "m" signifies metastable.

There is a direct mechanistic analogy between the formation of the enol in the chemical reaction and the formation of the equilibrium phases in the phase reaction. In both cases a fast initial reaction occurs (proton transfer in enolization, congruent melting in the phase reaction), which is followed by a slower rate-determing process (loss of a C−H proton to form the enol in the chemical reaction, nucleation and crystal growth in the phase reaction). As a result, the rate-determining step is in each case different from the reaction equation which describes the overall course of the reaction.

2.7 The colloidal aspect of physical science

Even after the structural, thermodynamic, and kinetic and mechanistic aspects of a surfactant system have been accounted for, one is often still confronted with colloidal phenomena during phase transformations. Colloidal phenomena are not encompassed by any of the above kinds of information. Because treating colloidal phenomena requires still more information (above and beyond that required to treat the thermodynamic and kinetic aspects), it is evident that information regarding the colloidal aspect is more complex than is any of the above-mentioned kinds of information. Many cases exist in which comprehensive knowledge of the physical science of surfactant systems simply must invoke the discipline of colloid science.

While not phrased in this language, this fact has been recognized since ancient times with respect to the nature of the product formed by freezing liquid eutectic mixtures of inorganic salts [30]. This process typically produces a finely dispersed mixture of very small crystals. In striking contrast, freezing mixtures that differ significantly in composition from the eutectic composition often yields larger and much better-formed crystals. Advantage of this fact is taken by geologists during investigations of the history of a rock formation.

The word "eutectic" (literal translation, "finely woven" [30]) may have been derived from the fine texture of crystal mixtures formed by freezing the eutectic mixture. This texture reflects the existence of many small crystals having a particular size, size distribution, shape, and packing arrangement. These are the parameters that define the colloidal structure of such mixtures.

If colloidal structure is formed, and particularly if it is long-lived, then knowledge of this structure, of the rates and mechanisms of the processes by means of which it is formed, and of the rates and mechanisms of the processes by means of which it collapses, are important dimensions of the physical science of the system.

It is critically important to treat the rate of change with respect to colloidal structure separately from the rate of equilibration with respect to phase structure. It is entirely possible for equilibrium to have been attained with respect to phase structure, within colloidal dispersions, long before the colloidal structure itself collapses. The two processes are not governed by the same factors, may proceed in parallel with one another, and must not be confused.

It is also worth noting that it is inappropriate to use the word "equilibration" with respect to changes in irreversible colloidal structure. Mixtures having irreversible colloidal structure are inherently nonequilibrium states.

2.7.1 The stability of colloidal structure

The time required for an irreversible colloidal structure to collapse varies from seconds to centuries (Section 2.4.4.2), which represents a dynamic range of $>10^9$. If one stores an aqueous mixture of a nonionic surfactant at a controlled temperature above its cloud point, the emulsion structure that is initially formed typically collapses into two clear liquids within minutes to hours. Vesicle dispersions of polar lipids and related compounds, on the other hand, may remain unchanged for years. One such dispersion of dioctadecyldimethylammonium chloride has been kept for more than 9 years without displaying grossly visible change in its colloidal structure, and without collapsing to single crystals (as happens rapidly with nucleation) [18].

"Ripening" ("coarsening") is the dissolution of small particles coupled with the growth of large ones, and is one important class of mechanisms for the collapse of colloidal structure (Section 2.7). It has been shown that the rate of ripening is directly related to the solubility of the dispersed material [31]. Flocculation and coalescence is the other major class of mechanisms for the destruction of colloidal structure, and it too occurs in biphasic surfactant dispersions [20]. Both of these processes, but especially the latter, have been the subject of massive investigations by colloid chemists.

2.8   The theoretical aspect of physical science

Gibbs' justification for his lifelong concern with the theoretical aspects of physical chemistry was that "it is the office of theoretical investigations to give the form in which the results of experiments may be expressed" [32]. Gibbs was keenly interested in the graphical display of fundamental concepts, and this approach has been adopted in presenting in graphical form the qualitative basis for immiscibility using free energy of mixing concepts (Chapter 4). The theoretical aspect of phase science provides a rationale for the experimental thermodynamic results, but this area is not addressed.

One reason for an interest in theory is the innate human desire to understand natural phenomena in terms of fundamental principles. Theories introduce models, and models serve to correlate, explain, and anticipate experimental phenomena. Models are extremely important, because they constitute a pervasive aspect of the human thought process. People seem to require models in order to function and make decisions – almost irrespective of whether or not the models used are sound and well-founded. A particularly valuable aspect of sound theoretical analysis is therefore to bias the models that people use towards ones that have a solid basis in experimental fact and sound theory, and in so doing to redirect attention away from lesser models. It may be presumed that the better the model the higher will be the probability that judgements made using it (before the fact) will be correct. This is extremely important, because it is decisions made before an experimental study is performed (rather than the interpretation of the data produced) which are critically important to charting the directions of research and development programs.

2.9   Interrelationships among the different aspects

The above categorization of the various aspects of the physical science of surfactants does *not* imply that they are unrelated – only that they may be cataloged, organized, and analyzed separately. Each aspect is typically investigated using different methodologies, but it is very important to recognize all the key aspects and not to confuse one aspect with another.

Strong relationships exist among several – perhaps all – of these various aspects of physical science. For example, the influence of solubility on the kinetics of ripening (Section 2.7.1) constitutes a relationship between thermodynamics (solubility is a phase parameter) and colloidal phenomena. Systems are defined by the molecular structures of their components, and (in principle, if not in fact) all aspects of their physical science could be inferred from knowledge of molecular structure (Section 2.4). The rates of phase reactions, and whether or not an intrinsic kinetic barrier to equilibration exists (above and beyond the engineering factors, Section 6.2), are probably related to the numerical value of the reaction entropy. This would be a relationship between kinetics and thermodynamics, of which numerous examples exist in the literature.

The important point is that the above categorization is not intended to dissect and fragment the subject of the physical science of surfactants, but to organize the various kinds of information with which one is concerned into a useful framework.

2.10   The collective value

Information on all the above aspects of a particular system constitutes a comprehensive body of knowledge that is of immense practical value, and is also of considerable academic interest. It is rare for a particular surfactant system to have been studied in all these aspects, and still less common for the information to have been collected and organized. Nevertheless, collecting such information in a concise format can be very useful. Except for the theoretical aspect, a serious attempt to do just that was made during the study of the dioctadecyldimethylammonium chloride (DODMAC)–water system [33,29,18].

TABLE 2.1
An outline of surfactant physical science

SURFACE AND COLLOID SCIENCE
Reversible colloidal structure
    Thermodynamics, kinetics, structures
Irreversible colloidal structure
    Sols – discrete particles
        Energetics, kinetics, structures
    Gels – networks
        Energetics, kinetics, structures

CHEMISTRY
Kinetic and mechanistic aspects
    Rate of approach to equilibrium of state
    Path taken to equilibrium (mechanism)
Thermodynamic aspects
    Analytical description – *phase science*
    Numerical description – thermodynamic studies
Structural aspects
    Phase structure
    Conformational structure
    Molecular structure

PHYSICS
Chemical physics
    Applications of fundamental laws to molecules
Basic physics
    The fundamental laws of nature
    Matter, force, motion, and energy

2.11 Summary

The various aspects of the physical science of surfactants are systematically arranged in Table 2.1, with the more fundamental aspects at the bottom. The relative "complexity" of each aspect (as defined within the text) increases as one ascends the table.

It may be seen from this analysis that the principal role of phase science is to define (analytically) the thermodynamics of mixing of a system. Thermodynamic parameters such as free energy, entropy, chemical potential, etc., do *not* exist in phase diagrams, but the regions within which a single phase is the equilibrium state are defined as a function of temperature, pressure, and composition. This information provides a solid basis for the analysis of more complex aspects of the physical science of the system, such as kinetic and mechanistic phenomena, or the formation, stability, and nature of colloidal structure.

Experience has repeatedly shown that knowledge of the equilibrium phase diagram is valuable. It tells one when a mixture should change further with time and when it should not, and it is the principal (if not the only) means of recognizing fundamental mistakes in the data. A particularly clear illustration of this emerged from investigations of the DODMAC–water system.

In agreement with earlier studies [34], it was found that if the crystal monohydrate $X \cdot W$ of DODMAC is mixed with water and heated, the thin, white, slurry of crystals is suddenly transformed at 39°C into a strong, translucent gel. However, the independently determined equilibrium phase diagram (Fig. 2.2) shows that no phase reaction occurs at this composition and temperature [33]. Moreover, it was shown that the

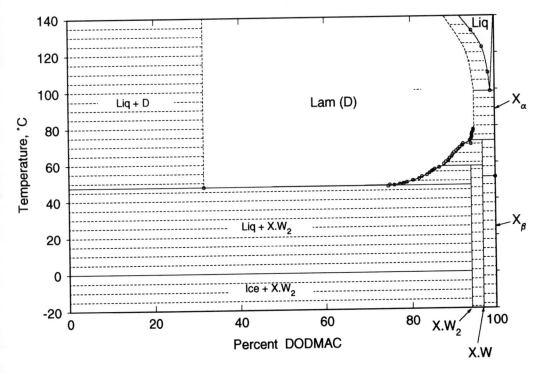

Fig. 2.2.   The dioctadecyldimethylammonium chloride (DODMAC)–water system [33].

equilibrium phase (which is the crystal dihydrate $X \cdot W_2$) does not display this gelation reaction at this temperature. (It does react with water, as expected, to form a liquid crystal phase at the Krafft discontinuity [33].) Further investigations revealed that the phenomenon is due to an unusual hydration reaction of the monohydrate [18], which produces a highly structured colloidal state *without the intervention of mechanical shear*. It is highly improbable that this information would have been revealed had not the equilibrium diagram existed. Knowledge of the equilibrium state was also critical to sorting out the mechanisms of the peritectic phase reactions of the two crystal hydrates (Sections 2.6, 6.5).

The whole of physical science may be important, and phase science is a cornerstone of this body of information.

References

1. Resnick, R. and Halliday, D. (1977). *Physics*, Part I, 3rd ed., John Wiley, New York; Halliday, D. and Resnick, R. (1978). *Physics*, Part II, 3rd ed., John Wiley, New York.
2. Dewar, M. J. S. (1969). *The Molecular Orbital Theory of Organic Chemistry*, McGraw-Hill, New York; Flygare, W. H. (1978). *Molecular Dynamics and Structure*, Prentice-Hall, Englewood Cliffs, NJ.
3. Moelwyn-Hughes, E. A. (1961). *Physical Chemistry*, 2nd ed., p. 1, Pergamon Press, London.
4. Laughlin, R. G. (1991). *Langmuir* **7**, 842–847.
5. Tokiwa, F. and Ohki, K. (1968). *Bull. Chem. Soc. Japan* **41**, 1447–1451.
6. Wyss, H. R. and Falk, M. (1970). *Can. J. Chem.* **48**, 607–614.
7. Klewe, B. and Pedersen, B. (1974). *Acta Cryst.* **B 30**, 2363–2371.
8. Hamilton, W. C. and Ibers, J. A. (1968). *Hydrogen Bonding in Solids*, pp. 188–237, W. A. Benjamin, New York.
9. McBain, J. W. (1918). *J. Soc. Chem. Ind.* **37**, 249–252.
10. Ekwall, P. (1988). *Colloid Polym. Sci.* **266**, 279–282.
11. Small, D. M. (1986). *The Physical Chemistry of Lipids. From Alkanes to Phospholipids*, Vol. 4, pp. 233–343, Plenum, New York.
12. Eliel, E., Allinger, N. L., Angyal, S. J. and Morrison, G. A. (1965). *Conformational Analysis*, Interscience, John Wiley, New York.
13. Würger, A. (1991). *Z. Phys. B: Condens. Matter* **84**, 263–267.
14. Snyder, R. G. (1992). *J. Chem. Soc. Faraday Trans.* **88**, 1823–1833.
15. Emsley, J. W. (1985). *Nuclear Magnetic Resonance of Liquid Crystals (NATO ASI Series)*, D. Reidel, Dordrecht.
16. Kléman, M. (1983). *Points, Lines, and Walls in Liquid Crystals, Magnetic Systems and Various Ordered Media*, John Wiley, New York.
17. Hadjipanayis, G. C. and Prinz, G. A. (1991). *Science and Technology of Nanostructured Magnetic Materials*, *NATO ASI Series B*, Vol. 259, Plenum, New York.
18. Laughlin, R. G., Munyon, R. L., Burns, J. L., Coffindaffer, T. W. and Talmon, Y. (1992). *J. Phys. Chem.* **96**, 374–383.
19. (1949). *Colloid Science, Reversible Systems*, Vol. II (H. R. Kruyt ed.), p. 483, Elsevier, New York.
20. (1952). *Colloid Science, Irreversible Systems*, Vol. I (H. R. Kruyt ed.), Elsevier, New York.
21. Aniansson, E. A. G. (1978). *Ber. Bunsenges. Phys. Chem.* **82**, 981–988.
22. Kahlweit, M. (1982). *J. Colloid Interface Sci.* **90**, 92–99.
23. Gormally, J., Gettins, W. and Wyn-Jones, E. (1981). *Mol. Interact.* **2**, 143–147.
24. Knight, P., Wyn-Jones, E. and Tiddy, G. J. T. (1985). *J. Phys. Chem.* **89**, 3447–3449.
25. Taylor, H. S. and Taylor, H. A. (1937). *Elementary Physical Chemistry*, 2nd ed., p. 384, Van Nostrand, New York.
26. March, J. (1968). *Advanced Organic Chemistry. Reactions, Mechanisms, and Structure*, pp. 458–459, McGraw-Hill, New York.

27. Lapworth, A. (1904). *J. Chem. Soc.* **85**, 30–46.
28. Bell, R. P. and Yates, K. (1962). *J. Chem. Soc.*, 1927–1933.
29. Laughlin, R. G., Munyon, R. L. and Fu, Y.-C. (1991). *J. Phys. Chem.* **95**, 3852–3856.
30. Prigogine, I. and Defay, R. (1954). *Chemical Thermodynamics*, (D. H. Everett ed.), Longmans, Green & Co., London.
31. Kabalnov, A. S., Makarov, K. N., Pertzov, A. V. and Shchukin, E. D. (1990). *J. Colloid Interface Sci.* **138**, 98–104.
32. Marsh, J. S. (1937). *Principles of Phase Diagrams*, pp. 122–170, McGraw-Hill, New York.
33. Laughlin, R. G., Munyon, R. L., Fu, Y.-C. and Fehl, A. J. (1990). *J. Phys. Chem.* **94**, 2546–2552.
34. Kuneida, H. and Shinoda, K. (1978). *J. Phys. Chem.* **82**, 1710–1714.

Part II

PHYSICAL CHEMISTRY

Chapter 3

The Thermodynamics of Immiscibility

3.1 The state variables and functions

A brief review of thermodynamics is appropriate before addressing the particulars of phase science. Those aspects of physical chemistry that directly relate to the thermodynamics of mixtures and their stability will be reviewed in this chapter.

Thermodynamics is that branch of physical chemistry which is concerned primarily with the quantitative dimensions (the state) of a mixture – particularly when it is at equilibrium – and with how the state is influenced by changes in conditions during a process.

State is defined by the magnitudes of four observable quantities, which are often termed the "system variables" [1]. These are:

n – quantity,
V – volume,
T – absolute temperature, and
p – absolute pressure.

"Observable" variables can not only be perceived; they can be measured, controlled, and may serve as independent variables during experimentation.

Changes of state typically involve the addition or subtraction of heat (q) and the performance of work (w), as well as changes in the system variables. From these data, other variables may be derived which (unlike heat and work) are independent of the path taken during a process. These fundamental thermodynamic quantities are termed "state functions", because their magnitudes are characteristic of and serve to define a particular state. They include:

E – energy (which is defined as a function of T and V),
A – Helmholtz free energy (which is defined as a function of T and V),
H – enthalpy (which is defined as a function of T and p), and
G – Gibbs free energy (which is defined as a function of T and p).

One other rather different state function exists,

S – entropy.

The above four state energy functions can never be measured in absolute terms; only the changes that occur during a process can be measured. Entropy, in contrast, can be uniquely defined. It is a function of all the "observable variables".

Quantity. Quantity refers to the mass of a particular component in a mixture. It is often expressed as the number of moles present, n. When more than one component is present, specifying the value of n for each component defines composition. Most thermodynamic parameters are expressed per unit value of n, in both one-component and multicomponent mixtures.

Volume. Volume, V, is a measure of the space occupied by a particular quantity of material. V depends on the identity (molecular structure) of the components, the numerical value of n, and the state.

Temperature. Temperature, T, is a measure of the ambient level of thermal energy, which is related to the motion through space (kinetic properties) of molecules. Temperature has a pervasive influence on all aspects of state.

The constants of proportionality which relate temperature to thermal energy are the Boltzmann constant, k, and the gas constant, R. The former is appropriate when considering individual molecules, and has the units energy/molecule/degree. The latter is appropriate for mole quantities (Avogadro's number) of molecules, and has the units energy/mol/degree.

RT (or kT) is also encountered in numerous other places, such as the ideal gas equation ($pV = nRT$). kT is *not* numerically equal to thermal energy; the kinetic energy of an ideal gas equals $3kT/2$. Nevertheless, it is usual to compare non-thermal energy terms with kT as a means of relating their magnitude to that of the ambient thermal energy. Also, it is often assumed that the thermal energy of states other than the gas phase is related to temperature in the same way as is found in ideal gases [2].

Pressure. Pressure, p, is a measure of the force that is exerted per unit area. (The lower case "p" will be used for pressure, as the upper case "P" will be reserved for the "number of phases" parameter of the Phase Rule.) In a mixture held at a particular pressure, a balance exists between the inward pressure exerted on the mixture by the container and the outward pressure exerted on the container by the mixture. Pressure tends always to be very uniform through space within fluid phases. If a local disturbance momentarily creates a pressure gradient, this gradient is rapidly dissipated (at a rate governed by the speed of sound in the mixture), and a uniform pressure is quickly reestablished [3]. The situation may be different in solids. (Sound is a longitudinal oscillation in pressure which travels at a characteristic speed. The velocity of sound is relatively slow in gas phases, and decreases with decreasing pressure. It is much higher in solids or liquids.)

For ideal gases, pressure is related to temperature and volume by the equation

$$p = \frac{RT}{V} \tag{3.1}$$

which has the dimensions of energy per unit volume (the "density" of energy).

Heat and work. Heat (q) and work (w) are observable parameters – but they differ fundamentally from those mentioned above in that they are *not* state parameters. Their value depends on how one moves from one state to another; that is, it depends on the path (for example in $p - V$ space) that is followed. The numerical difference between q and w, however, *is* one of the state functions (see later). (Incremental changes in q and w, dq and dw, are often described using a crossed "d" and termed "inexact differentials", because of their dependence on path.)

Energy. Energy, E, is the state function which is determined by T and V. As with all energy parameters, absolute values of E cannot be experimentally determined – only the changes that occur during a process. More will be said of E, q, and w in Section 3.2.

Enthalpy. Enthalpy, H, is the state function which is determined by T and p. Enthalpy is determined from data on q, w, and T, and is the appropriate energy parameter when a volume change occurs – that is, when mechanical work of expansion is done. Enthalpy is therefore the state function that is relevant to most actual experimental processes – including the thermodynamics of phase transitions.

Entropy. Entropy is, in the first approximation, a quantitative measure of the level of disorder. As mentioned earlier, entropy has a finite value in any particular state. Thermodynamically favored states are those of low energy – but of *high* entropy. Entropy is ordinarily increased as a result of decreasing the pressure. Entropy is also increased simply as a consequence of the existence of two or more different molecules in the same phase. In an ideal mixture, separating the components requires the expenditure of work or energy because the favorable entropy of the mixture must be overcome (Section 3.10).

The entropies of surfactants are influenced by the number of configurations the molecules present may assume and by the number of different conformational structures that exist, among other things. The number of different conformations varies widely among the different phases. Typically only one conformational structure exists in crystals at low temperatures, but millions of conformational structures may exist in fluid phases (Section 2.4.2). Other things being equal, if only a few conformational structures exist entropy is low while if many such structures exist entropy is high. Molecules which have the same conformational structure in both the crystal and the liquid state (such as bicyclic or aromatic compounds) display anomalously low entropies of melting and, as a result, anomalously high melting points (Section 3.7).

Gibbs free energy, G. It was recognized early by thermodynamicists, and especially by Gibbs, that some additive and/or multiplicative functions of the above variables are particularly useful. The two most important state functions for the present purpose are the enthalpy, H, and the Gibbs free energy, G. H may be defined as a function of E, p, and V as

$$H = E + pV \tag{3.2}$$

G, in turn, may be defined as a function of H, T, and S as

$$G = H - TS \tag{3.3}$$

Similarly, the Helmholtz free energy A may be defined as a function of E, T, and S as

$$A = E - TS \tag{3.4}$$

Great care must be exercised in manipulating these definitions, as constraints on their use result from the way in which they are defined. Recourse to a good thermodynamics text will be necessary for those who are seriously interested in this subject [4,5,6].

3.2 Heat, work, and heat capacities

Heat capacities are highly relevant to phase science. They can be determined using a variety of methods, but in the context of phase science are often determined using differential scanning calorimetry (DSC). DSC not only provides numerical thermo-dynamic data, but is sometimes the principal method used to determine phase diagrams (Appendix 4.4.3) [7].

During a differential scanning calorimetric study using modern instruments, the temperature of the environment within which an experimental sample (and also a reference material) exists is varied within a furnace (or furnaces) that is massive relative to the sample. The temperature difference between the two samples is monitored during the process; if a difference develops due to a phase transition within the sample, these data may be handled in one of two ways. In one approach the numerical value of the temperature difference is simply recorded and the numerical value of the heat effect is extracted by direct calibration using standard materials. In the other (the "null-bal-ance" principle), sufficient thermal energy is supplied to either the sample or the reference to maintain a temperature difference of zero and the amount of heat required to do so serves as a measure of the heat effect.

The raw data produced by DSC studies are the difference in the amount of heat transferred to the sample (relative to the reference), q, as a function of temperature. The heat capacity is then equal to dq/dT.

Two limiting heat capacities exist – one for processes that occur at constant pressure, C_p, and another for processes that occur at constant volume, C_v. These are defined by the equations

$$C_p = \left(\frac{\partial q}{\partial T} \right)_p \tag{3.5}$$

$$C_v = \left(\frac{\partial q}{\partial T} \right)_v \tag{3.6}$$

To extract thermodynamic information from these experimental data necessitates invoking a thermodynamic model.

Heat capacities at constant pressure. It is assumed, following Joule, that

$$dq = dE - dw \tag{3.7}$$

This equation signifies that the incremental change in heat content, dq, equals the change in the work done by the mixture on the environment, $-dw$, plus the energy added that is not used to do work, which is defined as dE. Since the work done during a reversibly executed process that occurs at constant pressure (dw) equals $-p\,dV$, it is evident that under this constraint dq equals dH

$$dq = dE + p\,dV = dH \tag{3.8}$$

Therefore, heat capacities measured at constant pressure for processes conducted in a reversible manner may also be expressed as

$$C_p = \left(\frac{\partial H}{\partial T} \right)_p \tag{3.9}$$

During heat capacity studies at constant pressure on condensed phases, the pressure is typically low and dV is small. (The term "condensed" phases denotes all phases except the gas phase. The term is useful because the enormous difference in density between the gas phase and all condensed phases clearly sets these two kinds of phases apart (except near critical conditions).) As a result, the contribution of the pdV work term is also very small. During heat capacity studies at constant pressure on gas phases, much larger changes in volume occur and the pdV term can be significant even at modest pressures.

Heat capacities at constant volume. If one starts again with eq. 3.7 and applies now the different constraint that the process occurs reversibly at constant volume ($dV = 0$), then no work of expansion is done and

$$dq = dE \tag{3.10}$$

It will be recalled that E has been defined as a function of T and V, so that

$$C_v = \left(\frac{\partial E}{\partial T}\right)_v \tag{3.11}$$

During heat capacity studies of dilute monatomic gases over a temperature range within which no change in electronic state occurs, dE is governed entirely by the change that occurs in the kinetic energy of the atoms. Applying eq. 3.11 to the expression for the kinetic energy of such a gas [8] one obtains

$$C_v = \tfrac{3}{2}R \tag{3.12}$$

which is consistent with experimental data.

At constant pressure not only does the kinetic energy of the gas change, but the volume changes and work of expansion is done. This work is equal to pdV, or equivalently RdT, so that $C_p = C_v + R$. The expected value of C_p is therefore

$$C_p = \tfrac{5}{2}R \tag{3.13}$$

and this is also found experimentally [8].

It is extremely difficult to measure C_v on condensed phases [9]. At constant pressure the change in volume which occurs as the temperature is increased is small, but the increase in pressure required to maintain volume strictly constant is extremely large. The magnitude of this pressure change is related to the inverse of the compressibility, β, which is large for gases but small for condensed phases [10].

DSC studies of aqueous surfactant mixtures are typically measured in closed containers that contain void space, and the expansion of condensed phases can easily be accommodated with a minimal increase in pressure under these circumstances. Changes in pressure resulting from the increased vapor pressure of water are also typically small (a few atmospheres at most). Such experimental results may be precisely defined as heat capacities measured at saturated vapor pressures [11], but are probably similar numerically to those that would be obtained during constant pressure studies.

If the temperature of a mixture of condensed phases is increased under conditions that approximate the constraint of constant volume, enormous pressure increases and dire consequences may result. Pressure changes are smooth between phase transitions, but discontinuous (and often relatively large) at phase transitions. If the mixture is held in a rigid container having no void space, the increase in pressure may rupture the container unless it is exceptionally strong. If rupture occurs a volume change results and considerable mechanical work is done.

A familiar example of this phenomenon (at phase transitions) is the cracking of a thick-walled glass milk bottle by freezing the milk. Further, the freezing of water has profound geological effects which include frost heave, modification of the structure of soils, the fracture of rock formations (and roadways), the loosening of soil towards landslides, etc. [12]. The phenomenon of expansion during melting (which occurs with most materials) was encountered during unsuccessful attempts to utilize zone refining to purify triacontanol [13]. (Triacontanol is the straight chain C_{30} alcohol n-$C_{30}H_{61}OH$.) Degassed samples sealed *in vacuo* within long glass or plastic tubes were used during these experiments, every one of which failed because the tubes ruptured during the process. These experiments could only be carried out in horizontal boats so that lateral volume expansion could easily occur within the melted zone.

3.3 Partial molar quantities: the chemical potential

Heat capacities are representative of a large group of thermodynamic functions which describe the dependence of one state variable on another, or upon composition. All these parameters are rigorously defined by partial derivatives of one quantity with respect to the other within stated boundary conditions. In a graph, the magnitude of these partial derivative functions equals the slope of the curve rather than its coordinates.

Of particular importance within this class are those functions that describe the dependence of a state variable or state function upon the quantity, n, of a particular component. These are termed "partial molar quantities"; that partial molar quantity which is particularly important to phase science (as well as to most other aspects of chemistry) is the "chemical potential", μ. Under specified conditions, each component, i, of a mixture has a chemical potential, μ_i, which is defined as

$$\mu_i = \left(\frac{\partial G_t}{\partial n_i} \right)_{T,P,n_{j \neq i}} \tag{3.14}$$

This equation signifies that the chemical potential equals the incremental change in *total* free energy, G_t, that results from an incremental change in the quantity of component i, at specified temperature, pressure, and composition. It is understood that the quantities of all the components (j) except the one under consideration (i) are held constant. If one were to plot G_t vs. n_i, μ_i would equal the slope of this curve at any particular composition. In the usual plot of G vs. mole fraction x_i, μ_i is *not* simply the slope (Section 3.15).

Often one is concerned with mixtures containing two or more components, in which case the above definition of chemical potential is straightforward. If only one component exists composition is invariant; then, the chemical potential simply equals the molar free energy. This is generally stated as a fact without explanation, but can be readily understood by recognizing that the total free energy, G_t, of a sample that contains n moles is the molar free energy G times the number of moles

$$G_t = G \times n \qquad (3.15)$$

Differentiation of eq. 3.15 with respect to n, at constant temperature and pressure, leads to

$$\left(\frac{\partial G_t}{\partial n} \right)_{T,P} = G = \mu \qquad (3.16)$$

The chemical potentials of pure compounds will be used in describing their phase behavior.

Chemical potentials within a phase are related to the free energy of that phase by the general equation

$$G^k = \sum n_i^k \mu_i^k \qquad (3.17)$$

where k indicates a particular phase. This equation states that the free energy is the summation of the terms $n_i \mu_i$ for all of the components in the phase. For a phase that is a binary mixture of water (W) and surfactant (S)

$$G = n_W \mu_W + n_S \mu_S \qquad (3.18)$$

The term $n_W \mu_W$ may be thought of as the "water free energy", and the term $n_S \mu_S$ as the "surfactant free energy" components of the total free energy. This breakdown of the free energy is very useful in diagnosing the thermodynamic factors that influence phase stability (Section 5.1), and in clarifying the relationship between water activity and aqueous surfactant phase behavior (Section 7.1).

3.4 The "extensive" and "intensive" classes of thermodynamic variables

It is important to recognize that qualitative differences exist among all these thermodynamic variables, and that their further classification is useful. Some are termed "extensive" variables, while others are termed "intensive" [4,5,6]. The difference is that the magnitude of "extensive" variables depends on the mass of the sample (n), while that of "intensive" variables does not. Using volume as an example, doubling the mass clearly also doubles the volume. Similar considerations apply to free energy, enthalpy, etc. Expressing these parameters as quantity per mole in no way alters this intrinsic property that they possess.

It is equally apparent that altering the mass of a sample, without otherwise changing the conditions, does not change the temperature or the pressure. Mass likewise has no influence on the magnitude of chemical potentials, or of any partial molar quantity.

Those parameters whose magnitude are independent of mass have been termed "intensive" parameters. Some important members of these two classes are:

Extensive: V, E, S, G, H, A, C_p, and C_v.
Intensive: T, p, μ and other partial molar quantities, and β.

β is the compressibility, which is defined as

$$\beta = -\frac{1}{V}\left(\frac{\partial V}{\partial P}\right)_T \tag{3.19}$$

3.5 The "density" and "field" classes of thermodynamic variables

In 1970 Griffiths and Wheeler suggested significant improvements in the basis for classification, and new labels for the classes [14]. Those previously designated as "extensive" variables were redesignated "density" variables, while those previously designated as "intensive" variables were redesignated "field" variables. (In this discussion, following Griffiths and Wheeler, no distinction will be made between the five basic "variables" and the state or derivative "functions". All will be termed "variables".) These suggestions went far beyond merely relabeling, however.

The analysis of Griffiths and Wheeler presumes that the system, the number of components, C, and the number of phases, P, have been specified. For such a mixture, n independent thermodynamic variables must be specified to fully define its state. "n" equals the number of degrees of freedom, F, in the Phase Rule.

Griffiths and Wheeler start with a set of $n+1$ field variables, which include the required n independent variables plus an additional one that is dependent on the others:

Selected field variables $= h_0, h_1, \ldots h_n$

The differentiation of one field variable with respect to another results in a density variable. The dependent field variable, h_0, is therefore chosen so that for each h_i a corresponding density variable, ρ_i is implicitly defined. (More precisely, ρ_i is not usually a simple density variable, but a particular function of density variables.) ρ_i is related to h_0 and h_i by the equation

$$\rho_i = \frac{dh_0}{dh_i} \tag{3.20}$$

To illustrate using a binary system within a single phase region ($n = 3$), one example of a set of field variables that might be chosen is

$$h_0 = -p$$
$$h_1 = \mu_1$$
$$h_2 = \mu_2 \tag{3.21}$$
$$h_3 = T$$

Applying eq. 3.20 to this set leads to

$$\rho_1 = \frac{d(-p)}{d\mu_1} = d_1$$

$$\rho_2 = \frac{d(-p)}{d\mu_2} = d_2 \qquad\qquad (3.22)$$

$$\rho_3 = \frac{d(-p)}{dT} = S/V$$

In this particular instance the variables d_1 and d_2 are literally the molar densities of components 1 and 2 (n_1/V and n_2/V), while S/V is the entropy divided by the volume (the "density" of entropy). A number of such sets may be defined.*

By use of this procedure, correlated sets of field and density variables may be defined that are related to each other by eq. 3.20. These sets possess important properties, one of which is, to quote Griffiths and Wheeler, that:

> the fields (in contrast with the densities) have the property that they take on identical values in two phases which are in thermodynamic equilibrium with each other.

Griffiths and Wheeler were concerned primarily with the analysis of critical phenomena, but their approach is broadly useful in phase science. It not only provides a means of expressing the condition of equilibrium that is both rigorous and simple, but also emphasizes the equally important characteristic of coexisting phases that, with certain exceptions (Section 3.7), they differ in their density variables. In most texts this latter aspect of the thermodynamics of phase equilibria, which is extremely important, is hardly mentioned.

The Griffiths and Wheeler analysis also provides an excellent basis for defining – on a purely thermodynamic basis – just what is a phase (Section 4.18). In so doing it also clarifies (in the first approximation) the concept of an "interface". Finally, in defining phase and interface, it provides a straightforward means of distinguishing interfaces between phases from structural surfaces within phases. Since all of these basic concepts are presently very much in need of clarification throughout surfactant science, the terminology and concepts of Griffiths and Wheeler will be adopted in this book.

The concept that an interface is a surface across which a spatial discontinuity in density variables exists (Section 4.18) is an extremely useful first approximation. However, the matter does not stop there. Extensive inquiry into the particulars and details of interfaces, their structures at the molecular level, their description using the methods of statistical mechanics, and their properties is presently under way [15]. These details are far beyond the scope of this book.

Griffiths and Wheeler also developed a method of graphically depicting phase behavior within "field-space" [16]. If for a binary system one field parameter is plotted against another the one-phase regions remain regions, two-phase miscibility gaps

---

*It was noted in Section 3.1 that pressure is related to the "density" of kinetic energy, or the total kinetic energy divided by the volume. Nevertheless, the behavior of pressure dictates that it be regarded as a field variable rather than a density variable in the context of phase science.

become lines (separating the one-phase regions), and three-phase discontinuities are the points of intersection of these lines. Such diagrams are particularly useful as a means of visualizing the variance that exists under these differing conditions of phase behavior (Chapter 5). If one of the field variables selected is temperature, they also depict the physical appearance of DIT cells during phase studies using this method (Section 4.18.1). However, they are of limited practical value for most phase work because the quantitative relationship between chemical potentials and composition is usually unknown.

Field-space diagrams have long been used to describe unary systems; the pressure–temperature phase diagrams of these systems are, in fact, field-space diagrams.

3.6 Defining systems and mixtures

Having reviewed the thermodynamic variables and described their classification, we now turn to their use. The first step in any analysis of phase behavior is that of defining the "system". This is accomplished simply by stating the number of compounds that will be considered and, for purposes of identification, their molecular structures.

System = number of components + molecular structures

Defining the sodium chloride – water system, in other words, amounts simply to stating that these two compounds, which have the molecular formulae NaCl and H_2O, are the ones with which we will be concerned. Defining a system excludes from consideration all other molecules, and thus imposes broad limits onto the analysis. It must be recognized, however, that those molecules which result from facile chemical reactions among the components must also be considered (Section 2.2.1). Defining a system implies nothing at all with respect to other circumstances.

Next, a phase diagram of the selected system is acquired. Phase diagrams typically impose additional restrictions, since they usually provide information spanning only a finite range of temperatures and pressures. However, a respectable binary phase diagram *must* span the entire composition range (Appendix 4.4). It should define the phase behavior of each of the pure components, as well as of all possible mixtures thereof, within the ranges of temperature and pressure investigated. Ideally either pressure or temperature is held constant, but this is rarely (if ever) the case. Fortunately, this is not a serious problem (as noted in Section 3.2) provided the analysis is restricted to condensed phases at moderate pressures. Very high pressures *will* change phase behavior [17].

Finally, the additional constraints related to mass and heat flow that are imposed for a particular process need to be recognized. During the analysis of calorimetric data, for example (Section 3.21), one is concerned with "closed mixtures". Mass is neither gained nor lost during a process that occurs with closed mixtures, while heat flow between the mixture and the environment is allowed. In considering "dilution" or "concentration" processes one is concerned with "open mixtures", which signifies that both mass and heat flow in and out of the mixture may occur. The third kind of process that may be encountered is the "adiabatic" process, which signifies that neither mass nor heat flow occurs between the mixture and the environment.

3.7   The thermodynamics of unary systems

The simplest systems contain only one component, and are termed "unary" systems [18]. Since composition is invariant, the form of the free energy functions that span the phase transitions of unary systems differs from those of binary and higher systems where composition may vary (see later). Because chemical potentials are numerically equal to molar free energies in unary systems, it suffices to plot the free energy of each phase state that exists against a particular field variable, for example, temperature. Another field variable, for example, pressure, is typically held constant. Which phase has the lowest free energy may then be determined simply by inspection.

In water at a pressure of one atmosphere (Fig. 3.1) one curve describes the free energy of the crystal phase (Ice I), another that of liquid water, and still another gaseous water. (The crystal chemistry of water at high pressures is complex. A series of phases of progressively increasing density exist as the pressure is increased which also depend on the temperature. The phase diagram of water at both high pressures and high temperatures is reproduced in many physical chemistry texts.) At most temperatures one or the other of these phases has the lowest free energy and, as a consequence, is the equilibrium phase at that temperature. Where the curves cross the free energies (and chemical potentials) of two phases are the same, and these two phases may coexist at these points. As the temperature is increased past a crossing point, the low temperature phase is replaced as the equilibrium state by the high temperature phase. To illustrate,

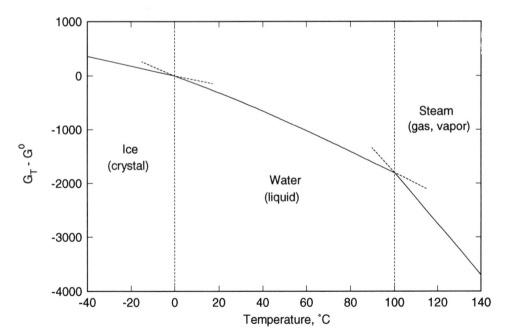

Fig. 3.1.   The free energies of the various states of water (ice, liquid water, and steam) that exist at one atmosphere pressure. The data for this figure were supplied by R. S. Hansen. Data on the thermodynamic states of water and steam are found in the Steam Tables [19] and on ice in handbooks [20], but the standard states taken in these references differ from that taken in the above figure.

ice is the equilibrium phase below 0°C, and ice may coexist with liquid water at this temperature. Liquid water is the equilibrium phase between 0 and 100°C, and this phase may coexist with the gas or vapor phase (steam) at 100°C. Above 100°C the gas is the equilibrium phase. A similar analysis may be undertaken as a function of pressure, at constant temperature.

In unary systems the field variables are the same within phases in equilibrium (Section 3.7). The enthalpy H, entropy S, and other density variables are different, but the free energy (because of its numerical equivalence to chemical potential) is the same. (The same will be true of azeotropic transitions, which are a special case of isothermal transitions of multicomponent mixtures during which composition also does not change.) If one were to overlay a plot of enthalpies on Fig. 3.1, the enthalpy of the low-temperature phase at each crossing point would be lower than that of the high-temperature phase. Heat flow to or from the environment is therefore required at each crossing point for the transformation of one phase to the other to occur. Since a volume change typically occurs as well, work is done and the applicable state function is enthalpy H (rather than energy E). Because these changes in heat content occur isothermally, they are called "latent" heats. (The complementary term is "sensible" heat.)

The relationship between free energy, enthalpy, and entropy at the crossing points are defined by eq. 3.3. For example, at the melting point of ice

$$G_{ice} = G_{liq} \tag{3.23}$$

The thermodynamic changes associated with the transformation of ice to liquid water are

$$H_{ice} - T \times S_{ice} = H_{liq} - T \times S_{liq} \tag{3.24}$$

from which one sees that

$$H_{liq} - H_{ice} = T \times S_{liq} - T \times S_{ice} \tag{3.25}$$

The temperature of the phase transition equals the ratio of the enthalpy change during the phase transformation to the entropy change. (In characterizing the thermodynamics of an isothermal transition, the temperature and enthalpy are directly measured and the entropy is calculated. The relationship defined by eq. 4.28 nevertheless holds.)

$$T_{melting} = \frac{(H_{liq} - H_{ice})}{(S_{liq} - S_{ice})} = \frac{\Delta H_{melting}}{\Delta S_{melting}} \tag{3.26}$$

This equation is valid for all reversible processes in which the compositions of the reactant and the product phases are the same. These include all phase transitions in unary systems, plus azeotropic discontinuities in multicomponent systems. This relationship applies to *all* reversible phase reactions – provided the stoichiometry of the reaction is taken into account (Section 3.21).

Metastable phases. Each of the curves describing the free energies of a given phase may be extrapolated past the temperature range within which it is the equilibrium state (Fig. 3.1). Such phases represent one kind of "metastable" phase state; they possess free energies which are higher than that of the equilibrium phase.

It is often possible to measure the properties of metastable phases in the same manner that one investigates equilibrium phases – provided the rate of equilibration of state is sufficiently slow. This is often the case when the entropy change during the phase transition is negative (Section 6.3). For example, it is often possible to investigate supercooled liquids below their freezing point – but not superheated crystals above their melting point. Since supercooling is known to occur in passing from one thermotropic liquid crystal state to another [21], these metastable states too are amenable to investigation. Investigations of superheated liquids are difficult.

Metastable polymorphic phases. Both equilibrium and metastable polymorphic states exist among crystal phases (Section 5.5.4). Equilibrium polymorphs in a unary system would be represented in the plot of free energy vs. temperature simply by another line and, within a particular range of temperatures, the polymorphic state would be the equilibrium phase.

Metastable polymorphic phases of surfactants may often be created by crystallization from different solvents, or by quickly cooling the liquid phase. The free energy curves of these states lie above that of the equilibrium phase under all conditions, so that these metastable polymorphic states differ fundamentally from the above-mentioned metastable phases. Metastable polymorphic crystal states are very important in the physical science of many nonsurfactant long-chain compounds, such as triglycerides [22].

Metastable polymorphic states that are sufficiently long-lived to be seen by casual observation are unknown for gases, liquids, and for lyotropic liquid crystals. (Slow transitions from one thermotropic liquid crystal to another have been recorded [24], but to the author's knowledge have not been observed in lyotropic systems.) If a polymorphic phase structure were to be created within these phases, it would be expected to relax quickly to the equilibrium state (Section 6.3.3). Metastable phases could likely be observed in all phases within a sufficiently short scale of time [23].

3.8 Composition units

We now consider systems containing more than one component, at which point composition enters as an important variable. It is useful first to look closely at what is meant by this term.

The "composition" of a mixture is designated by stating the fraction of each component that is present. The term "concentration" will not be used for this purpose; it will be restricted to designating the compositions of mixtures that are a single phase. This restriction is commonly understood, but is rarely stated. When concentrations are stated under the assumption that a mixture is a single phase, but it is not, then the degrees of freedom assumed is incorrect as well. Serious errors may result under these circumstances. While "concentration" is widely used with reference to liquid phases, the term may also be properly used for liquid crystal or crystal phases.

A "mixture" is any sample that has a defined composition. Neither temperature, pressure, nor any other aspect of the physical science of the sample, is implied in using this term. The word "system" will not be used in this sense, because of the ambiguity that this usage introduces. The word "system" is used in most physical chemistry texts both to define the components present (as will be used here), and to define a mixture of a particular composition. Ricci (who is one of the few authors to confront this ambivalence) resolved the issue by using the upper case "System" to designate the components present and the lower case "system" to designate a particular mixture, but his convention will not be used.

Ideally components are numbered and designated using a subscript, while phases are lettered and designated using a superscript. (Well-established designations of phases such as the "L_α" phase of polar lipids exist, however, and have been retained.) It is conventional to use the number 1 to designate the solvent (water) and higher numbers to designate the solute (the surfactant). To illustrate, the total weight fraction of component i in a mixture would be denoted by w_i. The weight fraction of component i in phase a (within a heterogeneous mixture) is w_i^a. The weight fraction of phase a in such a mixture is w^a.

Components and phases may also be identified using more explicit symbolism. These conventions will be used with the thermodynamic variables and functions as well as with composition.

As a reminder, for a mixture of a binary system (one having two components) the weight fractions of components 1 and 2 are defined as

$$w_1 = \frac{Q_1}{Q_1 + Q_2}, \quad w_2 = \frac{Q_2}{Q_1 + Q_2} \tag{3.27}$$

where Q_1 is the weight of component 1 that exists in the mixture and Q_2 is the weight of component 2. Ternary and higher order systems require additional terms in the denominator. In most phase diagrams composition is expressed either in units of weight fraction (weight per each unit weight of mixture), or more commonly in weight per cent (weight per hundred units of weight of mixture).

An excellent reason for expressing composition in weight units is that these are ordinarily the units of the raw data, and no mathematical transformations to other units (with the attendant risk of errors) are necessary. Another is that weight divided by density equals volume, but volume is a basic thermodynamic variable. Mass is not; it is insensitive to conditions. Volume units, especially when expressed as a fraction of the whole, are particularly relevant to phenomena such as optical scattering or rheology.

Finally, weight (a force) is proportional to mass and expressing composition as weight fractions eliminates the influence of gravity on defining compositions. Mixtures defined by weights determined in the same gravitational field will have the same mass ratio – even though the numerical value of the weight will vary with the gravitational field strength. In an orbiting satellite the weight of a mixture would be approximately zero, but the mass fraction determined from weights measured at the surface of the earth would remain valid.

It is often useful to convert weight fractions into mole fractions. The same conventions with respect to subscripts and superscripts are used. To illustrate for a two-component system containing n_1 moles of component 1 and n_2 moles of component 2, mole fractions are defined by

$$x_1 = \frac{n_1}{n_1 + n_2}, \quad x_2 = \frac{n_2}{n_1 + n_2} \tag{3.28}$$

Since the number of moles of component i equals the quantity of component i divided by its molecular weight, it is clear that defining w_i also rigorously defines x_i, and vice versa. Either composition unit can be used to suit the purpose at hand, but the quantitative relationship between weight fraction and mole fraction obviously depends on the particular molecular weights of the components.

Composition units of mass/volume are not to be used in phase science. While these may be convenient for analytical purposes (as in titrimetry), they are useful for these purposes only because both temperature and pressure are constant (or nearly so) during both the standardization of reagents and their later use. The problem with such units is that while mass is not a thermodynamic variable, volume is. The magnitude of volume depends on the system variables, and the volume of a mixture equals the sum of the volumes of the components only in rarely encountered ideal mixtures. Defining composition using mass units both unambiguously defines composition, and frees one of the necessity to independently determine innumerable densities.

3.9 The ideal mixing approach

The physical chemistry of dilute solutions is best approached by considering the behavior of ideal solutions [25]. A fair number of solutions (including surfactant solutions, Section 7.5) do indeed behave ideally or very nearly so, and may be quantitatively described using the laws governing ideal (or nearly ideal) behavior (Raoult's Law or Henry's Law). Moreover, those systems that do not behave ideally are often well described quantitatively by determining the extent to which they deviate from ideal behavior. For this purpose "thermodynamic activities" are used in place of compositions; the activities are related to compositions using "activity coefficients" (γ_i) by expressions such as $a_i = \gamma_i x_i$. This approach is useful because the relationship between activities and thermodynamic quantities (such as chemical potential) conforms to that of the relatively simple ideal laws. Interpreting activity coefficients requires extra-thermodynamic information, and is another matter.

This perspective is equally useful as a means of qualitatively visualizing the thermodynamic origin of immiscibility. Once the ideal mixing curve is understood and defined, the kinds of anomalies that lead to immiscibility may be introduced and visualized. Curiously, this approach is taken in only a few phase science texts. (This approach is taken by Oonk [26], but the form of the free energy of mixing functions proposed differ substantially from those described below.)

In taking this approach it is presumed that homogeneous mixtures which span the entire composition range can be imagined to exist. The critically important issue of whether or not a particular mixture is, or is not, thermodynamically stable is, for the moment, deferred.

This approach is not at all unrealistic; the components of many binary systems (e.g. of solvents) are indeed "infinitely miscible".* While only a few pairs of closely matched molecules are strictly ideal, a far larger number are miscible in all proportions [25].

*The term "infinitely miscible" is an extremely awkward one to defend, but is nevertheless widely used. What is meant by this term is that the components under consideration are miscible in all proportions, usually with reference to particular values of temperature and pressure. If a single adjective is to be used, "completely" or "perfectly" is preferred to "infinitely".

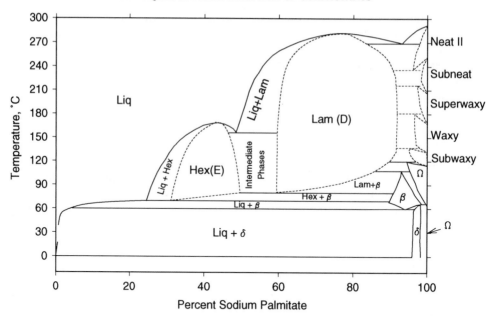

Fig. 3.2. The sodium palmitate–water system, compiled by F. B. Rosevear from the information available as of 1951 [26]. The "intermediate" phases, later found to exist between the hexagonal and lamellar phases [28], are not indicated in this diagram.

It is less widely appreciated that probably all surfactants are miscible with water in all proportions under the right conditions. An illustration of this principle can be seen in the phase diagram of the sodium palmitate–water system (Fig. 3.2) [26]. This diagram (determined from studies performed in sealed tubes at elevated pressures) shows that, at temperatures above 300°C, the liquid phase traverses the entire composition range. This signifies that this soap is indeed miscible with water in all proportions within this range of temperatures.

More surprisingly, this generalization appears to be true even for surfactants whose aqueous solutions display those miscibility gaps which lead to "cloud points" (Section 5.10). It has been suggested that such gaps likely constitute the lower portion of a "closed loop of coexistence" [16]. Closed loops have been shown to exist in the $C_{10}E_4$–water (Fig. 5.17) and $C_{10}E_5$–water systems (Fig. 3.3), and also in the decyl- and dodecyldimethylphosphine oxide–water systems. It is possibly important that in all four of these systems the upper critical temperature is close to the critical temperature of water itself (Section 10.6.1) [27].

Soaps contain the powerfully hydrophilic metal ion and carboxylate groups (Section 10.5.1), while the others have weakly hydrophilic hydroxy polyether (Section 10.5.5.2) or phosphine oxide (Section 10.5.4.1) hydrophilic groups. Nevertheless, all of these surfactants are, at sufficiently high temperatures, miscible with water in all proportions.

In the ideal mixing approach surfactants and water are regarded as being naturally miscible, and where this is not true one asks why not. The alternative is to regard these compounds as being naturally *im*miscible, and seek to explain the existence of miscibility. This latter view is incompatible with the general result that all materials

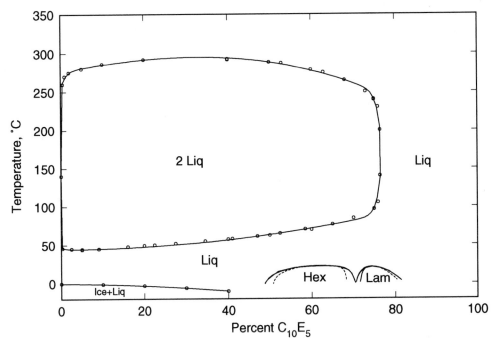

Fig. 3.3. The pentaoxyethylene glycol decyl ether ($C_{10}E_5$)–water system [16].

have a natural tendency to mix (due to entropy factors), and only fail to mix when an overriding enthalpy factor prevents mixing. Further, no good model for immiscibility exists while a simple model for ideal mixing does.

3.10 The ideal free energy of mixing

We will consider the mixing of n_1 moles of component 1 and n_2 moles of component 2, to form a homogeneous mixture – a single phase.

$$n_1 \text{ moles } 1 + n_2 \text{ moles } 2 \rightarrow \text{homogeneous mixture} \qquad (3.29)$$

The free energy change for this process, ΔG_t, equals the free energy of the mixture minus that of the pure components, and is given by the equation

$$\Delta G_t = (n_1\mu_1 + n_2\mu_2) - (n_1\mu_1^0 + n_2\mu_2^0) \qquad (3.30)$$

where the first two terms refer to the mixture. The superscript "0" is appropriate for the last two terms because the pure components have been taken to be the standard state.

Ideal mixtures may be defined as those within which the chemical potentials obey equation 3.31 [25]. This postulate also signifies that these mixtures obey Raoult's Law. In the general expression for chemical potentials, $\mu_i = \mu_i^0 + RT \ln(a_i)$ where a_i is the thermodynamic activity of component i. The basic postulate of Raoult's Law is that the thermodynamic activity equals the mole fraction ($a_i = x_i$)

$$\mu_i = \mu_i^0 + RT \ln x_i \tag{3.31}$$

Inserting this expression for μ_i into eq. 3.30 leads to

$$\Delta G_t = n_1 \mu_1^0 + n_1 RT \ln x_1 + n_2 \mu_2^0 + n_2 RT \ln x_2 - n_1 \mu_1^0 - n_2 \mu_2^0 \tag{3.32}$$

The standard state terms cancel, leading to

$$\Delta G_t = n_1 RT \ln x_1 + n_2 RT \ln x_2 \tag{3.33}$$

and dividing by $(n_1 + n_2)$ gives the free energy difference per mole of mixture

$$\frac{\Delta G_t}{(n_1 + n_2)} = \Delta G = x_1 RT \ln x_1 + x_2 RT \ln x_2 \tag{3.34}$$

It has been noted that only the entropy, and thermodynamic parameters which include entropy, are *not* linear functions of composition in ideal mixtures [25]. The remaining density variables *are* linear. For example,

$$V = x_1 V_1^0 + x_2 V_2^0 \tag{3.35}$$

3.11   The ideal "reduced free energy of mixing" function

The actual value of ΔG in eq. 3.34 is a function of temperature, but if both sides are divided by RT a dimensionless function, $\Delta G/RT$, may be defined. We will designate this the "reduced free energy of mixing", and denote it by ΔG_r

$$\Delta G_r = \Delta G/RT = x_1 \ln x_1 + x_2 \ln x_2 \tag{3.36}$$

Expressing ΔG as the reduced function ΔG_r is useful because ΔG_r may be discussed without the need to specify temperature. ΔG can be readily calculated from ΔG_r, if desired, by multiplying ΔG_r by RT.

$$\Delta G_r \times RT = G \tag{3.37}$$

Similarly, the entropy of mixing, S_m, is obtained by multiplying ΔG_r by $-R$

$$\Delta G_r \times (-R) = S \tag{3.38}$$

3.12   The form of the ideal free energy of mixing function

Figure 3.4 is a plot of ΔG_r vs. x_2 for ideal mixtures, which shows that this function is less than zero for all values of x_2. That is, ΔG_r falls below the line connecting the standard state chemical potentials of the two components. This important quantitative result is consistent with the above-mentioned qualitative principle: the ideal free energy

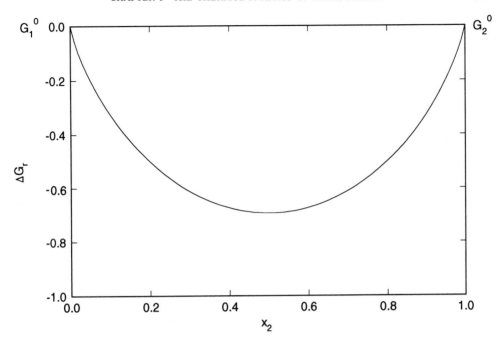

Fig. 3.4. The reduced free energy of mixing, G_r, vs. mole fraction of the solute, x_2, for ideal mixtures (Section 3.11).

of mixing is always negative (favorable). The entropy of mixing term, by itself, constitutes a sufficient driving force to insure that mixing occurs in ideal systems. Heats of mixing are zero. Thus, no enthalpic driving force for mixing exists, nor is it required.

Since $\Delta S = \Delta G_r \times (-R)$, it can be seen that a plot of ΔS vs. x_2 would lie above the line connecting the standard states for all values of x_2. This too is as it should be, since favored processes are characterized by positive changes in entropy (Section 3.1).

The minimum value of ΔG_r equals -0.69315. The value of the gas constant is $8.3130 \, \mathrm{J \, m^{-1} \, K^{-1}}$ ($1.9872 \, \mathrm{cal \, m^{-1} \, K^{-1}}$), so that the maximal free energy of mixing per mole of mixture, which is found at this composition (at 27°C or 300 K) is $-1.7286 \, \mathrm{kJ}$ ($-0.4132 \, \mathrm{kcal}$). The intercepts of the free energy of mixing function at the limits $x_1 \to 1$ and $x_2 \to 1$ correspond to the respective standard state free energies. From these calculations, it can be seen that the magnitude of the entropic driving force for mixing is small. As the composition of mixtures approaches either composition limit the numerical value of the free energy of mixing becomes extremely small – but it is still negative and sufficient to guarantee mixing. It need only have a finite (non-zero) negative value to insure that mixing occurs. Conversely, only a small positive value will preclude mixing.

3.13   Nonideal mixing with miscibility in all proportions

We now consider nonideal mixing processes with small deviations from ideality that are insufficient to cause immiscibility. This would result if the curve, regardless of its form, remains everywhere below the line connecting the standard state free energies and displays no inflections. Such mixtures have been exhaustively treated, both experimentally and theoretically [25].

3.14 Partial miscibility

We now come to the kind and degree of nonideality that results in miscibility over only part of the composition range. This situation is extremely common – and thus very important. If the free energy of mixing curve qualitatively resembles that in Fig. 3.4 the homogeneous phase is the state of lowest free energy at all compositions. If, however, an anomaly in the thermodynamics of mixing produces a bump in the curve, as shown in Fig. 3.5, then this is not true. Starting at the left (with pure component 1) in Fig. 3.5, one may add component 2 (moving along the curve to the right) up to composition a, and the homogeneous mixture *is* the equilibrium state. If the composition is further increased to c, however, it is not. Since the curve has taken an upward turn, this mixture may attain a lower free energy by separating into two phases. The two important questions are "which two phases?", and "how may they be recognized?".

3.15 The condition of equilibrium for coexisting phases

If c separates into two phases and one of these is d (Fig. 3.5), the constraint of mass balance requires that the composition of the remaining phase lie to the left of c. (It is very useful in phase science to speak of the "dilute" phase and the "concentrated" phase in connection with the process of phase separation. "Dilute" refers to the phase having a relatively high proportion of solvent, while "concentrated" refers to the phase having a relatively low proportion of solvent. The absolute concentrations are immaterial.) The separation of c into d and any other phase may be regarded as a "disproportionation phase reaction". If the phase formed from c were to have composition d, however, the resulting mixture is not the state of lowest free energy – regardless of the composition of the other phase.

This may be seen by invoking the condition of equilibrium, which states that the field variables must be the same in both phases (Section 3.5).* These include (besides temperature and pressure) chemical potentials. It has long been recognized that the chemical potentials of phases having different compositions can only be the same if the line tangential to the free energy of mixing curve at the composition of one of the phases is also tangential at the composition of the second. This has been termed the "double-tangent construction" [29], and it is worth considering why it is true.

For any curve having the form shown in Fig. 3.5, it is intuitively understandable that only one double-tangent line exists. At all compositions less than a, for example, lines tangent to the curve will not touch this curve at any other point. Just to the right of a, lines tangent to the curve will *cross* the curve at another place rather than be tangent to

---

*Equilibrium may exist at different levels within such mixtures. Within the metastable homogeneous liquid c, a pseudo "equilibrium" may be said to exist when the field variables are uniform within this phase. Since however this one-phase state is metastable with respect to the two-phase state, a still lower overall free energy can be achieved by the creation of two phases. The "condition of equilibrium" described in Section 3.5 may then be applied to define the state of lowest free energy within the two-phase state.

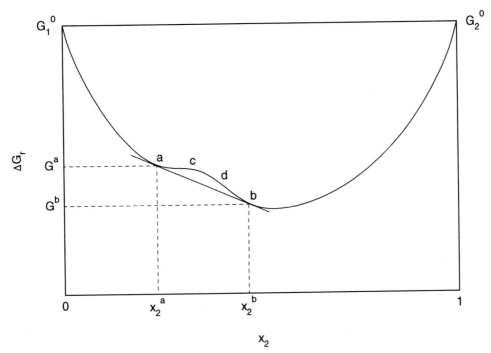

Fig. 3.5. The reduced free energy of mixing function, G_r, for a system having a single miscibility gap within which the coexisting phases are a and b. The fact that the free energies of phases a and b are different is clearly evident, as is also the fact that the slope of the free energy curve at these unique compositions is the same.

it. Only that line which is tangent at a is also tangent at another composition, and that composition is b. Once the free energy of mixing function is defined (by stating the system and the system variables), then a unique double-tangent line is also defined. Phases a and b may exist in equilibrium with each other, and are termed "coexisting" phases.

It is evident from Fig. 3.5 that the free energies of phases a and b (G^a and G^b) are different (Section 3.5). Phases a and b are simply the component phases which, together, constitute the state of the biphasic mixture whose gross composition is c (and also d).

This result illustrates the important principle mentioned in Section 3.5 that *while the field variables within coexisting phases are necessarily the same, the density variables (including the free energies) are typically different.* The free energies of coexisting phases in a biphasic mixture can only be the same when the phases have the same composition (Section 3.7).

The total free energy of a mixture of phases a and b, produced from a mixture having the gross composition c, is the weighted average of G^a and G^b. This free energy corresponds graphically to the value of G at the point where a vertical line passing through c crosses the double tangent line a–b (Sections 3.21, 4.11).

The span of compositions between coexisting phases a and b (x^b to x^a) is termed a "miscibility gap". The term is quite descriptive and is very useful. The components in this example are miscible over the entire composition range – except within the composition interval x^a to x^b. While this term is sometimes restricted to gaps between like phases (e.g. liquid/liquid gaps), it will be used in this book without regard for the structures of the coexisting phases.

Slope of the free energy of mixing function. The slope of the free energy vs. mole fraction plot is the difference between the chemical potentials of the two components

$$\frac{d(\Delta G)}{dx_2} = \mu_2 - \mu_1 \tag{3.39}$$

This may be shown by first dividing eq. 3.34 by RT, substituting $(1 - x_2)$ for x_1, and differentiating with respect to x_2. This yields

$$\frac{d(\Delta G/RT)}{dx_2} = \frac{d[(1 - x_2)\ln(1 - x_2) + x_2 \ln x_2]}{dx_2} \tag{3.40}$$

or

$$\frac{d(\Delta G/RT)}{dx_2} = -1 - \ln(1 - x_2) + 1 + \ln x_2 \tag{3.41}$$

which reduces to

$$\frac{d(\Delta G/RT)}{dx_2} = \ln x_2 - \ln x_1 \tag{3.42}$$

Multiplying by RT, and from eq. 3.31 for the chemical potential,

$$\frac{d(\Delta G)}{dx_2} = (\mu_2 - \mu_2^0) - (\mu_1 - \mu_1^0) \tag{3.43}$$

Equation 3.43 represents the strict definition of the slope, since it includes consideration of both the actual and the standard state chemical potentials. If the standard state chemical potentials are set equal to zero, then in simplified form

$$\frac{d(\Delta G)}{dx_2} = \mu_2 - \mu_1 \tag{3.44}$$

It is evident from eq. 3.44 that those compositions which fall at the two tangent points of the double-tangent line do indeed meet the criterion of equilibrium of state.

3.16 Multiple miscibility gaps

Figure 3.5 represents a hypothetical curve for mixtures which display a single miscibility gap, a situation which is often encountered in partially miscible solvents (e.g. benzene and water). The anomaly was produced in drawing this curve by summing the ideal mixing curve with a Gaussian function plus a linear correction term. (The use of a Gaussian function to produce the smooth curve in Fig. 4.5 has no theoretical basis whatsoever. This function simply represents a convenient means of generating an attractive curve having a plausible form.) In real systems several such gaps may be

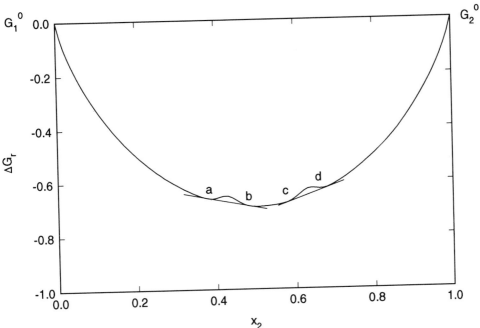

Fig. 3.6. The reduced free energy of mixing function vs. mole fraction of solute for mixtures showing two miscibility gaps between phases $a + b$ and $c + d$. The two miscibility gaps are to be treated independently of one another.

encountered along the same isotherm. These may be similarly visualized by inserting more than one anomaly-producing function, centered at different compositions. Each anomaly may be individually analyzed in a similar manner. Shown in Fig. 3.6 is a hypothetical mixing curve that would result in such behavior. The two different double-tangent lines are $a - b$ and $c - d$, and the two pairs of coexisting phases are $a + b$ and $c + d$.

An example of a surfactant–water system that displays two miscibility gaps along the same isotherm over a range of temperatures is the $C_{10}E_6$–water system (Fig. 3.7) [30]. The phase sequence in the direction dilute → concentrated mixtures along the 25°C isotherm is liq/hex/liq, and in each case the coexisting liquid and hexagonal phases are separated by small miscibility gaps.

In this example the two coexisting phases differ in phase structure. This is usual throughout phase science. In practically all instances of immiscibility coexisting phases do differ in *both* composition and structure, although liquid/liquid and crystal/crystal miscibility gaps represent exceptions to this rule. In these cases the two phases differ in composition and the details of structure, but their structures are qualitatively similar.

One may well ask how the free energy of mixing curve can remain a smooth function within a miscibility gap when the qualitative nature of the coexisting phases differ. This question cannot be answered, because in general the free energy of mixing curves within miscibility gaps is not known and cannot be determined. It is conceivable that two different curves exist (one for each structure) within such gaps.

Fortunately, this uncertainty does not invalidate the above descriptive analysis of the thermodynamics of immiscibility. Whether there are two curves or one, the

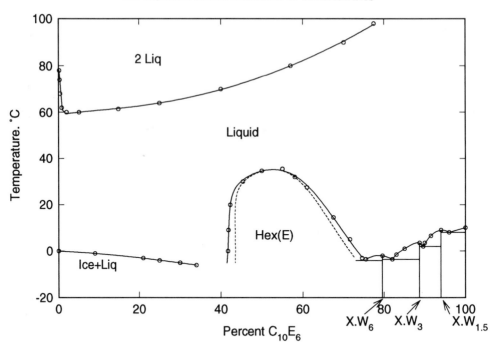

Fig. 3.7. The hexaoxyethylene glycol monodecyl ether ($C_{10}E_6$)–water system [30].

double-tangent line should still define the coexisting phases, and the above approach remains a satisfactory way of qualitatively visualizing the thermodynamic basis for immiscibility.

3.17 A pure component as a coexisting phase

In the general case (Fig. 3.5) both components were presumed to exist in both phases. In the sodium chloride–water system at $>0°C$ (Fig. 1.1) liquid mixtures coexist with the dry salt crystal. Similarly, in mixtures more dilute than 23.3%, pure ice may coexist with a liquid salt solution below $0°C$. Experimentally, there is no evidence to support the idea that either component is miscible with the crystalline phase of the other component in this system.

In the older literature one often encounters the view that all compounds are miscible with each other to some degree [31]. Yet, the notion that the dry crystal of salt in equilibrium with the liquid phase necessarily contains water is bothersome. Given sufficient time, one would expect that salt crystals in equilibrium with the coexisting liquid should become nearly perfect crystals. This is because the presence of the liquid phase provides a ready path (dissolution and reprecipitation) by means of which defects and excess surface energy can be eliminated, and if this happens the interior of such crystals will certainly contain no water. Similarly, ice is generally presumed not to be miscible with salts or other solutes. (The well-known ice clathrates do not constitute an exception to this rule, in that the local symmetry of the water within these structures, which correspond to icosahedral geometry, is qualitatively different from the local

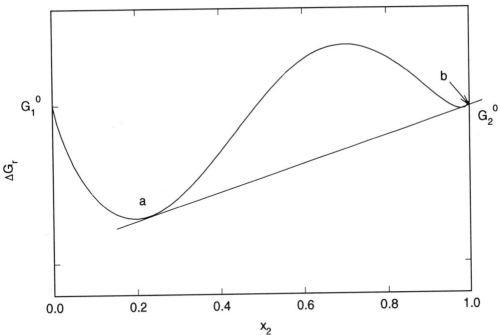

Fig. 3.8.   The reduced free energy of mixing function for mixtures showing a miscibility gap where one of the phases is a pure component.

symmetry of water molecules within the ordinary ice I structure [34].) At the crystal/liquid interface, where water and salt are forced to interact, the situation is obviously different.

Is the concept that two coexisting phases may be completely immiscible sensible? Answering this question definitively on a purely experimental basis will be extremely difficult, but it is not at all difficult to visualize free energies of mixing curves which lead to this result.

Starting with Fig. 3.5, we may imagine that the compositions of the two phases may be smoothly varied by adjusting the system variables. There exists no *a priori* reason why the composition of either one or both of the coexisting phases may not correspond precisely to that of a pure component (Fig. 3.8). In such cases the double tangent construction would apply and the condition of equilibrium would be met, but one of the phases would be a pure component. The above analysis is valid regardless of which component separates from the mixture, or of the structure of the phase that separates. In surfactant–water systems it may apply not only to the precipitation of a pure surfactant crystal (or of ice) from a liquid, liquid crystal, or crystal hydrate phase, but also to the vaporization of water from a liquid, liquid crystal, or crystal phase. As surfactants are ordinarily non-volatile, the coexisting gas phase is typically pure water.

3.18   Three coexisting phases at eutectics

In unary systems, three phases may coexist at unique values of the field variables (Section 4.5). This occurs at the triple point, which for water lies at 0.01°C and

0.0061173 bar (611.73 Pa, 4.58 Torr) [32]. At the triple point the free energies of all the phases are equal, which means simply that the free energy curves of all three phases cross at these coordinates.

Three phases may also coexist in binary and higher systems. At the eutectic discontinuity in the salt–water system (Fig. 1.1) ice, the eutectic liquid, and the salt dihydrate ($X \cdot W_2$) crystal may coexist. It is worthwhile to consider the form of the free energy of mixing curve that leads to this situation, and the manner in which it may evolve. Both may be visualized by approaching the eutectic temperature from above (Fig. 3.9). (In this analysis we will be concerned only with the ice $\rightleftarrows$ liquid and the liquid $\rightleftarrows X \cdot W_2$ (crystal dihydrate) miscibility gaps; while the $X \cdot W_2 \rightleftarrows X$ miscibility gap is also present it does not change with temperature.)

At −15°C the compositions of the ice and the coexisting liquid phases are 0% and 18.8%, respectively, while those of the liquid and $X \cdot W_2$ phases are 24.3% and 61.9%. At −19°C the two crystal compositions remain unchanged, but the liquid phase only spans the composition range from 21.9 to 23.7%. Finally, at the eutectic temperature (−21.1°C) the liquid phase composition is uniquely defined (23.3%).

Above the eutectic, the free energy of the liquid phase must constitute a local minimum which lies *below* the double tangent line that would otherwise connect the two outer (crystal) phases. Should this minimum rise as the temperature is lowered, it will fall precisely on this line at the eutectic temperature. At this point a "triple-tangent construction" exists: a single line is tangent to the free energy of mixing curve at three different compositions. Should the minimum continue to rise as the temperature is further lowered, it will lie *above* the double-tangent line of the outer phases. When this is true, the central liquid phase is metastable.

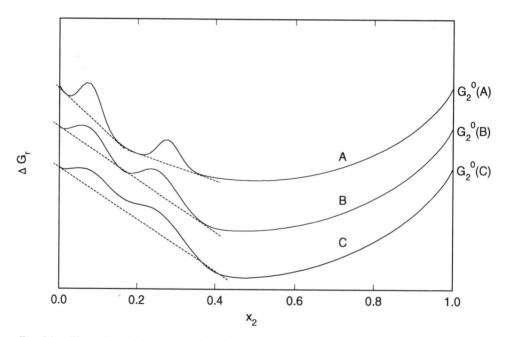

Fig. 3.9. The reduced free energy of mixing function at temperatures above (A), at (B), and below (C) a eutectic discontinuity.

It may seem improbable, at first glance, that conditions will exist which lead to the triple-tangent construction. However, it is not difficult to visualize this if it is accepted that the elevation of the central minimum (relative to the balance of the curve) may be smoothly varied by changing a field variable (in this case temperature). If the curves change in this manner, the triple-tangent condition is inevitably encountered at some temperature.

3.19   Three coexisting phases at peritectics

At eutectic discontinuities the central minimum lies below the double-tangent line of the outer phases at high temperatures, and is pushed above this line as the temperature is lowered. For peritectic discontinuities at which the central phase spans a finite composition range (Section 5.8), exactly the same progression in the free energy curves may exist except that the dependence on temperature is reversed. That is, the central phase exists as an equilibrium phase at low temperatures but becomes metastable as the temperature is raised. An example of a system in which this likely occurs is monoolein–water (Fig. 5.15).

Peritectic discontinuities of stoichiometric phase compounds. The above situation does not seem likely for the peritectic reactions of stoichiometric phase compounds (such as crystal hydrates). An example is the peritectic decomposition of the $X \cdot W_2$ hydrate of salt (Fig. 1.1). In these cases the isopleth at the hydrate composition represents a discontinuity in phase behavior which divides the diagram into discrete regions – one more concentrated than this isopleth and the other less concentrated. Each region may be treated separately for purposes of diagnosing the phase behavior, as if the $X \cdot W_2$ composition were a pure component and two-phase diagrams were joined side-by-side [33].

At temperatures below this discontinuity (0.1°C), two independent free energy of mixing curves likely exist – one for the more dilute region and one for the more concentrated region (Fig. 3.10). These curves may cross at the $X \cdot W_2$ isopleth. The usual double-tangent construction will exist within each curve, with the composition of one of the coexisting phases being either a pure component (salt) or a liquid phase, and that of the other being the $X \cdot W_2$ crystal (61.9%). As the temperature is raised, the angle that tangents to the curves form at the $X \cdot W_2$ composition will diminish until, at the peritectic temperature, a triple-tangent construction exists. The coexisting phases are the liquid (26.3%), the $X \cdot W_2$ crystal (61.9%), and the dry salt crystal (100%). At higher temperatures $X \cdot W_2$ is no longer an equilibrium phase, and a miscibility gap separating the liquid and the dry crystal X exists.

Isothermal vaporization of water. The isothermal vaporization of water from a mixture within a miscibility gap entails a phase reaction that is superficially analogous to the peritectic reaction of stoichiometric phase compounds (Section 3.21). The forms of the free energy of mixing curves, as the heat of vaporization is added and the water-poor coexisting phase is formed, may be described in a manner identical to that used for the peritectic in Fig. 3.10.

Polytectic discontinuities. The other isothermal three-phase discontinuity that is encountered is the polytectic discontinuity [34]. This degenerate form of the eutectic or peritectic discontinuities results when the compositions of two of the three coexisting phases are identical (Chapter 6). Plausible free energy curves spanning the temperature

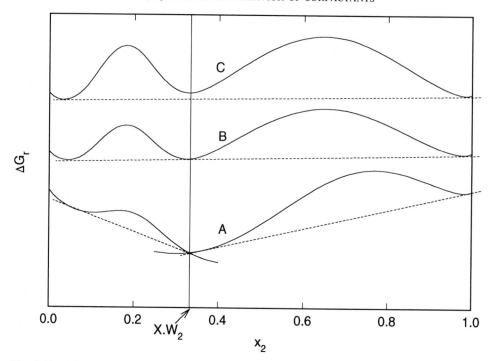

Fig. 3.10. The reduced free energy of mixing function at temperatures below (A), at (B), and above (C) the peritectic discontinuity of a stoichiometric crystal hydrate.

of polytectic discontinuities may be constructed by combining the curve for two coexisting phases (Section 3.14 or 3.17) with the free energy vs. field variable curves for unary systems (Section 3.7). Just as in unary systems, there will be two free energy curves at the same temperature for phases having the same composition, but one curve will apply to the low temperature situation, the other to the high temperature situation, and the enthalpies and entropies will differ.

3.20 Total immiscibility: the container issue

It is worthwhile to consider the likelihood that two components are completely immiscible. Free energy curves describing this situation may be easily visualized in two ways. One is by extending the analysis of immiscibility in which one phase is a pure component (Section 3.17) to include both phases. Alternatively, free energy of mixing functions are conceivable which lie above the standard state chemical potential line at all compositions. This too would signify that the total free energy increase required to form a homogeneous mixture is greater than the stabilization provided by its formation. This situation is not at all difficult to imagine, since the magnitude of the entropic stabilization is not very large and probably cannot increase indefinitely. The enthalpy of a phase, on the other hand, can probably vary over an extremely wide range. The hypothetical form of this particular free energy curve would differ from that described in Section 3.17 in that the double tangent construction does not exist at either composition limit. In either of these two cases, no one-phase mixture containing both components would exist as an equilibrium state.

Most compounds are immiscible with the container that holds them. A pure component must be held within a container, and the container is a chemical compound (or compounds). How, then, can one justify ignoring the container in considering phase behavior?

One possible answer is that if the container and the components are in fact completely immiscible, then this restriction makes it legitimate to ignore the container. In analyzing the situation from the perspective of the Phase Rule (Chapter 5), the container may be included – thus increasing the number of components of the system – but adding the constraint of immiscibility correspondingly reduces the number of degrees of freedom that remain. One may either include the container as another component and then (on the basis of the constraint of immiscibility) remove it from consideration, or ignore it altogether; the result is the same.

This issue is neither a pure exercise nor a trivial matter. In many commercial systems, components are often present (such as polymers) that are probably immiscible with all the other components of the mixture. When this is true, one may safely ignore them insofar as the phase behavior of the balance of the system is concerned, and thereby simplify the analysis. In so doing one must *not* ignore these materials when dealing with other aspects of the physical science of the system, such as its rheology or colloid science.

3.21   The thermodynamics of three-phase discontinuities

In principle, reversible, stoichiometrically well-defined phase reactions may occur as one crosses isothermal discontinuities along isoplethal trajectories in closed mixtures (Section 4.10). Those terms which dictate the thermodynamic changes that occur during these reactions may be perceived from the free energies of mixing curves.

The general form of the eutectic phase reaction is

$$w^a a + w^b b + q^{eu} = c^{eu} \tag{3.45}$$

This equation signifies that phase "a" reacts reversibly with phase "b" with the absorption of heat (q^{eu}) to form the eutectic liquid, c^{eu}. The stoichiometry of this reaction is rigorously defined, and is indicated by the coefficients (w^a and w^b) which indicate the quantities of phases a and b required to form unit quantity of c^{eu}. These stoichiometric factors are dictated by the form of the free energy of mixing curve, and may be calculated from the phase diagram of the system using the lever rule (Section 4.9). Unlike the analogous coefficients in chemical reaction equations, the stoichiometric coefficients of phase reactions are not typically integers. Although phases of differing composition are both reactants and products during this process, no mass is gained or lost; mass balance is precisely maintained during the eutectic phase reaction (Section 3.18).

Just as within miscibility gaps, the density variables of all three of the phases that may coexist at a eutectic typically differ from one another (Section 3.18). It is apparent from the curve, by inspection, that the weighted sum of the free energies of the two reactant phases must equal the free energy of the product phase. The net free energy change during this reaction process (ΔG^{eu}) therefore equals zero

$$\Delta G^{eu} = G^{c^{eu}} - (w^a \times G^a + w^b \times G^b) = 0 \tag{3.46}$$

The enthalpy and entropy changes that occur during the eutectic reaction are non-zero and positive, and are described by the equations

$$\Delta H^{eu} = H^{c^{eu}} - (w^a \times H^a + w^b \times H^b) > 0$$
$$\Delta S^{eu} = S^{c^{eu}} - (w^a \times S^a + w^b \times S^b) > 0 \tag{3.47}$$

The temperature of this isothermal reaction is dictated by the ratio of the enthalpy change for the process to the entropy change

$$T^{eu} = \frac{\Delta H^{eu}}{\Delta S^{eu}} \tag{3.48}$$

A similar analysis applies to the phase reaction that occurs at peritectic discontinuities. The peritectic phase reaction is

$$c^{peri} + q^{peri} \rightleftarrows w^a a + w^b b \tag{3.49}$$

The net free energy change for this reaction (ΔG^{peri}) is zero

$$\Delta G^{peri} = (w^a \times G^a + w^b \times G^b) - G^{c^{peri}} = 0 \tag{3.50}$$

The non-zero enthalpy and entropy changes during the reaction are defined by the equations

$$\Delta H^{peri} = (w^a \times H^a + w^b \times H^b) - H^{c^{peri}} > 0$$
$$\Delta S^{peri} = (S^a \times S^a + w^b \times S^b) - S^{c^{peri}} > 0 \tag{3.51}$$

and the ratio of the enthalpy to the entropy defines the peritectic temperature.

$$T^{peri} = \frac{\Delta H^{peri}}{\Delta S^{peri}} \tag{3.52}$$

The form of the thermodynamic equations for polytectic phase reactions is different; it is identical to that of unary phase reactions (Section 3.7). This is true because the third phase is really a bystander that is not involved in the reaction.

Isothermal vaporization of water. One other important kind of reversible process that involves three coexisting phases of differing composition is the isothermal vaporization of water from a biphasic mixture, to form water vapor plus the coexisting condensed water-poor phase (Section 5.1). An example would be the evaporation of water from a saturated salt solution to form water vapor plus the dry salt crystal. Three coexisting phases exist during this process and, at specified temperature or pressure, the mixture is invariant. The phase reaction which pertains to this process is not related to any explicit isothermal discontinuity (such as a eutectic or peritectic), but may occur at any point within any miscibility gap.

If the water-rich reactant phase is designated c and the product phases are phase b and water vapor (H_2O^g), the reaction equation is

$$c + q^{vap} \rightleftarrows w^{H_2O^g} H_2O^g + w^b b \qquad (3.53)$$

As always, mass balance is maintained, ΔG^{vap} equals zero, ΔH and ΔS are non-zero, and the temperature of the process equals the ratio of the enthalpy change to the entropy change.

References

1. Taylor, H. S. and Taylor, H. A. (1937). *Elementary Physical Chemistry*, 2nd ed., p. 384, Van Nostrand, New York.
2. Israelachvili, J. (1992). *Intermolecular and Surface Forces*, 2nd ed., Academic Press, London.
3. Stanley, R. C. (1968). *Light and Sound for Engineers*, pp. 247–263, Hart, New York.
4. Atkins, P. W. (1978). *Physical Chemistry*, 2nd ed., W. H. Freeman, San Francisco.
5. Glasstone, S. (1946). *Textbook of Physical Chemistry*, 2nd ed., pp. 693–814, Van Nostrand, New York.
6. Taylor, H. S. (1930). *A Treatise on Physical Chemistry*, 2nd ed., Van Nostrand, New York.
7. Chapman, D., Williams, R. M. and Ladbrooke, B. D. (1967). *Chem. Phys. Lipids* 1, 445–475.
8. Resnick, R. and Halliday, D. (1977). *Physics*, Part I, 3rd ed., pp. 497–538, John Wiley, New York.
9. Guggenheim, E. A. (1967). *Thermodynamics*, 5th ed., pp. 88–89, North-Holland, Amsterdam.
10. (1991). *Handbook of Chemistry and Physics* (D. R. Lide ed.-in-chief), 72nd ed., pp. 6-108–6-110, CRC Press, Boca Raton, FL.
11. Reference 9, pp. 121–123.
12. Scheidegger, A. E. (1991). *Theoretical Geomorphology*, pp. 44–46, 67–75, Springer, Berlin.
13. Laughlin, R. G., Munyon, R. L., Ries, S. K. and Wert, V. F. (1983). *Science* 219, 1219–1221.
14. Griffiths, R. B. and Wheeler, J. C. (1970). *Phys. Rev. A* 2, 1047–1064.
15. Widom, B. (1985). *Chem. Soc. Rev.* 14, 121–140.
16. Lang, J. C. and Morgan, R. D. (1980). *J. Chem. Phys.* 73, 5849–5861.
17. Bridgman, P. W. (1937). *J. Chem. Phys.* 5, 964–966; Bridgman, P. W. (1932). *The Physics of High Pressure*, MacMillan, New York.
18. Roozeboom, H. W. B. (1901). *Die Heterogene Gleichgewichte*, Vol. 1, F. Vieweg u. Sohn, Braunschweig.
19. Haar, L., Gallagher, J. S. and Kell, G. S. (1984). *NBS/NRC Steam Tables*, Hemisphere, Washington, DC.
20. (1984). *Perry's Chemical Engineer's Handbook*, (D. W. Green and J. O. Maloney, eds), 6th ed., McGraw-Hill, New York.
21. Demus, D., Diele, S., Grande, S. and Sackmann, H. (1983). *Advances in Liquid Crystals*, Vol. 6 (G. H. Brown ed.), p. 71, Academic Press; Volmer, M. (1939). *Kinetik der Phasenbildung*, Th. Steinkopf, Dresden.
22. (1988). *Crystallization and Polymorphism of Fats and Fatty Acids*, Vol. 31 (N. Garti and Kiyotaka Sato, ed.), Marcel Dekker, New York.
23. Knight, P., Wyn-Jones, E. and Tiddy, G. J. T. (1985). *J. Phys. Chem.* 89, 3447–3449.
24. Demus, D., Diele, S., Grande, S. and Sackmann, H. (1983). *Advances in Liquid Crystals*, Vol. 6 (G. H. Brown ed.), pp. 1–108, Academic Press, New York.

25. Rowlinson, J. S. and Swinton, F. L. (1982). *Liquids and Liquid Mixtures*, 3rd ed., Butterworth Scientific, London.
26. Rosevear, F. B. (1968). *J. Soc. Cosmetic Chemists* **19**, 581–594.
27. (1984). *Handbook of Chemistry and Physics*, 64th ed., p. F-66, CRC Press, Boca Raton, FL.
28. Madelmont, C. and Perron, R. (1976). *Colloid and Polymer Sci.* **254**, 6581–6595.
29. Bett, K. E., Rowlinson, J. S. and Saville, G. (1975). *Thermodynamics for Chemical Engineers*, The MIT Press, Cambridge, MA.
30. Clunie, J. S., Corkill, J. M., Goodman, J. F., Symons, P. C. and Tate, J. R. (1967). *Trans. Faraday Soc.*, 2839–2845.
31. Rivett, A. C. D. (1923). *The Phase Rule*, p. 9, Oxford University Press, London.
32. (1992). *CRC Handbook of Chemistry and Physics*, 72nd ed., pp. 6–10, CRC Press, Boca Raton, FL.
33. Findlay, A. (and Campbell, A. N.) (1938). *The Phase Rule and its Applications*, 8th ed., pp. 111–121, Dover Publications, New York.
34. Laughlin, R. G., Munyon, R. L., Fu, Y.-C. and Fehl, A. J. (1990). *J. Phys. Chem.* **94**, 2546–2552.

Chapter 4

Phase Diagrams and The Phase Rule

4.1 The dimensions of phase behavior

In previous chapters extensive use has been made of the phrase "phase behavior"; the meaning of this phrase will now be defined. In describing phase behavior, answers are provided to three questions:

(1) How many phases exist?
(2) What is the composition of each phase?
(3) What is the structure of each phase?

If all these questions can be answered correctly, then phase behavior is known.

Phase behavior may be described at several levels. A unique behavior exists at specific values of temperature, pressure, and composition. A finite range of behavior exists over a range of temperatures (at specified pressure and composition), over a range of compositions (at specified temperature and pressure), or over a range of pressures (at specified temperature and composition). A still wider range of phase behavior is encountered in considering a system over its entire composition and a wide temperature and/or pressure range.

Process paths exist during which composition and/or temperature change, and these may be plotted as lines on an empty figure having the same coordinates as phase diagrams. Such lines may be regarded as "process trajectories". The physical process of mixing one component with another, while holding the temperature constant, is described by a horizontal line extending from the initial to the final composition at the specified temperature. The physical process of heating or cooling a mixture is described by a vertical line which extends from the initial to the final temperature, at a specified composition.

More complex processes also exist during which both temperature and composition change. If an aqueous mixture is heated in an open container and water is allowed to evaporate, the (otherwise vertical) path would curve to the right at the higher temperatures. Such paths are often encountered during engineering processes in which water is vaporized within a heat exchanger. Similarly, if heat were evolved (or absorbed) during a mixing process this path too would also be curved rather than linear. Pressure variation could be considered as well, but we will not focus on this variable (except in Chapter 7).

A given process path may be followed in any system, and once the system has been defined the path may be overlaid onto the phase diagram of that system. It is then possible to infer, by inspection, the changes in phase behavior that occur along the path. This is one of the most important uses of phase information.

4.2 Phase diagrams of binary systems

The typical phase diagram of a binary system is a graph-like plot of composition (along
the abscissa) against temperature (along the ordinate). The convention of placing water
to the left and the surfactant to the right will be followed. The term "graph-like" is used
advisedly, for phase diagrams differ importantly from graphs. The span of compositions is restricted to 0–100%, and the vertical boundaries at these limits have special
significance: they are phase diagrams in their own right.

Phase boundaries. The area within phase diagrams is divided into regions by lines.
Some of these are straight lines that are either precisely horizontal or vertical, while
others are smooth curves. The latter generally depict the limits of miscibility of the
components, and are the "phase boundaries". A one-phase region always exists to one
side of phase boundaries and a two-phase region to the other. The regions are labeled
so that both the number of phases present, and their structures, can be read from the
diagram.

Isothermal discontinuities. Horizontal lines (lines of constant temperature or
"isotherms") are usually found [1]. (An "isotherm" usually signifies a horizontal line
that spans the entire composition range, but fragments of isotherms are also "isothermal". Similar considerations apply to isopleths.) These reflect the existence of
discontinuities in phase behavior which occur at these temperatures. Isothermal
discontinuities are most clearly revealed along those isoplethal paths that fall within
the composition span of the discontinuity (Sections 3.18, 3.19). Such isotherms usually
touch three different one-phase regions, which one must learn how to identify. Designating the phases that coexist at these lines would clutter the diagram, and they are not
usually labeled.

One of the isothermal discontinuities in the salt–water system lies at $-21.1°C$ and is a
"eutectic", while the other is at $0.1°C$ and is a "peritectic". A third kind is found in
other systems in which two of the three phases have identical compositions; the term
"polytectic" has been suggested as a generic name for this kind of discontinuity [2].

Isoplethal discontinuities. Vertical lines (lines of constant composition or "isopleths")
also exist [1]. Two vertical lines are always found at 0 and 100% compositions which
are the unary diagrams of the components. Along the line at 0% in Fig. 4.1, the phase
structures of water within various ranges of temperature (including the transition
from crystal to liquid phase at $0°C$) are indicated. Similarly, the right boundary constitutes the phase diagram of salt. Since salt melts at $801°C$ and undergoes no polymorphic change before melting, it is sufficient to indicate that the phase present is a
crystal. No discontinuity of state exists in salt crystals within the temperature range of
this diagram.

A third isoplethal discontinuity exists below $0.1°C$ at 61.9%. A crystal phase exists
along this isopleth whose composition corresponds exactly to a 2 : 1 mole ratio of water
molecules to sodium chloride ion pairs. This phase is termed a "crystal dihydrate", and
is abbreviated "$X \cdot W_2$". Such mixtures are termed "phase compounds" [3]. They are
rigidly defined stoichiometrically, and the stoichiometry is an integral mole ratio of the
components. The decahydrate of sodium sulfate ($X \cdot W_{10}$) is an example of a "high"
crystal hydrate [4], while the 1/8th hydrate of sodium dodecyl sulfate ($X_8 \cdot W$) is an
example of a "low" hydrate [5]. In the diagnosis of phase behavior, these isopleths (as
well as the components themselves) should be treated as phase "regions" whose span of
composition is zero.

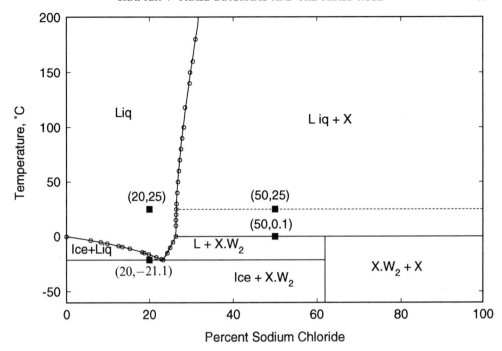

Fig. 4.1. The sodium chloride–water system, with the coordinates of the four particular mixtures considered in the text marked (as squares).

4.3   Reading and interpreting binary diagrams

A mixture of a particular composition held at a particular temperature may be located within Fig. 4.1 by its x, y coordinates. If one prepares a 20% solution of salt and holds it at a temperature of 25°C, this mixture lies at the coordinates (20,25).

The phase behavior at specific coordinates may be read from the diagram. The point (20,25) lies within a region that is labeled a one-phase region, and the phase in this region has a liquid structure. Because the mixture is a single phase, the phase composition is the same as the total composition.

For a 50% salt mixture at 25°C, *the diagram indicates that no phase having this composition exists at this temperature.* A mixture at these coordinates exists, instead, as two phases. Their compositions are read by following the horizontal dashed line at 25°C in both directions until one-phase regions are encountered. At the dilute end this line touches the liquid region at 26.43%, and at the concentrated end it touches the crystal "region" at 100%. Lines such as these, which define the compositions of coexisting phases within biphasic regions, are called "tie-lines". They are necessarily horizontal (isothermal) in binary diagrams, since otherwise the condition of equilibrium would be violated. (Another term for "tie-line" that may be encountered is "conode" [12, p. 19].)

A 20% mixture at −21.1°C displays a still different phase behavior. This mixture lies on the eutectic discontinuity, at which up to three phases may coexist. As with tie-lines, when three phases coexist their compositions correspond to the three points at which the eutectic line touches one-phase regions. To the left this eutectic touches the ice crystal phase at 0%, to the right it touches the crystal dihydrate phase at 61.9%, and in the middle it touches the liquid region at 23.3%. Eutectic lines are *not* tie-lines, and are not to be treated as such (Section 4.11).

A 50% mixture at 0.1°C falls on the peritectic discontinuity, at which a related but qualitatively different kind of phase behavior exists. As with eutectics, as many as three phases may coexist at peritectics. At this peritectic the three phases are the liquid of 26.30%, the dihydrate crystal of 61.9%, and the crystal of 100% composition. The fundamental difference between eutectics and peritectics is that the phase of intermediate composition at eutectics exists only *above* the temperature of the discontinuity, while this intermediate phase exists at peritectics only *below* the discontinuity. Polytectic discontinuities are degenerate in this regard, as only two phase compositions are involved and the issue is moot.

Finally, we may consider the phase behavior of a 20% mixture at −21.1°C that is held in a container having void space, and in which the pressure has been adjusted to 93 Pa (0.70 Torr). Under these explicit conditions four phases coexist: ice (0%), the 23.3% liquid, the 61.9% dihydrate crystal, and a gas phase whose composition is also 0%. As will be seen (Section 4.8.5), the existence of more than four phases at equilibrium is exceedingly unlikely. If found and proven, this result would invalidate the Phase Rule.

The above analysis amply illustrates the statement (Section 1.1) that phase diagrams constitute a dense form of information. With a little practice it is easily possible to read phase behavior at specific coordinates. With further practice progressive changes in phase behavior along isothermal or isoplethal process trajectories may readily be perceived (Section 4.8). Finally, phase diagrams make it possible to grasp quickly and comprehensively the phase behavior of a system as a whole. No better way to communicate visually all these kinds of information has yet been devised.

4.4 Phase diagrams of unary systems

The focus in this chapter is on binary diagrams, but other kinds are also important. The unary diagrams of the components are, at specified pressure, lines having the dimensions of temperature along which phase structures and discontinuities are designated. The diagrams of water and salt at one atmosphere are shown in Fig. 4.2.

Adding another component to a binary system produces a ternary system. Just as the unary diagram of water imposes limits on the phase behavior of all aqueous mixtures, the salt–water binary diagram imposes limits on the behavior of ternary systems containing salt, water, and any third component. The handling of ternary systems is described in Section 12.1.1.

4.5 The Phase Rule

In Chapter 3 the qualitative forms of thermodynamics of mixing profiles that result in various kinds of immiscibility were described. We now consider the relationships between the number of phases, the number of components, and the degrees of freedom (the variance) that are dictated by these thermodynamics. These are described by the law of science that has come to be known as "The Phase Rule" [6]. The Phase Rule

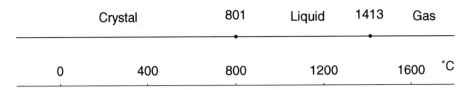

The Sodium Chloride System

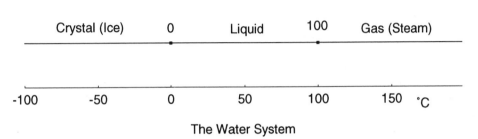

The Water System

Fig. 4.2. The unary phase diagrams of water and of salt at one atmosphere pressure. The scales of temperature (which are very different) are shown next to the two lines.

equation is

$$P + F = C + 2 \tag{4.1}$$

This equation implies that, in a system having C components, the number of phases present (P) plus the number of degrees of freedom that exist (F) is a constant for any mixture of these components that is held at particular values of temperature, pressure, and composition. It further states that the value of this constant equals the number of components plus two $(C + 2)$. While this equation may be rearranged at will, the above arrangement has the advantage that once the number of components is stated, those parameters $(P$ and $F)$ which depend on the numerical values of the system variables are collected to one side and their sum remains constant during the balance of the analysis. Each element of the Phase Rule equation will now be considered.

4.6 The number of components, C

In the vast majority of cases, the number of components in surfactant mixtures is simply the number of chemical compounds present. Samples which contain only a single nonionic surfactant molecule, such as $C_{12}PO$ or $C_{12}E_6$, contain only one component $(C_{12}PO = $ dodecyldimethylphosphine oxide; $C_{12}E_6 = $ hexaoxyethylene

TABLE 4.1
Terminologies for multicomponent systems

Number of components	Name	$P + F$
One	Unary	3
Two	Binary	4
Three	Ternary	5
Four	Quaternary	6
Five	Pentanary	7

glycol monododecyl ether. Acronyms used to designate chemical structure are defined in Appendix 1). Mixtures of $C_{12}E_6$ and $C_{12}E_7$, or of either of these surfactants with water, contain two components, and so on. The terminology for systems having varying numbers of components is shown in Table 4.1.

A few situations exist where the number of components does not equal the number of compounds mixed together. This complication arises when chemical reactions occur among the compounds present that result in the formation of new chemical compounds. When this occurs, it is extremely important to phase science. The distinction between chemical and physical processes, and the basis upon which they may be recognized, are described in Section 2.2.

Components may also react to form "phase compounds" such as the stoichiometric crystal hydrates (Section 2.2.2). Molecular structure remains intact within many phase compounds, in which case they constitute a feature of the physical science of the system. In other cases (such as the sulfonic acid monohydrates) a chemical reaction occurs (Section 10.5.1). Both kinds of phase compounds are readily handled within the rules of phase science.

In applying the Phase Rule, it is only the *number* of the phases and components that is important. The structures of the phases (whether gas, liquid, crystal, liquid crystal) is immaterial, as long as their number is correctly specified. Also, when chemical reactions occur it is the change in the number of components that is important in applying the Phase Rule. Which molecules are designated as components is (within reason) also immaterial.

It is better to regard the number of components as being the number of "compounds" present rather than the number of "molecules" (Section 2.2.1). While this fine point is purely a matter of definition, definitions are important if communication is to be clear and precise. This issue arises for ionic surfactants like sodium dodecyl sulfate (SDS). The compound SDS contains two discrete ions, each of which may be regarded as formally charged molecules. In microscopic samples of SDS the number of one ion present is precisely equal to the number of the other; the ratio of sodium cations to dodecylsulfate anions present cannot be independently varied. Exceptions exist in the mass spectrometer and within gas discharges, where individual ions can and do exist apart from oppositely charged ions. The selective adsorption of ions at interfaces produces important electrochemical effects, but the quantities required to produce these effects are small and the mixture as a whole remains electrically neutral. It is thus correct to describe SDS (and other salts) as a chemical compound and a single component, even though such materials are best viewed as containing two (or more) molecules.

4.7   The number of phases, *P*

Once C is stated, the Phase Rule imposes limits on the number of permissible combinations of number of phases (P) plus degrees of freedom (F) that may exist. The value of P is inextricably linked to the basic concept of "phase", which is considered in Sections 4.18 and 4.19. For the moment we will presume that the concept of phase is clearly understood and that the phase diagram is known, and consider the number of phases that exist at various points within the diagram.

When one is determining phase diagrams, the situation is reversed. During this process one must experimentally determine the number of phases present, or more precisely the change in this number that occurs as conditions are varied (Appendix 4). Doing so is often not straightforward.

4.8   The degrees of freedom or variance, *F*

We come now to the "degrees of freedom" parameter, F. "Variance" is a synonym for "degrees of freedom". Once the number of components is stated and the number of phases determined, F is also defined – but what is it?

The degrees of freedom, F, *is the number of system variables that must be specified in order to define fully the state of a mixture that is held under a particular set of conditions.* F is often stated with respect to a region within which qualitatively similar phase behavior exists. If (as above) C has been specified so that $C + 2$ is defined, then

$$F = (C + 2) - P \qquad (4.2)$$

For a binary system the Phase Rule allows up to four phases (no more) to exist (Section 4.5), in which case $F = 0$ and the system is said to be "invariant". If three phases exist $F = 1$ and the mixture is "univariant", if two phases exist $F = 2$ and it is "bivariant", and if one phase exists $F = 3$ and it is "trivariant".

The variance is determined by counting the number of different "system variables" (temperature, pressure, or composition) that must be specified in order to reduce the variance to zero. In such an invariant mixture all aspects of the thermodynamics of state are fixed, so that unique phase behavior exists.

As with C and P, it is the *number* of degrees of freedom that is important – not which system variables are regarded as degrees of freedom. Specifying the numerical value of any system variable removes it from further consideration, of course. In some texts the reduction in the number of degrees of freedom imposed by such restrictions are explicitly designated as "r", so that the Phase Rule equation becomes $P + F = C + 2 - r$ [7].

The number of composition variables stated must be sufficient to define rigorously total composition. (At this point we are concerned only with the composition of the mixture as a whole, and not with the compositions of any phases that may have separated from this mixture.) This is one less than the number of components, C, because the remaining composition parameter is not independently variable; it is dictated by the requirement of mass balance. For a binary system one composition is sufficient, for a ternary system two are necessary, for a quaternary system three, etc.

The system variables T and p are also state variables (Section 3.1). Composition is not, but stating composition defines the numerical values of field variables such as chemical potential. In principle, any thermodynamic field or density variable may serve as a degree of freedom (see Section 3.6). In practice, one must either use temperature, pressure, and composition, or independently determine the relationship between these observable parameters and a thermodynamic variable – which amounts to the same thing.

4.8.1   Variance in a one-phase region in a binary system

In the liquid region of the salt–water diagram (Fig. 4.1), $P = 1$. Since $P + F = 4$, $F = 3$. The predicted variance of three is consistent with the observation that temperature, pressure, and composition must all be specified to define fully this liquid. Viewed from a different perspective, any or all of these parameters may be varied by small increments while qualitatively preserving the phase behavior (a single liquid phase). However, changing any of these variables quantitatively modifies all the properties of the liquid phase.

The value of F calculated from the Phase Rule equation is the *maximum* value that may exist. By specifying additional system variables, the variance under these constraints or restrictions is reduced below this maximum. For example, if the temperature of the liquid is held constant only pressure and composition must be defined in addition, and such mixtures are bivariant. If any two variables are held constant only one must be specified and the mixture is univariant, and if three are defined it is invariant. If a mixture has a finite variance, which of the yet undefined system variables may be specified to render the mixture invariant is immaterial.

Most binary aqueous surfactant diagrams have no pressure dimension. For these diagrams pressure is presumed to be constant (or nearly so) over the temperature span of the diagram. Typically no gas phase is shown, which implies that the pressure is greater than the vapor pressure of the most volatile component (water) at the maximum temperature of the diagram. Pressure could (in principle) be designated along a third coordinate, in which case solid figures instead of planar regions would exist. Realistically, pressure does vary during most phase studies but its variation is assumed not to influence the result and it is therefore neglected.

4.8.2   Variance in a two-phase region in a binary system

We now consider mixtures which lie within the region to the right of the solubility boundary. At 50% and 25°C (50,25), where two phases exist, what are the predicted and the actual variance?

Since $P = 2$ the predicted value of F is also 2 – one less than in the one-phase region. We may again determine the actual variance by inquiring as to whether temperature, composition, and pressure each remain degrees of freedom.

If temperature is allowed to vary while holding composition at 50% and pressure at one atmosphere, the coordinates of such mixtures move up or down the 50% isopleth. The diagram indicates that small changes in temperature do not qualitatively change the phase behavior, but *do* quantitatively alter the properties of the phases present. At 25°C a liquid of 26.43% concentration and the dry crystal phase exist. The tie-line connecting these phases is shown as a dashed line in Fig. 4.1. At 30°C, however, a liquid whose concentration is 26.48% and the crystal exist. Both the composition of the liquid and all the properties of both phases differ at these two temperatures. Temperature must therefore be specified in order to fully define phase behavior within this region, and may be regarded as a degree of freedom (specified pressure and composition).

The same is true of pressure. Doubling the pressure on the 50% mixture at 25°C is not likely to change qualitatively the phase behavior, but at the higher pressure the liquid phase composition will be slightly different, the densities of both phases will be slightly higher, and other properties will vary as well. Pressure, too, remains a degree of freedom (specified temperature and composition).

We come next to composition. At 25°C, 1 atmosphere, and 50% composition the two phases present are as previously indicated, and the weight fractions of the liquid and the crystal phase are 0.680 and 0.320, respectively. (These fractions are calculated using the lever rule, Section 4.9.) If we add sufficient water to form, for example, a 45% mixture at 25°C, the new mixture lies on the same tie-line. Because this is true, *precisely the same two phases exist*. The compositions and all the properties of the liquid and crystal phases that exist at 50% are identical to those of the liquid and the crystal phases that exist at 45%. If salt is added so as to increase the composition to 55% the quantity of crystals present is increased by exactly the amount of salt added, but the composition and the properties of the liquid phase do not change. From these observations, it is concluded that composition is *not* a degree of freedom within this region (at specified temperature and pressure).

Something must obviously change as total composition is varied, and that is the fraction of the mixture in each phase. The fraction of the liquid increases as water is added, until at 26.43% all of the crystals disappear. The fraction of crystals increases as salt is added, but the quantity of liquid phase that previously existed remains unchanged. Its fraction will decrease, but will never reach zero.

To summarize, the consequence of moving from the one-phase (liquid) region to the two-phase (liquid plus crystal) region is to eliminate the influence of composition on phase behavior, except insofar as the fraction of the mixture in each of the two phases is concerned. Temperature and pressure must still be specified to define the state of the mixture so the observed value of F is 2, as predicted by the Phase Rule.

4.8.3   Variance at three-phase discontinuities between condensed phases

We now consider a 20% mixture at −21.1°C. This mixture lies on the "eutectic" line and the diagram indicates that three phases may coexist at this discontinuity. Their compositions are 0, 23.3, and 61.9%, and their structures are ice crystal, liquid, and dihydrate crystal $(X \cdot W_2)$. When three phases exist the predicted variance is one $(P + F = 4, P = 3,$ therefore $F = 1)$; what is the actual variance?

The variance may again be recognized by considering each of the system variables separately while fixing the remainder, starting with temperature (pressure being specified). The temperature of a eutectic line is precisely defined; it does *not* span a finite range of temperatures. If temperature is changed one of the phases present disappears. Which phase disappears depends on total composition and on whether the temperature is increased or decreased. For a mixture to the left of the eutectic liquid (e.g. 10 or 20%), an increase in temperature causes $X \cdot W_2$ to disappear. For a mixture to the right (e.g. 30%), an increase in temperature causes ice to disappear. If the temperature is lowered at any composition along this eutectic the liquid disappears. One cannot vary temperature without changing the number of phases; therefore, temperature is *not* a degree of freedom (at specified pressure).

As regards composition, the compositions of the three coexisting phases also do not vary with total composition. The compositions of the phase regions that this line touches are characteristic of the system, and do not depend on the gross composition of a particular mixture. The extremities of tie-lines are best viewed as not really being part of the line, but belong to the respective one-phase regions. (If one has ice at −21.1°C, for example, the other two phases need not even exist.) The same perspective applies to eutectic lines at the three points where these lines touch one-phase regions. From these observations, one may conclude that composition also is not a degree of freedom (at fixed pressure or temperature).

If pressure is changed at constant temperature, the same behavior is observed as when temperature is changed at constant pressure: one of the phases disappears. If the change in pressure is not too large then an appropriate change in temperature will reestablish a three-phase mixture, but all the properties of each phase would differ from those of the original state. Exactly the same result is found if small changes in temperature are introduced; within limits, a change in pressure (at the new temperature) will reestablish the original three phases.

The observed variance at eutectic discontinuities is thus one, in agreement with the prediction of the Phase Rule. Either temperature or pressure may be specified, but then the other is defined.

Peritectic and polytectic discontinuities. At the peritectic discontinuity at 0.1°C a liquid of 26.30%, the $X \cdot W_2$ crystal of 61.9%, and the dry X crystal of 100% composition may coexist. Precisely the same analysis as above may be applied to this discontinuity – except that the central $X \cdot W_2$ phase disappears on *heating* the mixture past the discontinuity rather than on cooling (as with eutectics).

The analysis of polytectic discontinuities is simpler than in the above cases, because compositions do not vary in either of the phases involved in polytectic phase reactions. None exists in the salt–water system, but an example may be seen in the diagram of the dioctadecyldimethylammonium chloride–water system where X_α, X_β, and the crystal monohydrate $X \cdot W$ may coexist at 52°C (Fig. 2.2).

4.8.4   Variance in mixtures containing an equilibrium gas phase

In the above sections we have been concerned only with equilibria among condensed phases, but it is also important to consider those situations where a gas phase coexists in equilibrium with condensed phases (Chapter 7). In aqueous surfactant systems the gas phase is, almost without exception, pure water substance.

Pure liquid water has a natural equilibrium vapor pressure at which the gas and liquid coexist, which is dictated solely by the temperature. If salt is dissolved in water the vapor pressure of water is lowered. If the hydrostatic pressure is adjusted to this value, the gas phase may coexist with the liquid as an equilibrium phase. Since $P = 2$ when both phases exist, F must also be 2. The various choices for F might include temperature and composition (in which case pressure is fixed), temperature and pressure (in which case composition is fixed), or pressure and composition (in which case temperature is fixed).

If sufficient salt is added to exceed the solubility and a second crystal phase exists, this biphasic mixture is now invariant with respect to composition (Section 4.8.2) and has the characteristic vapor pressure of the saturated solution. If the pressure over such a biphasic mixture is adjusted to its equilibrium vapor pressure, the gas phase may exist.

Now, $P = 3$ and $F = 1$. If any variable (such as temperature or pressure) is also fixed, the system is invariant.

As water is vaporized from such mixtures of saturated liquid and salt crystals (assuming one is dealing with a closed system), the salt in the liquid phase is transformed into the dry crystal. After the liquid phase has vanished the pressure may be further reduced, but then only the water vapor phase and the dry crystal exist.

When only a single condensed phase (liquid or crystal) and a water vapor phase coexist, these two phases are the constituent phases which comprise the state of the system (Section 3.18). The conditions that are necessary for the reversible vaporization of water from the mixture to occur (at constant temperature and pressure) do not exist under these circumstances. A reversible phase reaction (in a closed mixture) only exists when three coexisting phases are present (Section 3.21).

4.8.5   Variance in four-phase mixtures in a binary system

The Phase Rule indicates that up to four phases may exist in a binary system. If $P = 4$ then $F = 0$, and such a mixture is invariant. In principle one may have four phases of any structure, but in practice only one four-phase situation exists which can easily be visualized. As seen earlier (Section 4.8.4) three-phase mixtures at either the eutectic or the peritectic discontinuities are univariant, and the mixture will have a defined vapor pressure at the temperature of the eutectic. If the pressure on the mixture were to be adjusted to this pressure, a gas phase would exist. If the pressure were increased the gas phase would disappear, and if it were decreased one of the condensed phases would disappear. No system variable can be altered without losing a phase, which signifies that such a mixture is indeed invariant.

The fraction of each phase that exists within invariant four-phase mixtures is determined by the enthalpy of the mixture, just as is true of invariant three-phase mixtures. Phase fractions would likely vary in a complex manner in such a state, but as heat is added the fraction of water vapor phase present would likely increase until one of the water-containing phases disappears. The lowest enthalpy state is probably that in which the fraction of the gas phase is zero, and the highest enthalpy state is probably that in which the fraction of one of the condensed phases is zero.

4.8.6   The existence of variance within phases of fixed composition

In biphasic mixtures of salt and water consisting of the saturated solution and dry salt crystals, the condition of equilibrium requires that the chemical potentials of both water and salt be the same in both phases. This is easily understandable in the case of salt, which exists in both phases, but it must also be true of water. This raises the interesting question "how can a finite chemical potential of water exist in a phase (the dry salt crystal) which contains no water?"

Various elements of phase science, taken together, provide a plausible resolution of this puzzle. One is recognition that the condition of equilibrium requires only that the *chemical potential* of each component be the same in both phases; it is not concerned with the quantity of any component in any of the phases present. The compositions of coexisting phases usually differ, and a composition of zero is simply a limiting value that compositions may assume.

Further, it is important to recognize that the total free energy of a phase equals the sum of the contributions of the terms $n_i\mu_i$ for all of the components present (eq. 3.17). For water (w) in a mixture of salt and water, the water free energy term equals $n_w\mu_w$ and in the aqueous liquid this term has a finite value (as does the salt term). In the coexisting salt crystal, however, the value of $n_w\mu_w$ equals zero even though μ_w is finite, because n_w equals zero. The existence of a finite chemical potential of a particular component in a phase does *not* necessarily mean that that component contributes to the free energy of the phase.

Another relevant point is that the chemical potential of water influences the pressure of the mixture. A mixture of liquid and crystal phases has a finite vapor pressure at a particular temperature, and a third gas phase may coexist with these two phases when such a mixture is held at this pressure (Section 4.8.4). If heat were added to a mixture of liquid and crystal phases, the quantity of the gas phase and the dry crystal would increase. A point would be reached at which the liquid vanishes, and only the dry crystal and water vapor remain. At this point *both* phases are pure components, and the situation is doubly mysterious. Now, one must also explain how pure sodium chloride crystals can exert a finite chemical potential on a gas phase that consists of pure water.

One may view the water as exerting its chemical potential on the dry crystal via its pressure, and the sodium chloride as exerting its chemical potential on the water by the pressure it exerts on the gas phase. (The equilibrium pressure of pure sodium chloride is zero.) According to the analysis in Section 7.1, no water free energy component exists in the salt crystal (since it contains no water) and no salt free energy component exists in the water vapor phase. The condition of equilibrium is met, both phases have the proper chemical potentials and free energies, and all is well with the Phase Rule.

If the chemical potential of water were to be reduced below the equilibrium vapor pressure when the liquid solution, water vapor, and salt crystal phases coexist, the liquid will vanish and the pressure of the mixture would be reduced. Varying the chemical potential of water amounts to changing the pressure that each phase exerts upon every other.

A similar issue (and a similar analysis) applies to stoichiometric crystal hydrates, except that both the chemical potential of water, and also the water free energy contribution to the crystal free energy, are finite. Varying the water activity within these phases (over the finite range within which they exist, Section 7.2) does not change the composition of the crystal hydrate. It *does*, however, affect the pressure that is mutually exerted by all the phase states that are present upon each other.

It is evident from the above just how the chemical potential (or activity) of water may, in fact, serve as a degree of freedom within dry crystals and crystal hydrates – even though the crystal composition is independent of this parameter. The chemical potential of water influences the state of the system, under these constraints, by way of its influence on pressure.

4.9   Phase fractions in two-phase regions: the lever rule

When two phases coexist at specified temperature and pressure, composition may be varied without altering either the field or the density variables. When this is true, the fraction of each phase is defined solely by consideration of mass balance (*not* by thermodynamics), and may be calculated using the lever rule. In applying this rule, it is presumed that a particular tie-line has been selected. Doing so defines the composi-tions of the coexisting phases, and the fraction of each phase in any mixture which falls on this tie-line may then be calculated.

If the two phases have compositions a and b and the mixture composition is c, the tie-line may be lifted from the diagram, and redrawn as the line $a–b$, with the point c in between. The lever rule states that the fractions of the two phases at composition c are

$$w^a = \frac{(b - c)}{(b - a)}, \quad w^b = \frac{(c - a)}{(b - a)} \tag{4.3}$$

Inspection of these equations reveals that the fraction of a particular phase is propor-tional to the lever arm opposite that phase. For example, as the composition c approaches b the length of the lever arm $a–c$ increases, and when c equals b then w^b equals one.

4.10   Phase reactions at isothermal discontinuities

As the temperature is raised to the melting point of ice, the "melting phase reaction"

$$\text{Ice} + q^{\text{melting}} \rightleftarrows \text{Liquid water} \tag{4.4}$$

occurs. As a eutectic discontinuity is crossed during an isoplethal process, a similar (but more complex) "eutectic phase reaction" occurs. At this particular eutectic, the reaction equation is

$$0.624\,\text{g Ice} + 0.376\,\text{g } X \cdot W_2 + q^{\text{eu}} \rightleftarrows 1\,\text{g Liquid (23.3\%)} \tag{4.5}$$

During this reaction, two crystals (ice and $X \cdot W_2$) react with one another to form the eutectic liquid. This is a reversible physical process which, as with chemical reactions among molecules, has a rigidly defined stoichiometry. The stoichiometry differs from that of most chemical reactions, however, in being a scalar ratio of phases (containing many molecules), rather than an integral ratio of molecules. The coefficients before the reactants indicate the quantity of each phase that are required to form unit quantity of liquid. To be driven forward heat (q^{eu}) must be added, but the reaction occurs isothermally. The enthalpy and entropy of the liquid reaction product are greater than the combined enthalpies and entropies of the two crystal reactants, and the reaction temperature is dictated by the ratio of these two quantities (eq. 4.5).

If an excess of either reactant phase exists, it is not involved in the reaction and will be left over after the eutectic reaction is complete. By analogy with molecular reactions, the maximum quantity of liquid product phase that may be formed is governed by the quantity of the limiting reactant phase that is present; the excess reactant phase is an inert ingredient.

Being reversible, the eutectic reaction may be driven backward by removing heat from the eutectic liquid phase. The quantities of ice and $X \cdot W_2$ that may be formed (from a given quantity of eutectic liquid) are also dictated by the above equation. If either ice or the $X \cdot W_2$ phase coexists with the liquid at the outset, this phase is not involved and is (as above) carried through the process unchanged. In the forward direction the above reaction may be regarded as a "combination" reaction, while in the reverse direction it may be viewed as a "disproportionation" reaction.

At peritectic discontinuities the peritectic phase reaction occurs. For $X \cdot W_2$ the equation is

$$1 \text{ g } X \cdot W_2 + q^{\text{peri}} \rightleftarrows 0.517 \text{ g Liq } (26.3\%) + 0.483 \text{ g } X \qquad (4.6)$$

At many surfactant peritectic discontinuities the reactant is a crystal hydrate, and the products are a liquid crystal plus a crystal of lower hydration. This reaction may be analyzed in a manner identical to that of the eutectic reaction.

Finally, as noted in Section 4.8.4, the process of isothermally vaporizing water from a biphasic mixture may also be treated as a reversible isothermal process (Section 3.21). In the case of the saturated salt solution and the dry crystal state at 25°C, the phase reaction which occurs is

$$1 \text{ g Liq } (26.43\%) + q^{\text{vap}} \rightleftarrows 0.2643 \text{ g NaCl}^{\text{crystal}} + 0.7357 \text{ g } H_2O^{\text{vapor}} \qquad (4.7)$$

The vaporization and peritectic phase reactions are in some ways similar. In both reactions a single reactant phase (which contains both components) reacts (upon absorption of heat) to form a dilute product phase (which is pure water vapor in the vaporization process and a dilute condensed phase at a peritectic) plus a concentrated product phase. Both phase reactions are disproportionation reactions.

Phase reaction equations have proven to be of considerable value for two principal reasons. First, they express phase transformations in a form that is familiar to all chemists. Second, they considerably simplify analysis of the thermodynamics, kinetics, mechanisms, and other aspects of reversible processes (Section 3.21). For example, once the eutectic and peritectic phase reaction equations are defined, determining the thermodynamics of this reaction allows the thermodynamics of all other mixtures spanning the composition range of these discontinuities to be inferred (at the temperature of the discontinuity).

4.11   Phase fractions at isothermal discontinuities: the Tammann triangle

The lever rule must be applied with great care at eutectics and peritectics. Three qualitatively different kinds of phase behavior are observed along isopleths spanning the eutectic discontinuity in the salt–water system; these are exemplified by following the 20, 23.3, and 30% isopleths (Fig. 4.3). Starting from below the eutectic temperature at 20%, two crystal phases coexist up to the eutectic temperature, and the lever rule may be applied in a straightforward manner until the eutectic temperature is reached. As further heat is added the eutectic phase reaction commences and three phases coexist; when it is finished the $X \cdot W_2$ phase has vanished, because the ice was in excess and $X \cdot W_2$ was the limiting reactant phase. The extent of the phase reaction, which dictates the fraction of each phase present, is governed by the enthalpy of the mixture.

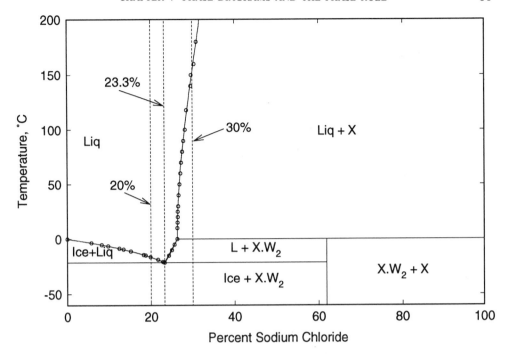

Fig. 4.3. The sodium chloride–water diagram with the 20, 23.3, and 30% isopleths drawn as vertical dashed lines.

The lever rule is not to be used during the course of a phase reaction (while three phases coexist).

Once the eutectic phase reaction is finished, two phases again coexist and the lever rule may be used. The appropriate tie-line is the segment of the eutectic line that lies between ice and the eutectic liquid.

At 30% the limiting reactant phase is ice, and $X \cdot W_2$ is left over after the eutectic reaction is complete. Along the 23.3% isopleth the exact stoichiometric ratio of the two crystal phases exists, and the phase reaction consumes all of these two phases.

The extent of the phase reaction is dictated by the quantity of the limiting reactant phase present, and this is a linear function of composition within the intervals between the eutectic liquid composition and the ends of the discontinuity. This was recognized some time ago by Tammann, who applied it to the analysis of calorimetric data at various compositions spanning such discontinuities [8]. Tammann's analysis is depicted using a "Tammann triangle" (Fig. 4.4). It is most commonly applied at eutectic discontinuities, because severe kinetic problems impair its use at peritectics.

In the Tammann triangle, the *isothermal* (or latent) heats that are required for the eutectic reaction to occur (at the eutectic temperature) are plotted against composition. The nonisothermal component of the thermogram is, if present, ignored. The isothermal heat effect is zero at each end of the eutectic line, because no phase reaction occurs at these limits. The maximum heat effect is observed at the eutectic composition, and the heat effects at intermediate compositions vary linearly with composition between these limits. Data such as these have been used not only to determine the thermochemistry of these transitions, but also the compositions of coexisting phases.

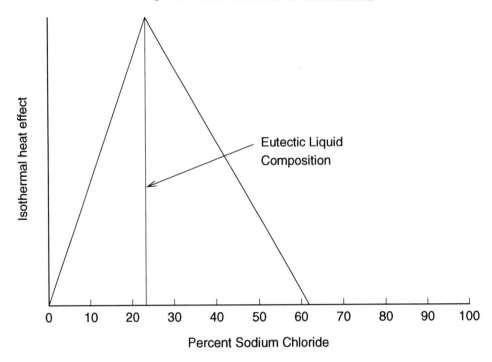

Fig. 4.4. The Tammann analysis of the isothermal heat effects expected at the temperature of the Ice–Liquid–$X \cdot W_2$ eutectic in the sodium chloride–water system.

Examples of the Tammannn analysis may be found in recent studies of the sodium dodecyl sulfate (SDS)–water [9], dipalmitoylphosphatidyl choline (DPPC)–water [10], and decyldimethylphosphine oxide ($C_{10}PO$)–water [11] systems.

4.12 The alternation rule

As one passes along the 20% and 30% isotherms (Section 4.8.3, Fig. 4.3), a transition from a two-phase to a three-phase state, then back to a two-phase state, occurs. At comparable compositions within the span of the peritectic discontinuity, the same pattern exists with respect to the number of phases below, at, and above the discontinuity. The same is true of polytectic discontinuities.

This even–odd–even sequence exists not only along isopleths, but also along isotherms. For example, if salt composition is increased at room temperature starting from 0%, one passes from a one-liquid phase to a two-phase liquid plus crystal region, and then back to a one-crystal phase region. The principle that *an alternation between odd and even numbers of phases occurs* (*in increments of one*) *along isoplethal and isothermal process trajectories* has been termed the "alternation rule" [12]. This principle is implicit in the Phase Rule, and is applicable to all classes of systems. It presumes that the phase transitions are first order (Section 7.5).

Exceptions to the alternation rule appear to exist along particular isopleths at isothermal discontinuities. At the composition of the eutectic liquid or along the $X \cdot W_2$ isopleth, for example, it may appear that three phases are formed from one phase.

However, it is well to recognize that either one, two, or three phases may exist, in fact, and that these apparent exceptions are not exceptions at all if viewed from the perspective of the phase reactions that occur (Section 4.10). At eutectics two crystal phases react to form a single liquid phase ($2 \rightarrow 1$), and at peritectics a crystal reacts to form a liquid plus another crystal ($1 \rightarrow 2$). At melting points and azeotropic points a one-phase state reacts to form two, then a phase vanishes to again form a one-phase state. The alternation rule is, in fact, rigorously followed at each discontinuity of state that occurs – including those associated with isothermal phase reactions.

The alternation rule is the most important device available for evaluating whether or not a particular phase diagram is consistent with the Phase Rule. For example, a boundary which separates two one-phase regions clearly violates this rule, and is therefore highly suspicious. The application of this rule to the analysis of phase behavior in the vicinity of liquid crystal regions is illustrated in Section 5.6.11.

4.13 Isothermal discontinuities as "thermodynamic stop-signs"

It is useful to think of isothermal discontinuities as "thermodynamic stop-signs". A new phase, when it appears, has a different enthalpy and usually a different structure than the phases from which it was created. When an existing phase disappears heat must be supplied or removed, its structure destroyed, and the material within it utilized to construct new phases.

These are complex, drastic changes and many different elementary processes which require heat and/or mass transport must occur if equilibrium is to be maintained. The pause in temperature allows these processes to be completed before the new condition of equilibrium is re-established. Once this changeover has been accomplished then temperature is again a degree of freedom, and smooth variations in state may occur as the temperature changes. While these too require heat and perhaps mass transport, they occur smoothly instead of discontinuously.

Viewed from this perspective, it is hardly surprising that isothermal phase reactions do not occur instantly (Chapter 6). A finite time is invariably required to reach equilibrium in the course of each of these processes.

4.14 The kinds of isothermal discontinuities

The various kinds of isothermal discontinuities that exist, and their essential features, may be classified on the basis of the number of phases in the regions that exist as one passes through these discontinuities along isoplethal paths, and the qualitative nature of the phases that exist.

The alternation rule requires that, with the specific exceptions noted above (which vanish if the phase reaction analysis is followed), the number of phases at a discontinuity is always one greater than the number above and below it. If one phase exists above and below a discontinuity, then two must exist at the discontinuity. A useful shorthand for these sequences is 1–2–1 and 2–3–2. No variants exist within the 1–2–1 class, but more complexity exists within the 2–3–2 class.

Three possible kinds of 2–3–2 discontinuities exist: the eutectic, the peritectic, and the polytectic. A reversible $1 \rightleftarrows 2$ phase reaction occurs as one passes through all these discontinuities, but the polytectic is different from the others in that both the reactant and the product phases have the same composition. The eutectic and peritectic are also differentiated by the fact that the state of higher enthalpy at eutectics is the 1-phase state, while the state of higher enthalpy at peritectics is the 2-phase state. At both eutectics and peritectics, a special situation exists along the isopleth at the composition of the intermediate phase because along this isopleth the $1 \rightleftarrows 2$ phase reaction occurs stoichiometrically.

The structures of the phases that exist at each of these discontinuities is inconsequential insofar as their thermodynamic features are concerned, but their structures give the discontinuity its name.

Number of variables required. The number of variables required to define fully an isothermal discontinuity depends on its classification. In all cases the class, the temperature, and (ideally) the pressure must be specified. (Pressure will, henceforth, be ignored.) In the 1–2–1 class two phase structures and one composition are required as well, so that a total of five parameters must be defined for discontinuities of this class. For eutectics and peritectics in the 2–3–2 class, three phase structures and three composition parameters must be specified plus class and temperature, for a total of eight. Since the compositions of two of the phases are the same at a polytectic, three phase structures but only two compositions must be specified, for a total of seven.

Polytectic discontinuities appear superficially to be just another tie-line within phase diagrams, because two of the phases fall at the same coordinates. Which end of the line

TABLE 4.2
Terminologies of isothermal discontinuities

Name	Phase structures		
1–2–1 (five variables)			
Melting point	X	L	
Boiling point	L	G	
Sublimation point	X	G	
Crystal polymorphism	X_β	X_α	
Crystal $\rightarrow$ liquid crystal	X	LX	
Azeotropic points			
$\quad$ Vapor/liquid	L	G	
$\quad$ Liquid crystal/liquid	LX	L	
$\quad$ Liquid/liquid crystal	L	LX	
2–3–2 (eight variables)			
Eutectics			
$\quad$ Salt/water	Ice	L	$X \cdot W_2$
$\quad$ Krafft discontinuity	L	LX	X
$\quad$ Krafft discontinuity (alternate)	L	L	X
Peritectics			
$\quad$ Salt $X \cdot W_2$	L	$X \cdot W_2$	X
$\quad$ DODMAC $X \cdot W_2$	LX	$X \cdot W_2$	$X \cdot W$
$\quad$ DODMAC $X \cdot W$	LX	$X \cdot W_2$	X
Polytectics			
$\quad$ DODMAC $X_\beta \rightarrow X_\alpha$	$X \cdot W$	X_β	X_α
Isothermal vaporization of water			
$\quad$ All biphasic regions	Any	Any	G

G, gas; L, liquid; X, crystal; LX, liquid crystal; W, water.

represents the two-phase state must therefore be designated, in some way, to identify the polytectic line.

The various dimensions of these discontinuities, and their common names, are listed in Table 4.2.

4.15   Some different ways of using the Phase Rule

In the above sections a system whose phase diagram is known was examined to determine the variance within different regions and at discontinuities. This is an important use of the Phase Rule, but is not the only one.

This Rule may also be used to infer the number of phases present from data on the variance of a mixture. An example is the determination of the number of phases that exist within liquid solutions that display micellar structure. From experimental data it is apparent that the variance of micellar solutions is three in binary surfactant–water systems, and this result demands that these solutions be regarded as a single phase [13]. This holds at compositions up to the true solubility boundary, but just past the solubility boundary a micellar solution of defined composition coexists with another defined phase (another liquid, a liquid crystal, or a crystal). (DIT experiments are fully consistent with this conclusion, in that discontinuities are never seen in the vicinity of critical micelle concentrations (cmcs).) This discontinuous change to an easily recognizable two-phase state poses no problem if the micellar solution is regarded as a single phase. This situation would constitute a flagrant violation of the Phase Rule and the derived alternation rule (Section 4.12), however, if the micellar solution were to be viewed as a two-phase region.

The micelle has been viewed as a second phase, in part, because of the sharpness with which changes in the properties of surfactant solutions often occur at cmcs. However, the sharpness of cmcs varies with chain length, and curvature at the cmc is progressively more evident as the chain is shortened (to C_8 or C_{10}). Careful studies have also demonstrated that curvature exists at the cmc of SDS, which displays a relatively sharp cmc.

Useful quantitative descriptions of the solution and surface chemistry of micellar solutions, based on the "pseudo-phase separation" model, have been developed [14]. This approach has also been particularly useful in characterizing the nonideality of mixtures containing two or more surfactants [15,16]. The conflict with the Phase Rule in using this model does not invalidate its use, but the approximation that has been made should always be kept in mind.

Determining variance from DIT studies. DIT studies permit a usually unambiguous determination of whether or not composition is a degree of freedom. ("DIT" stands for "diffusive interfacial transport", which is the name given the isothermal phase studies method during which the phases that exist are formed by swelling, and their structures and compositions determined *in situ* (Appendix 4.4.4).) If composition is observed to vary smoothly within a particular range, then variance clearly exists with respect to composition within that range. Conversely, composition is evidently *not* a degree of freedom within miscibility gaps, which span finite composition ranges but occupy no space within the cell.

Determining the number of components from variance and data on the number of phases present. If F and P have been determined, then C may be inferred. While this might appear to be a strange application of the Phase Rule, it helps one to determine whether or not chemical reactions between components are important features of a particular system.

If one attempts to perform a phase study of a pure soap in water, for example, data which are inconsistent with the Phase Rule for binary systems may be obtained. At compositions less than about 20%, the observed crystal solubility boundary has a negative slope (increases in temperature as the composition is reduced). This boundary cannot have a negative slope in a binary system, as it is then impossible to draw tie-lines which connect the coexisting liquid and crystal phases without passing through a one-phase liquid region. Results such as these signify that the mixture is not behaving as a binary system and, from other information, one may infer that hydrolysis of the soap to form fatty acid is important. The "binary" soap–water phase diagrams reported in the literature were, in fact, determined in the presence of sufficient added base to suppress hydrolysis [17]. The magnitude of this problem depends on composition; while it is serious at compositions less than about 20%, it is less noticeable at higher compositions.

The hydrolysis chemistry of salts of unassociated carboxylic acids is fully described by the equations

$$RCO_2^-, Na^+ + H_2O \rightleftharpoons RCO_2H + Na^+, OH^-$$

$$K_h = \frac{K_w}{K_A} \tag{4.8}$$

$$pK_h = 14 - pK_A \cong 9$$

where K_w is the autolysis (self-dissociation) constant of water and K_A is the dissociation constant of the carboxylic acid [18]. Since the pK_As of carboxylic acids are about 5, the pK_hs of their salts are about 9. The hydrolysis of soaps at concentrations where they are associated, or exist as a liquid crystal phase, cannot be described using this model; the apparent pK_A drifts significantly (by several pK_A units) as the ratio of soap to fatty acid changes [19]. This chemical information is fully consistent with physical data from the phase studies and the existence of crystalline phase compounds of fatty acids and soaps (acid–soaps) (Section 2.2.1).

It is interesting to compare the behavior of soaps with that of amine salts. Dodecyl-ammonium chloride, for example, can dissociate as follows:

$$C_{12}H_{25}NH_3^+, Cl^- \rightleftharpoons C_{12}H_{25}NH_2 + H^+$$

$$pK_A \cong 9 \tag{4.9}$$

The pK_A of this salt [20], which accurately describes its hydrolysis behavior under conditions where it is unassociated, is numerically similar to the hydrolysis constant pK_h of unassociated soaps [19]. In associated solutions, the same sort of distortion of hydrolysis equilibria occurs with ammonium salts as is observed with soaps. Yet, mixtures of dodecylammonium chloride and water do *not* display the complexity in their phase behavior that do soaps. A satisfactory explanation for this interesting discrepancy has not been suggested.

4.16   A comparison of isothermal and isoplethal process paths

The consequences as to phase behavior of following either isothermal or isoplethal process paths have been diagnosed in the above analysis. While these two paths may appear superficially to be similar, there are important differences. The consequences of following isoplethal paths are, as a rule, far more complex than are those of following isothermal paths. These differences stem from fundamental differences in the nature of the temperature and pressure variables, on the one hand, from composition variables on the other.

It is impossible for a mixture to remain invariant during any process in which the temperature (or pressure) is varied, because the properties of phases inevitably change as either of these two parameters is varied. In striking contrast, it is entirely possible for a mixture to remain invariant during a process in which composition is varied. Examples are described in Section 4.8.2. Within the span of compositions for which this is true, the state of the mixture does not change – only the phase ratios. Thus, there are no circumstances under which changes in temperature and pressure do *not* alter the thermodynamic state of a mixture.

There are at least two important consequences of these differences between composition and temperature or pressure, one of which is how best to view one process in terms of the other. While this is a trivial issue within segments of any trajectory that pass through single-phase regions, it is not at all trivial within multiphase regions.

To illustrate, one may view an isoplethal process as a series of closely spaced isothermal processes, or one may view an isothermal process as a series of closely spaced isoplethal processes. If isothermal processes are intrinsically simpler than isoplethal, then the former view represents an analysis of a relatively complex process as a series of relatively simpler processes. The latter view, in contrast, represents the analysis of a comparatively simple process as a sequence of more complex processes. Clearly, the former approach is to be preferred.

A second consequence of the above premise relates to the selection and use of phase study methods (Appendix 4). In executing a phase study one should use every tool available, but it is well to remember that the inherently greater complexity of isoplethal paths leads to intrinsic problems in all isoplethal methods, and that these problems are side-stepped by using isothermal methods. The foundation of surfactant phase science was laid using isoplethal methods out of necessity; the only isothermal method that was available during this period is not generally applicable to surfactant systems. The development of isothermal swelling phase study methods is expected to alter this situation, in time [21].

4.17   Boundary conditions for use of the Phase Rule

In applying the Phase Rule it has been assumed that certain boundary conditions have been met; these will now be spelled out.

4.17.1   The condition of equilibrium of state

First and foremost is the presumption that equilibrium of state has been attained (Section 3.5). This condition is implicit in the derivation of the Phase Rule, and will be regarded as a *sine qua non* for its application. This means, first, that the existing mixture has the lowest free energy among the various kinds of states that are accessible (Section 3.15). Further, it implies that the field variables are uniform throughout the space occupied by the mixture. They must be uniform not only within phases, but also among phases. If gradients in field variables exist within a phase, that phase is *not* at equilibrium and spontaneous fluctuations will, in time, eliminate these gradients. If the field variables are uniform within each phase but are not the same in all phases, then the system as a whole is not at equilibrium. This condition of equilibrium of state has been stated in a rigorous mathematical form (Section 3.5).

In the case of surfactant molecules the concept of equilibrium of state is in most cases straightforward, since the compounds are thermally and chemically stable under the conditions of a phase study (with the exceptions noted in Section 5.2). It is worth noting, however, that the Phase Rule may be applied to systems which are recognizably metastable, provided the kinetic barrier to change is sufficiently large.

Hydrazine and sodium sulfate may be used to illustrate this situation. The free energy of formation of hydrazine is positive, which signifies that (at equilibrium) hydrazine is unstable relative to a mixture of nitrogen and hydrogen. Such decomposition will indeed happen quickly if a catalytic metal surface is provided, but in the absence of a catalyst hydrazine is extremely long-lived. Its phase behavior may therefore be determined in a straightforward manner.

Sodium sulfate forms two hydrates (a hepta- and a decahydrate [4]), and the solubilities and peritectic discontinuity temperatures have been determined for both hydrates. The solubility is far higher and the peritectic temperature lower for the heptahydrate, from which it may be inferred that the heptahydrate is metastable with respect to the decahydrate. The Phase Rule may still be applied to analysis of the phase behavior for the heptahydrate, so long as a path for the conversion of one phase to the other does not exist. Should formation of the decahydrate occur, however, the heptahydrate is to be regarded as a metastable state.

4.17.1.1   Equilibrium of state and colloidal structure

In mixtures having more than one phase for which it can be generally agreed that both phases are in every sense of the word macroscopic, the application of the Phase Rule is straightforward. If one of the phases present is subdivided and dispersed within the other, however, the surface area to volume ratio increases and, eventually, colloidal structure is introduced (Section 2.4.4). Under these circumstances additional factors must be considered. From a purely structural perspective, and in many properties such as light scattering, a clear distinction between a mixture of two phases and a highly structured single phase containing both components cannot be made. Energetically, one must recognize that the free energy of the mixture becomes progressively larger as the area/volume ratio is increased. Further, an increase in the curvature of interfaces usually accompanies an increase in area/volume ratio and this too may have to be taken into account.

The debate over how to treat colloidally structured mixtures, and the transition from a macroscopically structured mixture to a colloidally structured mixture, has extended over more than a century [7,22]. This issue was again raised in connection with the celebration of the centenary of the Phase Rule [23] during which (and also many years earlier [7]) it was suggested that the "degree of dispersity" of colloidally structured mixtures should be regarded as a degree of freedom in its own right (along with the system variables), and that the Phase Rule equation should be modified accordingly.

This approach is not taken in this book. The philosophy adopted herein is that the Phase Rule is to be applied only to analysis of the equilibrium state of a mixture; the extremely important subject of properly treating colloidal mixtures (and other non-equilibrium aspects of their physical science) is regarded as a separate matter to be treated using appropriate methods. This position is not new or unusual; it has been adopted by the vast majority of authors in phase science [24,25,26]. Its evident and considerable value lies in the fact that the equilibrium state serves as a solid bedrock of information for a particular system that is fully defined once the system variables are stated. Such information is not only of value in its own right; it also provides a solid basis for the further analysis of more complex nonequilibrium phenomena (including colloidal phenomena).

4.17.2   The dynamic aspect of equilibrium

A dynamic exchange of components across interfaces exists in phases at equilibrium. If a thin membrane which barred the transport of either component were to be inserted between coexisting phases (for example, between a salt crystal and the coexisting liquid phase), these phases could no longer be described as being at equilibrium – even if they possessed equilibrium compositions, structures, and thermodynamics.

In Section 1.1, the salt crystals that coexist with the saturated liquid phase were described as remaining "seemingly" unchanged. This qualifying adjective is mandatory because, in fact, the matter within phases in equilibrium is in a state of constant flux. An exchange of salt constantly occurs between the crystal and liquid phases, and if water existed within the salt crystal the same would be true of this component. If one were to add radiolabeled sodium chloride crystals to such a mixture, a measurable exchange of labeled molecules between the two phases would be observed. At the outset a non-equilibrium distribution of labeled molecules would exist, but at equilibrium the entropy of mixing would insure that the labeled molecules will be distributed uniformly throughout the mixture. (It is presumed in this instance that the mixing of these heavy isotopes is in fact ideal. If there were a measurable "isotope effect", then the distribution of the isotope at equilibrium would not be uniform.)

4.17.3   Fluctuations in space and time within equilibrium phases

Local fluctuations in composition, which span a finite range of space and time, occur within all equilibrium phases. If these fluctuations represent a departure from the equilibrium state, however, they do not persist. If the mixture is *not* at equilibrium and these fluctuations move it closer to equilibrium, they will persist. It is just such processes that enable a nonequilibrium mixture to attain equilibrium [27]. One cannot infer the equilibrium state of a phase from snapshots of composition, structure, or thermodynamics that are taken over too small a distance or within too short a span of time.

Local fluctuations are particularly large, and thus most easily observed, in the vicinity of critical points. Solutions of surfactants such as $C_{10}E_4$ at temperatures just below the lower consolute boundary (Fig. 4.5) display a uniform and persistent haziness to the naked eye, termed "critical opalescence" [28]. Its intensity fades as the temperature is

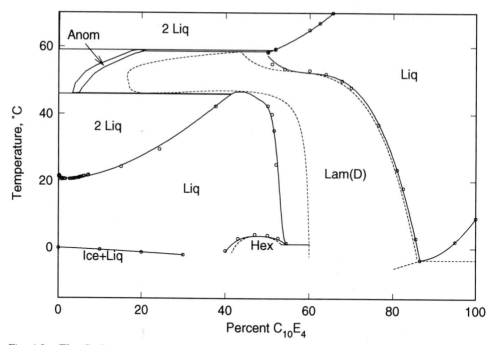

Fig. 4.5. The $C_{10}E_4$–water system between -15 and $70°C$. The published diagram has been modified at the lower limit of the anomalous phase corridor in accordance with recent DIT results [30].

lowered, and it is usually no longer grossly perceptible 5–10°C below the critical temperature. Further, as one passes along isotherms that lie just below the critical temperature, the intensity of critical opalescence passes through a maximum near the critical composition. This correlation has been used to determine experimentally critical compositions in both binary and ternary systems [29].

Similar phenomena are observed in the vicinity of the upper consolute boundaries of aqueous zwitterionic surfactant mixtures, except that the dependence on temperature is reversed.

Critical opalescence is visible to the naked eye because the spatial dimensions of the fluctuations in composition and refractive index (which may be described by a characteristic correlation length) are similar to the wave length of visible light (400–700 nm) [28]. As temperature and composition move away from the critical point the correlation lengths diminish; turbidity may no longer be visible even though the fluctuations must exist. One must be aware of critical opalescence during phase studies, otherwise it can be mistaken for phase separation (Appendix 4.4.3).

4.17.4   The influence of surface energies

For surfactants, an important boundary condition imposed on the Phase Rule is that the surface energy within a mixture must be sufficiently small, relative to the energy of the bulk phase, that it does not measurably perturb the equilibrium state. After all, the capacity to modify efficiently the mechanical properties and energies of interfaces is a pervasive characteristic of surfactant molecules (Section 11.2.1).

The importance of this boundary condition is reflected by the historic fact that McBain did not at first assume, *a priori*, that the Phase Rule would in fact apply to surfactant systems (Chapter 2). During the succeeding seven decades, however, massive evidence has been produced which supports the position that the Phase Rule does indeed apply, in a straightforward manner, to aqueous surfactant systems. Further, these data support the premise that phase transitions in these systems are typically first order (Section 7.6).

Since it is impossible to have a phase that does not have an interface as well, the area/volume (A/V) ratio and specific interfacial areas (A/mass) can never be precisely zero. (The term interface is usually used with reference to the surface separating any two phases, while "surface" is more commonly used with respect to the interface with air. The differences are purely semantic.) Since A/V varies linearly with the reciprocal of the linear dimension in particles of similar geometry, both A/V and surface energy are increased by dispersing one phase in another to produce colloidal structure (Section 2.4.4). The extent of the influence of surface energies is thus intimately related to the qualitative nature (sol vs. gel, Section 2.4.4.2), and especially to the quantitative dimensions, of the colloidal structure that exists. The surface energy component is in the first approximation the product of the interfacial tension and specific surface area, and will be largest when both of these factors are large. However, the excess energy of colloidal states relative to the equilibrium state may include other components, for example curvature energy [31,32].

4.17.5 The influence of external fields

It was recognized by Gibbs that external gravitational, electrical, and magnetic fields could influence the state of equilibrium [33]. Under most circumstances, one is concerned with mixtures that exist in a gravitational field of 1 g, in a magnetic field of about 0.5 gauss [34], and in very weak electrical fields. If the magnitude of any of these fields is changed, then the state of equilibrium will also be changed. The dependence of surfactant phase behavior on gravitational field strength has been investigated [35].

During isothermal phase studies dispersions of one phase in another are often encountered, and the two phases must be cleanly separated to perform the study using these procedures. As the phases usually differ in density, this separation may possibly occur spontaneously by the action of ambient gravity. However, when the separation is slow, it is tempting to increase the gravitational field strength by centrifugation. If separation does not result within a clinical centrifuge, the mixture may be placed in an ultracentrifuge. The gravitational field within a preparative ultracentrifuge at high speeds may range from 254 000 g at the top to 485 000 g at the bottom of the sample tube.

It is entirely possible for gravitational fields of this magnitude to distort phase behavior [35,36]. In addition to altering quantitatively the compositions of coexisting phases, they may on occasion qualitatively change the phase diagram. During the $C_{10}E_4$–water phase study, for example, it was observed that attempts to isolate the anomalous (L_3) phase by the ultracentrifugation of approximately 2% mixtures resulted, instead, in separation of the L_1 and the lamellar phase. (Mixtures within the anomalous–lamellar miscibility gap did separate into these two phases.) The presence of the L_3 phase at 1 g was amply documented and its existence later confirmed by swelling methods, so that its existence may be regarded as having been firmly established.

Electrical fields. External ambient electrical fields are ordinarily weak during phase studies, but special circumstances exist under which this may not be true. Mixtures held between closely spaced charged plates, as for example during electrophoresis or within solid state devices, may be exposed to much higher electric fields. The equilibrium phase state can be expected to be altered under these circumstances, just as happens with the thermotropic liquid crystal phases utilized in electronic displays [37].

Magnetic fields. Mixtures held within an NMR spectrometer are exposed to very high magnetic fields. Within a 500 mHz NMR spectrometer operating at the proton resonance frequency, the sample is exposed to a magnetic field of 11.7 tesla (117 000 gauss). This is about 240 000 times larger than the ambient magnetic field at the surface of the earth [38]. This may be of concern during investigations of lyotropic nematic liquid crystal phases, because viewing these phases within the magnetic field of an NMR spectrometer is one of the means by which they have been characterized [39]. Fields of less than one tesla can orient such phases [40]. Therefore, care must be exercised in this area to insure that the phase behavior determined under these circumstances is not distorted from that which applies at the weak ambient magnetic field outside the spectrometer.

4.18   The concepts of "phase" and "interface"

Having reviewed the thermodynamics of immiscibility (Chapter 3) and the rules and boundary conditions governing phase behavior (above), it is timely to probe more deeply into the basic concepts of "phase" and "interface". The spatial profiles of compositions and of thermodynamic variables within phases and across interfaces are especially relevant to these concepts.

4.18.1   Spatial composition profiles

It can be seen from the $C_{10}E_4$–water phase diagram (Fig. 4.5) that a 2% mixture at 20.8°C will exist, at equilibrium, as two liquid phases whose compositions are 0.62 and 4.32%. If this mixture is allowed to stand quietly in a test tube at this temperature, these phases will separate from one another as clear upper and lower layers. The interface between them will be easily visible and razor sharp, each layer will be uniform throughout its extent, and the mass fractions of the two phases will be 0.627 and 0.373, respectively.

An entirely different situation would exist during a DIT (Appendix 4.4.4) study of this system at this temperature [21]. In this experiment the DIT cell (which is a long, thin capillary) is partly filled with $C_{10}E_4$, and the remainder of the cell is filled with water. An interface is created between the two components; precisely at the moment of its creation a step function in concentration exists along a trajectory which spans this interface. After a finite time has lapsed the countercurrent diffusion of water into the $C_{10}E_4$ (and of $C_{10}E_4$ into the water) will have occurred, and three transverse interfaces spontaneously will appear which divide the cell contents into four phase bands (Fig. 4.6). These interfaces will spread apart with the passage of time, but persist for a day or more. The $C_{10}E_4$ concentrations to either side of each interface do *not* vary with time [21].

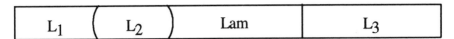

Fig. 4.6. A schematic diagram of the phase bands encountered during DIT studies of the $C_{10}E_4$–water system at 30°C. The sequence of phases observed is consistent with the phase behavior of this system at this temperature (Fig. 4.5).

The phase band at the water end of the cell has a liquid structure (L_1). At short times pure water exists at this end of the cell, and a gradient in $C_{10}E_4$ concentration extends up to the first interface. The concentration within this band at this interface is 0.62%.

The second band is also a liquid (L_2). A gradient in $C_{10}E_4$ concentration likewise exists within the L_2 band; at its dilute end (at the first interface directly opposite the L_1 phase) the $C_{10}E_4$ concentration is 4.32%, while at its concentrated end the concentration is 52.64%. The third and fourth bands are the lamellar liquid crystal and a third liquid (L_3), respectively. Concentration gradients also exist within these bands, discontinuities in concentration exist at each interface, and (at short times) dry $C_{10}E_4$ exists at the far end of the L_3 band. The contents of the cell span the entire composition range, but what would be a smooth composition gradient (if the surfactant was miscible with water in all proportions) is interrupted by these discontinuities in composition. They are visible as interfaces because discontinuities in refractive indices also exist.

At 30°C a qualitatively similar sequence of phases is found, but the compositions at the first interface are <0.1% and 24.24%, respectively. A plausible composition profile along the length of the cell at 30°C is shown in Fig. 4.7.

Parallel experiments using sodium chloride and water would yield similar, but not identical, results. In the test tube experiment at 30°C, a salt crystal of uniform composition (100%) would coexist in equilibrium with a salt solution that is also of uniform composition (26.48%). In the diffusion experiment the crystal phase would remain uniform in composition, but a gradient would exist within the liquid phase. Such gradients have been visualized around a dissolving salt crystal by using interference methods, and the rates of dissolution quantitatively described using this approach [41].

Some very important attributes of phases are revealed by examining and comparing the equilibrium test tube and the nonequilibrium DIT experiments. In the test tube the regions adjacent to the interface are single phases – but so are all the bands in the DIT experiment. In the test tube compositions are uniform throughout each phase, while in the DIT cell finite spatial gradients in composition exist within each phase band (Fig. 4.8). In the dissolving sodium chloride experiment spatial gradients exist within the liquid phase, but not in the crystal.

From general experience, as well as from specific observations such as the above, it is evident that uniformity of composition is *not* a mandatory criterion that a mixture must meet in order to be regarded as a single phase. Each of the two layers in the test tube and all the bands in the DIT experiments are clearly single phases, but compositions are uniform within the phases in the test tube while smooth spatial gradients in composition exist within the phase bands in the DIT cell. Uniformity of composition *is*, however, a prerequisite of phases at equilibrium with one another. If composition varies within a phase then so do chemical potentials (and other field variables), and spatial gradients in

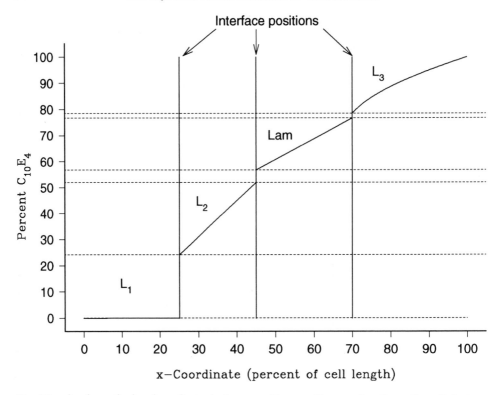

Fig. 4.7. A schematic drawing of a typical composition profile spanning the entire cell during a DIT study of $C_{10}E_4$ at 30°C.

field variables do not exist within phases at equilibrium. A peculiarity of the DIT and other swelling experiments is that equilibrium phases exist only at the interfaces; mixtures within the phase bands are *not* at equilibrium [42].

Along spatial profiles spanning the interface between equilibrium phases (as in the test tube), the field variables not only remain uniform within the phases but are also precisely the same within both phases. A discontinuity in the field variables does *not* exist at the interface separating equilibrium phases. A smooth gradient in the field variables exists within each of the phase bands along spatial profiles spanning the interfaces in the DIT cell, but the gradients within adjacent bands must intersect at the interface and the field variables must be the same at this position in space.

The spatial profiles of compositions and of the density variables are entirely different from the spatial profiles of the field variables, in both the test tube and the DIT cell. Within the test tube compositions are uniform within each phase, but are different in the two phases. A discontinuity *does* therefore exist at the interface in the spatial profile of compositions. Within the DIT cell smooth gradients in composition exist within each phase band, but discontinuities in composition also exist at the position of each interface in this experiment as well.

If compositions differ to either side of an interface, so must the entire ensemble of density variables also differ. Therefore, the spatial profiles of density variables within phases and across interfaces (in both the test tube and the DIT cell) must qualitatively

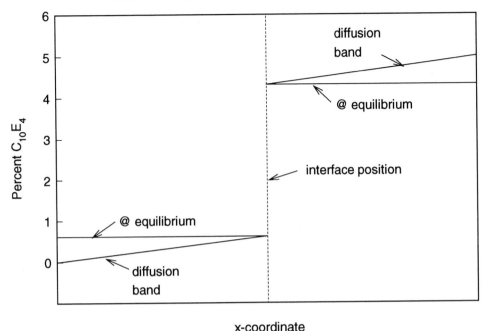

Fig. 4.8. Composition profiles at the liquid/liquid interface in $C_{10}E_4$–water at 20.8°C within any mixture in the miscibility gap (equilibrium condition, zero spatial gradients in composition), and within the phase bands that exist in a diffusion band during a swelling experiment (finite gradients).

resemble the spatial profiles of composition. Gradients in density variables do not exist within phases at equilibrium (the test tube), but do exist within phases that are not at equilibrium (the DIT bands). Discontinuities in the density variables along spatial profiles spanning the interface exist at the interface in both cases. It was shown in Chapter 3 how each of these profiles of field and density variables may be qualitatively understood from consideration of free energy of mixing functions in sufficiently non-ideal mixtures.

Structures to each side of interfaces. The structures of the phases that exist to each side of an interface are typically different. In the DIT cell at the interface between the two most dilute phases at both 20.8 and 30°C, all four phases have a liquid structure. At 20.8°C the concentration of $C_{10}E_4$ is greater than the cmc in both phases. (Critical micelle concentration (cmc) is the range of concentrations in liquid surfactant solutions within which micellar structure first appears. For $C_{10}E_4$ the cmc at 20°C is 6.4×10^{-4} M (0.021%). The cmc concentration must exist at a particular position within the first liquid band during the diffusion experiment, but no perceptible discontinuity is ever observed within this band.) At 30°C the $C_{10}E_4$ concentration is very close to (or below) the cmc in the dilute phase, but far above it in the concentrated phase. From these observations, it is apparent that micellar structure either may, or may not, be present within coexisting liquid phases.

Still more dramatic differences in phase structure exist to either side of the interface at the L/D and the D/L interfaces (the dilute and the concentrated sides of the lamellar region, respectively), since in these instances one phase is liquid while the other is liquid crystal. The structure of the liquid to the concentrated side of the lamellar phase has not been determined; it is uncertain as to whether or not it has micellar structure.

The scale of uniformity in phases. Throughout the phase literature, mixtures that are single phases are described as "homogeneous" and mixtures containing two or more phases as "heterogeneous". By any of the methods available for determining the structure of phases in Gibbs' time, these distinctions would have been eminently reasonable.

Now, however, the situation is different. Multiphase dispersions known to be "heterogeneous" that are crystal clear (due to extremely small particle size) have been prepared [43,44], and "homogeneous" phases that display perceptible turbidity exist (Sections 4.17.3, 5.10, 8.4.3). The capacity to determine phase structure has become highly developed in recent years. The meanings of the words "homogeneous" and "heterogeneous" therefore need to be re-examined with this new information in mind, and a more robust definition of what is a "phase", and of what constitutes an "interface" between phases, is needed.

The first step towards refining our concepts of phase and interface is to again recognize that all so-called "homogeneous" phases are actually heterogeneous, if examined over the molecular scale of distance [45]. The molecular heterogeneity of micellar solutions is more easily perceived than is that in salt solutions, but only because the scale of distance spanned by the structural elements present is larger. Unusually intense light scattering exists in micellar solutions, and when the "sphere-to-rod" transition occurs within micellar solutions the heterogeneity that exists is revealed with striking clarity by both optical and rheological data [46].

Nevertheless, heterogeneity also exists in salt solutions. There are 9.6 water molecules per NaCl ion pair in the saturated solution. Free ions, ion pairs and ion multiplets likely exist within this liquid, and perhaps also aggregates that resemble salt crystals in some degree. Some water molecules exist within the inner hydration sphere of the sodium ions, others in the outer sphere, still others are hydrogen bonded to chloride ions, and a few may exist as "free" water. At the molecular level, this liquid phase is anything but "homogeneous".

The same picture holds for long-chain molecules. Considerable evidence presently exists to suggest that within the liquid at the melting point, extensive crystal-like short-range structure exists even though the phase is optically isotropic [47]. Substantial disorder also exists in the coexisting crystal at this transition temperature (Section 8.3). Heterogeneity exists even within gas phases at low pressures. The existence within these phases of clusters (van der Waals molecules) whose abundance depends on pressure, temperature, and molecular structure, is thoroughly documented [48].

A different kind (and a far higher level) of local heterogeneity of structure exists within liquid crystal and crystal phases, where phase structure is pervasive and highly developed. The most perfectly structured phase known is the single crystal at zero K. At ordinary temperatures thermal energy introduces uncertainty in the positions of molecules within the lattice and the lattice dimension depends on temperature, but the lattice persists (Section 8.3 [49]). The "thermal ellipsoids" which define these fluctuations are routinely documented during single crystal X-ray structural studies.

One encounters, then, a wide range of structures within the phases that aqueous surfactant systems form. Liquid solutions below the cmc display no perceptible long-range structure beyond that ordinarily found in liquids. Micellar liquid solutions display structure within which aggregates of surfactant molecules exist, and the surfaces of micelles can be characterized using a variety of methods. Liquid crystal phases display still more highly developed structures, and surfaces are readily perceived within these phases. Crystal phases display the most highly developed and sharply defined structures of all, and many surfaces can be perceived within crystal phases. The issue before us, then, is how to distinguish those structural surfaces that exist within a phase (but are not interfaces) from those unique surfaces which separate coexisting phases and are true interfaces.

4.19 Suggested definitions of "phase" and "interface"

Consideration of the spatial profiles within macroscopic phases and across interfaces separating such phases, as described above for both equilibrium and nonequilibrium situations, suggests the following definition of the concept of "phase". *A phase may be regarded as a volume element of a mixture within which smooth variations in space of the density variables exist.* If the spatial variation within a particular volume element is zero, this volume element is not only a single phase but is a phase in equilibrium as well. If the spatial variation in density variables is smooth but has a finite value, then this volume element qualifies as a single phase but this phase is not in equilibrium.

A definition of the "interface" between two phases is implicit in the above definition of a phase: *an interface is a surface across which a spatial discontinuity in density variables exists.* Recognizable structural surfaces exist within most surfactant phases, but if a spatial discontinuity in density variables does not exist across a particular surface then it is not a true interface.

To illustrate, if one starts at the interface between macroscopic phases at equilibrium (the test tube situation, above) and moves away from it in one of the phases along the normal to the interface a sufficiently large distance (see later), a particular numerical value of the density variables that is characteristic of the macroscopic phase will be encountered. If one returns to the starting point and moves in the opposite direction into the other phase, a particular numerical value of the density variables will be encountered along this path as well, and this value will differ from that encountered in the first phase.

If one starts from any point at a structural surface within a macroscopic phase at equilibrium and moves along similar opposed paths (without crossing an interface), the numerical values of the density variables to one side of the starting point will be identical to their values to the other side. Within structurally isotropic phases the same values are encountered in all directions. Within structurally anisotropic phases (Section 8.2) the numerical values of the observed density variables depend upon the direction chosen, but the symmetry in values along opposed paths will remain (presuming that the structure as a whole is not curved). Along spatial profiles within phases not in equilibrium (the DIT experiment, above), smooth variations in both field and density variables (but no discontinuities) will be encountered.

The above definitions of "phase" and of "interface" are based on characteristic features of the spatial variation of specific thermodynamic variables. From Gibbs' time to the present phases have been defined primarily in terms of uniformity of composition [50], but often they are defined in terms of uniformity of "properties" [51]. Both of these attributes constitute a shaky basis for defining the concept of phase, since (as noted earlier) no phase is uniform in composition at molecular scales of distance, and some properties are the same in both phases. Moreover, it is widely recognized that the properties of bulk phases (thermodynamic and otherwise) are distorted as the dimensions of the volume element that they occupy reach the colloidal size domain.

Partial miscibility has its origin in the existence of anomalies (or nonideality) in the free energies of mixing. It would therefore seem preferable in defining both "phase" and "interface" to focus on the spatial discontinuities that exist in the relevant thermodynamic variables (as above), rather than to include consideration of spatial discontinuities in compositions, or in properties in general. The quantitative description of phases and details of the structures of interfaces are evidently not addressed by this simple view of phases and interfaces; these matters are presently the subject of intense scrutiny [52].

What span of distance should be regarded as being macroscopic? The above concepts are straightforward for "macroscopic" or "bulk" phases, but are less clear for volume elements of a phase that are sufficiently small to be regarded as having "colloidal dimensions". At a suffiently large dimension (say a millimeter), it would probably be generally agreed that a particle having such a dimension is macroscopic. If the dimensions of a 1 mm particle were to be reduced by a factor of two, the properties of the resulting 0.5 mm particles would closely resemble those of the initial particle. This subdivision process could be repeated many times starting at these values, and the same result would be found. However, a point would eventually be reached at which further reduction in the dimensions of the particle exerts a measurable influence on its properties. This would not likely occur suddenly, but would instead occur to a progressively larger degree as the particle size is reduced. Eventually the distinction between the small particle and all the larger ones would become sufficiently large that the particle would generally be regarded as being within the colloidal size domain. Phase science is clearly applicable to all those particles of a phase whose dimensions may be regarded as being macroscopic, but its application to mixtures having particles of colloidal dimensions is less clear. The most important classes of colloidal structures that are encountered in surfactant systems are the vesicle and liposomal colloids.

Vesicles, liposomes, and the issue of "phase". The particles that exist in colloidal dispersions of insoluble surfactants (such as polar lipids) tend to have a vesicular or a liposomal structure. "Vesicles" are hollow, their outer membrane is a single bilayer, and they usually have a spherical shape. "Liposomes" may be similar in their outer diameter to vesicles (or perhaps larger), but differ in that many concentric bilayers exist within these particles. For a polar lipid that occupies c. 0.65 nm^2 per molecule at interfaces, a moderately small unilamellar vesicle 50 nm in diameter contains about 24 100 molecules and about 200 molecules are found within a great circle. These numbers indicate that while the dimensions of vesicles are bimolecular in the direction across the shell, they are much larger along the other two dimensions within the membrane surface. The quantitative relationship of the properties of vesicular and liposomal dispersions to the equilibrium properties of aqueous polar lipid mixtures is an important issue.

Dispersions of vesicles undergo isothermal phase transitions during isoplethal studies (Section 11.3.2.7), just as do dispersions of liposomes [53]. The transition temperatures observed in vesicular dispersions are close to those exhibited by liposomal dispersions when the vesicles are large, but typically occur at slightly lower temperatures.

Deviations from the liposomal value become progressively larger as the vesicle size is reduced, but the size of vesicles cannot be reduced indefinitely. Vesicles having a diameter as small as 20 nm have been prepared and investigated, but smaller ones (which are presumably formed during intensive sonication) are not sufficiently long-lived to be studied [54,55]. In some related systems, both open bimolecular leaflets and three-dimensional particles that are not closed exist [56].

If one assumes that the physical behavior of liposomal dispersions resembles that of the equilibrium state, then the physical behavior of vesicles demonstrates that one dimension of particles of a dispersed phase can have bimolecular dimensions without qualitatively altering phase behavior – provided the other two dimensions are very much larger. The quantitative perturbations of phase structure and transitions that exist in vesicles (relative to liposomes) may be attributed to their colloidal overall dimensions and to the bimolecular dimensions of their membranes. All these factors contribute to the existence of the excess free energy and intrinsic instability that is characteristic of irreversible colloidal states. Perhaps the major uncertainty in this area is the properties of biphasic mixtures of macroscopic phases that *are* at thermodynamic equilibrium. While liposomal dispersions appear to resemble a macroscopic phase more closely than do vesicles, it cannot be assumed *a priori* that liposomal dispersions do, in fact, represent the equilibrium state.

4.20 Tensions within phases and at interfaces

Equilibrium surface tensions at interfaces reflect the existence of a spatial imbalance in the density variables to either side of the interface, and hence in the forces that act upon this special surface. The existence of an equilibrium force at an interface is consistent with the concept that a difference in density variables exists to each side of interfaces. At structural surfaces remote from interfaces, however, density variables are uniform along any two opposing directions and the forces in any given direction are precisely balanced by those in the opposite direction. This is true no matter how extensively structured the phase may be. As a result, equilibrium tensions do not exist at the structural surfaces within phases at equilibrium.

The possibility exists that fluctuations in space and time may occur within phases which lead to transient nonequilibrium tensions at structural surfaces, as is also true of the thermodynamic variables (Section 4.17.3). Nevertheless, persistent equilibrium tensions cannot exist. True interfaces, which are the surfaces that separate coexisting phases, are the only surfaces at which equilibrium tensions exist.

References

1. Tamas, F. and Pal, I. (1970). *Phase Equilibria Spatial Diagrams*, Iliffe Books (Butterworth), London.
2. Laughlin, R. G., Munyon, R. L., Fu, Y.-C. and Fehl, A. J. (1990). *J. Phys. Chem.* **94**, 2546–2552.
3. Glasstone, S. (1946). *Textbook of Physical Chemistry*, 2nd ed., pp. 755–761, Van Nostrand, New York.
4. (1965). *Solubilities*, 4th ed., Vol. II, pp. 1121–1122, American Chemical Society, Washington, DC.

5. Rawlings, F. F. and Lingafelter, E. C. (1955). *J. Am. Chem. Sci.* **77**, 870–872.
6. Rock, P. A. (1969). *Chemical Thermodynamics*, MacMillan, London.
7. Defay, R., Prigogine, I., Bellemans, A. and Everett, D. H. (1966). *Surface Tension and Adsorption*, pp. 94–95, John Wiley, New York.
8. Tammann, G. (1924). *Lehrbuch der Heterogenen Gleichgewichte*, Vieweg, Braunschweig; Tammann, G. (1903). *Z. Anorg. Chem.* **37**, 303–313.
9. Kekicheff, P., Grabielle-Madelmont, C. and Ollivon, M. (1989). *J. Colloid Int. Sci.* **131**, 112–132.
10. Grabielle-Madelmont, C. and Perron, R. (1983). *J. Colloid Interface Sci.* **95**, 471–482.
11. Chernik, G. G. and Fillipov, V. K. (1991). *J. Colloid Interface Sci.* **141**, 415–424; Chernik, G. G. and Sokolova, E. P. (1991). *J. Colloid Interface Sci.* **141**, 409–414.
12. Masing, G. (1944). *Ternary Systems*, Translated by B. A. Rogers, Reinhold, New York.
13. Hall, D. G. and Pethica, B. A. (1966). *Nonionic Surfactants*, 1st ed., Vol. 2, pp. 516–557, Marcel Dekker, New York.
14. Hato, M. and Shinoda, K. (1973). *Bull. Chem. Soc. Japan* **46**, 3889–3890.
15. (1992). *Mixed Surfactant Systems*, ACS Symposium Series, Vol. 501 (P. M. Holland and D. N. Rubingh eds), American Chemical Society, Washington, DC.
16. Zhu, B. and Rosen, M. J. (1984). *J. Colloid Interface Sci.* **99**, 435–442.
17. Personal communication, O. T. Quimby.
18. (1976). *Handbook of Biochemistry and Molecular Biology*, 3rd ed., Vol. I, pp. 157–269, CRC Press, Cleveland, OH.
19. Ekwall, P. (1938). *Kolloid Z.* **84**, 284–291; Harva, O. and Ekwall, P. (1948). *Acta Chem. Scand.* **2**, 713–726.
20. Somasundaran, P. and Ananthapadmanabhan, K. P. (1979). *Solution Chemistry of Surfactants*, Vol. 2 (K. L. Mittal ed.), pp. 777–800, Plenum, New York.
21. Laughlin, R. G. and Munyon, R. L. (1987). *J. Phys. Chem.* **91**, 3299–3305.
22. Rusanov, A. I. (1978). *Phasengleichgewichte und Grenzflächenerscheinungen*, Akademie, Berlin.
23. Rusanov, A. I. and Fridrikhsberg, D. A. (1976). *Russian J. Phys. Chem.* **50**, 1809–1814.
24. Glasstone, S. (1946). *Textbook of Physical Chemistry*, 2nd ed., pp. 693–814, Van Nostrand, New York.
25. Ricci, J. E. (1951). *The Phase Rule and Heterogeneous Equilibrium*, Van Nostrand, New York.
26. Findlay, A. (and Campbell, A. N.) (1938). *The Phase Rule and its Applications*, 8th ed., pp. 1–14, Dover, New York.
27. Ricci, J. E. (1951). *The Phase Rule and Heterogeneous Equilibrium*, pp. 5–8, Van Nostrand, New York; Gibbs, J. W. (1928). *The Collected Works of J. Willard Gibbs*, Vol. I (W. R. Longley ed.), p. 96, Longmans, Green, & Co., New York.
28. Berne, B. J. and Pecora, R. (1976). *Dynamic Light Scattering*, pp. 257–261, John Wiley, New York.
29. Kuneida, H. and Friberg, S. (1981). *Bull. Chem. Soc. Jpn.* **54**, 1010–1014; van der Donck, J. C. J. and Stein, H. N. (1993). *Langmuir*, **9**, 2270–2275.
30. Laughlin, R. G. (1990). *Food Emulsions and Foams: Theory and Practice*, No. 277, Vol. 86 (P. J. Wan, J. L. Cavallo, F. Z. Saleeb, and M. J. McCarthy eds, E. L. Gaden, Series ed.), pp. 7–15, American Institute of Chemical Engineers, New York.
31. (1952). *Colloid Science, Irreversible Systems*, Vol. I (H. R. Kruyt ed.), p. 79, Elsevier, New York.
32. Leermakers, F. A. M. and Scheutjens, J. M. H. M. (1989). *J. Phys. Chem.* **93**, 7417–7426; Helfrich, W. (1973). *Z. Naturforsch.* **28**, 693–703.
33. Gibbs, J. W. (1928). *The Collected Works of J. Willard Gibbs*, Vol. I (W. R. Longley ed.), p. 62, Longmans, Green, & Co., New York.
34. Abragam, A. (1961). *The Principles of Nuclear Magnetism*, p. 64, Clarendon Press, Oxford.
35. Rossen, W. R., Davis, H. T. and Scriven, L. E. (1986). *J. Colloid Interface Sci.* **113**, 248–268.
36. Lang, J. C. and Morgan, R. D. (1980). *J. Chem. Phys.* **73**, 5849–5861.
37. Schiller, P. and Schiller, K. (1990). *Liq. Cryst.* **8**, 553–564.
38. Derome, A. E. (1987). *Modern NMR Techniques for Chemistry Research*, Vol. 6 (J. E. Baldwin ed.), p. 7, Pergamon Press, Oxford.

39. Rao, N. V. S. (1984). *Mol. Cryst. Liq. Cryst.* **108**, 231–243; Photinos, P., Melnik, G. and Saupe, A. (1986). *J. Chem. Phys.* **84**, 6928–6932.
40. Reizlein, K. and Hoffmann, H. (1984). *Prog. Colloid Poly. Sci.* **69**, 83–93.
41. Wilhelm, R. H., Conklin, L. H. and Sauer, T. C. (1941). *Ind. Eng. Chem.* **33**, 453–457.
42. Laughlin, R. G. (1992). *Adv. Colloid Interface Sci.* **41**, 57–79.
43. Ottewill, R. H., Sinagra, E., MacDonald, I. P., Marsh, J. F. and Heenan, R. K. (1992). *Colloid Polym. Sci.* **270**, 602–608.
44. Markovic, I., Ottewill, R. H., Cebula, D. J., Field, I. and Marsh, J. (1984). *Colloid Polym. Sci.* **262**, 648–656.
45. McBain, J. W., Vold, R. D. and Vold, M. J. (1938). *J. Am. Chem. Soc.* **60**, 1866–1869.
46. Hoffmann, H., Platz, G., Rehage, H., Schorr, W. and Ulbricht, W. (1981). *Ber. Bunsenges. Phys. Chem.* **85**, 255–266; Weers, J. G. and Scheuing, D. B. (1990). *ACS Symp. Series. Fourier Transform Infrared Spectrosc. Colloid Interface Sci.*, Vol. 447, American Chemical Society, Washington, DC.
47. Small, D. M. (1986). *The Physical Chemistry of Lipids. From Alkanes to Phospholipids*, Vol. 4, pp. 215–217, Plenum, New York.
48. (1984). *ACS Symposium Series, Vol. 263: Resonances in Electron-Molecule Scattering, van der Waals Complexes, and Reactive Chemical Dynamics*, (D. G. Truhlar ed.), American Chemical Society, Washington, DC.; (1990). *Dynamics of Polyatomic Van der Waals Complexes*, Vol. 227 (N. Halberstadt and K. C. Janda eds), Plenum, New York (NATO ASI Series. Series B: Physics); (1982) *van der Waals Molecules*, Faraday Discussion of the Chemical Society No. 73, Royal Society of Chemistry, London.
49. Cruickshank, D. W. J. (1956). *Acta Cryst.* **9**, 747–753.
50. Gibbs, J. W. (1928). *The Collected Works of J. Willard Gibbs*, Vol. I (W. R. Longley ed.), pp. 63, 96, Longmans, Green, & Co., New York.
51. Adamson, A. W. (1979). *A Textbook of Physical Chemistry*, 2nd ed., pp. 413–414, Academic Press, New York.
52. Widom, B. (1985). *Chem. Soc. Rev.* **14**, 121–140.
53. Ladbrooke, B. D. and Chapman, D. (1969). *Chem. Phys. Lipids* **3**, 304–356.
54. Mason, J. T., Huang, C. H. and Biltonen, R. L. (1983). *Biochemistry* **22**, 2013–2018.
55. Cornell, B. A., Fletcher, G. C., Middlehurst, J. and Separovic, F. (1982). *Biochim. Biophys. Acta* **690**, 15–19.
56. J. L. Burns, personal communication; Vinson, P. K., Talmon, Y. and Walter, A. (1989). *Biophys. J.* **56**, 669–681.

Chapter 5

The Characteristic Features of Surfactant Phase Behavior

5.1 Thermal energies, water free energies, and chemical potentials

The influence of temperature, pressure, and composition on surfactant phase behavior will now be treated. It will be recalled (Section 3.1) that both temperature and pressure are intimately related to kinetic energy. Absolute temperature provides a measure of the ambient level of kinetic energy, while pressure reflects the ambient density of kinetic energy (energy per unit volume). During processes performed on a closed system, holding temperature constant (while varying pressure and volume) amounts to holding the total amount of kinetic energy constant while varying its density. Holding pressure constant (while varying the temperature and volume) amounts to holding the density of kinetic energy constant while varying the total amount. During processes in which both temperature and pressure change, both the total amount and also the density of kinetic energy vary.

The partial molar quantities – including chemical potentials – describe the dependence of free energy on composition (Section 3.3). In aqueous mixtures the chemical potential of water is related to its thermodynamic activity by the equation

$$\mu_W = \mu_W^0 + RT \ln(a_W) \tag{5.1}$$

where the term μ_W^0 is the standard state chemical potential and a_W is the thermodynamic activity of water. During the analysis of phase thermodynamics it is usual to take the pure substance at a particular temperature and pressure as the standard state, in which case the activity of the pure substance is set equal to unity. For a volatile component such as water, the activity of water is reflected by its partial pressure, and is defined as p_W/p_W^0 where p_W^0 is the vapor pressure of pure water. Within any state in which water and a surfactant are miscible, the value of μ_W (at a given temperature) is invariably less than is the standard state value, so that the equilibrium partial pressure of water over the mixture is always less than that of pure water.

The total free energy of a water-containing surfactant phase is given by the expression

$$G = n_W\mu_W + n_S\mu_S \tag{5.2}$$

in which the term $n_W\mu_W$ represents that component of the total free energy of the phase which is dependent on water composition, and $n_S\mu_S$ represents that component which is dependent on surfactant composition. The term "$n_W \mu_W$" will be termed the "water free energy" (it is tempting to label this term the "hydration free energy" component, but the term "hydration" is used in so many other ways and carries so many implicit connotations that this label will not be used), and "$n_S \mu_S$" the "surfactant free energy". The relative contributions of these two components is related to the composition of the phase. The magnitude of the $n_W \mu_W$ term, and its dependence on both n_W and μ_W, is of special importance. It is worth noting that the chemical potential of water in a particular phase may be finite while at the same time the amount of water (n_W) may be zero (Section 4.8.6). When this is true it is obvious, from eq. 5.2, that water does not contribute to the free energy of that phase. The same situation may exist for the surfactant.

5.2   The thermal and chemical stability problem

Before proceeding, it must be recognized that a factor unrelated to the physical science of surfactants may exist that precludes the full analysis of phase behavior, or even the determination of phase diagrams. That is thermal or chemical instability of the molecule. If the molecular structure of a surfactant breaks down during a study, whether by the action of heat alone or by chemical reaction with water, definitive physical studies cannot be performed. This is a serious issue, as it is not at all unusual for crystal or liquid crystal states to extend to temperatures where surfactant molecules are unstable.

Most ionic surfactant salts do not melt reversibly as the dry crystal, so their melting points cannot be determined. The alkali metal soaps are exceptions to this general rule, in that they appear to be stable (for short times, at least) at temperatures beyond 300°C. Sodium alkyl sulfates, sulfonates and quaternary ammonium surfactants are ordinarily not stable at their melting points. For example, the thermal stability limit of dioctadecyldimethylammonium chloride (DODMAC) is about 135°C so that physical data recorded above this temperature are meaningless [1]. Cationic surfactant salts formed by neutralizing primary, secondary, or tertiary amines with mineral acids are considerably more stable than are quaternary ammonium salts; for example, dioctadecylammonium chloride and dioctadecylmethylammonium chloride both melt reversibly at very high temperatures (Appendix 4.3). The thermal stability depends on the nucleophilicity of the anion, so that bromides are less stable thermally than are chlorides.

Most zwitterionic compounds are also unstable at the melting point. These compounds (and also amine oxides) are stabilized thermally to a significant degree simply as a result of being in a crystal phase, so that melting and decomposition occur in concert. In preparing such materials, synthetic chemists can often obtain a reproducible melting/decomposition temperature during the first determination on a sample, but repeated determinations on the same sample give progressively lower melting temperatures. The decrease can be attributed to chemical instability and conventional freezing point lowering due to creation of the impurity molecules. It has been shown that nucleophilic displacement reactions of the negative group on the carbon atoms attached to the positive group occurs in ammonio carboxylates, which lead to aminoester decomposition products [2]. The reaction occurs both intra- and intermolecularly [3]. Aminoesters differ dramatically in their phase behavior from that of zwitterionic compounds, and constitute additional components that severely modify physical behavior. Elimination reactions to form olefins are also possible.

In the above-mentioned cases, mixtures of surfactants with water are often significantly more stable than the dry compound [1,4]. Sometimes this stabilization is dramatic, as in the case of phosphonium hydroxides. These compounds are stable in aqueous solution at the boiling point, but decompose at room temperature in the dry state [5]. Moreover, since water often dramatically reduces the melting point of the compound, fluid phases may be accessible in concentrated water-containing mixtures that do not exist with the dry compound. This melting-point-reducing phenomenon (Fig. 3.2) was critically important to the successful execution of the DIT phase study of the DODMAC–water system [1].

The polyoxyethylene oxide monoether and phosphine oxide surfactants are highly stable compounds thermally, as indicated by the fact that many of them can be analyzed using gas chromatography during experiments in which the temperature may be programmed to $> 300°C$. The interior of a gas chromatography (gc) column is one of the cleanest environments that exists, however, so that the apparent thermal stability indicated by gc data may not reflect their stability under other circumstances. The presence of dissolved air, or of acidic or basic surfaces on the wall of the container, may lead to instability at lower temperatures in ordinary samples than the gc data would suggest.

Reactive functional groups. The stabilizing effect of water does not extend to molecules that possess a functional group that is capable of reacting with water by hydrolytic cleavage. (Because both are important, the term "hydrolysis" is used to describe proton transfer reactions to and from water, while the term "hydrolytic cleavage" is used for the breaking of a chemical bond by the action of water.) In these cases instability results from chemical reactivity, even though the compound may be thermally stable. The instability of salts of alkyl sulfate esters towards hydrolysis, and the profound effect of the alcohol hydrolysis product on their physical behavior, is well documented [6]. These salts are relatively stable in neutral and basic solution, but they undergo hydrolysis in acidic solutions or in neutral mixtures at elevated temperatures. Hydrolysis rates depend strongly on the temperature, however, so that studies in acidic solution (or even of the alkylsulfuric acid itself [7]) are possible with care. During recent phase studies of sodium dodecyl sulfate (SDS), precautions were taken to avoid this hydrolytic cleavage reaction, and to verify the expected stability after each series of experiments [8].

Carboxylate ester functional groups are liable to hydrolytic cleavage reactions, and uncertainty exists as to the stability of ester groups with the polar lipids. Ether analogs have attracted considerable attention for this reason [9]. Amide groups are considerably less reactive than are esters. As with the alkyl sulfates, hydrolysis does not inevitably occur with ester or amide groups but the possibility must be recognized and the stability constantly checked.

The kinetics of hydrolysis of reactive groups within phases that coexist with liquid water solutions can be dramatically different from the kinetics of hydrolysis within the aqueous liquid phase itself (Section 11.5.2). It is possible that the differences in the free energies of coexisting phases (Section 3.15) may be responsible for these differences in chemical stability.

5.3 Solubility boundaries

A "solubility boundary" is the dilute boundary of the first gap in miscibility that is encountered as one component of a system is added to the other (Section 1.1). Those solubility boundaries that are encountered as surfactants are added to liquid water are particularly important. Solubility boundaries are named after the phase structure of the coexisting phase that exists once the solubility is exceeded. A "crystal solubility boundary" is one at which a crystal separates, a "liquid solubility boundary" is one at which a liquid separates, and so on. The solubility boundary is that of the solvent phase (not the coexisting phase), however. Most systems have several qualitatively different solubility boundaries at different temperatures.

One tends to think of solubility only with respect to the liquid phase, but it is useful not to impose this arbitrary constraint on use of the term. For example, inspection of the salt–water diagram (Fig. 1.1) reveals that while the solubility of salt in liquid water is substantial (*c.* 25%), the solubility of salt in ice crystals is zero. The solubility of water in salt crystals is also zero above 0.1°C, but below 0.1°C water reacts with dry salt crystals to form the stoichiometric crystal hydrate $X \cdot W_2$. Such phases, which incorporate both components within a phase whose composition is rigidly defined stoichiometrically, are best regarded as "phase compounds"; this distinctive form of "miscibility" is not usually described as solubility.

The qualitative aspect of solubility. It is extremely important to recognize that solubility has both a qualitative and a quantitative aspect. The reason for this is clear from the perspective provided by the thermodynamic analysis of partial miscibility (Chapter 3). The composition of the saturated solution is defined by the circumstances under which the condition of equilibrium is met, which depends on the thermodynamic properties of both phases. This being the case, it is evident that the thermodynamic properties of the coexisting phase are just as important in determining solubility as are those of the solution phase.

In treating solubility, both the numerical value of the solubility and also the composition and structure of the coexisting phase should always be stated. The importance of doing so is recognized implicitly in the reference books to solubilities [10]. Invariably, the formulae of coexisting crystal phases (and by implication, their structures) are stated along with numerical solubility/temperature data. It is comparatively straightforward to translate these tabular data into a graphical phase diagram, since both kinds of information are provided. The salt–water phase diagram utilized throughout this book was derived from this source.

As a result of the uniquely large variety of phase states that surfactants form, a comparably wide variety of qualitatively different solubility boundaries exists. The coexisting concentrated phase, past the limit of solubility, is often termed the "saturating phase", and may be any of these states. Which state is found depends mainly on the temperature. In the familiar short chain anionic and nonionic surfactants either a crystal, a cubic liquid crystal, or a hexagonal liquid crystal phase is frequently the saturating phase, but the lamellar phase is also found in some systems. In nonionic surfactants the saturating phase may be any of these, or another liquid phase. In zwitterionic and some cationic surfactants a crystal, one of the cubic phases, and occasionally a liquid phase are the usual saturating phases.

Generally, enthalpic (ΔH) contributions to the free energy are relatively more important with respect to which states exist at low temperatures, while entropic

$(-T\Delta S)$ contributions are dominant at higher temperatures. If the standard state of water is taken to be ice at 0°C (Fig. 4.1), the free energy (and chemical potential) of water (at one atmosphere) *decreases* with increasing temperature in all its phase states, because the change in the entropic contribution with temperature is always numerically larger than is the change in the enthalpic contribution.

Since the crystal phase is always the surfactant phase of lowest enthalpy and entropy, this phase is invariably the coexisting phase at the surfactant solubility boundary at low temperatures – unless the freezing of water intervenes. Intervention of the separation of ice (rather than the surfactant crystal) on cooling dilute liquid mixtures is more likely if the surfactant is weakly crystalline (has a relatively high enthalpy). Surfactant molecules that have an awkward shape (conformational structure), for example due to substituents in the chain, cannot pack densely within a crystal and do not form energetically stable (low enthalpy) crystals (Section 11.3.2). In these cases the crystal solubility boundary may simply not exist (Section 5.4.4).

Because surfactant crystals are low-melting, for example in comparison to most inorganic crystals, a temperature is almost always reached in aqueous surfactant systems at which the saturating phase changes from a crystal to a fluid phase. The form of the crystal solubility boundary just below this transition is often distinctive. The crystal solubility boundary, and the phase discontinuity at its upper temperature limit, will now be treated.

5.4 Crystal solubility: the Krafft boundary and the Krafft eutectic

Many surfactants are highly soluble in liquid water at high temperatures, but at lower temperatures separate from solution as a crystal phase. The crystal solubility boundary of such surfactants is called the "Krafft boundary", in honor of its discoverer (Section 14.2). This boundary has a distinctive shape, and terminates at its upper temperature limit at a eutectic discontinuity; this eutectic is called the "Krafft discontinuity" or "Krafft eutectic". The saturating phase at the solubility boundary at temperatures above the Krafft eutectic is usually (but not always) a liquid crystal phase.

Far below the temperature of the Krafft eutectic the solubility of surfactants is small and the crystal solubility boundary is steep, which signifies that a low temperature coefficient of solubility exists. As the temperature is increased a turnover or "knee" develops, however, and just above this knee a "plateau" exists (Fig. 5.1). Along the plateau the slope is small and the temperature coefficient of solubility is very large; it may exceed 10%/°C. The plateau terminates at the Krafft eutectic, at which the crystal solubility boundary intersects that solubility boundary which exists at higher temperatures. The liquid phase at the intersection of these two solubility boundaries is also the most dilute of the three phases that may coexist at the Krafft eutectic.

A cusp exists in the boundary of the liquid region at the intersection of these two solubility boundaries. Since thermodynamic variables typically change in a smooth manner with temperature within liquid phases, *cusps in a liquid boundary invariably signify a discontinuity of state within the coexisting phase.* During the determination of phase diagrams, finding such cusps alerts the investigator to examine closely the coexisting phase at the temperature of the cusp. Very often a peritectic decomposition of a crystal hydrate, or a polymorphic phase transition in the crystal, is found at this

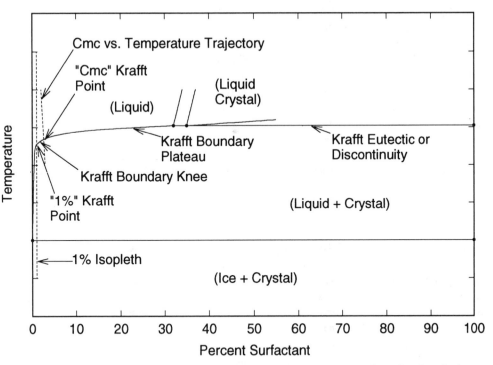

Fig. 5.1.   The general form of the Krafft (crystal solubility) phase boundary showing the lower nearly vertical part, the knee, and the plateau region. The Krafft boundary terminates at its upper limit at the Krafft eutectic. The "cmc Krafft point"is the point of intersection of the locus of critical micelle concentration (cmc) vs. temperature, and the "1% Krafft point" is the temperature of the boundary at a specified composition (1%).

temperature. Crystal hydrates were discovered in the dodecylammonium chloride–water [11] and dimethyl sulfone diimine–water systems [12] in just this manner.

5.4.1   The Krafft boundary, Krafft points, and micellization

Not surprisingly, the existence of the knee and plateau in the Krafft boundary has attracted considerable attention. Krafft himself associated the temperature of the knee in soap–water systems with the melting point of the corresponding fatty acid [13] – a correlation which was soon recognized to be coincidental [14].

More recently, it has been noted that the composition of the boundary at the knee is similar to the critical micelle concentration (cmc) of surfactants. (It has even been suggested that the situation is akin to a "triple point" in that three "phases" coexist [15].) The dependence of the cmc on temperature, at temperatures above the Krafft boundary, has been determined for a number of soluble alkyl sulfate salt surfactants. The cmc vs. temperature function has been extrapolated to its intersection with the independently determined Krafft boundary, and the point of intersection (which clearly falls within the knee) has been termed the "Krafft point" (Fig. 5.1) [15].

An interesting consequence of these data is that micellar solutions of these compounds do not exist (as equilibrium states) below the temperature of the "cmc Krafft point"; the liquid phase below this temperature is a molecular solution. In other cases (e.g. dodecyl methyl sulfoxide) evidence has been obtained that the cmc lies at compositions below that of the vertical arm of the Krafft boundary [16], but these data are not so well documented as in the case of the alkyl sulfates (above).

It is worth noting that experimental determinations of cmcs in the vicinity of solubility boundaries are treacherous. Phase separation necessarily introduces a plateau in the surface tension vs. ln[C] function, which closely resembles that which occurs for micellization in solution [17]. Care must therefore be used not to misinterpret a discontinuity in such data as a cmc, when it in fact corresponds to a solubility boundary.

In other work the Krafft boundary has been characterized by determining its temperature at a single arbitrarily selected composition (often 1%, sometimes 15%), and this too has been termed the "Krafft point" (Fig. 5.1).

Several features of the Krafft boundary are worth noting with respect to its relation to micellization in the solution phase.

Occurrence of the knee and plateau in other systems. First, it is important to recognize that the presence of a knee and a plateau in a crystal solubility boundary is *not* unique to aqueous surfactant systems, and that the existence of these features does *not* require that micellar phenomena exist within the liquid phase. They are found in the crystal solubility boundary in the phase diagrams of all solute–solvent systems that are useful for purification by recrystallization, and micellization within the liquid phase does not exist in the vast majority of these systems.

Three distinct regions may be recognized to exist in the generalized crystal solubility boundary that is invariably encountered during recrystallization. The presence of small quantities of a good solvent typically depresses the "melting point" of any crystalline compound; more correctly, the boundary temperature of the liquid region (the liquidus) is lowered by the presence of solvent. At the melting point no solvation is required to form the liquid from the crystal; this occurs solely by the (isothermal) addition of heat. At lower temperatures, solvation is required in order to break down the crystal and form a liquid phase. The amount of solvent required increases as the temperature is lowered and, as a result, the solubility decreases. This region of the crystal solubility boundary may be described as the "high-temperature arm".

The melting-point lowering associated with the "high-temperature arm" occurs inevitably, if miscibility exists in the liquid phase but does not exist in the crystal state [18]. However, it does not occur universally (Fig. 10.3), and the magnitude of the lowering is reduced if the components are miscible within *both* the liquid and the crystal phase [19]. Also, if the dry crystal does not melt in a simple fashion to form a liquid at its upper limit of stability (Section 5.5.3), the phase behavior may be different.

A second region of the crystal solubility boundary is found at temperatures far below the melting point. At such temperatures the high stability (low enthalpy) of the crystal phase dominates the thermodynamics of mixing, and within this temperature regime the level of solvation energy (and thus quantity of solvent) required within the liquid phase to overcome the considerable stability of the crystal phase is extremely high. When this is true the coexisting liquid phase is very rich in solvent and the solubility is very small.

At its lower temperature limit, this low-temperature arm may be expected to terminate at a eutectic which will lie just below the melting point of the solvent. At this eutectic the crystal phase of the solvent, the eutectic liquid, and the crystal phase of the solute will likely coexist.

The third region of the crystal solubility boundary is found at temperatures between the high-temperature and the low-temperature arms of the solubility boundary. As the temperature is lowered (starting from the melting point of the pure solute) the high-temperature high-solubility arm and the low-temperature low-solubility arm must somehow be joined in a continuous manner since they are part of the same boundary. Numerous experimental studies have shown that the joining of these two arms typically occurs within a narrow range of temperatures – thus producing a more-or-less well-defined plateau in the crystal solubility boundary. The slope of this plateau may vary, but it necessarily lies only a short distance in temperature below the melting point of the solute. A knee inevitably results where the plateau merges with the low temperature (nearly vertical) arm of this same boundary. Representative examples of such phase behavior are seen in diagrams of $C_{12}PO$ in selected organic solvents (Fig. 5.2).

Related, but somewhat different, behavior is observed in alkanol solvents. In these solvents (where specific solvation via hydrogen bonding exists) the melting point depression region (high-temperature arm) is extended over a large composition span, and the plateau is pushed to lower temperatures and a narrower composition range than was observed in the other systems. Often the plateau is not visible, as shown in Fig. 5.3 and also in the aniline diagram in Fig. 5.2.

The phase boundaries of the alkanol systems fall in a regular pattern in that the rate of lowering of the boundary temperature diminishes as the molecular weight of the alcohol increases. The diagrams are virtually identical (in concentrated mixtures) if plotted on a mole fraction basis, and water falls nicely in line (as befits its status as the parent molecule from which the other alkanols are derived) until sufficiently water is present to form liquid crystal phases.

Continuity of structure within the coexisting crystal phase at the knee. A second consideration with respect to the form of the Krafft boundary and its relationship to

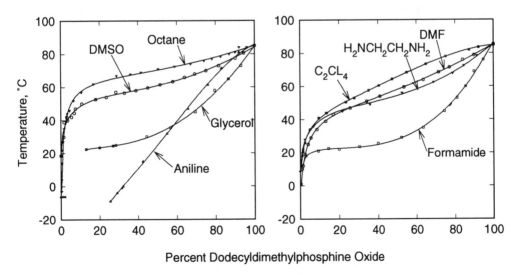

Fig. 5.2. The phase diagrams of dodecyldimethylphosphine oxide in various organic solvents. In all cases the area below each boundary is a biphasic liquid + crystal region.

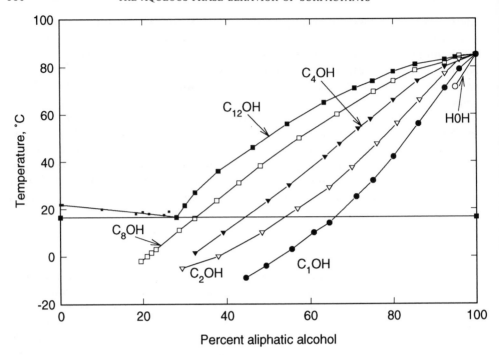

Fig. 5.3. The phase diagrams of dodecyldimethylphosphine oxide in a homologous series of aliphatic alcohol solvents.

micellization is the fact that, within the temperature range of the knee and plateau of this boundary, the phase structure of the coexisting crystal phase typically does *not* change discontinuously at the temperature of the knee. Although the Krafft boundary is strongly curved at the knee, it must (and does) vary smoothly within this temperature region. This result is hardly surprising, in as much as curvature also exists in other bulk and surface properties of the liquid phase in the vicinity of the cmc (Section 8.5) [20].

   Influence of crystal and liquid crystal thermodynamics on the temperature of the Krafft boundary. It has been suggested on the basis of the cmc Krafft point data that the form of the Krafft boundary in the vicinity of the knee and the plateau region is related to the changes that occur in the solution chemistry that are associated with micellization [15]. An alternative view is that the temperature range at which the Krafft boundary is found is defined by the temperature of the Krafft eutectic, and that its form is defined by the same factors that influence the general form of crystal solubility boundaries (above).

   The temperature of the Krafft eutectic (and therefore of the Krafft boundary itself) is defined by the thermodynamic properties of the coexisting liquid, liquid crystal, and crystal phase at this eutectic (Section 5.6.2). The structural and thermodynamic changes that occur within the liquid phase at the cmc, which lies at concentrations far below that of the liquid phase at the Krafft eutectic (for soluble surfactants), would appear to have no direct bearing on the temperature of the Krafft eutectic.

   It seems likely that both the knee and the plateau features of the Krafft boundary result simply from the necessity to smoothly connect the low-temperature arm of the surfactant crystal solubility boundary with the coexisting liquid phase at the Krafft eutectic – just as occurs in the case of recrystallization diagrams (above). The difference lies in the fact that in recrystallization diagrams the composition of this liquid is the pure solute, while in soluble surfactant diagrams the composition of this liquid is that of the aqueous micellar liquid phase at the Krafft eutectic. Focusing attention on the very dilute solution chemistry (at the cmc) does not factor into the analysis consideration of those factors which actually determine the temperature of the Krafft eutectic.

Finally, it may be noted that the both the free energies and the enthalpies of micellization [21] are extremely small, in comparison with the substantial amounts of heat required to melt crystals along isopleths that span the Krafft boundary plateau. Regardless of whether formation of the surfactant liquid phase occurs solely by the action of thermal energy (as at the melting point), or by the combined action of thermal energy and solvation energy, the amount of heat typically required is far larger than is that which is associated with micellization in dilute solution.

5.4.2 The slope of the Krafft plateau

The slope of the Krafft boundary is much smaller in the plateau region than at lower temperatures, but at no point may its slope be precisely zero. It must be possible, along the entire length of this boundary, to draw tie-lines that connect the boundary with the coexisting crystal phase. Further, these tie-lines may *not* pass through another one-phase region (Sections 4.15, 10.7.2.1). The slope within the plateau must always be positive, but its magnitude may vary considerably.

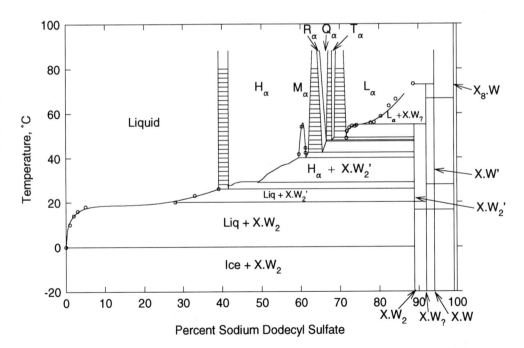

Fig. 5.4. An edited version of the published sodium dodecyl sulfate–water diagram, illustrating the relatively large difference between the temperature of the knee and the Krafft eutectic in an anionic surfactant–water phase diagram [8]. The complex array of liquid crystal phases that exists in this system is also evident.

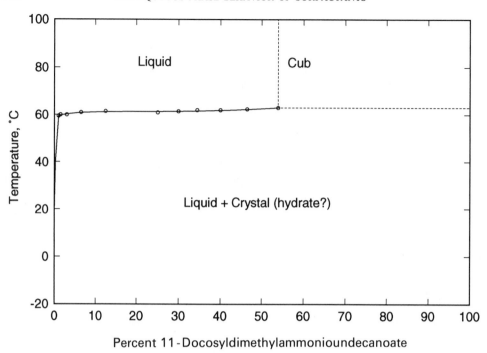

Fig. 5.5. Partial diagram of the docosyldimethylammonioundecanoate (C$_{22}$AU)–water system, illustrating the small slope of the Krafft boundary plateau in zwitterionic surfactant–water phase diagrams.

Within the low-temperature arm of the Krafft boundary, the slope is under no such constraint. It may (insofar as the Phase Rule is concerned) be either positive or negative, but such data as are available suggest that it is usually positive.

The Krafft plateaus of ionic surfactants characteristically display a relatively steep slope, and the knee is not sharply defined. In soap–water (Fig. 3.2) and sodium alkyl sulfate–water systems (Fig. 5.4), the knee displays substantial curvature and may lie as much as 15–20°C below the Krafft eutectic.

The slope of the Krafft plateau varies widely among nonionic surfactant–water diagrams. This boundary is rarely found in polyoxyethylene oxide monoether surfactants (C$_x$E$_y$), because the dry compounds are typically low-melting and the boundary is metastable (Section 10.5.5.2). A substantial slope exists in the *N*-dodecanoyl-*N*-methylglucamine–water diagram (Fig. 10.21), which resembles that often seen in the diagrams of ionic surfactants. The smallest slope encountered, by far, is found in the diagrams of zwitterionic surfactants. The slope of the Krafft boundary in these systems may be so small as to be almost imperceptible (Fig. 5.5). The slope of this plateau greatly influences the significance of single-point (e.g. "1%") Krafft boundary determinations (Section 5.4.1). Whereas a determination at a single composition defines the position of the Krafft boundary of a zwitterionic surfactant such as C$_{22}$AU with considerable accuracy, the same data on most surfactants leaves the temperature of both the Krafft boundary and the Krafft eutectic uncertain by as much as 15–20°C. Clearly determination of the full boundary is to be preferred, especially since (in the case of the Krafft boundary) this determination is not very difficult (Appendix 4.4.3).

5.4.3 The solubility at the Krafft eutectic

The composition of the liquid phase at the intersection of the Krafft boundary with the Krafft eutectic varies over an extremely wide range (from 0 to >60%). In short-chain surfactants that display an equilibrium Krafft boundary (Section 5.4.4), this liquid is very concentrated (Section 11.3.1). The diagram of sodium decane sulfonate (Fig. 5.6) provides an example in which the liquid composition is about 40%.

If the solubility of the surfactant in the liquid phase is decreased (as for example by increasing the lipophilic group chain length), the composition of the liquid phase at the Krafft eutectic also necessarily decreases (Section 11.3.1). The concentration of this liquid for C_{12} or C_{14} ionic surfactants is typically 25–40%. A similar value would likely exist for $C_{12}AO$, except that its Krafft boundary is metastable. However, the concentration at the Krafft eutectic in $C_{16}AO$ is only 1.5% (Fig. 5.7). The phase behavior of this compound is extremely unusual, in this regard.

Lengthening the chain from C_{12} to C_{16} shrinks the composition span of the Krafft plateau, and if this trend were to continue the plateau feature should completely disappear. That is exactly what is found in the diagrams of surfactants that are poorly soluble in liquid water. Only the vertical low solubility arm of the Krafft boundary exists, but it still terminates (as it must) at a Krafft eutectic.

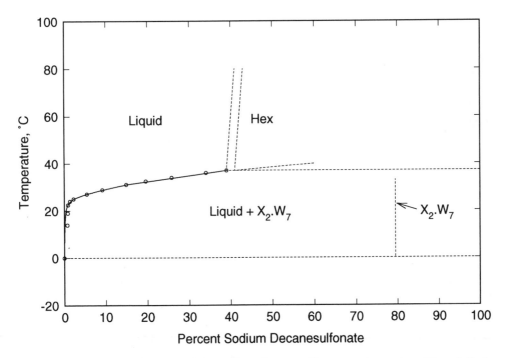

Fig. 5.6. Partial phase diagram of the sodium decane sulfonate–water system. The relatively steep slope of the Krafft plateau is also evident in this diagram.

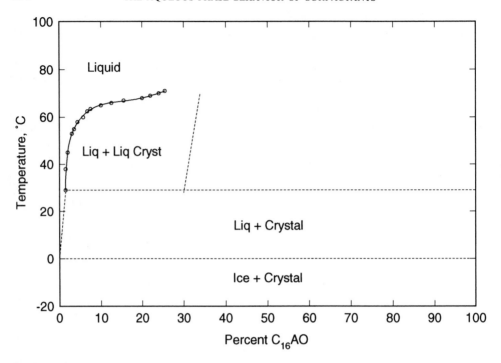

Fig. 5.7. Partial phase diagram of the hexadecyldimethylamine oxide–water system. Finding the liquid crystal solubility boundary at dilute but easily measurable concentrations, as seen in this diagram between 30 and 60°C, is extremely unusual.

Well-documented examples of this behavior are found in the monoglyceride–water (Fig. 5.15) and DODMAC–water (Fig. 2.2) phase diagrams. Although our knowledge of the phase behavior of polar lipid surfactants is meager (Section 11.3.2.7), qualitatively similar behavior almost certainly exists in the case of these compounds.

It is important to note that a small solubility in water is absolutely *not* synonymous with immiscibility with water. DODMAC, for example, would conventionally be regarded as being poorly soluble in water because the composition of the saturated liquid phase is extremely small. However, DODMAC and water are miscible over approximately two-thirds of the composition span within the lamellar liquid crystal phase region. Viewing "miscibility" from the perspective of phase behavior provides a broader, more general, and sometimes more appropriate perspective than does the more narrowly defined concept of "solubility". Solubility in liquid water represents but one dimension of miscibility, and this dimension is very important. There are many other dimensions of miscibility, however, and these may be equally important for same purposes.

5.4.4   Metastable Krafft boundaries and Krafft eutectics

Changes in molecular structure typically influence the temperature of the Krafft eutectic. The restraints on how far the discontinuity can be raised have not been well defined, but it has been found to lie above 100°C for a number of surfactants (including C_{12}dimethylammonioethanesulfonate [22], C_{22}dimethylammonioacetate, and C_{42}dimethylammoniohexanoate).

On the other hand, the Krafft eutectic cannot be lowered indefinitely without encountering interference from the freezing of water as dilute solutions are cooled. Shortening the length of the lipophilic group typically reduces the temperature of the Krafft eutectic (Section 11.3.1), and eventually its temperature will be projected to fall below the freezing point of water. When this happens, neither the Krafft boundary nor the Krafft eutectic exist as equilibrium phenomena. If they are not too low, the Krafft boundary may still be observable as a metastable boundary. The freezing of water to form ice requires nucleation but is usually extremely fast. (Water freezes via a nucleation and growth path, and freezing may be arrested if nucleation sites are absent. It has been the author's experience that it is very difficult, during a surfactant phase study, to prevent the nucleation process from occurring. Once nucleation occurs, the growth of ice crystals is extremely fast.) Because of this, the formation of ice may easily override other phase phenomena below 0°C, and dictates the kinetically favored state of the system as well as the equilibrium state.

This is one of the important ways in which the unary phase behavior of water (the fact that it freezes at 0°C) imposes rigid limitations upon the phase behavior of all aqueous surfactant systems. If the freezing point of water were either −100°C or +100°C, for example, our perspective as to surfactant phase behavior and its dependence on structure would be entirely different than it is.

When the Krafft eutectic is metastable, a different kind of eutectic is observed on cooling dilute liquid mixtures. If an equilibrium Krafft boundary exists, the phases at the discontinuity are typically $L-LX-X$. If the Krafft boundary is metastable, the eutectic first encountered on cooling dilute liquid mixtures is often an Ice$-L-LX$ eutectic. Such behavior is seen along the isotherm at about −2°C in the phase diagram of the $C_{12}E_6$–water system (Fig. 5.8). The liquid phase at this eutectic, which would correspond to the most dilute phase at the usual $L-LX-X$ Krafft eutectic, is instead the intermediate eutectic liquid phase at this Ice$-L-LX$ eutectic.

At still lower temperatures, another eutectic is found at the lower limit of the hexagonal phase (Fig. 5.8). It appears that the eutectic liquid crystal composition at this Ice$-LX-X$ eutectic is centered within the liquid crystal region, rather than being at its dilute limit as is normally found (Section 5.6.3). The data on such features are sparse, however.

Metastable Krafft discontinuities are found in the case of both soluble surfactants (Fig. 5.8) and insoluble surfactants. In the DODMAC–water system (Fig. 3.2) the Krafft eutectic lies well above the freezing point of ice and is thus an equilibrium phenomenon. In shorter chain-length dilong chain dimethyl quaternary ammonium chlorides (such as didodecyldimethylammonium chloride), in unsaturated long chain polar lipids (such as dioleyoyllecithin), and in many similar compounds the liquid phase (which is nearly pure water) may be expected to freeze to form ice at a temperature that is very close to 0°C, while the coexisting surfactant phase remains a liquid crystal. It has been suggested (but not proven) that in such cases a polytectic exists at 0°C rather than a eutectic at lower temperatures [1]. If true, the coexisting liquid phase at the discontinuity is, in fact, pure water. (This possibility was also suggested (but also without proof) in the case of the 0°C discontinuity found in the DODMAC–water system.)

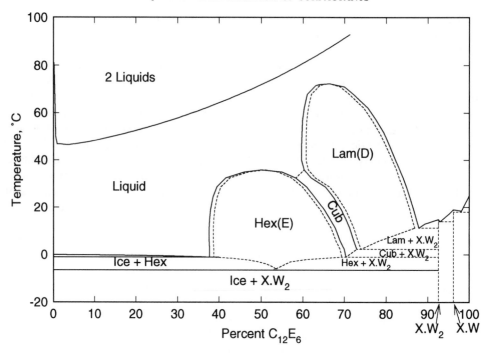

Fig. 5.8. The $C_{12}E_6$–water system, which has a metastable Krafft boundary [23]. The crystal hydrates in this system are indicated to melt congruently.

5.4.5   A definition of the Krafft boundary and Krafft eutectic

The Krafft boundary may be defined as the crystal solubility boundary of surfactants, and the Krafft eutectic as the eutectic that exists at the upper temperature limit of the Krafft boundary. In spite of the fact that it was the sudden increase in crystal solubility within a narrow temperature range that first attracted Krafft's attention to this boundary [24], it was seen above (Section 5.4.1) that the existence of the knee and the plateau in this boundary is not unique to aqueous surfactant systems. The near vertical lower arm and the eutectic must exist regardless of the solubility of the surfactant, but the plateau region is found only in the diagrams of soluble surfactants whose Krafft eutectic lies above 0°C (Section 5.4.4).

The coexisting (saturating) phase at the Krafft boundary is *always* a crystal, but it may be either the dry crystal or a crystal hydrate. In all too many cases one cannot say with certainty which is the case, simply for lack of information.

The Krafft discontinuity is – without exception – a eutectic phase transition (Section 4.14). The dilute phase is always a liquid phase and the concentrated phase is always a crystal phase, but the structure of the intermediate phase at this transition varies.

5.4.6   The significance of the Krafft eutectic

The Krafft eutectic is one of the most important phase phenomena encountered in aqueous surfactant systems. It was first recognized by Vold [25] that this eutectic phase discontinuity (unlike the knee in the Krafft boundary) is a true thermodynamic discontinuity. The reason for its importance resides in the fact that it divides surfactant–water diagrams into distinctive upper and lower temperature regions. Within the upper temperature region, liquid crystal phases may exist as equilibrium phases within specified ranges of temperature and composition. Within the lower temperature region, liquid crystal phases do not exist as equilibrium phases at any composition. However they *may*, and often do, exist as metastable phases within the lower temperature region (Section 5.4.4).

The surfactant crystal phase persists as an equilibrium phase to temperatures well above that of the Krafft eutectic (in concentrated mixtures), but the area within which it is found as an equilibrium state becomes progressively smaller as the temperature is increased. Above the melting temperature of the dry crystal, it of course vanishes. This is a limitation on surfactant phase behavior that is imposed by the unary surfactant diagram.

5.5   The phase behavior of surfactant crystal phases

5.5.1   Dry crystals

Many surfactants in the dry state exist as well-formed crystals which usually possess a bilayer structure (Section 8.3.1). Surfactant crystals undergo polymorphic and melting phase transitions in much the same manner as do most other crystals [26], except that sometimes they do not melt directly to a liquid phase but pass through a dry (thermotropic) liquid crystal phase before reaching the isotropic liquid phase (Fig. 10.21 [27]).

During a phase reaction among polymorphic phases, the crystal that exists at low temperatures (X_β) may be transformed on heating into a different structure (X_α) [1]. Such phase reactions are isothermal and (in principle) reversible. The high-temperature phase necessarily has the higher enthalpy and entropy of the two (Sections 4.10, 3.21). When the kinetics of the transformations of one polymorphic surfactant crystal to another are sufficiently fast, a calorimetric discontinuity is observed [1]. Polymorphism may therefore be recognized either by programmed temperature X-ray data (structural information), or by calorimetric studies (heat capacity data).

Direct observation (using either bright field or polarized light microscopy) is *not* a reliable means of recognizing polymorphism in surfactant crystals because a disconti-nuity in optical texture may (or may not) be observed at such transitions. In the DODMAC system, the phase transition of X_β to X_α (at 52°C) is visible using both calorimetrical and X-ray methods, but is invisible by direct observation [1].

5.5.2   The dry soap phases

The most thorough investigations of dry surfactant crystals have been conducted on the soaps – sodium palmitate in particular. As the low temperature Ω phase of soap is heated, it undergoes a progressive sequence of changes to other phase states before finally melting to form the isotropic liquid phase at 288°C (Fig. 3.2) [28]. Each of these states displays a distinctive and well-defined X-ray pattern (Fig. 5.9). Water is miscible to a finite degree with each state, so it is not unreasonable to describe them as highly structured fluids. Whether or not they should be regarded as liquid crystal phases

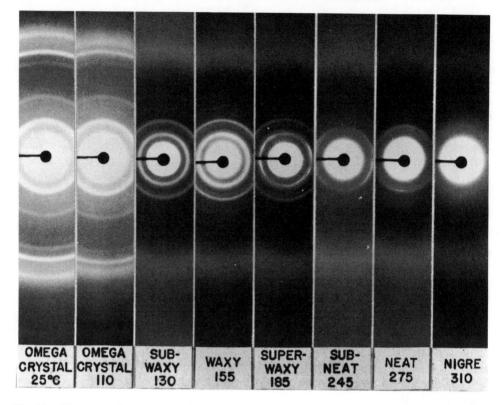

| OMEGA CRYSTAL 25°C | OMEGA CRYSTAL 110 | SUB-WAXY 130 | WAXY 155 | SUPER-WAXY 185 | SUB-NEAT 245 | NEAT 275 | NIGRE 310 |

Fig. 5.9. X-ray powder patterns of the dry (or nearly dry) sodium palmitate phases at various temperatures, documenting the existence of a variety of phase structures (see Fig. 3.2).

hinges entirely on how one defines "liquid crystals". It is possible that they are better regarded as "plastic crystal" phases, which are disordered crystal-like states that display a more random disorder than do liquid crystals [29]. Regardless of their classification, the existence of these states (if not their structures) is firmly established.

5.5.3 Dry (thermotropic) liquid crystal phases

As noted in Section 5.5.1, the dry surfactant crystal is on occasion transformed into a liquid crystal phase at a temperature below that at which the isotropic liquid exists. This was suggested to be the case for polyglycerol ester mixtures [30], and has been observed in a number of dilong chain cationic surfactant salts [31] and in various polyol surfactants [27,32,33]. The impact of this phenomenon on the phase behavior of binary systems will be considered in connection with the phase behavior of liquid crystal states (Section 5.6.5).

Dry surfactant liquid crystal phases are "thermotropic", in the strict sense of the word, since they result solely from the action of thermal energy on the crystal. Their existence does not hinge on the combined action of thermal and solvation energy, as with "lyotropic" liquid crystals. While these phases differ in their miscibility with water from the familiar thermotropic liquid crystal states (e.g. the cholesteric phases formed by elongated dipolar semirigid molecules (Section 5.6.5) [34]), they may be similar structurally.

A polytectic discontinuity may exist in the binary diagram with water at the temperature of any phase discontinuity within the dry surfactant. Polytectic phase state discontinuities have been described elsewhere as regards their thermodynamic analysis (Section 3.21) and their general properties (Section 4.8.3).

5.5.4 Polymorphic states: equilibrium and metastable

Polymorphic crystals (or polymorphs) may be defined as *crystals that are formed from the same molecules and have the same composition, but differ in crystal structure.* Two different kinds of polymorphs exist – equilibrium and metastable. If the crystal X_β is transformed reversibly at a particular temperature to another crystal having a different phase structure X_α (Section 5.5.4)

$$X_\beta + q \rightleftarrows X_\alpha \qquad (5.3)$$

and both phases may (under different conditions) exist as equilibrium states, then X_β and X_α may be regarded as "equilibrium polymorphs". Equilibrium polymorphs may coexist at particular values of the system variables, but (except at these discontinuities) they do not otherwise coexist. The transformation of one to the other is a thermodynamically reversible isothermal phase transition. Equilibrium polymorphs may exist in stoichiometric phase compounds (such as crystal hydrates) [35], as well as in crystals of a pure component [1].

Nonequilibrium polymorphs. A fundamentally different kind of polymorphism exists if, for example, the freezing of a liquid phase results in the formation of a non-equilibrium or metastable crystal phase, X_m, whose structure differs from that of the equilibrium crystal.

$$\mathrm{liq} - q \rightarrow X_m \qquad (5.4)$$

X_m may be termed a "metastable (or nonequilibrium) polymorph" of the equilibrium crystal, X. Metastable polymorphs are not usually shown in phase diagrams. They are "kinetically stable" states whose existence hinges on the presence of a kinetic barrier to attaining equilibrium of state. Metastable polymorphs may be found at the same system variables as the equilibrium phase, but the two do not – under any condition – coexist in equilibrium. Rather, the transformation of a metastable polymorph to the equilibrium polymorph has a finite negative free energy and is an irreversible process.

Because the metastable polymorph X^m has the higher free energy, it is necessarily more soluble (and lower melting) than is the equilibrium polymorph X. This principle is beautifully documented in the case of sodium sulfate heptahydrate (Section 4.17.1) [10], which is an isolable (but metastable) crystal hydrate that is far more soluble than is the equilibrium decahydrate. (These two crystals differ in composition and are not polymorphs.)

Metastable polymorphic crystals are widely encountered and are particularly well-documented in triglyceride systems [36]. Liquid tristearin, on being rapidly cooled, initially forms the α crystal phase. If the α phase is rapidly heated it first melts, then recrystallizes, to form the β' phase (it has been suggested that two β' phases exist). The β' phase, in turn, is similarly converted (at a higher temperature) to the equilibrium β phase. The exact path followed depends on the heating rate and other experimental details. The equilibrium phase melts reversibly (so long as it is not completely melted and nucleation exists), to form the equilibrium liquid. Once formed, the β phase does *not* revert to other polymorphs at lower temperatures; instead, the metastable polymorphs (given a path) are transformed into the β phase.

Both the α and the β' phases may therefore be regarded as metastable polymorphs of the β phase, since only the β crystal phase exists below its melting point if equilibrium of state is attained. Recrystallization of tristearin from a solvent promotes formation of the equilibrium polymorph, but recrystallization does *not* guarantee that the equilibrium crystal will result. A nonequilibrium polymorph may crystallize from one solvent, while the equilibrium polymorph may be formed from another. An example of this phenomenon in a surfactant system was encountered in the case of *N*-dodecanoyl-*N*-methylglucamine. This compound forms a metastable polymorph when recrystallized from acetonitrile, but the equilibrium polymorph when recrystallized from hexane [37]. Metastable polymorphic crystals likely exist in a great many surfactants, but relatively little attention has been paid to this area.

The issue of "polymorphism" in liquid crystals. It is noted elsewhere (Section 6.3) that metastable liquid crystal states attain equilibrium within a very short time. This suggests that polymorphism of the sort described above for crystals does not exist among the liquid crystal states. The term "polymorph" will therefore not be used herein to describe the various lyotropic liquid crystals formed by surfactants. Reversible transitions from one liquid crystal to another are well known in thermotropic liquid crystal-forming compounds, and the term "polymorph" is perhaps more appropriate in connection with these states. The use of this term will be restricted to crystalline phases herein.

There are several reasons for taking this position. First, the term polymorph has historically been reserved for crystal states [38]; to use it for other kinds of phases as well will predictably introduce confusion. Second, there are no recorded instances to date of one aqueous surfactant liquid crystal being reversibly transformed to another of differing structure but the same composition. When multiple liquid crystal states exist (Section 5.6.10) then one state is typically transformed by cooling (at its lower eutectic limit) into the other, but the reactant and the product phases in this phase reaction differ in composition. Different liquid crystal regions are invariably separated by a biphasic region (however narrow), and the transition from one state to the other at the boundary of the liquid crystal region is non-isothermal. (Azeotropic transitions do exist at which a liquid crystal phase is converted reversibly to a *liquid* phase [39].) Finally, the conversion of one liquid crystal to another along an isopleth is restricted to only a narrow range of compositions. Along other isopleths (within the same phase region) the liquid crystal does not undergo the same phase transition.

5.6 Liquid crystal regions

Liquid crystal phases exist as equilibrium states only at temperatures above the Krafft eutectic. The structures of these phases will be described in Chapter 8, while the form of the regions they occupy, the influence of system variables on these regions, and liquid crystal solubility will be considered in this section.

Liquid crystals are invariably restricted to a finite range of both composition and temperature. Being fluids, their compositions are not stoichiometrically defined – as is often the case with hydrated crystals. (It has been suggested that stoichiometric hydrates exist in fluid states [79], but not in the sense that they exist in crystal phases. The approximately stoichiometric hydration of polyoxyethylene nonionics by two waters per ether oxygen has also been noted [80].) Being liquid crystals, they differ from liquids in having a well-developed (if also disordered) phase structure. Since liquid crystals differ from one another structurally, they not only occupy defined regions within the phase diagram but may coexist with each other, with liquid phases, or with crystal phases.

The general form of an idealized liquid crystal phase region is depicted in Fig. 5.10. This form may be altered when several such phases are crammed together within a narrow span of compositions (Section 5.8), or when discontinuities of state interfere (Section 5.6.6). Nevertheless, it is representative of the dominant liquid crystal regions – those that occupy the larger areas. The following discussion is focused on the boundary of the liquid crystal region proper (line e–f–g–h–i–e) (not that of the coexisting liquid phase). The boundary of this liquid phase, which generally lies above the liquid crystal region boundary, is considered in Section 5.6.8.

5.6.1   The upper temperature limit of liquid crystal regions

The upper part of the boundary of liquid crystal regions (Fig. 5.10) often has a symmetrical bullet or dome shape. The lower part of the boundary of this same region is unsymmetrically curved, however, so that the lowest temperature limit of liquid

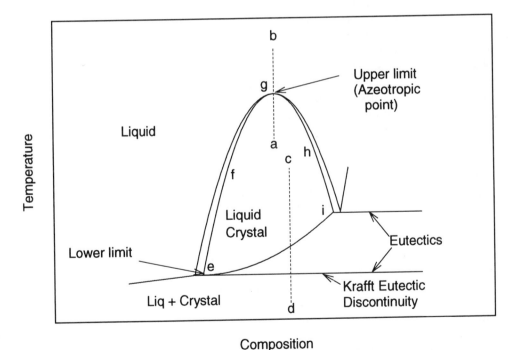

Fig. 5.10. The general form of liquid crystal regions, and of the contiguous boundaries and discontinuities.

crystal phases invariably lies at the most dilute composition limit of the region. Since the upper and the lower parts of the liquid crystal phase boundary are dictated by qualitatively different thermodynamic parameters, they are described by different mathematical functions and cusps exist where they intersect. One cusp lies at the lowest temperature (and composition) limit, while another cusp lies at the highest composition limit and falls at an intermediate temperature. Additional cusps may exist as well (Section 5.7), and the form of the liquid crystal region may be severely modified (Section 5.10.6).

In surfactants that are highly soluble in liquid water (Section 5.10.6), the boundary of the liquid phase that coexists with the liquid crystal surrounds a large fraction of the upper boundary of the liquid crystal region. This liquid boundary therefore has a shape which parallels that of the liquid crystal boundary. In the case of surfactants that are insoluble in liquid water, the boundary of the coexisting liquid has an altogether different shape (Section 5.6.8).

The liquid boundary to the dilute side of the most dilute liquid crystal region (line e–f–g, Fig. 5.10) is the solubility boundary that lies above the temperature of the Krafft eutectic – irrespective of the numerical value of the solubility of the phase (Section 5.6.8). The quantitative dimensions of the miscibility gap between the liquid and the liquid crystal are small and are usually ill-defined in soluble surfactants (Appendix 4), but the span of this gap is much larger in poorly soluble surfactants (see, however Section 5.6.8).

It is required thermodynamically, and is highly probable, that at the upper temperature limit of liquid crystal regions (point g in Fig. 5.10) the liquid and the liquid crystal boundaries touch asymptotically at a unique composition and temperature. A liquid and a liquid crystal phase of the same composition coexist at this point, and the reversible phase reaction $LX + q \rightleftarrows L$ may occur at this temperature along the isopleth a → b which passes through this composition (g). Neglecting the obvious differences in phase structures, this phase transition closely resembles thermodynamically the upper azeotropic point encountered in vapor–liquid equilibria [40]. It therefore seems appropriate also to describe this discontinuity of state in surfactant systems as an "azeotropic" discontinuity. Definitive calorimetric studies in the $C_{10}PO$–water system have shown that the transition from liquid crystal to liquid along isoplethal paths such as a → b (Fig. 5.10) is, indeed, strictly isothermal or "azeotropic" [39]. Along isopleths to either side of this composition within the same phase region, the phase transitions between the liquid crystal and liquid phases involve changes of phase composition during the process, and are thus "zeotropic" [41]. These latter transitions are inherently non-isothermal; even if conducted reversibly, they would span a finite range of temperatures.

As with phase transitions in unary systems, the azeotropic transition of a liquid crystal to a liquid phase as at point g is a congruent 1–2–1 phase transition which occurs isothermally along the a–b isopleth. Azeotropic transitions are termed "congruent" (in the jargon of phase science) because the reactant phase and the product phase have the same composition. Zeotropic transitions are termed "incongruent" because the reactant phase differs in composition from the product phase.

The upper temperature limit of liquid crystal regions vs. the melting point. The relationship between the upper temperature limit of liquid crystal phases (which is a measure of their thermal stability) and the melting point of the dry crystal varies widely and is not easily predictable. This result is hardly surprising, since the phases involved are qualitatively different. In soap–water diagrams (Fig. 3.2) the liquid crystal region of greatest stability rises well above the "melting point" of the dry soap, but it will be recalled that the phase in equilibrium with the dry liquid is *not* the same crystal that exists at low temperatures (Sections 5.5.2, 5.5.3). The liquid crystal is usually more stable thermally than is the crystal in surfactants having 12 or more carbons, regardless of the hydrophilic group. Among very short chain surfactants, however, the thermal stability of liquid crystal phases is often considerably less than is that of the crystal. In these compounds the melting point lies well above the upper temperature limits of the liquid crystals, and the region of miscibility within the liquid phase is extremely large. A well documented example of this behavior is found in the octyldimethylphosphine oxide–water system, where the surfactant melting point is 67.8°C, the upper temperature limit of the hexagonal liquid crystal phase (the most stable phase) is 17°C, and the crystal solubility at 25°C is 87% [42].

5.6.2   The lower temperature limit of liquid crystal regions

The discontinuity that exists at the lower temperature limit of liquid crystal states is determined by fundamentally different parameters from those that determine their upper limit. When liquid crystal phases are cooled beyond their lower limit, they typically undergo disproportionation to form a more dilute liquid crystal and a more concentrated crystal phase. Further, while the upper temperature limit is usually governed by an azeotropic discontinuity, the lower limit occurs at a eutectic. For surfactants that display an equilibrium Krafft boundary, the general form of the phase reaction which occurs on cooling along the isopleth passing through the eutectic composition (the "eutectic isopleth") is

$$\text{liq cryst} \rightleftarrows \text{dil fluid} + \text{crystal} + q^{\text{eu}} \tag{5.5}$$

The "dilute fluid" product may be either a liquid (in which case the phases involved are L–LX–X), or another liquid crystal (in which case the phases are LX–LX'–X). The concentrated phase at either of these eutectics is invariably a crystal.

The azeotropic point at the upper temperature limit of liquid crystal regions is approximately centered within the region. However, at the lower temperature limit the eutectic composition invariably lies near the most dilute composition limit of the region. This aspect of the form of the lower boundary is quite important. On cooling along isopleths near the concentrated limit of a liquid crystal (such as path c–d in Fig. 5.10), one passes (at the phase boundary) from the one-phase liquid crystal (LX) region into the two-phase liquid crystal-plus-crystal ($LX + X$) region. Once this biphasic region is entered crystals separate, and the liquid crystal composition becomes progressively more dilute as the temperature is lowered. The path that the liquid crystal composition follows (assuming equilibrium is maintained) corresponds exactly to the form of the liquid crystal boundary between the composition of the isopleth c → d and the composition at the eutectic temperature. Experimentally, supercooling is invariably observed along trajectories such as c → d and it is very difficult to maintain equilibrium (Section 6.3.1).

5.6.3   The lower temperature limit of liquid crystal regions when the Krafft boundary is metastable

If the Krafft boundary is projected to lie below the freezing point of water, it no longer exists as an equilibrium boundary and surfactant crystals do *not* separate on cooling dilute mixtures. Ice separates instead, and the composition of the coexisting liquid is increased as a result until the temperature of a eutectic (which is *not* the Krafft eutectic) is reached. At this transition ice, the eutectic liquid, and a liquid crystal phase coexist. The phase reaction which occurs at this Ice–L–LX eutectic (on cooling) is

$$\text{liq}^{\text{eu}} \rightleftarrows \text{Ice} + \text{liq crystal} + q^{\text{eu}} \qquad (5.6)$$

This eutectic does *not* represent the lowest temperature at which the liquid crystal exists. As the temperature is lowered further more ice forms, and this drives the composition of the liquid crystal to still higher values. Eventually yet another eutectic transition is reached; the phase reaction which occurs at this Ice–LX–X eutectic (also on cooling) is

$$\text{liq crystal}^{\text{eu}} \rightleftarrows \text{Ice} + \text{crystal} + q^{\text{eu}} \qquad (5.7)$$

The composition and structure of both the liquid crystal and the crystal phases vary widely, and often the crystal is a hydrate. The temperature of this eutectic is the lowest temperature at which equilibrium liquid crystals exist when the Krafft boundary is metastable. In this sense it is of comparable importance to the Krafft eutectic itself.

A familiar system which has a metastable Krafft boundary is $C_{12}E_6$–water (Fig. 5.8). In this case the temperature of the Ice–L–LX eutectic is about $-2°C$ and that of the Ice–LX–X eutectic is about $-6°C$.

When liquid $C_{12}E_6$–water mixtures of $< 38\%$ composition are cooled, the surfactant crystal does not exist as an equilibrium phase until the temperature falls below the eutectic at the lower limit of existence of liquid crystals ($-6°C$). Put another way, a driving force for formation of the crystal state does not exist until the temperature is below this lowest eutectic. This fact must be recognized in testing for stability in liquid phases. Conditions that favor the formation of seed crystals must be achieved if equilibrium is to be readily attained, and for this to occur the mixture should be cooled below this lowest eutectic. In nonionic surfactants having metastable Krafft boundaries this temperature is generally above $-20°C$ (as in Fig. 5.8), but in the sodium 8-hexadecylbenzene sulfonate–water system it has been found to lie near $-70°C$ [43].

5.6.4   The lower azeotropic temperature limit of liquid crystal regions

On rare occasions, a liquid crystal region terminates at an azeotropic point on cooling instead of at a eutectic discontinuity. This unusual behavior was first discovered in the dodecylammonium chloride–water system (Fig. 5.11) [44], but has also been found in the decyl and octyl homologs [45]. One important consequence of this behavior is that this liquid crystal region is isolated from other highly structured phases as an island which is completely surrounded by the liquid phase. The phase involved (in this particular system) is the hexagonal phase. If salt is present, the liquid crystal regions are fused and the behavior is more usual as to form. These systems are also unusual in that the lyotropic nematic phase is reported to exist [46].

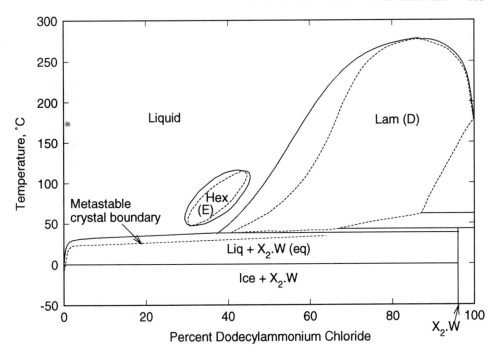

Fig. 5.11.   The dodecylammonium chloride–water system, showing the unusual hexagonal phase island [44].

The thermodynamic analysis of phase behavior at this lower azeotropic point would exactly parallel that of low-boiling azeotropic mixtures, such as exist in the ethanol–water system. At the lower azeotropic points of liquid crystals a reversible phase reaction occurs which has the form

$$\text{liquid} + q^{\text{az}} \rightleftarrows \text{liquid crystal} \qquad (5.8)$$

The highly unusual aspect of this particular phase reaction is that the enthalpy and entropy of the liquid crystal are *higher* than are those of the liquid phase. Another system where this unusual reversal (as to the structure of the upper state) is found is $C_{10}E_4$–water [47], except that in this system it occurs at an *upper* (instead of a lower) azeotropic point. Both of these transformations are superficially analogous to the "melting" of a liquid to form a "crystal"! Their existence suggests that the thermodynamic balance between these liquid and liquid crystal states is extremely delicate.

5.6.5   The factors that govern the temperature limits of liquid crystal regions

The form of the upper boundary of liquid crystal regions may be qualitatively understood by recognizing that the liquid typically has a higher entropy than does the liquid crystal (notwithstanding the unusual behavior of dodecylammonium chloride and $C_{10}E_4$, Section 5.6.4). The form of the lower boundary, on the other hand, is governed by the balance between the sum of thermal energy plus water free energies, on the one hand, and crystal free energy on the other. Disruption of the crystal is opposed by its low enthalpy, while thermal and solvation energies favor its disruption.

The upper temperature limit. It is instructive to start within the liquid crystal region (points a or c in Fig. 5.10), and consider the consequences of first increasing, and then decreasing, the temperature. The thermodynamic relationship between the liquid state and the liquid crystal state is most clearly evident at the azeotropic point, where the line a → b crosses the point of tangential contact of both phase boundaries. At this point the liquid and the liquid crystal coexist, and the higher enthalpy of the liquid phase is just balanced by its higher entropic energy. The calorimetric thermogram along this isopleth is very sharp and extremely simple because the transition is strictly isothermal [39], and the ratio of these two phases at this transition is governed by the heat content of the mixture. Since the compositions of the two phases are the same, both the water and the surfactant free energy terms ($n_W\mu_W$ and $n_S\mu_S$, Section 5.1) as well as the chemical potentials of both components, are necessarily the same in both phases.

Similar thermodynamic relationships exist along isopleths during which zeotropic transitions occur (at compositions to either side of the a → b isopleth), but the calorimetric thermograms are more complex and reflect the non-isothermal nature of these transitions. Analysis of the thermodynamic data is also far more complex [39].

The azeotropic composition is centered within the phase region, but it is not entirely clear why this is true. This result indicates that the thermal stability of the liquid crystal falls off to both higher and lower compositions. Also, it is interesting that the enthalpies of liquid crystal → liquid phase transitions along isopleths within concentrated regions of liquid crystals are substantially greater than are those within dilute regions [48].

The lower temperature limit. As all liquid crystal phases are cooled, a temperature is inevitably reached at which the surfactant crystal phase precipitates from the liquid crystal. (This is true irrespective of whether or not the Krafft boundary is metastable or not, but when it is metastable then ice (rather than the surfactant crystal) will precipitate from the more dilute parts of the most dilute liquid crystal (Section 5.4.4).) This crystal is invariably more concentrated than is the liquid crystal, so that the remaining liquid crystal becomes progressively more dilute as the crystal separation proceeds. However, this process can proceed only so far. When the temperature reaches the eutectic temperature the liquid crystal concentration is at its most dilute limit, and this phase does not exist at lower temperatures.

Liquid crystals *may* exist, however, well below their lower eutectic limit as nonequilibrium states. It is interesting that these nonequilibrium phases display a finite limit of swelling – just as do equilibrium phases [29]. Further, their dilute boundary is an extension of the equilibrium boundary that lies above the eutectic. Metastable liquid crystals are observed on cooling the liquid along isopleths that pass through the region of existence of nonequilibrium phases below the temperature of the Krafft plateau. These liquid crystals are not observed along more dilute isopleths, however; supercooled liquids or glasses are formed instead. While qualitatively similar behavior has long been known to exist in inorganic salt–water systems [81], this behavior was first observed in a surfactant system only recently (Fig. 10.21).

These results are consistent with the hypothesis that *the form of the lower part of the liquid crystal boundary is dictated by the balance between the combined influence of*

thermal energy and water free energies, on the one hand, and the enthalpic free energy of
the crystal phase on the other:

$$\text{crystal} + \begin{pmatrix} \text{thermal energy} \\ + \\ \text{water free energy} \end{pmatrix} \rightleftarrows \text{liquid crystal} \tag{5.9}$$

This is clear from consideration of the equation for the irreversible isothermal phase reaction that occurs between liquid water and the surfactant crystal to form the liquid crystal

$$nW' + X \rightarrow (X \cdot W_n)^{LX} \tag{5.10}$$

The water free energy of the liquid crystal corresponds to $n_W \mu_W$, where n_W is the stoichiometric coefficient of water in this equation and μ_W is the equilibrium chemical potential of water that exists within the coexisting phases. A particular value of n_W will be required at a particular temperature; at a higher temperature, where the level of thermal energy is greater, the amount of water (and therefore of water free energy) that is required to form the liquid crystal is relatively lower. At the lower eutectic temperature the level of water free energy is the maximum that the liquid crystal can tolerate, and the liquid crystal concentration at this eutectic is the most dilute that exists. Adding more water (or increasing its chemical potential) beyond this point does not produce the liquid crystal state, but instead a more dilute fluid phase of differing structure. At the Krafft eutectic temperature this dilute phase is the micellar liquid, but at other eutectics it may be a more dilute liquid crystal having a different structure.

The stoichiometric ratio of liquid phase to crystal phase that is required in order to form the liquid crystal during the eutectic phase reaction is normally very high. To illustrate, if one assumes that the concentrations of the liquid, liquid crystal, and crystal phases at the Krafft eutectic are 25, 27, and 100% surfactant, then 0.973 g of liquid but only 0.027 g of crystal are required to form 1 g of liquid crystal. The stoichiometric weight ratio of liquid to crystal phase required in this particular example is 36 : 1; this ratio is extremely sensitive to the compositions of the coexisting phases.

On heating mixtures along isopleths more concentrated than the eutectic composition (path d → c, Fig. 5.10), the excess crystal reactant phase that remains at the eutectic temperature (after the isothermal eutectic reaction is complete) is dissolved in the liquid crystal to a progressively higher degree as the temperature is increased. At the liquid crystal boundary, the dissolution is complete.

5.6.6   The composition limits of liquid crystal regions

The lower boundary of liquid crystal regions typically spans the widest range of composition of this phase region. As with the liquid phase boundary at the Krafft eutectic (Section 5.4.3), the cusp in the liquid crystal boundary at its upper composition limit (point i in Fig. 5.10) signifies the existence of a discontinuity of state within the coexisting phase. Below the temperature of this cusp the coexisting phase is a crystal, while above this temperature it may be either a different crystal, a liquid crystal (Figs 5.13 and 5.14, Section 5.6.8), or (in some cases) a liquid.

Varying composition at constant temperature is equivalent to holding kinetic energy constant while varying the water free energy. The fact that liquid crystal regions display a lower composition limit indicates that an upper level of water free energy exists above which they are unstable. The fact that they display an upper composition limit indicates that a minimal level of water free energy is required in order for them to exist (Section 5.6.7). The composition limits thus also define, indirectly, the range of water free energies within which a particular phase is stable. This range varies with the temperature, and typically becomes progressively narrower as the temperature is increased until the azeotropic point is reached.

The idealized form of the lower boundary of the liquid crystal regions shown in Fig. 5.10 resembles the actual form in many surfactant systems (e.g. Fig. 10.21) [49]. This form may be altered, however, as in both the hexagonal and the lamellar phase regions of the SDS–water system (Fig. 5.4) [8]. In this system both of these boundaries are convex upward to the dilute side of the phase region, but convex downward to the concentrated side. This gives the boundary a sigmoid shape. In the case of the lamellar phase the temperature of the plateau corresponds to the peritectic temperature of the crystal dihydrate, $X \cdot W_2$, but the reason for the complex form of the hexagonal boundary is uncertain. Perceptible changes in phase structure do occur with composition in both phases, and may be responsible in part for these odd shapes [50].

Within the narrow temperature range of the plateau region of the SDS hexagonal boundary, the composition of the liquid crystal changes very rapidly with temperature. It is thus possible that the calorimetric endotherm that was observed may *not* reflect the existence of a true thermodynamic discontinuity, but instead an unusually rapid rate of dissolution of crystals with increasing temperature. Calorimetric studies of the *N*-dodecanoyl-*N*-methylglucamine–water system revealed the existence of an endotherm at the temperature of the knee of the Krafft boundary (Fig. 10.21).

5.6.7 Phase reactions that occur during isothermal mixing

The phase reactions that occur along isothermal mixing paths and the factors that influence the composition span of liquid crystal regions will now be considered. Suppose one selects the isothermal path e → l (Fig. 5.12), and follows it from right to left with stops at points f, g, h, etc. This corresponds physically to adding aliquots of liquid water so as to form mixtures of the desired composition, and equilibrating the mixture (e.g. by stirring or other forms of mixing) after each addition.

The consequence of adding the first aliquot (to reach point f) is the formation of a liquid crystal phase of composition g. A stoichiometric phase reaction occurs during this process which is

$$W + X \rightarrow LX^g \tag{5.11}$$

This process is irreversible; the driving force for it is

$$\Delta G = G_{LX^g} - (G_W + G_X) \tag{5.12}$$

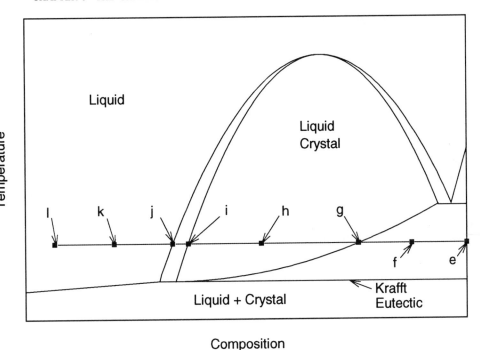

Temperature

Composition

Fig. 5.12. The compositions of mixtures (f through l) formed by adding water to mixture "e", along the dilution trajectory e → l at a temperature just above the Krafft eutectic.

Adding a second aliquot so as to reach point g invokes precisely the same phase reaction, and at the same time eliminates the crystal phase altogether. Within the composition interval e → g, the ratio of liquid crystal to crystal phases changes after each addition (in accordance with the lever rule), but exactly the same phase reaction occurs after each addition.

A qualitatively different process occurs as more water is added (past g) to form mixture h in that the added water simply dissolves (again irreversibly) in the liquid crystal phase and dilutes this phase. As still more water is added to form various mixtures within the span g to i the composition, details of the structure, and all of the properties of the liquid crystal phase vary smoothly and continuously (Section 4.8.1). This was *not* true within the interval e → g.

This smooth dilution process ceases abruptly at point i; past point i, a different stoichiometric phase reaction process (analogous to that described within the interval e → g) again occurs. The added water, within the interval i → j, reacts stoichiometrically with the liquid crystal to form a liquid phase of composition j until point j is reached. As still more water is added to form k, the liquid phase is again simply diluted. From k to l and beyond, discontinuities in phase behavior are not encountered.

These observations illustrate the general principle that *an alternating sequence of irreversible stoichiometric phase reactions (within miscibility gaps) and dilution processes (within one-phase regions) occurs along isothermal mixing paths.* When water is mixed with the crystal at temperatures just above the Krafft eutectic, along an isotherm which passes through a single liquid crystal region ($e \rightarrow 1$ in Fig. 5.12), the following processes occur:

(1) $X + W \rightarrow LX$
(2) $LX + W \rightarrow$ more dilute LX
(3) $LX + W \rightarrow$ Liq
(4) Liq $+ W \rightarrow$ more dilute Liq

The reason for the alternation is that during steps 1 and 3 composition is not a degree of freedom (Section 4.8.2), while during steps 2 and 4 it is (Section 4.8.1). One can never escape the influence of the Phase Rule.

Such alternation between stoichiometric phase reactions and dilution processes along isothermal mixing paths is universal. The addition of water to salt crystals (Fig. 1.1) also results at first in a stoichiometric reaction (to form the saturated liquid phase), but when the composition of the mixture reaches the solubility boundary the further addition of water results simply in dilution of this liquid.

5.6.8   The liquid crystal solubility boundary

Emphasis has to this point been placed on the boundaries of the liquid crystal region itself, but the boundary of the coexisting liquid region is also important. One reason is that it is this liquid boundary that is actually determined during most phase studies. It is difficult (and unusual) for the liquid crystal boundary itself to be accurately determined, whereas determination of the liquid boundary may be accomplished very simply (Appendix 4.4.3) [51].

It was noted in Section 5.6.1 that if a surfactant is highly soluble in liquid water, the form of the coexisting liquid boundary parallels that of the liquid crystal boundary proper. The composition span of the miscibility gap between these two phases is typically small (from $<1\%$ to a few per cent) under these circumstances.

If a surfactant is insoluble within the liquid phase, however, the form of the liquid phase boundary is altogether different. This situation is well known in monoglyceride–water systems (Fig. 5.15) [52], and the term "monoglyceride-like phase behavior" has been used to describe the behavior of such systems (Section 5.10.6). It is also found in many di-long chain surfactant–water systems.

Monoglycerides are very weakly hydrophilic diol ester surfactants (Section 10.5.5.1) that are miscible with water within liquid crystal phases but are poorly soluble in liquid water. (Water *is* miscible with these compounds within the concentrated liquid and a hexagonal phase, if formed, has the "inverted" structure; the "normal" hexagonal phase does not exist in these systems.) Liquid crystal or other phases separate from such solutions at very low monoglyceride compositions, and the entire dilute part of the diagram consists of biphasic regions within which the saturated liquid and another phase coexist. At low temperatures the coexisting phase may be a crystal, at intermediate temperatures a liquid crystal, and at sufficiently high temperatures it is a concentrated liquid (Section 5.6.1). Those isothermal discontinuities which separate these different biphasic regions are typically eutectics which are closely related to those found (above) in soluble surfactants. (It is worth noting that these discontinuities are typically described as "melting" transitions in the instruction manuals of calorimeters, and that this description is simply wrong if the term "melting" is used in the strict sense of the word from a phase science perspective.)

The only important difference is that the dilute liquid phase at the lowest (Krafft) $L-LX-X$ eutectic is almost pure water, instead of being a concentrated micellar solution.

Qualitatively similar phase behavior is also found in aqueous systems of most polar lipids [53] and other di-long chain surfactants. A well-documented example is provided by the DODMAC–water phase diagram (Fig. 3.2). In this system the boundary of the liquid region is pushed so closely against the water border that it is essentially vertical. (The Krafft boundary of soluble surfactants at temperatures far below the Krafft plateau has a similar shape (Section 5.4.3).) The exact form of such liquid solubility boundaries is not readily discerned, and cusps or other irregularities (should they exist) would be almost imperceptible.

The form and composition range of the liquid crystal region proper is not severely altered by lack of miscibility of the surfactant with liquid water – but this is not true in the $C_{10}E_4$–water and in other pivotal systems (Section 5.10.7). The liquid crystal phase is, after all, a legitimate region of miscibility in its own right.

5.6.9 The factors that govern the composition limits of liquid crystal regions

At the eutectic at the lower temperature limit of a liquid crystal, the available thermal energy is just barely sufficient for the phase to exist and the amount of water required is the maximum that is allowed. At somewhat higher temperatures less water is required to form the liquid crystal, and the water composition of the coexisting liquid crystal phase (within the liquid crystal/crystal miscibility gap) is lower. Above the temperature of the cusp at the concentrated limit of the phase (point i, Fig. 5.10), hydration of the crystal produces initially another phase having a different structure before the liquid crystal region with which we are concerned is formed.

Below the eutectic at the lower limit of the liquid crystal (point e, Fig. 5.10), the amount of water required to break down the crystal is sufficiently great that the liquid crystal is not actually formed. Instead, a more dilute liquid phase (which can tolerate far higher levels of water than can any liquid crystal) results. As will be seen in Chapter 7, the application of these principles leads to useful correlations between phase behavior and relative humidity.

"*Infinite swelling*". The phrase "infinite swelling" has been used to describe the swelling behavior of polar lipid molecules [54]. This phrase deserves to be evaluated critically from the perspective of phase science because, taken literally, a phase that undergoes "infinite swelling" may presumably be diluted isothermally with water, in any amount, without encountering a discontinuity of state. As defined in this way, "infinite swelling" does indeed occur during the dilution of (for example) liquid ethanol with water. As water is added the ethanol is continuously "swollen" to form a liquid mixture of lower ethanol composition. Micellar solutions of soluble surfactants can similarly be diluted with water in any amount without encountering a phase boundary. (An exception exists when the liquid region is interrupted by a liquid/liquid miscibility gap.)

However, polar lipids having naturally occurring chain lengths are insoluble in liquid water, and the coexisting phase is a liquid crystal or "gel" phase. The liquid crystal phases that exist in such systems do *not* undergo "infinite swelling", as the term is defined above. A limit to swelling exists for all liquid crystal phases, perhaps because the phase structures of liquid and of liquid crystal phases are qualitatively different. These phases differ with respect to the symmetry elements that they possess, order parameters, and other features. They are also distinct with respect to their thermodynamic parameters. A liquid crystal phase can never, for these reasons, be part of the same continuous phase region as a liquid.

5.6.10   Multiple liquid crystal regions

More than one liquid crystal state exists in practically all surfactant–water systems, and the various kinds of phase behavior that result when this occurs will now be considered. In Fig. 5.10 the liquid crystal region shown is also the separating phase at the liquid crystal solubility boundary that lies just above the Krafft eutectic. The fact that this liquid crystal (which will be designated A) exists at the temperature of the Krafft eutectic signifies two things:

(1) that A is the liquid crystal state which requires the least thermal energy in order to exist, and
(2) that A is also the liquid crystal state which is most resistant to destruction by the further addition of water.

Liquid crystals more concentrated than A require higher minimal levels of thermal energy to exist, and also are not so resistant to destruction by the addition of water, as is A.

Often the form of the next more concentrated liquid crystal region (B) is qualitatively similar to that of A (Fig. 10.21). B must also terminate at its lower temperature of existence at a eutectic, but constraints imposed by the Phase Rule introduce complexities into the phase behavior in this temperature region. Because of these constraints, the lower eutectic temperature of B either may – or may *not* – exist at the temperature as the cusp at the upper composition limit of A (point i, Fig. 5.10).

The alternation rule requires that in passing along an isopleth or isotherm the number of phases present must alternate between even and odd (Section 4.12). It also requires that phases be inserted or deleted one at a time, and in a systematic and logical fashion. To illustrate, in passing from the two phase state below a eutectic through the eutectic, a third phase is inserted at a composition that falls between the two phases initially present. After the eutectic phase reaction has occurred, either one or the other of the terminal phases is deleted at higher temperatures. In passing from the biphasic region above a eutectic to the biphasic region below it, a third phase is "inserted" whose composition actually lies beyond the limits of the two phases initially present.

In traversing the Krafft eutectic along isopleths between the composition of A and the composition of X and starting from below its temperature, one may (according to the above rules) pass through the sequence

$$L + X \rightarrow L + A + X \rightarrow A + X$$

The abbreviation "$L + X$" corresponds to the two-phase region below the eutectic within which the liquid (L) and the crystal (X) phases coexist, and "$A + X$" symbolizes the region above it where liquid crystal A and X coexist. The "$L–A–X$" signifies the isothermal eutectic line at which one or more of these three phases may be found.

This terminology signifies that in the first step liquid crystal A is inserted between liquid L and crystal X, and that in the second the terminal L phase is deleted. Both processes occur via the eutectic phase reaction $L + X + q_{eu} \to A$. L is completely consumed because X is in excess, and part of this phase is left over after the reaction is complete. *The substitution of one phase for another is not allowed along isopleths passing through such 2–3–2 phase state discontinuities* – only the insertion or deletion of a phase.

Because of these restrictions, when multiple liquid crystals exist one may *not* pass directly from an $A + X$ region (next to the A liquid crystal) to a $B + X$ region (next to the B liquid crystal). Additional steps are required, which must occur in multiples of two. One reasonable hypothesis, illustrated in Fig. 5.13 along the isopleth a → b, is the sequence:

$$A + X \to A + L + X \to L + X \to L + B + X \to B + X \to B$$

This path requires first insertion of the liquid L phase and deletion of A, then insertion of B and deletion of L. Two eutectic discontinuities ($A + L + X$ and $L + B + X$) must exist; first A, and then L, are deleted as each eutectic is passed.

Another plausible sequence is:

$$A + X \to A + B + X \to B + X \to B$$

in which liquid crystal B (rather than L) is directly inserted in the first step (to form an $A + B + X$ eutectic), and then A is deleted. This behavior is shown in Fig. 5.14, along

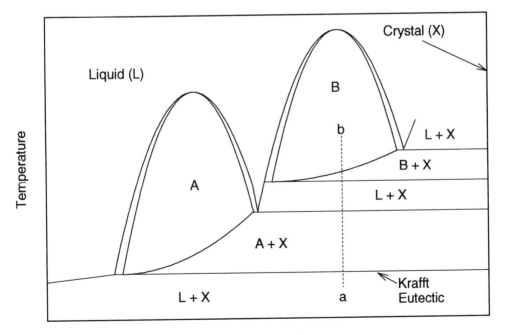

Fig. 5.13. Multiple liquid crystal phases in which the coexisting liquid phase surrounding A extends down to the temperature of the cusp at the upper composition limit of A.

the isopleth a → b. The liquid L, which coexists with A over most of the upper boundary of region A, is not encountered along this path because in this model L does not extend down to the temperature of the $A + B + X$ eutectic. Instead, it terminates at an $A + L + B$ eutectic which lies at a higher temperature (Fig. 5.14). When this happens an additional cusp may be expected to exist in the boundary of region A at the temperature of the $A + L + B$ eutectic. This behavior can be seen in the diagrams of the sodium palmitate–water system near the upper limits of both the lamellar and the hexagonal phases (Fig. 3.2).

The upper temperature limit of phase B may be an azeotropic discontinuity, as shown in Figs 5.13 and 5.14 (Section 5.6.1), but other kinds of behavior exist if the phase is squeezed between two other phase regions (Section 5.8). If a third liquid crystal (C) exists, C may be inserted following either of the two models described above. In many surfactant systems only two or three liquid crystal regions exist, but *six* are found in the SDS–water system (Fig. 5.4) [8]. In soap–water systems "intermediate" liquid crystal phases were found to exist between the hexagonal and the lamellar phase during the 1970s, but exactly how they are inserted remains unclear [55]. These phases went unrecognized for more than 50 years after the first soap–water phase study was performed, which illustrates dramatically the premise that the answer to the question "how many phase states exist?" may easily be uncertain in aqueous surfactant systems.

The fundamental difference between the behavior shown in Fig. 5.13 and that in Fig. 5.14 is that the liquid phase extends down to the temperature of the concentrated

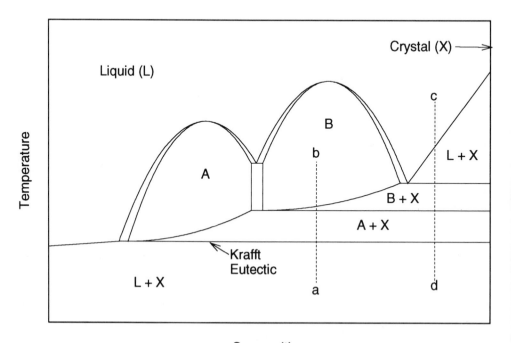

Fig. 5.14. Multiple liquid crystal phases in which the second liquid crystal phase (B) is inserted at the temperature of the cusp at the upper composition limit of A, while the liquid phase surrounding A does not extend down to this temperature.

cusp of A in Fig. 5.13, but does not do so in Fig. 5.14. The latter behavior actually appears to be more common, but either is possible. Other presently unrecognized kinds of behavior may also exist, as these details of surfactant phase behavior have historically been difficult to ascertain.

5.6.11   Phase reactions along concentrated isoplethal process paths

A particularly important consequence of the existence of several liquid crystal states is that eutectic discontinuities are introduced at the lower temperature limit of each phase. Each of these discontinuities are necessarily encountered along isopleths that pass through them, so that at very high compositions all these eutectics will be encountered. As the composition is reduced those that lie at high temperatures are not encountered, so that the number of eutectics encountered becomes smaller as the composition is reduced. A particularly significant example of this behavior was encountered during the SDS–water study [8]. It is critically important to recognize the existence of these aspects of surfactant phase behavior, if one is to interpret correctly calorimetric thermograms determined at compositions greater than that of the liquid phase at the Krafft eutectic (point e, Fig. 5.10).

This principle may be illustrated in a model system by cooling mixture c in Fig. 5.14, and following the isopleth c → d. As this mixture crosses the liquid boundary into the $L + X$ miscibility gap, the crystal phase separates. The liquid composition is reduced by the separation of the concentrated crystal phase, and follows the liquid phase boundary profile down to the temperature of the $B + L + X$ eutectic. At this discontinuity an isothermal change of state occurs. A quantity of heat equal to the latent heat of the eutectic reaction must be removed, and the eutectic liquid phase is transformed (by the eutectic phase reaction) into the more dilute B liquid crystal (plus more crystal phase). As the temperature is lowered still further, more crystals separate and the composition of B follows the lower boundary of region B until the $A + B + X$ eutectic is encountered. Again the latent heat must be removed in order to drive the eutectic reaction forward, liquid crystal A results, phase B disappears and still more crystals are formed isothermally. Lowering the temperature still further causes still more crystals to separate and the composition of A follows the lower boundary of region A until the Krafft ($L + A + X$) eutectic is reached, at which point all these processes are once again repeated. Lowering the temperature further causes still more crystals to separate and the concentration of L to follow the Krafft boundary plateau to progressively more dilute liquid compositions. Once the knee is passed, most of the surfactant has separated from the mixture as the crystal phase. Finally, ice forms and the components are no longer miscible. From this point downward the thermodynamic properties (and perhaps the structures) of the phases vary, but presumably their compositions do not.

Throughout this process the fraction of the crystal phase in the mixture steadily increases, while the fraction of the coexisting fluid phase decreases. *It is essential to realize that "merely" changing the temperature of a mixture along isopleths that pass through biphasic regions and isothermal discontinuities, such as are illustrated in Fig. 5.14, dictates the existence of extensive mass transport (along with heat flow and structural rearrangements) if equilibrium is to be maintained.* If one starts at low temperatures and heats mixture d along the path d → c, these processes are exactly reversed. The latent heats at the various eutectics may, in principle, be revealed along this path by calorimetric studies, but often they are not because the scanning rate is too fast. During the SDS–water study, for example, it was established during preliminary experiments that a scanning rate of 0.2°C/minute failed to reveal all of the isothermal discontinuities that exist. The entire study was performed at 0.08°C/minute, for this reason [8]. Similar observations have been made during calorimetric studies of polar lipid–water mixtures [56].

The phase behaviors depicted in Figs 5.13 and 5.14 are typical of a large number of the surfactant–water systems encountered in industrial processes. Some other factors that also influence surfactant phase behavior will now be considered.

5.7 The influence of crystal hydrates on phase behavior

It has been assumed to this point, for the sake of simplicity, that no crystal hydrates exist. This is true for a number of surfactants such as the phosphine oxides [57], but practically all surfactants form stoichiometric crystal hydrate phase compounds. It is therefore necessary to consider the considerable impact that these states may exert on surfactant phase behavior.

Modes of decomposition. Crystal hydrates invariably decompose on being heated to a certain temperature. ("Decomposition" is used only with reference to phase structure, as typically molecular structure is not altered during such "decomposition".) Two modes of decomposition are known: congruent decomposition, during which another (usually liquid) phase of the same composition is formed, and incongruent decomposition, during which two phases of differing composition result. The crystal hydrates formed by $C_{12}E_6$ are reported to decompose congruently (Fig. 5.8). However, most surfactant crystal hydrates decompose incongruently via the peritectic phase reaction. The thermodynamic (Section 3.21) and phenomenological (Section 4.8.3) attributes of the peritectic decomposition of the $X \cdot W_2$ crystal of sodium chloride (Fig. 1.1) have been described.

The general form of these peritectic phase reactions is

$$X \cdot W_n + q^{\text{peri}} \rightleftarrows w^{\text{fluid}}\text{Fluid} + w^{X \cdot W_m}X \cdot W_m \quad (m < n) \tag{5.13}$$

This equation signifies that the crystal phase $X \cdot W_n$ reacts, upon absorbing an amount of heat, q^{peri}, to form a phase more concentrated than $X \cdot W_n$ (which is invariably another crystal, $X \cdot W_m$), and a fluid phase that is more dilute than $X \cdot W_n$. The fluid phase will be the phase of highest enthalpy and entropy among these three states, and its formation likely serves as the principal driving force for the reaction. The fluid phase formed during peritectic reactions of nonsurfactant crystal hydrates is invariably a liquid (Fig. 1.1). However, the fluid phase formed during peritectic reactions of surfactant crystal hydrates is invariably a liquid crystal.

Further, one must recognize that the peritectic reaction has been found to be mechanistically complex so that this reaction probably never actually occurs as written in eq. 5.13 (Section 6.5.1). If so, the heat observed during a calorimetric study in which peritectic reactions occur thus more nearly reflects the heat of the congruent "melting" of the crystal hydrate (to form a metastable reaction product), than it does the equilibrium heat of the peritectic phase reaction.

Crystal hydrates as the coexisting phase. Over the span of temperatures within which a crystal hydrate exists, this phase compound replaces the dry crystal as the phase that coexists with liquid and liquid crystal phases. The existence of a crystal hydrate does not in any way alter the qualitative aspects of phase behavior. However, it *does* alter the compositions of the phases present, and because of this it also strongly influences phase ratios. When crystal hydrates are involved, the fraction of crystal phases that exist (at a particular gross composition) within crystal/fluid miscibility gaps is higher than would be the case if the dry crystal were involved. This impact on phase ratios strongly influences the physical properties of biphasic mixtures, especially their rheology. Their influence on these properties will be exerted along both isoplethal and isothermal process paths.

More than one crystal hydrate may exist, as in SDS (Fig. 5.4) [58] and DODMAC (Fig. 3.2) [1]. The crystal phase that coexists with fluid phases is the most highly hydrated phase that exists at the temperature under consideration, but lower hydrates may also exist that come into play at higher temperatures (above the peritectic temperature of the highest hydrate).

The crystal hydrate having the most water invariably displays the lowest peritectic temperature, the next more concentrated hydrate a higher temperature, and so on. The peritectic lines are thus arranged in a stair-step fashion within the phase diagram until, finally, only the dry crystal phase exists as an equilibrium crystal phase (Fig. 3.2).

Crystal hydration reaction chemistry. In addition to influencing phase behavior, the presence of crystal hydrates introduces new phase chemistry. If a dry crystal is mixed with water under conditions such that a crystal hydrate is the equilibrium phase, the water must react with the dry crystal to form the hydrate in addition to engaging in other phase reactions that may be required (Section 5.6.7). Heat may be absorbed or liberated during this process. The formation of crystal hydrates often occurs in parallel with other physical processes.

The hydration of crystals likely occurs by diffusion through a layer of the solid reaction product which forms at the surface of the reacting crystal. The rates of such reactions vary enormously. Such experience as is available suggests that single chain surfactants form crystal hydrates comparatively rapidly (during periods of weeks), while di-long chain surfactants (DODMAC dihydrate) may require months or years to form, depending on conditions [59].

The dehydration of crystal hydrates is usually rapid, which is understandable in that a positive entropy change is associated with this process. The rapid decomposition of hydrates can be highly disruptive to a crystal structure. (The term "efflorescence" has been used to describe the transformation of a well-formed reflective crystal to a powdery particle as water is lost during a peritectic reaction [56].) In one instance a metastable crystal phase resulted from such a process [1]. Thus, the existence of crystal hydrates is likely to introduce a complex array of both equilibrium and non-equilibrium phase reactions that occur as mixtures are heated or cooled, and these phase reactions may give rise to significant calorimetric heat effects. Even the existence of these phase compounds, let alone details of the phase chemistry involved, is often poorly documented.

If a mixture of coexisting liquid crystal plus crystal hydrate phases is heated to the peritectic temperature of the crystal hydrate, the hydrate is replaced at this temperature (at equilibrium) by another crystal of lower hydration. Because of this discontinuity of state in the crystal phase, a cusp exists in the boundary of the liquid or liquid crystal region at peritectic temperatures (Section 5.4). It has never been easy to document these cusps historically, but they were clearly evident during the phase study of the

DODMAC–water system [1]. The temperatures of the cusps (determined using the isothermal DIT method) coincided with the temperatures of the $X \cdot W$ and $X \cdot W_2$ peritectic decomposition reactions (determined using calorimetry). Similar phenomena were earlier encountered during phase studies of the dodecylammonium chloride– [44] and SDS–water systems [8].

Summary. The existence of crystal hydrates represents a significant, but straightforward, perturbation of the phase chemistry of a system. While their existence does not alter the fundamental features of surfactant–water phase behavior, it does significantly influence important quantitative aspects such as phase ratios. Those properties of mixtures which are directly influenced by phase ratios, such as rheology, are particularly strongly affected. The existence of crystal hydrates may also introduce new phase reaction chemistry, and (under some circumstances) provides a driving force for heat and mass transport that would not otherwise exist.

5.8   The squeezing together of liquid crystal regions

The form of liquid crystal regions illustrated in Fig. 5.10 may be altered by squeezing one region between two others. This occurs in many systems, and when it does the azeotropic transition at the upper temperature limit of the phase may be obliterated. Details of the phase behavior of such phase regions are not easily documented, but this issue was addressed during a study of the monoolein–water system (Fig. 5.15) [47]. In

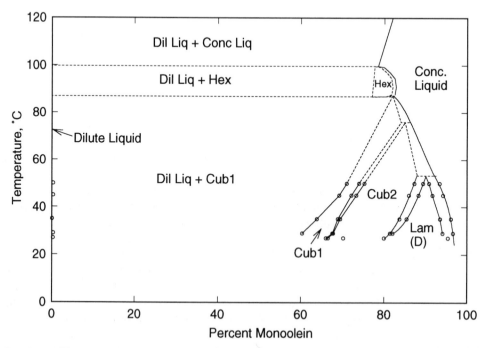

Fig. 5.15.   The monoolein–water phase diagram, showing the lamellar phase squeezed between a dilute cubic phase and a concentrated liquid phase, terminating at a peritectic at its upper limit [47].

this system the lamellar liquid crystal region lies under the more concentrated of the two cubic phase regions, is compressed to a point at its upper limit, and decomposes at this temperature (via a peritectic phase reaction) into a cubic and a liquid phase (Section 3.19). Possibly, this is the fate of the "intermediate" phases now recognized to exist between the principal hexagonal and lamellar liquid crystals in many systems [8,56,60].

A reasonable hypothesis as to the general form of squeezed phase regions is that they decompose via a peritectic transition at their upper limit of existence, and via a conventional eutectic transition at their lower temperature limit (as in Figs 5.13 or 5.14). A plausible illustration of such behavior is suggested in the modified diagram of the dodecyldimethylamine oxide–water system (Fig. 10.12).

5.9 The influence of dry liquid crystal states

Many nonsurfactants, and a few surfactants, display a dry liquid crystal state that exists at temperatures between the range of existence of the crystal and liquid phases (Section 10.5.5.1). Water is miscible with these states if the molecules from which they are formed possess operative hydrophilic groups (are surfactants). In such cases the thermotropic liquid crystal region simply extends into the binary composition space, in much the same manner as does the dry liquid phase region itself in the more usual diagrams. The general form of the liquid crystal phase regions of aqueous surfactant systems having thermotropic phases thus appear to be conventional, except that they are truncated at the 100% surfactant boundary. A well-defined example of this behavior is found in the *N*-methyl-*N*-dodecanoylglucamine–water system shown in Fig. 10.21.

5.10 "Cloud points" and the liquid/liquid miscibility gap

Attention has so far been focused on the various states of matter that exist within aqueous surfactant systems. We will now consider the *absence* of a phase – the gap in miscibility that may exist within the liquid phase. This gap is responsible for the "cloud points" that are widely encountered in aqueous mixtures of weakly polar (usually, but not always, nonionic) surfactants. Miscibility gaps of all sorts are encountered in surfactant–water phase diagrams, but this particular gap is (almost) unique in that the coexisting phases have qualitatively similar structures. It is for this reason both intriguing and important, and has attracted considerable attention.

Miscibility gaps within which like phases coexist, together with the associated critical phenomena, are (in principle) possible in phases other than the liquid phase. It has recently been claimed that a miscibility gap and critical point exist within the lamellar phase of the sodium 5-dodecylbenzenesulfonate–water system [84]. It has also been reported that coexisting lamellar liquid crystal phases of widely differing compositions coexist in the $(C_{12})_2NMe_2^+,Br^-$–water system [62], but the existence of a critical point at which this gap vanished could not be established in this system. It has also been suggested that a miscibility gap exists within the hexagonal/monoclinic liquid crystal region of the SDS–water system [8].

These preliminary results suggest that the intrusion of miscibility gaps and the associated critical phenomena may not be restricted to the liquid phase, as experience to this point would have suggested.

It has also been hypothesized that miscibility gaps may exist within crystal phase regions in binary mixtures of crystalline hydrocarbons. Solid solutions have been reported to exist in solid phases I and II in the 2,2-dideuteroheneicosane/nonadecane ($C_{21}H_{42}D_2$/$C_{19}H_{40}$) system (Fig. 5.16). This diagram indicates that these two compounds are miscible in all proportions in solids I and II, which lie below the liquidus boundary. The fraction of *gauche* bonds in the hydrocarbon chains has been shown to be substantial within these solid phase regions [63].

An azeotropic-like discontinuity exists below the solid I phase region, and the general form of this boundary signifies that solid solutions also exist in phase II. At still lower temperatures an upper consolute boundary is suggested to intrude into the phase II region, which (if true) would result in critical behavior within a crystal phase [61].

5.10.1 Lower consolute boundaries and lower critical points

If a clear aqueous solution of a pure surfactant having a cloud point is heated, the mixture becomes cloudy at a particular temperature. This clouding occurs suddenly, is easily reproducible, and is readily reversible. Cloudiness may appear for several reasons in liquid surfactant mixtures, but often it is due to the separation of small droplets of a second liquid phase in the form of an emulsion. If a mixture that is cloudy for this reason is allowed to stand at a particular temperature, it will separate (usually within a few minutes to a few hours) into two clear liquids of well-defined composition. Such

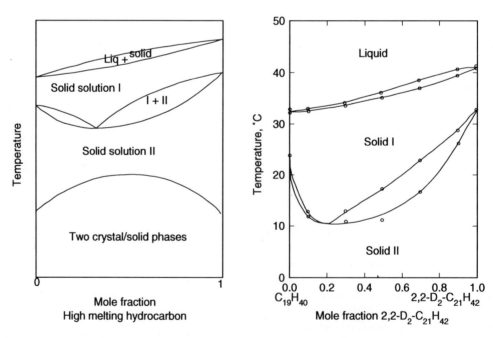

Fig. 5.16. The generalized form of binary hydrocarbon phase diagrams (left), and the 2,2-dideuteroheneicosane($C_{21}H_{42}D_2$)/nonadecane($C_{19}H_{40}$) system (right) [61].

mixtures are invariant with respect to composition (temperature and pressure specified).

Such phenomena indicate that a gap in miscibility exists within the liquid region of the system. The width of this gap is often found to decrease as the temperature is lowered, and it vanishes at a particular temperature (the "critical temperature") and composition (the "critical composition"). At temperatures far above the critical temperature neither of the two coexisting liquid phases is unusual with respect to its structure or properties. However, phases that lie below and close to the minimum (which is termed the "critical point") display perceptible opalescence and differ structurally from liquids far below the minimum. A particularly important difference is that the characteristic dimensions of the spontaneous fluctuations in composition (and refractive indices) that occur are exceptionally large. These fluctuations are responsible for the "critical opalescence" that such phases display (Section 8.4.3) [64]. If one determines the turbidity of mixtures within the liquid phase as composition is varied along isotherms near to but below the critical point temperature, a maximum in turbidity occurs at the critical composition.

In commercial mixtures, many compounds which in binary systems have widely differing phase behavior will exist. Such mixtures may, in principle, be rigorously treated as ternary or higher systems. If regarded as a binary system seeming violations of the Phase Rule will be observed, such as variation in the compositions of the coexisting phases as gross composition is varied within the miscibility gap. Also, determination of the boundary is fuzzier and depends on the history of the mixture (such as time, method of mixing, etc.).

Critical points also exist at a particular point along isothermal liquid/liquid miscibility gaps in ternary systems; these are called "plait points" (Section 12.1.1.4).

5.10.2 Upper consolute boundaries and upper critical points

The most common consolute boundary found in surfactant systems, by far, is the lower consolute boundary of polyoxyethylene and other weakly polar surfactants. However, aqueous liquid mixtures of selected zwitterionic and ionic surfactants also display miscibility gaps in which clouding occurs on *cooling* their solutions, rather than on heating [65]. Phase studies reveal that these cloud points too correspond to the boundary temperature of a liquid/liquid miscibility gap, but this particular gap is inverted as to temperature with respect to those described in Section 5.10.1. The boundary of the gap in which separation occurs on cooling is termed an "upper consolute boundary", and the critical point an "upper critical point". The upper and lower consolute boundaries are in some respects qualitatively similar, but differ in several important respects in aqueous surfactant systems.

Many aqueous systems of small amphiphilic molecules (such as butanol) that do not display surfactant behavior according to the criteria developed later (Chapter 9) display upper consolute boundaries [66].

Critical points exist in many unary systems, and would presumably exist in all systems if chemical decomposition did not intervene. These may be regarded as being more closely related to upper than to lower consolute boundaries since the coalescence of phases occurs on increasing the temperature. In unary systems the critical point is the temperature and pressure at which the distinction between the liquid and the gas phase vanishes. Those fluid phases which exist above these critical points are termed "supercritical fluids"; many compressed gases at room temperature qualify as "supercritical fluids" and such fluids are presently widely exploited during chromatographic analyses [67], in large-scale separation processes [68], and possibly during toxic waste management [69].

5.10.3 The closed loop of coexistence

Liquid/liquid miscibility gaps are widely encountered in nonsurfactant systems. Innumerable binary systems of partially miscible solvents display one phenomenon or the other, and some display both [70,71]. In some systems the lower and upper consolute boundaries are connected, so that a "closed loop of coexistence" exists. In other cases an upper consolute boundary lies *below* a lower consolute boundary, so that the two are separated by a band of liquid phase. One suspects that in such cases the lower consolute boundary (at the higher temperature) is the lower portion of a closed loop of coexistence whose upper limit is inaccessible. These phenomena have been investigated both experimentally and theoretically for more than a century. In the closed loop, a lower critical point exists at the lower temperature limit and an upper critical point at the upper limit, but these are features of the same continuous phase boundary.

In 1966 it was discovered that the closed loop of coexistence exists in surfactant–water systems [42], and this phenomenon was later investigated in depth [72]. The phase diagrams of four systems which display this phenomenon were determined, and the shape of the boundary in the vicinity of the lower critical points (along with the critical exponents which describe its shape) was accurately determined for a number of such systems [73] (Fig. 5.17). Because of the extremely high temperatures to which these

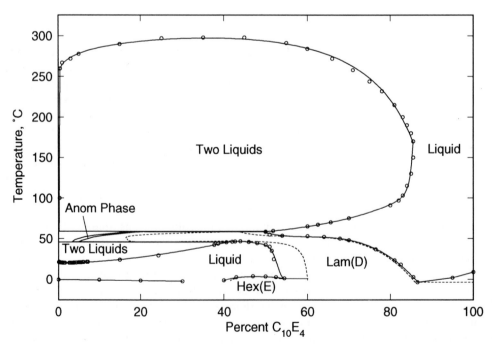

Fig. 5.17. The $C_{10}E_4$–water system, showing the entire closed loop of coexistence [72].

loops extend, these investigations are both experimentally difficult and restricted to the limited number of compounds that are sufficiently stable (both thermally and hydrolytically) to survive the study. These phase phenomena are of considerable interest to those who are concerned with the theoretical aspects of this field.

It is worth noting that isothermal "closed loops of coexistence" have also been found in a few ternary solvent systems, such as the phenol/acetone/water system between 30 and 56.5°C [74]. Miscibility in all proportions exists in all three of the component binary systems at these temperatures, but a closed miscibility gap is found in mixtures containing all three components. Two isothermal critical points (plait points, Section 12.1.1.4) are required of such a loop, which is typically found only within a narrow range of temperatures.

5.10.4   A comparison of lower and upper consolute boundaries

It is found experimentally that the upper and the lower consolute boundaries of aqueous surfactant mixtures usually differ strikingly in their shape. Comparisons of the two kinds of gaps across a large number of systems reveals that the lower consolute boundary resembles a rounded and distorted "V", while the upper consolute boundary of aqueous zwitterionic systems is more symmetrical and resembles an upside-down "U". This subjective observation is reflected by differences in the critical compositions. The critical compositions of lower consolute boundaries are typically near 1–2%, while those of upper consolute boundaries are about 25%. They are difficult to estimate accurately since the boundary is rather flat at the critical point. In this respect the general form of the zwitterionic boundary resembles that of the upper part of the closed loops of coexistence, but the compositions spanned by these two phenomena differ somewhat.

The shape of the lower consolute boundary found in diagrams of the phosphine oxides and polyoxyethylene nonionics is characteristic of those of many surfactant–water diagrams, but this boundary in the dodecyl methyl sulfone diimine–water system (Fig. 5.18) has a particularly striking "V" shape [75]. The coexisting liquid phases in this system are both unusually dilute, and even some distance above the critical point are remarkably similar in composition (Section 10.5.4.3).

At the other extreme, the lower consolute boundary displayed by "capped nonionic" surfactants can be extremely broad. In the $C_{10}E_4NMe_2$–water system (Fig. 5.19), for example, this miscibility gap spans most of the composition range. The critical composition is not exactly known, but 20°C above the critical temperature (which is 27°C) the compositions of the coexisting phases are $c.$ 0 and 95%. Liquid crystal phases do exist in this system, but are weakly stable thermally and lie below the critical temperature of the miscibility gap.

The shape of the upper consolute boundary in zwitterionic–water systems does not vary much as a function of molecular structure. The most thoroughly studied system of this type is the nonyldimethyammoniopropyl sulfate–water system [65]. A partial phase diagram was determined, the position of the miscibility gap was determined, and the structure of the micellar liquid phase that surrounds this gap was investigated using NMR methods. These studies showed, unexpectedly, that no unusual association phenomena exist – even in very concentrated solutions. Only small spherical micelles were found.

Calorimetric behavior. The heat effects that are observed using scanning calorimetry, as one passes into liquid/liquid miscibility gaps, are relatively small. Measurable effects

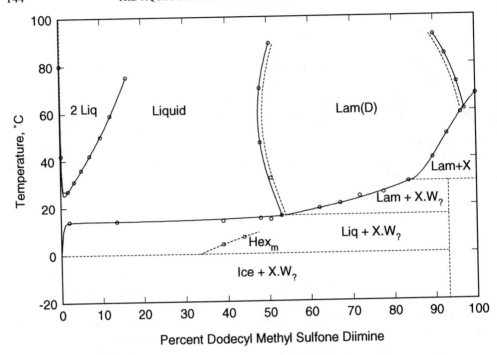

Fig. 5.18. The dodecyl methyl sulfone diimine–water system, showing the unusually narrow liquid/liquid miscibility gap.

have been observed along isopleths which span the concentrated part of the gap, but the gap may be calorimetrically "invisible" along more dilute isopleths [76]. It is intriguing to note that the same pattern is found within liquid crystal regions [77]. The numerical values of the observed heat effects along more concentrated isopleths is greater than that observed along more dilute isopleths in both instances.

5.10.5 Interference between the liquid/liquid miscibility gap and the Krafft boundary

If one views surfactant phase behavior from a broad perspective with respect to molecular structure, it appears that the various phase phenomena (miscibility gap, liquid crystal regions, Krafft boundary, etc.) move about in response to changes in structure more or less independently of one another. An early example of this phenomenon is McBain's observation, from comparing stearates and oleates, that the Krafft boundary temperature is lowered by this particular structural change (introducing a Δ9-*cis* double bond) without strongly affecting liquid crystal phenomena [78].

The position of the liquid/liquid miscibility gap may be similarly altered, relative to other phase phenomena, by suitable structural modifications. Some kinds of structural modification appear to cause this gap to "interfere" with these other phenomena, and when this occurs the phase behavior of the system is dramatically modified.

Both the upper and the lower consolute boundaries may be caused to interfere with the Krafft boundary. Interference of the upper consolute boundary with the Krafft boundary is particularly important among ammonio sulfate zwitterionic surfactants

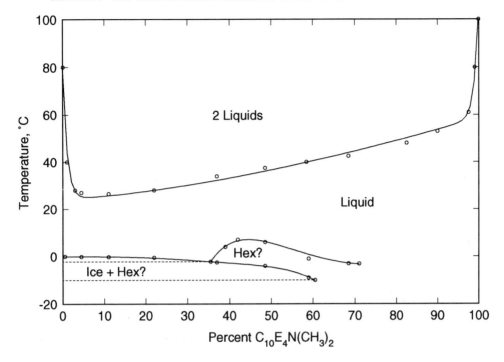

Fig. 5.19.   The dimethylamino-capped-$C_{10}E_4$ ($C_{10}E_4NMe_2$–water) system, showing the unusually broad miscibility gap and weakly stable liquid crystal phases.

(Fig. 5.20). As noted in Section 10.5.3.1, these compounds are relatively weakly hydrophilic members of this subclass of very strongly hydrophilic functional groups. A complete phase diagram of such a system has not been determined, but it is known that the liquid/liquid miscibility gap is very prominent and rises above the Krafft eutectic in these systems [65]. The coexisting phase at the solubility boundary that lies above the Krafft eutectic is not the usual liquid crystal phase, as a result (Section 5.4), but is a *liquid* phase instead. The Krafft eutectic (normally a $L + LX + X$ eutectic) is a $L + L + X$ eutectic in such cases. Liquid crystal regions still exist at high compositions.

Interference of the lower consolute boundary with the Krafft boundary is evident in the dodecyl-bis(2-cyanoethyl)phosphine oxide–water system (Fig. 5.21). Here, a change similar to that described above occurs in the phases that coexist at the Krafft eutectic (from $L + LX + X$ to $L + L + X$). This diagram differs from the zwitterionic surfactant diagrams, however, in that the liquid/liquid gap becomes broader as the temperature is increased rather than shrinking. Liquid/liquid miscibility decreases with increasing temperature in this system rather than increasing, and does not terminate at an upper critical point as in Fig. 5.20. Perhaps a closed loop would be observed if the study could be extended to sufficiently high temperatures.

5.10.6   Interference between the liquid/liquid miscibility gap and liquid crystal regions

A closed loop of miscibility exists in the $C_{10}E_4$–water system. This closed loop is qualitatively similar in overall appearance to those found in the $C_{10}E_5$–, $C_{10}PO$–,

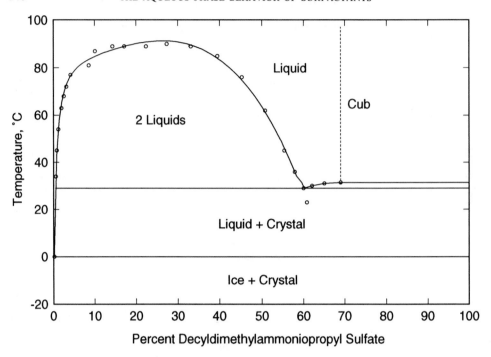

Fig. 5.20. The 3-(decyldimethylammonio)propyl sulfate–water system, showing interference of the upper consolute boundary with a Krafft boundary.

and $C_{12}PO$–water systems (Section 5.10.3), but the miscibility gap and the liquid crystal regions are widely separated by a liquid region in the latter three systems. As the hydrophilicity of these surfactants is reduced or their lipophilicity is increased, however, both the miscibility gap and the liquid crystal regions may be enlarged. Such changes may be accomplished by changing either the length of the lipophilic chain (Section 11.2) or the length of the polyoxyethylene group (Section 10.5.5.2). Eventually the miscibility gap and the liquid crystal regions interfere, and when they do the impact on phase behavior is profound. These two phase phenomena interfere with one another in the $C_{10}E_4$–water system, and this interference has two major consequences:

(1) a corridor which cuts through the miscibility gap is introduced, and
(2) a new phase state is created.

The new phase, which was termed the "anomalous" phase [72], is found over a narrow span of compositions. The span remains nearly constant as the temperature is increased, but the actual compositions increase. The composition span of the anomalous phase is far more dilute than is that of most liquid crystals; it is a very water-rich phase. The phase is turbid and bluish when observed in white light. Though apparently isotropic, it displays readily visible shear birefringence and may be regarded as a highly structured liquid. It also appears to be a delicately balanced phase in that it is sensitive to and may vanish altogether in high gravitational fields (Section 4.17.5). Dilute mixtures, which at 1 g contain this phase plus the still more dilute liquid, separate in an ultracentrifuge into the dilute liquid and the lamellar phase. The anomalous phase is

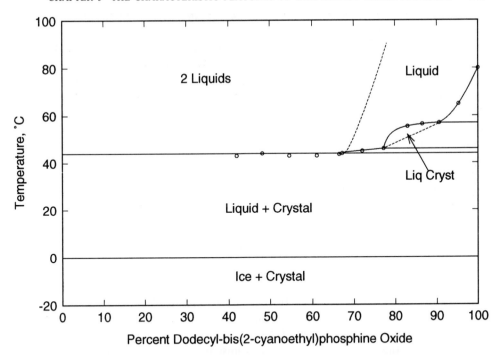

Fig. 5.21. The dodecyl-bis(2-cyanoethyl)phosphine oxide–water system, showing interference between the lower consolute and the Krafft boundary.

only weakly capable of dissolving hexadecane, as indicated by the fact that this phase region extends only a short distance into the ternary $C_{10}E_4$–C_{16}–water diagram (Fig. 12.11, Section 12.3.3) [73]. The anomalous phase is possibly related structurally to the "sponge" phase [79] or the "L_3 phase" [80] that has been found in numerous three- and four-component surfactant mixtures.

The boundaries of the anomalous phase corridor are delineated at their upper and lower limits by pairs of closely spaced three-phase discontinuities [72]. In the published diagram both are peritectics at the upper limit of the corridor while one is a eutectic and the other a peritectic at the lower limit. A small modification of this diagram is suggested by more recent studies, which require that both of the lower discontinuities be eutectics (Fig. 4.5) [47]. The phase behavior will be described below assuming the revised diagram to be correct.

In passing through the anomalous phase corridor (starting from within the low temperature liquid phase), one encounters (at varying compositions to the right of the anomalous phase) the following sequence of two-phase regions and three-phase discontinuities with increasing temperature. L is the dilute liquid, L' the liquid of intermediate composition, L'' the most concentrated liquid, An the anomalous phase, and D the lamellar liquid crystal phase):

$$L \rightarrow L + L' \rightarrow L + An + L' \rightarrow An + L' \rightarrow An + D + L' \rightarrow An + D \rightarrow$$

$$An + D + L'' \rightarrow An + L'' \rightarrow L + An + L'' \rightarrow L + L''$$

While extraordinarily complex, this behavior is consistent with the Phase Rule through-out and (except for the second discontinuity) is well documented. Recent careful studies of the phase behavior of the $C_{12}E_5$–water system [81] have shown it to be qualitatively similar to the $C_{10}E_4$–water system at temperatures near the bottom of the closed loop.

The appearance of the $C_{10}E_4$–water diagram suggests that the lower and the upper parts of the closed loop constitute elements of the same phase boundary, and that this boundary is interrupted at intermediate temperatures by intrusion of the anomalous phase corridor. Qualitatively similar behavior has been observed in isothermal com-position space in ternary systems (Section 12.3.3). In the $C_{10}E_4$–hexadecane–water system at 19°C, for example, the lamellar liquid crystal phase region appears to intrude deeply into the large principal liquid/liquid miscibility gap, and divides this gap into two smaller ones. Phenomena that are perhaps related are also seen in the C_8E_4–heptane–water system (Section 12.3.4).

5.10.7   The "pivotal phase behavior" concept

As the length of the polyoxyethylene chain is shortened from $C_{10}E_5$ to $C_{10}E_4$ the relatively simple phase behavior observed in $C_{10}E_5$ becomes very complex. What, one may ask, are the consequences of further shortening the length of the oxyethylene group? While this question cannot be answered with assurance, partial information regarding the phase behavior of the $C_{10}E_3$–water system is shown in Fig. 5.22. These data show that this surfactant is much less miscible with liquid water than is $C_{10}E_4$,

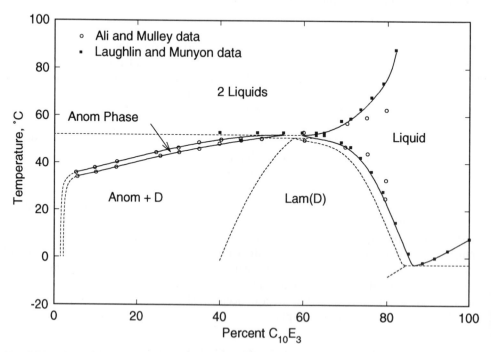

Fig. 5.22.   A preliminary diagram of the $C_{10}E_3$–water system, incorporating data from Ali and Mulley [82] and from Laughlin and Munyon [83]. The qualitative form of the diagram has been established by use of DIT studies, but uncertainties remain as to quantitative aspects.

since the lower part of the lower consolute boundary and the extensive micellar solution region below it no longer exist. The very dilute (nearly pure water) liquid phase and the concentrated liquid (which is a solution of water in the surfactant) remain. In addition, a discrete isotropic phase region is found whose span of composition is very narrow and varies significantly with temperature. This phase is found within the composition interval between the dilute liquid and liquid crystal regions. It extends from below 0°C to 51.1°C, which is 1.3°C above the upper temperature limit of the lamellar liquid crystal phase in this system. (The lamellar phase is also a "squeezed" phase (Section 5.8) which decomposes at its upper limit at a peritectic phase transition rather than at an azeotropic transition.) The form of the narrow isotropic phase region is strikingly similar to that of the anomalous phase in $C_{10}E_4$–water, but is shifted to higher compositions and extends to lower temperatures than the anomalous phase of $C_{10}E_4$.

Decreasing the number of oxyethylene groups in an AE_x surfactant reduces its hydrophilicity (Section 10.5.5.2), and in passing from $C_{10}E_4$ to $C_{10}E_3$ a significant reduction in miscibility with liquid water occurs. This is evident from the fact that the primary miscibility gap extends to below the temperature at which ice forms in the $C_{10}E_3$–water system, while it vanishes at a critical point at 19°C in $C_{10}E_4$–water.

In many surfactants having very weak hydrophilic groups, one expects to encounter the kind of phase behavior seen in monoglyceride–water systems. Monoglyceride phase behavior differs qualitatively from that of the $C_{10}E_4$–water system in two important respects. First, the lower consolute boundary and remnants of the micellar liquid region below this boundary (that are visible in the $C_{10}E_4$–water diagram) are absent in monoglyceride–water diagrams, just as they are in the $C_{10}E_3$–water diagram. Second, the anomalous liquid phase found in $C_{10}E_4$–water (and presumably also in $C_{10}E_3$–water) does not exist. Monoglyceride-like phase behavior may be expected to appear in C_nE_x–water systems having sufficiently short E_x groups, but has not been well documented.

Comparing phase behavior in the series $C_{10}E_3$, $C_{10}E_4$, and $C_{10}E_5$ reveals that the $C_{10}E_4$ structure occupies a focal or "pivotal" position in this series, in that interference between miscibility gap and liquid crystal regions as the number of oxyethylene groups is reduced first appears in this structure. Thus, the phase behavior of $C_{10}E_4$ (and of $C_{10}E_3$) is not only complex; it would appear to be delicately balanced since adding one oxyethylene group to the E_4 structure (to form the E_5 surfactant) dramatically alters the aqueous phase behavior. Removing one (to form the E_3 surfactant) significantly reduces the miscibility of the surfactant with water, while (probably) retaining the anomalous phase.

The addition of more oxyethylene groups to $C_{10}E_5$ does not qualitatively change the phase behavior from that of $C_{10}E_5$ insofar as interference between the miscibility gap and liquid crystal regions is concerned. It does alter the structures of the liquid crystal phases that exist, however, in that the stability of the hexagonal phase is promoted and that of the lamellar phase is reduced by additional oxyethylene groups. Eventually the lamellar phase is eliminated (Section 10.5.5.2).

HLB and the various types of phase behavior. It is useful to invoke the concept of "HLB" (the balance between hydrophilicity and lipophilicity, Section 11.4.2), as a basis for qualitatively correlating phase behavior with molecular structure in the above series of surfactants. All the above kinds of phase behavior are often found in many families of weakly hydrophilic surfactants within which either the hydrophilicity or the lipophilicity may be varied in a systematic manner; it is *not* restricted to the above compounds.

A broadly encountered sequence of phase behavior exists within these families, which can be described as follows:

low HLB	(not found in $C_{10}E_x$ series)	(monoglyceride-like)
intermediate HLB	$C_{10}E_4-$ and $C_{10}E_3-$water	(pivotal)
high HLB	$C_{10}E_5-$ and $C_{10}E_6-$water	(soluble surfactant)

In the $C_{10}E_y$ series $C_{10}E_4$ may be regarded as the "pivotal" surfactant, for the reasons described above. In the $C_{12}E_y$ series $C_{12}E_5$ is clearly the pivotal surfactant, in that $C_{12}E_6$ is a highly soluble surfactant with widely separated miscibility gap and liquid crystal regions while $C_{12}E_4$ and $C_{12}E_3$ appear to resemble $C_{10}E_3$ [80]. For each homolog lengthening the hydrophilic chain produces a soluble surfactant and destroys pivotal phase behavior, but pivotal behavior can be reintroduced by holding the hydrophilic chain constant and lengthening the lipophilic chain.

In alkyldimethylphosphine oxide surfactants the C_{14} homolog is clearly the pivotal (intermediate HLB) surfactant, while the C_{12} homolog displays soluble surfactant (high HLB) behavior and the C_{16} homolog displays monoglyceride-like (low HLB) behavior. This pattern is seen even more clearly among diethylalkylphosphine oxide homologs over the chain-length span of C_{12} to C_{16}, because the Krafft eutectic temperature is lower. Within the phosphine oxides HLB is altered by varying the lipophilic group rather than by changing the hydrophilic group (Section 11.4.2).

Numerous instances of pivotal behavior have been encountered among single-bond surfactants in addition to those mentioned above. Once the pivotal structure is identified, the phase behavior of structurally related surfactants may be shifted in either direction in a predictable manner by altering either the hydrophilicity or the lipophilicity of the compound.

5.11 Summary

When considering the influence of temperature on surfactant phase behavior, the melting point of the dry surfactant is one particularly important point of reference and the temperature of the Krafft eutectic is the other. At sufficiently high temperatures all surfactants (ionic, nonionic) are miscible with water in all proportions. Below the Krafft eutectic temperature liquid crystal phases do not exist as equilibrium states.

At temperatures below the knee in the Krafft boundary, surfactants and water are very poorly miscible. (In many surfactants the freezing of water intervenes so that the Krafft boundary does not exist as an equilibrium phase boundary, but even in these cases a comparable lower limit to miscibility is still evident.) As the temperature is increased from within this low temperature region towards the Krafft eutectic, a sudden increase in miscibility occurs at the knee and within the narrow temperature span of the Krafft plateau. The micellar liquid phase and crystal phases coexist only within this narrow transition temperature zone.

The existence of liquid crystal regions at higher compositions and at temperatures above the Krafft eutectic has a profound effect on the phase changes that occur along both isothermal and isoplethal process paths which traverse this region. Along isothermal paths (such as occur when the dry surfactant is mixed with water, or when water is evaporated from an aqueous mixture), alternating sequences of stoichiometric phase reactions and dilution/concentration processes occur. Along isoplethal paths (as when mixtures of defined composition are heated or cooled), a series of discontinuous change-of-state phase reactions alternate with continuous variations of state. These phase discontinuities are often encountered regardless of whether or not the path actually passes through any liquid crystal phase region (Section 5.6.11). Such phenomena strongly influence the physical properties of mixtures, and are extremely influential with respect to engineering processes.

From consideration of those phase reactions which occur at the upper and lower temperature limits of liquid crystal regions, and from analysis of the forms of liquid crystal regions, it may be suggested that the principal factors which govern the dependence of surfactant phase behavior on system variables are:

(1) the enthalpies of crystal phases,
(2) the numerical values of water free energies (which depend on both the thermodynamic activity of water, or relative humidity, and the water composition), and
(3) the ambient level of thermal energy (which is dictated by the temperature).

The upper temperature limit of liquid crystal phases depends strongly on the molecular structure of the surfactant and varies widely. Because both liquid crystal and liquid phases are states of relatively high entropy, the temperatures at which the one is transformed reversibly to the other may be considerably higher than is the melting point of the dry surfactant. In rare instances, a liquid phase is transformed isothermally by the addition of heat into a liquid crystal phase (Section 5.6.4).

Liquid/liquid miscibility gaps often exist, particularly in weakly hydrophilic surfactants. The lower consolute (cloud point) boundary, when it exists, is probably the lower part of a closed loop of coexistence. This phenomenon may be treated separately from the crystal and liquid crystal aspects of surfactant phase behavior, and may interfere with these features in some instances. If the molecular structure of the surfactant is adjusted so as to cause the liquid/liquid miscibility gap and liquid crystal regions to interfere, considerable additional complexity in phase behavior is introduced and new phases may result.

An upper consolute boundary is observed in selected zwitterionic surfactant–water systems. The existence of this boundary is not related to unusual association within the liquid phase.

References

1. Laughlin, R. G., Munyon, R. L., Fu, Y.-C. and Fehl, A. J. (1990). *J. Phys. Chem.* **94**, 2546–2552.
2. Cohn, E. J. and Edsall, J. T. (1943). *Proteins, Amino Acids, and Peptides as Ions and Dipolar Ions*, p. 75, Hafner, New York; Willstaetter, R. and Kahn, W. (1904). *Ber. deut. chem. Ges.* **37**, 401–417; Willstaetter, R. (1902). *Ber. deut. chem. Ges.* **35**, 594–584.
3. C. R. Degenhart, unreported work.
4. Sahyun, M. R. V. and Cram, D. J. (1963). *J. Am. Chem. Soc.* **85**, 1263–1268.
5. Hays, H. R. and Laughlin, R. G. (1967). *J. Org. Chem.* **32**, 1060–1063.
6. Kurz, J. L. (1962). *J. Phys. Chem.* **66**, 2239–2245.
7. Bravo, C., Hervés, P., Leis, J. R. and Peña, M. E. (1990). *J. Phys. Chem.* **94**, 8816–8820; Garcia-Rio, L., Leis, J. R., Peña, M. E. and Iglesias, E. (1992). *J. Phys. Chem.* **96**, 7820–7823.

8. Kekicheff, P., Grabielle-Madelmont, C. and Ollivon, M. (1989). *J. Colloid Int. Sci.* **131**, 112–132.
9. Dorset, D. L. and Pangborn, W. A. (1982). *Chem. Phys. Lipids* **30**, 1–15; Klenk, E. and Debuch, H. (1963). *Progr. Chem. Fats Lipids*, Vol. 6 (R. T. Holman, W. O. Lundberg and T. Malkin eds), pp. 1–29, Pergamon Press, New York; Hermetter, A. (1988). *Comments Mol. Cell. Biophys.* **5**, 133–149.
10. (1965). *Solubilities*, 4th ed., Vol. II, pp. 1121–1122, American Chemical Society, Washington, DC.
11. Broome, F. K. and Harwood, H. J. (1950). *J. Am. Chem. Soc.* **72**, 3257–3260.
12. Laughlin, R. G. and Yellin, W. (1967). *J. Am. Chem. Soc.* **89**, 2435–2443.
13. Krafft, F. (1899). *Ber. deut. chem. Ges.* **32**, 1596–1608.
14. Hartley, G. S. (1936). *Aqueous Solutions of Paraffin Chain Salts*, p. 38, Hermann, Paris.
15. Shinoda, K. (1967). *Solvent Properties of Surfactant Solutions*, Vol. 2 (K. Shinoda ed.), pp. 12–16, Marcel Dekker, New York.
16. Personal communication, R. C. Mast.
17. Schwartz, R. and Strnad, J. (1987). *Tenside, Surfactants, Deterg.* **24**, 143–145.
18. Oonk, H. A. J. (1981). *Phase Theory: The Thermodynamics of Heterogeneous Equilibria*, pp. 165–188, Elsevier Scientific, New York.
19. Timms, R. E. (1984). *Prog. Lip. Res.* **23**, 1–38.
20. Mukerjee, P., Mysels, K. J. and Dulin, C. I. (1958). *J. Phys. Chem.* **62**, 1390–1396.
21. Corkill, J. M. and Goodman, J. F. (1969). *Advances in Colloid and Interface Science*, Vol. 2, pp. 297–330, Elsevier, Amsterdam.
22. Barnhurst, J. D. (1961). *J. Org. Chem.* **26**, 4520–4522.
23. Clunie, J. S., Goodman, J. F. and Symons, P. C. (1969). *Trans. Faraday Soc.* **65**, 287–296.
24. Krafft, F. (1899). *Ber. der Deut. Chem. Gesell.* **32**, 1596–1608.
25. Vold, R. D. (1939). *J. Phys. Chem.* **43**, 1213–1231.
26. Haase, R. and Schoenert, H. (1969). *Solid–Liquid Equilibrium*, (E. S. Halberstadt, translator ed.), pp. 81–87, Pergamon Press, Oxford.
27. Fu, Y. C., Glardon, A. S. and Laughlin, R. G. (1993). Presented at the American Oil Chemists' Society Meeting, Anaheim, California, April 25–29.
28. Small, D. M. (1986). *The Physical Chemistry of Lipids. From Alkanes to Phospholipids*, Vol. 4, Plenum, New York; Ferguson, R. H., Rosevear, F. B. and Stillman, R. C. (1943). *Ind. Eng. Chem.* **35**, 1005–1012.
29. Timmermans, J. (1938). *J. chim. Phys.* **35**, 331–344.
30. Seiden, P., Lutton, E. S., Sanders, R. A. and Laughlin, R. G. (1990). *8th International Symposium on Surfactants in Solution*, June 10–15, Gainesville, FL.
31. Laughlin, R. G. and Munyon, R. L. (1991). Presented at the International Colloid and Surface Science Symposium, Compiegne, France, July 7–13, 1990.
32. Lutton, E. S., Stewart, C. B. and Fehl, A. J. (1970). *J. Am. Oil Chem. Soc.* **47**, 94–99.
33. Jeffrey, G. A. and Maluszynska, H. (1989). *Acta Cryst.* **B45**, 447–452.
34. Saupe, A. (1974). *Liquid Crystals and Plastic Crystals*, Vol. 1 (G. W. Gray and P. A. Winsor eds), pp. 18–47, Ellis Horwood, Chichester.
35. Rawlings, F. F. and Lingafelter, E. C. (1955). *J. Am. Chem. Soc.* **77**, 870–872.
36. (1988). *Crystallization and Polymorphism of Fats and Fatty Acids*, Vol. 31 (Garti, N. and Sato, Kiyotaka eds), Marcel Dekker, New York.
37. Wireko, F. C. and Mootz, M. R. (1992). Presented at the 41st Annual Denver Conference on Application of X-ray Analysis, Colorado Springs, Colorado, August 3–7.
38. (1953). *The Van Nostrand Chemist's Dictionary*, (J. M. Honig, M. B. Jacobs, S. Z. Lewin, W. R. Minrath and G. Murphy eds), p. 550, Van Nostrand, New York.
39. Chernik, G. G. (1991). *J. Colloid Interface Sci.* **141**, 400–408.
40. Findlay, A. (and Campbell, A. N.) (1938). *The Phase Rule and its Applications*, 8th ed., pp. 114–118, Dover, New York.
41. Rose, J. (1961). *Dynamic Physical Chemistry*, John Wiley, New York.
42. Herrmann, K. W., Brushmiller, J. G. and Courchene, W. L. (1966). *J. Phys. Chem.* **70**, 2909–2918.
43. Blum, F. D. and Miller, W. G. (1982). *J. Phys. Chem.* **86**, 1729–1734.

44. Broome, F. K., Hoerr, C. W. and Harwood, H. J. (1951). *J. Am. Chem. Soc.* **73**, 3350–3354.
45. K. Fontell, personal communication.
46. Rizzatti, M. R. and Gault, J. D. (1986). *J. Colloid Interface Sci.* **110**, 258–262; Sammon, M. J., Zasadzinski, J. A. N. and Kuzma, M. R. (1986). *Phys. Rev. Lett.* **57**, 2834–2837.
47. Laughlin, R. G. (1990). *Food Emulsions and Foams: Theory and Practice*, No. 277, Vol. 86 (P. J. Wan, J. L. Cavallo, F. Z. Saleeb and M. J. McCarthy eds., E. L. Gaden, Series ed.), pp. 7–15, American Institute of Chemical Engineers, New York; R. Strey, unreported results.
48. Chernik, G. G. and Sokolova, E. P. (1991). *J. Colloid Interface Sci.* **141**, 409–414.
49. Ekwall, P. (1975). *Advances in Liquid Crystals*, Vol. 1 (G. H. Brown ed.), pp. 1–142, Academic Press, New York.
50. Hendrixx, Y., Charvolin, J., Kekicheff, P. and Roth, M. (1987). *Liq. Cryst.* **2**, 677–687.
51. Laughlin, R. G. (1976). *J. Colloid Interface Sci.* **55**, 239–241.
52. Lutton, E. S. (1965). *J. Am. Oil Chemists Soc.* **42**, 1068–1070.
53. Ladbrooke, B. D. and Chapman, D. (1969). *Chem. Phys. Lipids* **3**, 304–356.
54. Rand, R. P., Fuller, N. L. and Lis, L. J. (1979). *Nature (London)* **279**, 258–260.
55. Madelmont, C. and Perron, R. (1976). *Colloid and Polymer Sci.* **254**, 6581–6595.
56. Grabielle-Madelmont, C. and Perron, R. (1983). *J. Colloid Interface Sci.* **95**, 471–482.
57. Marcott, C., Laughlin, R. G., Sommer, A. J. and Katon, J. E. (1991). *Fourier Transform Infrared Spectroscopy in Colloid and Interface Science, ACS Symposium Series*, Vol. 447 (D. R. Scheuing ed.), pp. 71–86, American Chemical Society, Washington, DC.
58. Rawlings, F. F. and Lingafelter, E. C. (1955). *J. Am. Chem. Soc.* **77**, 870–872.
59. Laughlin, R. G., Munyon, R. L. and Fu, Y.-C. (1991). *J. Phys. Chem.* **95**, 3852–3856.
60. Henriksson, U., Blackmore, E. S., Tiddy, G. J. T. and Soderman, O. (1992). *J. Phys. Chem.* **96**, 3894–3901.
61. Maroncelli, M., Strauss, H. L. and Snyder, R. G. (1985). *J. Phys. Chem.* **89**, 5260–5267.
62. Fontell, K., Ceglie, A., Lindman, B. and Ninham, B. (1986). *Acta Chem. Scand.* **A40**, 247–256; Chen, S., Evans, D. F. and Ninham, B. W. (1984). *J. Phys. Chem.* **88**, 1631–1634.
63. Maroncelli, M., Song, P. Q., Strauss, H. L. and Snyder, R. G. (1982). *J. Am. Chem. Soc.* **104**, 6237–6247.
64. Berne, B. J. and Pecora, R. (1976). *Dynamic Light Scattering*, pp. 257–261, John Wiley, New York; Kunieda, H. and Friberg, S. (1981). *Bull. Chem. Soc. Japan* **54**, 1010–1014.
65. Nilsson, P-G., Lindman, B. and Laughlin, R. G. (1984). *J. Phys. Chem.* **88**, 6357–6362.
66. Lawrence, A. S. C. and McDonald, M. P. (1967). *Liquid Crystals*, (G. H. Brown, G. J. Dienes and M. M. Labes eds), pp. 1–19, Gordon & Breach, New York.
67. (1990). *Analytical Supercritical Fluid Chromatography and Extraction*, (Lee, M. L. and Markides, K. E. eds), Chromatography Conferences, Inc., Provo, Utah.
68. Latta, S. (1990). *Inform* **1**, 810–816.
69. Shaw, R. W., Brill, T. B., Clifford, A. A., Eckert, C. A. and Franck, E. U. (1991). *Chem. & Eng. News* **69**, 26–39.
70. Francis, A. W. (1961). *Critical Solution Temperatures*, Vol. 31 (R. F. Gould ed.), p. 48, American Chemical Society, Washington, DC.
71. Francis, A. W. (1963). *Liquid–Liquid Equilibriums*, pp. 72–76, Interscience, John Wiley, New York.
72. Lang, J. C. and Morgan, R. D. (1980). *J. Chem. Phys.* **73**, 5849–5861.
73. Lang, J. C. (1985). *Proc. Intl. Sch. Phys. "Enrico Fermi", (Phys. Amphiphiles)*, Vol. 90, pp. 336–375.
74. Reference 71, pp. 85–90.
75. Laughlin, R. G. (1978). *Advances in Liquid Crystals*, Vol. 3 (G. H. Brown ed.), pp. 99–148, Academic Press, New York.
76. G. G. Chernik, personal communication.
77. Filippov, V. K. and Chernik, G. G. (1986). *Thermochimica Acta* **101**, 65–75.
78. Laing, M. E. and McBain, J. W. (1920). *Trans. Chem. Soc.* **117**, 1506–1528.
79. Olsson, U., Wuerz, U. and Strey, R. (1993). *J. Phys. Chem.* **97**, 4535–4539; Wennerstrom, H. and Olsson, U. (1993). *Langmuir* **9**, 365–368; Anderson, D., Wennerstrom, H. and Olsson, U. (1989). *J. Phys. Chem.* **93**, 4243–4253.

80. Mitchell, D. J., Tiddy, G. J. T., Waring, L., Bostock, T. A. and McDonald, M. P. (1983). *J. Chem. Soc., Faraday Trans. I* **79**, 975–1000.
81. Strey, R., Schomäker, R., Roux, D., Mallet, F. and Olsson, U. (1990). *J. Chem. Soc. Faraday Trans.* **86**, 2253–2261.
82. Ali, A. A. and Mulley, B. A. (1978). *J. Pharm. Pharmacol.* **30**, 205–213.
83. R. L. Munyon, unreported work.
84. Ockelford, J., Tminini, B. A., Narayan, K. S. and Tiddy, G. J. T. (1993). *J. Phys. Chem.* **97**, 6767–6769.

Chapter 6

The Kinetic and Mechanistic Aspects of
Surfactant Phase Behavior

6.1 General principles

The consequences of following isoplethal and isothermal process trajectories on the equilibrium state of aqueous surfactant systems were described in Chapter 5. In order to apply the principles of thermodynamics to the analysis of phase behavior, it is necessary to presume that equilibrium was maintained throughout each process during this analysis. All real processes are irreversible [1], however, so that equilibrium is *not* maintained (Section 5.6.11) and (strictly speaking) $q \neq \Delta H$ (Section 3.21) for the various phase reactions encountered. Time is always required for phase transformations to reach equilibrium, but the time required varies widely (from a fraction of a second to years) [2]. It has repeatedly been emphasized that it is extremely important conceptually to distinguish equilibrium phenomena from kinetic phenomena, because serious mistakes will result if a metastable state is taken to be an equilibrium state, or vice versa.

Knowledge of the equilibrium state is useful regardless of whether phase reactions are fast or slow [2]. Prior knowledge of the equilibrium phase diagram is required to interpret the rates and infer the mechanisms of phase reactions, and this is particularly true of slow reactions. Systems that attain equilibrium rapidly are nearly always in a state of equilibrium, and for these systems kinetic factors are important only within short time scales. Also, it is relatively easy to determine the equilibrium state of systems that attain equilibrium quickly.

Mixtures in systems that attain equilibrium of state slowly rarely exist in a state of equilibrium. Knowledge of rate processes over long time scales is particularly important in these systems, and the problems encountered in determining the equilibrium behavior of such systems are considerably more severe. It is very likely that most of the present uncertainties in the physical science of surfactants stem either from lack of knowledge of the true equilibrium state, from the mistaken identification of a metastable state as an equilibrium state, or from the mistaken identification of colloidally structured biphasic mixtures as homogeneous phases.

The relevance of equilibrium states to kinetics is evident from the fact that the magnitude of the deviation from equilibrium is an important parameter with respect to the rates of phase processes. A well-known illustration of this principle is found in the laws governing the growth of crystals from solution; the value of the supersaturation (C/C_{eq}) enters into the equations which describe rates of crystal growth [3,4]. The concentration *per se* is perhaps not so important as is the ratio of the actual concentration to the equilibrium concentration.

Considerable information exists regarding various dynamic aspects of liquid solutions of surfactants, such as self-diffusion coefficients and the dynamics of exchange between the micellar and molecular states of aggregation [5,6,7,8]. Far less quantitative information exists regarding the kinetics and mechanisms of their phase reactions. A few T-jump studies of isoplethal transformations from one phase state to another have been reported [9], but one must presently rely largely on qualitative information from casual observations to discern the kinetic aspects of phase behavior.

We are concerned with both reversible and irreversible processes during which a change of state occurs along isothermal and isoplethal paths. Characteristic changes in free energy, enthalpy, and entropy accompany these processes, but in considering rates it is particularly important to focus on the change that occurs in the entropy.

6.2 The role of heat and mass transport: the engineering factors

Some phase transformations are accompanied by changes in phase composition, while others are not. Except for isothermal mixing in ideal systems, heat transport to or from the environment may also be required during these processes. Mass transport within the mixture is required when phase compositions change (Section 5.6.11), but not otherwise. Both mass and heat transport may occur either by conductive or by convective processes, which are governed by factors such as viscosities, transport coefficients, mechanical shear rates, particle sizes and surface areas, etc. [3]. These factors are always present, they are treated in chemical engineering texts [10], and they are very important to the kinetics of all phase reactions. This is true both for laboratory-scale phase reactions and for engineering-scale industrial processes.

In a great many instances these "engineering factors" are rate-determining insofar as phase reactions are concerned, but in others intrinsic rate-limiting factors exist which reduce the rate of the process below that allowed by the engineering factors. It is these intrinsic rate-limiting factors with which we will be primarily concerned.

Mass transport during isothermal processes. It is obvious that mass transport is required during isothermal mixing processes. In dissolving salt in water, the salt must somehow move from within the crystal into the space occupied by the liquid phase, and the liquid must occupy the space originally occupied by the salt crystal. During the reaction of a crystal with water to form a crystal hydrate, however, the initial reaction may only occur at the crystal surface. Diffusion of water through a layer of reaction product formed at the surface is required for the reaction to proceed beyond the first molecular layer. Diffusion coefficients governing transport through molecular crystals are smaller than are diffusion coefficients through fluid phases by a factor of roughly 10^7 [11], so that transport processes which require seconds in a fluid phase may require a year's time in a crystal phase. The rough kinetic data on the rate of formation of the crystal dihydrate $X \cdot W_2$ of DODMAC from its monohydrate $X \cdot W$ are consistent with this premise [2].

Mass transport during isoplethal processes. It is less clearly evident, but nevertheless true, that mass transport is required during most isoplethal processes. This is because phase ratios change during such processes if the process passes through biphasic regions. The changes in phase structures, compositions, and ratios that occur along a concentrated isoplethal path in a prototypical aqueous surfactant system are described in Section 5.6.11. An analysis of isothermal mixing processes is presented in Section 5.6.7.

Heating or cooling a mixture without changing its composition is probably the most common process performed in all of science. It is extremely important to recognize that "simply" changing the temperature of a mixture usually creates a driving force not only for change of state – but also for both heat and mass transport. The consequences of changing the temperature are, in general, far more profound and complex than are those which result from changing composition.

6.3 Rates of phase transformations along isoplethal paths

An intrinsic kinetic barrier to a phase transformation can be expected to exist when the entropy change during the process is both negative and large, which signifies that the process is highly improbable. Otherwise, phase reaction rates are dictated by the engineering factors.

The phase state of lowest entropy is the crystal. The entropies of liquid crystal and liquid states are much larger, and the differences among these states are relatively small. From the fact that the liquid phase usually exists at higher temperatures than does the liquid crystal, it is apparent that liquids typically do have a somewhat higher entropy and enthalpy than do liquid crystals of the same composition. Exceptions to this rule do exist (Section 5.6.4), and on rare occasions (as in the polyoxyethylene glycol alkylphenyl monoethers), the liquid crystal "bellies out" so that within a narrow span of compositions the liquid crystal lies above the liquid phase. This is not usual, however.

It is to be expected from these considerations that the particular phase transformation most likely to be kinetically slow is the formation of a crystal phase from any disordered or fluid phase.

$$\text{crystal} \underset{\text{slow}}{\overset{\text{fast}}{\rightleftharpoons}} \text{fluid} \tag{6.1}$$

The reverse process is *not* likely to display intrinsically slow rates. The rates of phase transformations among fluid phases are expected to be governed by the engineering factors. All these expectations are consistent with the available information.

6.3.1 The formation of a crystal from a fluid state

During the determination of phase diagrams using isoplethal methods (Appendix 4.4.3), the temperature of the Krafft boundary may only be inferred from observations made while the mixture is being heated. Supercooling is inevitably observed when the liquid phase is cooled from above the Krafft boundary temperature to below it. That is, the temperature must be well below the equilibrium boundary temperature before crystals are formed at an observable rate [12]. The magnitude of supercooling required to induce crystal formation varies widely with both surfactant structure and composition. It tends to be greater in concentrated solutions, near the upper composition limit of the Krafft plateau, than in dilute solutions.

If the rate of crystal formation is sufficiently slow, a supercooled glassy state results from cooling concentrated liquid phases to below the Krafft plateau. If the isopleth passes through a metastable liquid crystal region, however, the metastable liquid crystal is quickly formed; this pattern of behavior was well documented during the N-dodecanoyl-N-methylglucamine study [13]. Once crystals are formed, they quickly redissolve when the mixture is heated to the equilibrium boundary temperature (if they are sufficiently small).

The phenomenon of "oiling out", which is widely encountered by organic chemists utilizing recrystallization as a means of purification, also reflects the kinetic difficulty with which crystal phases form. During oiling-out the solute separates from the hot solution on cooling as a metastable liquid phase, rather than as the equilibrium crystal phase. "Oiled-out" phases are usually transformed into crystals by waiting for a period of time, by allowing solvent to evaporate, by scratching the flask with a sharp-edged stirring rod, or by adding seed crystals of the same (or of a different but related) compound. The widespread existence of this phenomenon emphasizes that separation of the solute phase from a solution upon cooling does not depend upon forming the crystal phase structure. The miscibility is determined by fundamental factors which do not depend critically upon whether the separating phase is liquid or crystalline in structure.

A similar result is obtained when a liquid crystal is cooled past its lower boundary. The biphasic region below liquid crystal regions is always a liquid crystal plus crystal region (Section 5.6.2), and supercooling inevitably occurs on passing into this biphasic region by cooling. In striking contrast, crystals quickly disappear (at a rate governed by the engineering factors) when biphasic mixtures of liquid crystal and crystal are heated. If viscosities are high then convective mixing and transport processes may be retarded or eliminated, and in these cases the absolute rates of such processes may be very slow – even though no intrinsic kinetic barrier exists.

The rate of formation of crystals from liquid or liquid crystal states by cooling is not only slow, but highly variable. This is to be expected when it is recognized that crystal formation probably occurs by a nucleation and growth mechanism [3]. If nucleation sites exist the rate of growth is reproducible and can be more easily studied. If nucleation sites do not exist, their formation is required before growth can commence and the observed rate of crystal growth is more erratic.

6.3.2   The formation of a liquid or liquid crystal from another state

The separation of a liquid crystal phase from a liquid (e.g. by cooling along isoplethal paths) does not require an easily observable span of time, and perceptible supercooling is not observed. This is true of all the major classes of liquid crystal phases – hexagonal, lamellar, and cubic. This is *not* to say that these processes are instantaneous. A change from a liquid to a liquid crystal entails profound changes in all aspects of state, and requires a finite time. Nevertheless, the actual time required for these particular transformations is too short to be perceived by direct observation.

The separation of a liquid phase from another liquid ("clouding") is also too fast to be directly observed. This phenomenon is not only fast but is readily reversible – provided the phase which separates remains dispersed in the other. If settling occurs, assistance with mass transport (from mixing) is required to quickly re-establish equilibrium.

The formation of fluid states from crystals – whether by isothermal mixing or by temperature variation – also appear to be "fast". No barriers to equilibration appear to exist other than the engineering factors.

A topic of considerable importance to modern polymer science is that of "spinodal decomposition", which is concerned with the kinetics of phase changes. "Spinodal" lines are presumed to exist within the miscibility gaps in a phase diagram, and may be understood by reference to the free energy of mixing curve for mixtures that display partial miscibility (Section 3.14, Fig. 3.5). In Section 3.14 the focus of the discussion was on the actual phase boundaries of the coexisting phases (compositions a and b in Fig. 3.5), which are sometimes termed the "binodal" points on the curve. Within the miscibility gap inflections exist at two compositions along the curve; these inflections are the "spinodal points" at the temperature of the curve. The locus of these spinodal points as a function of temperature (or pressure) define the spinodal curve. This curve lies within the miscibility gap, and is significant with respect to the rates at which metastable phases lying within the miscibility gap attain equilibrium.

It has been recognized for some time that metastable phases which lie inside the spinodal curve reach equilibrium of state significantly slower than do metastable phases which lie between the spinodal curve and the binodal curve [14]. While it is highly probable that these concepts apply to aqueous surfactant systems, investigations of the "spinodal decomposition" of metastable surfactant phases have not so far been reported.

6.3.3 T-jump experiments from one state to another

The above qualitative observations are consistent with the results of T-jump experiments during which one phase state is transformed to another of the same composition [9]. The temperature of these mixtures necessarily passed through a biphasic region during the jump process, but the rate of change in temperature may have been sufficiently high that disproportionation did not have time to occur.

The following changes of state were investigated in this way:

Hexagonal → liquid (L_1) in a $C_{12}E_{16}$–NaCl–water mixture,
Liquid (L_1) → lamellar phase in a $C_{12}E_6$–NaCl–water mixture, and
Bicontinuous cubic → liquid (L_1) in a $C_{12}E_6$–CsCl–water mixture.

L_1 is the normal micellar solution phase (Section 12.4.1).

The time constant governing the rate of return to equilibrium for all of these processes was a fraction of a second, and the time for completion of the transformations ranged from 0.5 to 3 s. While they are in no sense instantaneous, they would not be readily observable by casual observation.

Changes of state in polar lipid and monoglyceride systems have been investigated by T-jump and P-jump experiments using synchrotron X-ray data [15].

6.4   The rates of phase transformations along isothermal paths

An alternating sequence of stoichiometric phase reactions and dilution processes occurs along isothermal mixing paths (Section 5.6.7). Engineering factors are dominant with respect to the kinetics of all of these processes, because intrinsic kinetic barriers are virtually nonexistent. This is a particularly important attribute of the DIT phase studies method, during which the phases that exist are produced by just such processes. Being isothermal, the DIT method sidesteps the numerous kinetic problems which result from the change of temperature that occurs during isoplethal studies. These include the formation of metastable (and often long-lived) phase states, the mass and heat transport required to maintain equilibrium, and (occasionally) the formation of colloidally structured mixtures.

The influence of engineering factors can be illustrated using the stoichiometric phase reaction between water and salt crystals. This reaction is strongly influenced by the specific surface area – and therefore the size – of the crystals. Small crystals (which have a large surface area) dissolve quickly, while large ones dissolve slowly. A common example of the importance of this factor is seen in comparing confectioner's sugar with granular sugar [16]. The very real differences between these two materials, as regards their interactions with aqueous and oily liquids, must stem in part from the difference in particle size and specific surface area (area/mass).

The interface as the site of reaction. It is generally recognized that the primary interactions between two phases only occur at interfaces. During the dissolution of salt, the water strips the surface layer of molecules away thus exposing the next layer. The deposition of salt from solution at the surface is occurring at the same time, if not at the same site. If the liquid is undersaturated the dissolution process proceeds at a faster rate than does deposition until equilibrium is attained – at which point the rate of dissolution equals the rate of deposition. If the liquid is supersaturated deposition occurs faster than does dissolution, until the forward and reverse rates are again the same.

It has been shown that a gradient in concentration develops within the liquid phase surrounding a dissolving salt crystal [17]. This is easily understandable since the dissolution process will quickly saturate a thin layer of liquid next to the crystal, and before more salt can dissolve the salt in this saturated layer must be transported away from the surface. The mathematical form of such transport is consistent with the model that transport occurs by a diffusive mechanism within an "unstirred boundary layer" next to the crystal, and by convection outside this layer [18].

Transport processes that occur within phases which contain both components are most easily observed and measured during DIT studies (Section 4.18.1). At the time the experiment is initiated, the concentration profile along the DIT capillary is a step-function. If the components are miscible, this profile relaxes as a result of diffusive transport. If the components are miscible in all proportions a smooth profile results, while if miscibility gaps exist discontinuities in composition develop that are visible as interfaces and divide the contents of the cell into bands.

The length of the bands, the position of interfaces within the cell, and the concentration profiles within the bands are governed by rates of transport. In the case of CTAB these parameters have been analyzed assuming that diffusion coefficients are independent of composition (which is incorrect) [19]. A more rigorous transport model has been developed [20], but it has not been possible to utilize this model because composition profile data have not been available.

Exploratory studies spanning a wide range of surfactant structures suggest that the variation in transport rates among these molecules is not very large. Short chain surfactants, such as $C_{10}E_4$, display relatively fast interfacial movement so that interfaces may move to the end of the cell and disappear within 24 hours. Longer chain length surfactants display slower transport, so that studies of these compounds may be performed over a span of days. Single-chain and double-chain surfactants do not differ dramatically as regards the rate of movement of the interfaces between phase bands.

So far, no evidence for an intrinsic barrier to transport at interfaces has been disclosed by these investigations, but a very strange composition profile was observed during a study of $C_{10}E_4$ within the liquid phase of intermediate composition (Fig. 5.6). The dilute part of the band displayed almost no composition gradient, while a steep gradient occurred near the concentrated end of the band.

From these miscellaneous observations, the rates of isothermal mixing processes do appear to depend largely on the "engineering factors".

6.5   The mechanisms of surfactant phase reactions

Most phase transformations occur in a single step, which can vary widely in rate. In some reactions, however, the path to equilibrium is indirect. A fast initial process occurs that is followed by slower processes, which leads to the build-up of a reaction intermediate. Analogous situations are common in molecular chemistry (Section 2.6); the study of the path by which reactions actually occur (the "reaction mechanism") is the focus of physical organic chemistry [21]. It can be seen that the study of reaction kinetics and the study of reaction mechanisms are strongly intertwined.

It appears that those phase reactions whose rates are governed by the engineering factors are mechanistically simple. The absolute rates of these processes may vary enormously, but they seem to occur directly and without evidence of mechanistic complexity. Other phase reactions are more complex, and occur via a sequence of fast and slow steps.

Not much attention has historically been paid to this area in surfactant systems. Recent studies of a pair of di-long chain cationic surfactant salts have provided insight regarding rates and mechanisms, however, which is worth reviewing. The two compounds studied were DODMAC and DOACS. DODMAC is dioctadecyldimethyl-ammonium chloride, and DOACS is dioctadecylammonium cumenesulfonate (p-2-propylbenzenesulfonate). The results were based on isoplethal studies and on qualitative observations made during DIT studies [2,22].

6.5.1   Mechanism of the peritectic decomposition of crystal hydrates

Two crystal hydrates exist in the DODMAC–water system ($X \cdot W$ and $X \cdot W_2$), and one is found in the DOACS–water system ($X \cdot W$). All of these crystal hydrates, on being heated, decompose via a peritectic phase reaction. The equation for the equilibrium process which occurs during this reaction for the monohydrate crystal of both systems is

$$X \cdot W + q^{\text{peri}} \rightleftarrows Lam^{\text{eq}} + X \tag{6.2}$$

where Lam is the lamellar liquid crystal phase. The dihydrate $X \cdot W_2$ of DODMAC decomposes via a similar reaction (at a lower temperature) to the lamellar liquid crystal and the monohydrate crystal. The DOACS monohydrate decomposes to the dry crystal and the dilute *liquid* phase.

None of these peritectic reactions actually occurs as written; all three occur, instead, in two steps. The first step is simple (congruent) melting, which leads in all three cases to a liquid crystal of the same composition as the crystal hydrate

$$X \cdot W + q^{\text{meta}} \rightarrow Lam \cdot W^{\text{meta}} \tag{6.3}$$

This liquid crystal is evidently metastable, because its composition is higher than is that of the equilibrium liquid crystal phase at the temperature of the reaction.

This fast initial reaction is followed by slower crystal nucleation and growth processes. It is presumed that the slow process is the formation of nucleation sites. The equilibrium products of the peritectic phase reaction are thus formed by the process

$$Lam \cdot W^{\,meta} \rightarrow Lam^{\,eq} + X^{\,eq} \tag{6.4}$$

The first step would be expected to be highly endothermic, but a large fraction of this heat would be recovered in the second step if equilibrium were rapidly attained (which it is not). The equilibrium heat of the phase reaction will be the difference. While this peritectic decomposition mechanism is unfavorable in terms of the heat required, it occurs rapidly because it has a large positive entropy. Since the first reaction is congruent mass transport is not required – only heat flow. This factor should also favor the reaction kinetically.

Observations made during studies of the decahydrate crystal phase of sodium sulfate $(Na_2SO_4 \cdot (H_2O)_{10}$, Glauber's salt) may be interpreted in exactly the same manner as above, although the observations were not at the time expressed using the same concepts. In this case too a liquid phase is observed to be formed rapidly, while equilibrium of state is attained much more slowly [23,24]. The fact that the temperatures of such isothermal discontinuities are depressed by the presence of third components, much in the same way as are melting points [25], suggests that the above-mentioned mechanism may be rather general.

One important consequence of the complexity of the above-suggested mechanism is to render the interpretation of calorimetric studies of the heat of peritectic reactions highly uncertain. The extent to which the measured heat corresponds to the equilibrium value depends on the extent to which equilibrium is reached during the slow step. The probability that equilibrium is actually attained during the process is close to zero.

DOACS. In the DODMAC–water system the peritectic reaction occurs at a temperature above the Krafft discontinuity, so that the initially formed liquid crystal phase need only undergo a shift in composition (by separation of the crystal phase) to attain equilibrium.

In the DOACS–water system, however, the peritectic discontinuity lies *below* the Krafft discontinuity and a very different situation exists [22]. The equilibrium products of this peritectic reaction are a dilute liquid and a crystal, just as in the peritectic reaction of the dihydrate crystal of sodium chloride:

$$X \cdot W + q^{\,peri} \rightarrow Liquid + X \tag{6.5}$$

Nevertheless, the reaction mechanism seems to be exactly the same as in the case of the DODMAC crystal hydrates. In the DOACS–water system the initially formed lamellar monohydrate liquid crystal phase not only has the wrong phase structure, but it is supersaturated with water. This liquid crystal phase was observed *not* to swell, in contact with water, until the temperature was raised to above the Krafft discontinuity temperature. Then, normal swelling behavior (which conforms to the phase diagram of the system) occurs.

The swelling behavior of DOACS is unique among the systems explored to date. The unusually narrow range spanned by the lamellar liquid crystal phase formed by

DOACS is also noteworthy. Presumably, the fact that both ions in this salt are amphiphilic is responsible for this phase behavior.

6.5.2 Possible mechanisms at eutectic discontinuities

As surfactant mixtures are heated past eutectics, a crystal must often react with another phase to form a liquid or liquid crystal product. Because the reaction product (in the direction of increasing temperature) is a high-entropy phase (the eutectic fluid), the rates of such reactions are probably controlled by the engineering factors.

When a eutectic mixture is cooled past this discontinuity, however, the net entropy change is unfavorable and the attainment of equilibrium can be expected to be extremely slow. The mechanism of the eutectic reaction in the cooling direction may therefore be expected to resemble that of other crystal-forming reactions: supercooling should readily occur, and equilibrium will likely be attained via nucleation and crystal growth. The conditions for reaching equilibrium quickly are worst at the eutectic composition, because neither of the crystal products already exists (Section 2.7).

6.6 Summary

It can be seen throughout that the entropy change that occurs during phase reactions is extremely important in determining the rates of these reactions. Phase reactions that are characterized by a large positive entropy change may be kinetically favored – even when opposed by a large enthalpy factor. The reaction of one fluid or high entropy phase to form another is also usually fast – but is not instantaneous.

Congruent phase reactions (during which no composition changes occur) also appear to be kinetically favored. The absence of the requirement for mass transport may favorably influence the rates of such processes.

As in many other areas of surfactant physical science, the kinetic aspects of the phase chemistry of these materials is a fruitful area for further study.

References

1. Kestin, J. (1976). *Benchmark Papers on Energy. The Second Law of Thermodynamics*, Vol. 5 (J. Kestin ed.), pp. 206–207, Dowden, Hutchinson, and Ross, Stroudsburg, PA.
2. Laughlin, R. G., Munyon, R. L. and Fu, Y.-C. (1991). *J. Phys. Chem.* **95**, 3852–3856.
3. Nyvlt, T., Sohnel, O., Matuchova, M. N. and Broul, M. (1985). *Kinetics of Industrial Crystallization*, Elsevier, Amsterdam.
4. Prigogine, I. (1968). *Introduction to Thermodynamics of Irreversible Processes*, John Wiley, New York.
5. Stilbs, P. (1987). *Prog. Nucl. Mag. Reson. Spectrosc.* **19**, 1–45.
6. Aniansson, E. A. G. (1978). *Ber. Bunsenges. Phys. Chem.* **82**, 981–988.
7. Kahlweit, M. (1982). *J. Colloid Interface Sci.* **90**, 92–99.
8. Gormally, J., Gettins, W. and Wyn-Jones, E. (1981). *Mol. Interact.* **2**, 143–147.
9. Knight, P., Wyn-Jones, E. and Tiddy, G. J. T. (1985). *J. Phys. Chem.* **89**, 3447–3449.
10. Bird, R. B., Stewart, W. E. and Lightfoot, E. N. (1960). *Transport Phenomena*, John Wiley, New York.
11. Barrer, R. M. (1941). *Diffusion in and Through Solids*, Cambridge University Press, Cambridge.

12. Volmer, M. (1939). *Kinetik der Phasenbildung*, Th. Steinkopf, Dresden.
13. Fu, Y. C., Glardon, A. S. and Laughlin, R. G. (1993). Presented at the American Oil Chemists' Society Meeting, Anaheim, California, April 25–29.
14. Spinolo, G. and Anselmi-Tamburi, U. (1989). *J. Phys. Chem.* **93**, 6837–6843; Shibanov, Y. D. and Godovskii, Y. K. (1989). *Prog. Colloid Polym. Sci.* **80**, 110–118.
15. Caffrey, M., Magin, R. L., Hummel, B. and Zhang, J. (1990). *Biophys. J.* **58**, 21–29.
16. Johnson, J. C. (1976). *Specialized Sugars for the Food Industry. Food Technology Review No. 35*, p. 27, Noyes Data Corporation, Park Ridge, NJ; Cakebread, S. H. (1971). *Confect. Prod.* **37**, 407–410, 412, 461–464, 470, 535–538; Sherwood, T. K., Pigford, R. L. and Wilke, C. R. (1975). *Mass Transfer*, McGraw-Hill, New York.
17. Wilhelm, R. H., Conklin, L. H. and Sauer, T. C. (1941). *Ind. Eng. Chem.* **33**, 453–457.
18. Bird, R. B., Stewart, W. E. and Lightfoot, E. N. (1960). *Transport Phenomena*, pp. 140-142, John Wiley, New York.
19. Hakemi, H., Varansi, P. P. and Tcheurekdjian, N. (1987). *J. Phys. Chem.* **91**, 120–125.
20. Gerritsen, H. C. and Caffrey, M. (1990). *J. Phys. Chem.* **94**, 944–948.
21. March, J. (1968). *Advanced Organic Chemistry. Reactions, Mechanisms, and Structure*, pp. 458–459, McGraw-Hill, New York.
22. Laughlin, R. G. and Munyon, R. L. (1991). Presented at the International Colloid and Surface Science Symposium, Compiegne, France, July 7–13, 1990.
23. Findlay, A. (and Campbell, A. N.) (1938). *The Phase Rule and its Applications*, 8th ed., p. 184, Dover, New York.
24. DeBray (1868). *Compt. rend.* **66**, 194. (From reference 23.)
25. Muller, H. J. (1933). *Ann. Chim.* **8**, 143–241.

Chapter 7

Surfactant Phase Behavior and Relative Humidity

7.1 Water activities, activity coefficients, and the Phase Rule

In Chapter 5 the relevance of the thermodynamic activity of water to various aspects of the phase behavior of aqueous surfactant systems was described. In this chapter some specific ways in which phase behavior influences water activity – and water activity influences phase behavior – will be described.

Water activity in food products is widely recognized by food technologists to be correlated with microbial growth: microbial growth is suppressed when the water activity in the food lies below a certain value [1]. The activity of water is more useful than is water composition, with respect to such correlations, for the same reason that pH provides a better measure of acidity than does titratable acidity. Food products are not only extremely complex in their chemical constitution, but have many phases. One could hardly expect the water in them to behave in an ideal manner, so that the relationship between water activity and composition is unknown and is likely to be very complex. The value in measuring and controlling water activity lies in the fact that it provides a useful parameter that allows for the existence of these nonidealities.

Water activities have not received much recent attention during analysis of the phase behavior of surfactants, although they played a role in confirming and consolidating early studies of the soap–water systems [2,3]. However, they are worthy of considerable attention for they are potentially of considerable value. (Considerable work has been done on the measurement of the osmotic forces within the liquid crystal phases of polar lipid surfactants [5], but the relationship of water activity to composition is not so well known as is the relationship to the long spacings of the phase.) For any situation in which a surfactant mixture is allowed to equilibrate with the atmosphere (at a particular temperature and water activity, or relative humidity), the initial water composition is irrelevant insofar as the equilibrium state of the mixture is concerned. Instead, the water composition (and thereby the phase behavior) at equilibrium is defined by the atmospheric water activity.

The equilibrium vapor pressure of pure water (p_w^0, in Torr) [4] displays a characteristic dependence on temperature which is well described using the algorithm

$$\ln(p) = \frac{(a + cT)}{(1 + bT)}$$

$$a = 1.484\,101\,7$$
$$b = 0.004\,430\,345\,2$$
$$c = 0.080\,883\,235$$
$$r^2 = 0.999\,999\,986$$

(7.1)

between -15 and $200°C$. (The data are taken from *The Handbook of Chemistry and Physics*, 64th ed., pp. D-192–D-194. Vapor pressure is expressed in Torr; 760 Torr equals one defined atmosphere, which equals 101 325 Pa (Pascal) or N (Newton) m^{-2}.)

The presence of the surfactant in aqueous surfactant phases reduces the vapor pressure of water below that of pure water. The ratio of the actual partial pressure, p_W, to p_W^0 is taken to be the "activity" of water

$$a_W = \frac{p_W}{p_W^0} \tag{7.2}$$

The activity, in turn, is related to chemical potential by the equation

$$\mu_W = \mu_W^0 + RT \ln(a_W) \tag{7.3}$$

Raoult's Law. The relationship between thermodynamic activity (from which chemical potentials can be inferred using eq. 7.3) and composition (which is measurable experimentally) is very important. The simplest relationship known is that described by Raoult's Law (Section 3.12) [6]. In applying Raoult's Law the pure substance is taken as the standard state. If both components of aqueous surfactant mixtures were to follow Raoult's Law, the activity of both the water and the surfactant would equal their respective mole fractions. For the water

$$a_W = x_W \tag{7.4}$$

and the chemical potential of water would be described by the equation

$$\mu_W = \mu_W^0 + RT \ln(x_W) \tag{7.5}$$

For the surfactant the activity of the surfactant too equals its mole fraction and

$$\mu_S = \mu_S^0 + RT \ln(x_S) \tag{7.6}$$

Systems that follow Raoult's Law are often described as "ideal" systems. Deviations from Raoult's Law commonly occur for one or the other of the components, in which case the activity is measurably different from the mole fraction. In such cases an activity coefficient, γ_W, may be defined such that the product of the mole fraction times the activity coefficient equals the activity:

$$a_W = \gamma_W \times x_W \tag{7.7}$$

The numerical value of the activity coefficient is the ratio of the measured activity to the mole fraction (a_W/x_W).

The numerical values of activity coefficients serve as a quantitative measure of deviations from Raoult's Law. An activity coefficient of one would correspond to ideal or Raoult's Law behavior. Measured activity coefficients may be either greater than or less than one, and vary greatly in magnitude. As will be seen in Section 7.5, the activity coefficients of water found in aqueous surfactant systems are reasonably close to one, which suggests that quantitative predictions of water activities using Raoult's Law are surprisingly good. The Raoult's Law values are, in any case, a good place to start in considering water activities.

Henry's Law. Only a tiny number of systems behave in a strictly ideal manner, but it is found experimentally for a great many systems that, in dilute solutions, the solvent obeys Raoult's Law while the solute obeys Henry's Law. Henry's Law (like Raoult's Law) states that the activity of a solute is proportional to its mole fraction in the mixture:

$$a_2 = K_H \times x_2 \qquad (7.8)$$

Henry's Law differs from Raoult's Law in that the constant of proportionality is not one. The experimentally determined Henry's Law constant, K_H, may be viewed as an "activity coefficient", but this is not ordinarily done as its value is often very far from one.

Henry's Law is obeyed only below a finite solute concentration; this limiting value is determined experimentally and varies widely. This Law takes many forms, and one should be aware that in some of these forms the constant of proportionality is not dimensionless (as above) [7]. Henry's Law is particularly useful for describing the activity of solutes that have a measurable vapor pressure, which (just as in the case of water) serves as a direct measure of their activity (eq. 7.2). The dimensionless Henry's Law constants of volatile lipophilic compounds in aqueous solutions differ grossly from one (Section 11.2.1).

7.1.1   The equivalence of water activities and relative humidities

Relative humidity is numerically similar to water activity. Relative humidity is defined as the weight fraction of water in the atmosphere relative to the weight fraction of water in air that is saturated with water [8]. Thermodynamic activities, on the other hand, are defined as the ratio of the partial pressure of water to the vapor pressure of pure water (eq. 7.2). Both variables may be expressed either as a pure fraction or as a per cent fraction. Numerical differences may exist between relative humidity and water activity since they are not defined in precisely the same way, but these are probably small and will be ignored for the present purpose.

It is useful to recognize that relative humidity is actually a thermodynamic activity parameter (or very nearly so), as well as being a dimension of weather information that is very familiar to the general public. Recognition of the dependence of water activity on composition within regions of miscibility, coupled with the restrictions imposed by the Phase Rule, leads to useful predictive correlations between phase behavior and relative humidity.

The activity of water varies smoothly with composition along those segments of isotherms in binary systems that fall within one-phase regions, while it is fixed and independent of composition within miscibility gaps (Section 4.8.2). If the water activity at equilibrium in one phase of a multiphasic mixture can be reliably determined or calculated, its value is necessarily the same in all the coexisting phases as well. This relationship holds regardless of the chemical complexity of the system.

7.2 Water activity profiles in the sodium chloride–water system

Mixtures of a salt with water that contain liquid, crystal, and gas phases in equilibrium are invariant if the temperature is specified. Under these constraints the pressure is fixed, and if the solute is non-volatile (as in surfactant systems) it equals the vapor pressure of the water present. Such mixtures are widely used to maintain a constant relative humidity environment within a closed container. The vapor pressure depends on the molecular structure of the salt, so that by selecting the proper salt the relative humidity can be varied [9]. It is critically important, in creating a controlled relative humidity environment using this method, to insure that the ratio of salt to water is such that the liquid, the crystal, and the vapor phases all exist and that equilibrium is maintained. Often temperature is not well controlled during such experiments, but probably the activity does not vary nearly so much with temperature as does the vapor pressure itself within a modest range of temperatures.

The coexisting phases within the liquid/crystal miscibility gap in the salt–water system at 25°C are the 26.43% salt solution and the dry crystal. In a graph of water activity vs. composition (Fig. 7.1), a smooth variation in water activity must occur between $a_W = 1$ and $a_W = 0.755$ within the composition span of 0–26.43%. Then, a horizontal line must exist at $a_W = 0.755$ which spans the composition range of 26.43–100%. Finally, since the water activity in pure salt is zero, the water activity

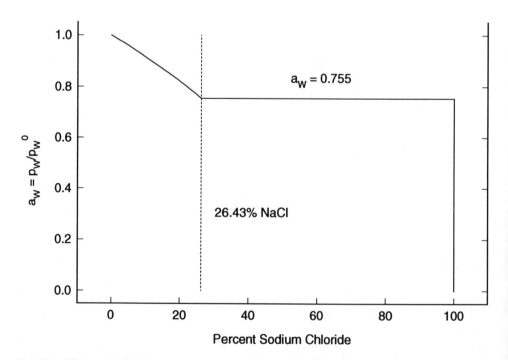

Fig. 7.1. The water activity (a_W) vs. composition profile spanning the 25°C isotherm in the sodium chloride–water system. At the plateau between the liquid and crystal phases, a_W equals 0.755.

must fall from 0.755 to 0 along the 100% isopleth. The issue of how water may have a finite activity within a dry salt crystal is discussed in Section 4.8.6.

The water activity profile along the 25°C isotherm thus consists of three distinct segments:

(1) a smooth curve, which extends from $a_W = 1$ at 0% to $a_W = 0.755$ at 26.43%,
(2) a horizontal segment, which extends from $a_W = 0.755$ at 26.43% to $a_W = 0.755$ at 100%, and
(3) a vertical segment, which extends from $a_W = 0.755$ at 100% to $a_W = 0$. at 100%.

The activity of water varies only within segments 1 and 3. In segment 1 the composition of the phase varies along with a_W. In the third segment the phase contains no water and its composition does *not* vary; nevertheless, a_W must be finite within this phase and its value must vary (Section 4.8.1). The second (horizontal) segment corresponds to the wide composition range within which the water activity is constant, within the liquid/crystal miscibility gap.

These a_W profiles resemble the composition profiles which would exist if DIT phase studies of the sodium chloride–water system were to be carried out (Appendix 4.4.4). If one rotates Fig. 7.1 about the 45° diagonal (so as to exchange the x and y axes), and replaces a_W by the position of the crystal/liquid interface along the x-coordinate, the resulting figure (Fig. 7.2) qualitatively resembles a composition profile that might exist along a cell during a DIT study of sodium chloride at 25°C. The form of the profile within the liquid phase would not be the same, but the

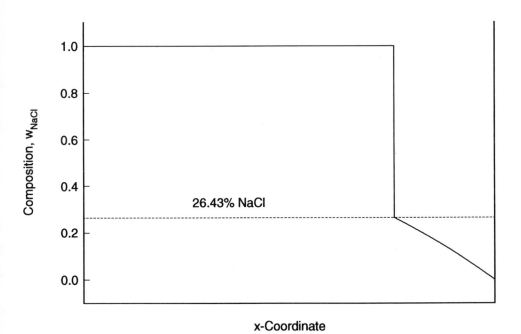

Fig. 7.2. The form of the composition vs. x-coordinate profile expected during a DIT phase study of the sodium chloride–water system at 25°C, obtained by rotating Fig. 7.1 about an axis passing through the origin at a 45° angle to the abscissa.

composition ranges of the phase bands (which are dictated by the phase diagram) must be identical.

A different water activity profile exists at each temperature. The qualitative nature of the profile is dictated by the phase behavior along the isotherm, and will therefore change smoothly between isothermal discontinuities but discontinuously at these temperatures. For this reason, all the activity profiles in the salt–water system between 200°C and 0.1°C will have a form similar to that in Fig. 7.1, but those which exist below 0.1°C will be different. At −10°C, for example, a profile such as that in Fig. 7.3 must exist.

Starting at the left in the phase diagram (Fig. 7.1) ice coexists with a 14.06% liquid phase, this liquid extends from 14.06 to 25.00%, the 25.00% liquid and the $X \cdot W_2$ crystal coexist up to the composition of $X \cdot W_2$ (61.9%), and $X \cdot W_2$ and X coexist between 61.9 and 100%.

It follows from this diagram that, in the most dilute segment of this activity profile, a horizontal line will exist at $a_W = 1$ between 0% and 14.06%. The value of $a_W = 1$ is required because one of these phases (ice) is pure water. (It is also true, incidentally, that since x_W is <1, then γ_W in the coexisting liquid must be >1; the observed value is 1.05.) The a_W must then decrease smoothly, within the span of the liquid phase, until the concentrated boundary of the liquid region is reached. Within the gap between this liquid and $X \cdot W_2$ another horizontal line must exist which reflects the fact that a_W is invariant within this span of compositions. A vertical line at the composition of $X \cdot W_2$ will connect this line to still another horizontal line (at a lower a_W) that extends from the composition of $X \cdot W_2$ to 100%. Finally, a vertical line must be drawn at 100% which connects this last horizontal line to $a_W = 0$ at 100%. As above, the vertical lines designate the regions of variable a_W at fixed compositions.

The form of the profile of water activities spanning the entire composition range may thus be anticipated from the diagram, and if the activities of water within the $L \rightleftarrows X \cdot W_2$ region were to be determined it would be possible to predict, with reasonable accuracy, the numerical values of a_W within the entire liquid region. If, in addition, the activity of water in a mixture of $X \cdot W_2$ and X phases were to be determined (using vapor pressure measurements), the entire profile would be known. Activity coefficients could be inferred as well from such data.

It was assumed in drawing Fig. 7.3 that the activity coefficient decreased linearly with mole fraction in the liquid phase (Section 7.5) from 1 to the same value found at this composition along the 25°C isotherm (above). Also, the plateau values of the activities were chosen arbitrarily in this figure. It is thus correct as to form, but is *not* to be utilized to infer numerical data.

The influence of temperature on hygroscopicity. Since both the qualitative and the quantitative details of the activity profile change with temperature in accordance with the phase diagram, so must the influence of water activity on phase behavior also vary. An example of this variability is seen from data on the hygroscopicity of dimethyl-dodecylphosphine oxide [10].

At 37.8°C this compound is not hygroscopic (absorbs no water from the atmosphere) at either 47 or 80% relative humidity. At 48.9°C, however, it remains non-hygroscopic at 47% but is strongly hygroscopic at 80% relative humidity. An 84.2% mixture (the lamellar liquid crystal phase) is found at equilibrium at this relative humidity (the same salt solutions were used to regulate water activity at the different temperatures, so that it is likely that the water activities at the different temperatures were similar but not exactly the same).

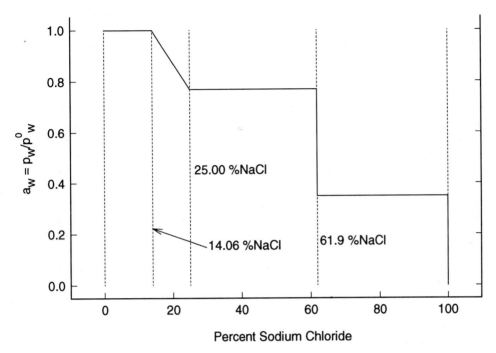

Fig. 7.3. A plausible water activity profile along the −10°C isotherm in the sodium chloride–water system.

These results suggest that the plateau in the activity profile which connects the dry crystal with the lamellar liquid crystal phase in this system lies below $a_W = 0.8$ at 37.8°C, but above 0.8 at 48.9°C. This result is consistent with the qualitative picture (Section 5.6.2) that the amount of water required to form the liquid crystal phase decreases as the temperature is increased.

7.3 The use of water activity profiles to predict phase compositions

If composition is specified, the water activity can be read from activity profiles. This is one way in which these profiles can be used, but an alternative use is to specify water activity and then infer the composition of the equilibrium mixture using the profile. Once the composition is determined, the phase behavior may be inferred by reference to the phase diagram.

This "reverse" use of activity profiles can be illustrated using the salt–water profile (Fig. 7.1). If a_W is taken to be 0.8 the composition of the mixture which has this activity is about 22%, and from the phase diagram this mixture is seen to be a liquid phase. If the activity is 0.7, however, the phase that exists is the dry salt crystal; the liquid phase does not exist at this water activity. While it would be highly improbable for the activity to coincide exactly with that of the plateau, both the liquid and the crystal would coexist if it did.

One would say in the first case (above) that salt is "hygroscopic" (absorbs water from the air to form a liquid), while in the second case it is not. This example clearly illustrates the fact that hygroscopicity depends on water activity (relative humidity), as well as on the nature of the material.

The same kind of predictions could be made using the profile in Fig. 7.3, if the profile were actually known. Taking the activity values in Fig. 7.3 as experimental results (*which they are not*), one would conclude that the ice phase will not exist if the activity is just below one – only the liquid phase. As the a_W is reduced starting from one, the composition of the liquid will increase until, at some value, it too disappears – to be replaced by the $X \cdot W_2$ crystal. Over a finite range of activities $X \cdot W_2$ will be the equilibrium phase, and finally at still lower activities X will result. X will persist as the equilibrium phase as a_W approaches zero. At an a_W of precisely zero water does not exist and one no longer has a binary system.

The extension of these concepts to ternary (or higher) systems is straightforward in principle, but the additional degrees of freedom that exist due to the larger number of components present must be taken into account (Section 12.1.1.2). One predictable effect of the presence of more than two components will be to lower and smear out the sharp isothermal discontinuities found in binary systems (Chapter 12).

7.4 The water activity profile at 25°C in the $C_{12}E_6$–water system

In 1969, both the phase diagram and the water activity profile along the 25°C isotherm in the $C_{12}E_6$–water system were reported [11]. This study is important in that the activity profile spanning a complete isotherm has been determined within a system whose phase diagram is also fully defined. This profile is shown in Fig. 7.4. (Considerable water activity data are also available in polar lipid–water systems [5], but while these data have been correlated directly with the long X-ray d spacing (a structural parameter) the relationship to composition is less certain. Also, the equilibrium states along the isotherm are ill defined (Section 11.3.2.7).)

The diagonal line in Fig. 7.4 describes the values for a_W that would have been observed had Raoult's Law been followed for this component. It was assumed during this study that the behavior of the mixture up to the cmc concentration is ideal with respect to both the surfactant and the water [12]. A sharp deviation from Raoult's Law behavior in the dependence of water activity on composition occurs at the critical micellar concentration (cmc). From earlier work [13] a positive deviation would be expected (due to micellar aggregation), and this was indeed observed. However, the composition range (in mole fraction units) spanned by this deviation is very narrow; when x_W falls below about 0.9 the curve turns and becomes approximately parallel to the Raoult's Law line. The deviation from Raoult's Law increases slightly as x_W approaches about 0.3, then the curve crosses below the Raoult's Law line so that for $x_W < 0.2$ (which corresponds to 99.18% surfactant) the deviation from ideal behavior is negative ($\gamma_W < 1$).

The activity coefficients of water in these mixtures (γ_W) may be calculated by dividing the observed activity by x_W, and are depicted in Fig. 7.5. These data show that the deviation from ideality of the thermodynamic activity of water in this system is modest over virtually the entire composition range. The largest positive deviation occurs near $x_W = 0.3$ (98.77%), and a negative deviation occurs at $x_W < 0.2$ (>99.18%). Between

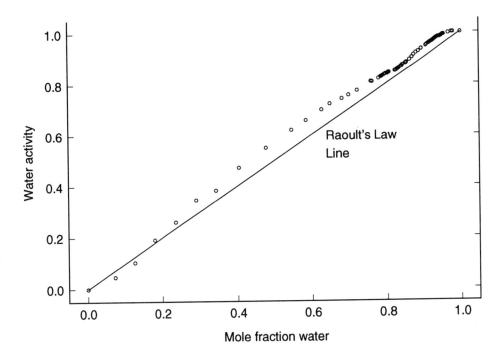

Fig. 7.4.   The measured water activity profile along the entire 25°C isotherm in the $C_{12}E_6$–water system [11].

the composition span of 0 to 97% (by weight) surfactant, the observed activity coefficients range from 1 to about 1.05! Larger deviations from ideality than this occur in sodium chloride and other salt solutions.

When one considers the wide range of compositions spanned and the structural complexity of the phases that exist, these deviations from nonideality are remarkably small. Along this composition span one encounters a micellar liquid phase, a hexagonal liquid crystal, a cubic liquid crystal, a lamellar liquid crystal, and finally a concentrated (probably non-micellar) liquid phase [11]. Yet, the activity of water is not far from ideal within all of these phase states. The most serious (negative) deviations occur within the concentrated liquid phase.

During the late 1930s the vapor pressures of soap–water mixtures in the sodium laurate–water [2] and sodium palmitate–water [14] systems were investigated, and more recent studies of surfactant–water systems have been reported [15]. The phase diagrams of the soap–water systems remain uncertain within crystal regions and the intermediate phases (Section 8.4.14) were unrecognized at the time, so the phase behavior at the temperature of these studies (90°C) is not completely known. Nevertheless, these data suggest that the general form of water activity/composition profiles in soap–water systems is qualitatively similar to that in the nonionic $C_{12}E_6$–water system. For sodium laurate mixtures of less than about 80 wt% composition (x_W c. 0.8) the activity coefficients lie below 1.1 (if mole fractions are calculated on a formula weight basis), and rise to about 2.4 at a mole fraction of 0.2 (98.5 weight %). This result was noted [3], but its considerable significance was not emphasized. If it is

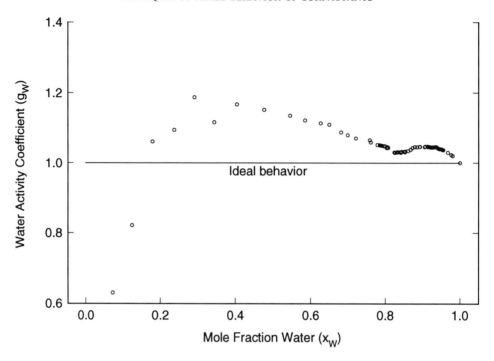

Fig. 7.5. The activity coefficient of water, γ_W, along the 25°C isotherm in the $C_{12}E_6$–water system [11].

true that the above-described behavior is found in both a weakly hydrophilic nonionic surfactant and also in the strongly hydrophilic soaps, then such behavior may well be characteristic of all surfactant–water systems. The data in Figs 7.4 and 7.5 suggest that it may be legitimate to apply Raoult's Law to estimate the activity of water, as a first approximation, over virtually the entire span of compositions within an aqueous surfactant phase diagram!

These results seriously contradict the earlier implication, drawn from Krafft's early data on the boiling point elevations of soap solutions, that surfactant solutions are strongly nonideal (Section 11.2.1). Krafft's measurements were made over a small composition range (at high values of the mole fraction of water) and near the cmc. In an expanded plot of the $C_{12}E_6$–water data within this region the departure from ideality will appear to be extremely large, but an altogether different impression is left when data which span the entire composition range are viewed.

7.5 The order of phase transitions in the $C_{12}E_6$–water system

When the preliminary data from the $C_{12}E_6$–water study were first obtained the miscibility gaps between the various phases were not evident [16], and the possibility that these transitions are second order was therefore carefully investigated. With the acquisition of data of sufficient accuracy and precision, it was found (as shown in Fig. 7.6) that the water activity function is, in fact, discontinuous with respect to

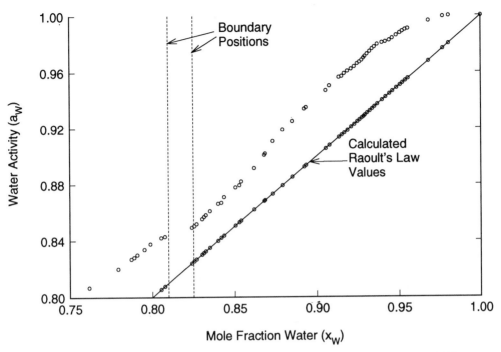

Fig. 7.6. The discontinuity in a_W observed at the liquid/hexagonal liquid crystal miscibility gap in the $C_{12}E_6$–H_2O system.

composition within the composition regions where two phases coexist. The existence of this discontinuity requires that a plateau in the activity-composition function exist, and provides evidence that there is a narrow miscibility gap between the liquid and liquid crystal phases. When such gaps are large (e.g. 50 or 75%) their existence is readily apparent, but when they are small (as in Fig. 7.6) documenting them is far more difficult. (The quantitative determination of the compositions of coexisting phases at miscibility gaps that span a composition range as small as 0.6% has recently been accomplished using the DIT-IR method [21].)

The above work constitutes perhaps the most careful investigation reported to date of the question of whether or not surfactant liquid or liquid crystal → liquid crystal phase transitions are first order or second order [17]. The experimental distinction between first and second order transitions rests on whether or not miscibility gaps exist; in first-order transitions the gap does exist while in second-order it does not. This study documents the fact that even this narrow gap exists, and (as noted in Section 14.3) may be regarded as the clinching evidence in support of the hypothesis that the Phase Rule applies directly, and without qualification, to aqueous surfactant systems.

Qualitative information. The result of this quantitative study is fully consistent with a large body of qualitative information gathered during many phase studies using the step-wise dilution or other methods (Appendix 4.4.3). During step-wise dilution studies, one or more isopleths were almost invariably encountered along which the partial transformation of a liquid phase into a mixture of liquid and liquid crystal phases was observed. Typically this biphasic situation persisted over a finite range of temperatures below the liquid boundary temperature, while at still lower temperatures the mixture was transformed into a one-phase liquid crystal phase. This qualitative behavior was observed with all classes of liquid crystals (hexagonal, lamellar, and cubic), and with all classes of surfactants. While these qualitative observations by themselves would perhaps not be convincing, taken together with the above quantitative isothermal studies they leave little doubt that most – perhaps all – transitions between aqueous surfactant phases of differing structure are first-order transitions.

If one considers the probable form of free energies of mixing curves for miscibility gaps that span a very narrow composition range, it would appear likely that the numerical value of the instability that exists within these gaps is very small – but it must nevertheless exist.

It has been suggested that the nematic $\rightleftharpoons$ lamellar phase transition that occurs as the temperature is increased along selected isopleths in the $C_7F_{15}CO_2^-,Cs^+$/water system is second-order [18]. The behavior of the SDS–water system at the R_α/Q_α liquid crystal boundary (Fig. 5.4) also suggests the possibility that this is a second order transition [19]. However, in neither case can the order of these transitions be regarded as having been firmly established. In thermotropic systems selected liquid crystal transitions are generally regarded as second-order transitions [20], but the situation may be different in these unary systems.

References

1. Troller, J. A. and Christian, J. H. B. (1978). *Water Activity and Food*, Academic Press, New York.
2. McBain, J. W. and Salmon, C. S. (1920). *J. Am. Chem. Soc.* **42**, 426–460.
3. Vold, R. D. and Ferguson, R. H. (1938). *J. Am. Chem. Soc.* **60**, 2066–2076.
4. (1984). *Handbook of Chemistry and Physics* (R. C. Weast, ed.-in-Chief), 64th ed., pp. D-192-194, CRC Press, Boca Raton, FL.
5. LeNeveu, D. M., Rand, R. P. and Parsegian, V. A. (1976). *Nature* **259**, 601–603; Elworthy, P. H. (1962). *J. Chem. Soc.* 4897–4900.
6. Lewis, G. N. and Randall, M. (revised by K. S. Pitzer and L. Brewer) (1961). *Thermodynamics*, 2nd ed., pp. 224–241, McGraw-Hill, New York.
7. Stumm, W. and Morgan, J. J. (1981). *Aquatic Chemistry*, pp. 40–42, John Wiley, New York.
8. (1983). *Handbook of Chemistry and Physics* (R. C. Weast ed.-in-chief), 64th ed., p. F-102, CRC Press, Boca Raton, FL.
9. (1991). *Handbook of Chemistry and Physics* (D. R. Lide ed.-in-chief), 72th ed., p. **15**–21, CRC Press, Boca Raton, FL.
10. R. G. Laughlin, unreported work.
11. Clunie, J. S., Goodman, J. F. and Symons, P. C. (1969). *Trans. Faraday Soc.* **65**, 287–296.
12. Corkill, J. M. and Goodman, J. F. (1969). *Advances in Colloid and Interface Science* **2**, 297–330.
13. Krafft, F. (1899). *Ber. deut. chem. Ges.* **32**, 1596–1608.
14. McBain, J. W., Vold, R. D. and Vold, M. J. (1938). *J. Am. Chem. Soc.* **60**, 1866–1869.
15. Altunina, L. K., Sokolova, E. P. and Churyusova, T. G. (1973). *Vestn. Leningrad Univ., Fiz. Khim.* 83–88.
16. J. F. Goodman, personal communication.
17. Fried, V., Hamek, H. F. and Blukis, V. (1977). *Physical Chemistry*, MacMillan, New York; Rutgers, A. J. (1954). *Physical Chemistry*, pp. 290–298, Interscience, New York.
18. Boden, N., Jackson, P. H., McMullen, K. and Holmes, M. C. (1979). *Chem. Phys. Lett.* **65**, 476–479.

19. Kekicheff, P., Grabielle-Madelmont, C. and Ollivon, M. (1989). *J. Colloid Int. Sci.* **131**, 112–132.
20. Demus, D., Diele, S., Grande, S. and Sackmann, H. (1983). *Adv. Liquid Crystals*, Vol. 6 (G. H. Brown ed.), p. 71, Academic Press, New York.
21. Laughlin, R. G. and Munyon, R. L. (1987). *J. Phys. Chem.* **91**, 3299–3305.

Part III

STRUCTURAL CHEMISTRY

Chapter 8

The Structures and Properties of Surfactant Phases

8.1 Introduction

In Chapter 2 the hierarchy of structural information (molecular, conformational, phase, colloidal) that is important to surfactant physical science was outlined. In the present chapter the phase structures of the phases that surfactants form will be described. Phase structure was defined in Section 2.4.3 as *"the manner in which molecules are arranged in space within a phase"*. Both phase and conformational structure will be considered, however, because these two levels of structure are so often intimately related (Section 2.4.3.1). Since the optical and rheological properties of surfactant phases are influenced by their phase structure, these properties of the various phases will also be considered.

It is imperative to distinguish phase structure from colloidal structure, which is concerned with the manner in which phases themselves (within multiphase mixtures) are arranged in space. If these two levels of structure are not distinguished, serious problems (including violations of the Phase Rule) may result. If colloidal structure exists within a particular surfactant mixture, it too (along with the phase structure) influences the optical and rheological properties of mixtures. However, that is another subject; the present focus is on the properties of one-phase mixtures.

The objective of this chapter is to provide a qualitative review of the structures of surfactant phases, insofar as they are presently known, and to indicate the properties of phases and the relationship of these properties to phase structure, where that relationship is evident. The quantitative description and analysis of structure, as well as the physical theory underlying these phases, will be left to others.

The nature of the equilibrium state has a strong bearing on the behavior of mixtures in which one phase is colloidally dispersed within the other. Those discontinuities of state which occur in macroscopic mixtures must also occur in colloidal mixtures, but at the same time many aspects of state are quantitatively perturbed by the excess energies (surface and bending) associated with colloidal states [1]. The temperatures of phase transitions may be somewhat lower, conformational structures may differ, and the arrangement of molecules in space may be altered from that which exists in the equilibrium state [2].

The primary focus in phase science is necessarily on the equilibrium phase state and its structure, but attention must be given to these perturbations of state that exist within colloidal mixtures. This issue is particularly important in the case of insoluble or poorly soluble surfactants, because colloidal structure is unusually long-lived when a surfactant is insoluble. *Both* the (equilibrium) phase science and the science of irreversible colloids are important aspects of the physical science of these particular mixtures.

Three classes of condensed phase structures are encountered in aqueous surfactant systems: crystals, liquid crystals, and liquids. The gas phase, when present, is typically pure water. While this phase is very important as regards the influence of relative humidity on phase behavior (Chapter 7), no consideration need be given here to its structure.

8.2   Optical and structural isotropy and anisotropy

The texture observed when phases are viewed in polarized light is especially important in surfactant phase science, because it may sometimes be used to determine – at a glance – the phase structure. This was made possible by careful studies of the optical physics of liquid crystal phases by Friedel [3], and by the correlations of the textures that are observed with phase structure which were later established by Rosevear [4]. Textural information has historically been purely subjective in nature and best interpreted by trained experts who are conversant with optical physics, but one can foresee the possibility that modern image processing could alter this situation.

Phases that appear dark when viewed between crossed polars are described as "isotropic" or "nonbirefringent", while those that appear bright are termed "anisotropic" or "birefringent" [5]. When a phase is described as being isotropic or anisotropic, reference is usually being made to its optical properties. However, these same words are also used to characterize the various phase structures themselves. Isotropic structures are identical along any three orthogonal directions in space, while anisotropic structures are not. (One direction is "orthogonal" to another if the vector component of the first, in the direction of the second, is zero. Sets of Cartesian coordinates are orthogonal in this sense.)

Caution must be exercised in using the terms "isotropic" and "anisotropic". While structurally isotropic phases are always optically isotropic, all structurally anisotropic phases appear, when viewed along certain directions, to be optically isotropic (nonbirefringent). The optical properties of a phase are dictated by its structure, but the reverse is absolutely not true.

Those phase states that are structurally isotropic are gases, liquids, cubic liquid crystal phases, and cubic crystal phases. All other phases are structurally anisotropic. Three crystallographic subclasses exist within the cubic class: primitive, body centered, and face centered [6]. All the cubic subclasses may (in principle) exist within both liquid crystal and crystal states. All three are known in crystals, but the primitive cubic structure is uncommon in liquid crystals [7,8].

The intrinsic action of a substance on light does not depend on the polarization of the light, but our ability to perceive this action does. For this reason, it is necessary to view a sample that is held between a polarizer and an analyzer which are "crossed" in order to determine its optical behavior. Being "crossed" means that the direction of vibration of the electric fields of the light waves that are transmitted by the polarizer are at right angles to the direction that is transmitted by the analyzer. (The direction of vibration of the electric field is transverse to (across) the direction of propagation.) The sample is illuminated using an unpolarized light source placed beyond the polarizer, and is observed from in front of the analyzer while looking in the direction of the light source. The optical properties are best diagnosed by rotating the sample about the axis of the direction of illumination as it is being observed.

When polarized light passes through a phase which is structurally isotropic, the plane of polarization of the emergent ray is unaltered relative to that of the incident ray. When polarized light passes through a phase that is anisotropic, however, the plane of polarization of the emergent ray *is* altered under most conditions. (An exception exists when the ray passes through the phase along its optic axis. In this case its polarization is *not* altered due to this special circumstance – even though the phase is anisotropic.) This arises from the fact that the refractive indices of anisotropic phases vary depending on the direction of polarization of the incident light, relative to the structure of the phase. The refractive index of an isotropic phase, in contrast, is independent of the direction of polarization of the incident light [9].

The refractive indices of anisotropic phases fall between fixed limits that are defined by molecular structure and by the numerical values of the system variables. The magnitude of their anisotropy is the numerical difference in refractive index between the upper and lower limits. In strongly anisotropic states this difference is large, in weakly anisotropic states it is small, and in isotropic states it is precisely zero (not merely small).

When light passes through an anisotropic phase in most directions, the emergent ray is split into two parallel rays. The planes of polarization of these two rays are at right angles to each other, and they are usually of unequal intensity. One ray is termed the "ordinary" ray, because it is refracted ("bent") as it passes through interfaces separating regions of differing refractive index in a normal (or ordinary) manner. This means that the relationship between the angle of bending (refraction) of the incident and the emergent rays is governed by Snell's Law [5].* The other ray is termed the "extra-ordinary" ray because it is refracted more severely than is the ordinary ray, and is said to display "anomalous refraction" because it appears to violate Snell's Law. A classic experiment which illustrates this phenomenon is the observation of a dot through a suitably oriented single crystal such as calcite [10]. The dot is seen as two dots, and for this reason the effect is also termed "double refraction".

Since the plane of polarization of light is not altered by passage through an isotropic phase, the light that passes the polarizer is absorbed by the analyzer as if no sample were in the light path. It is for this reason that isotropic mixtures appear dark. Because the plane of polarization is split into ordinary and extraordinary rays after it passes through a birefringent phase, a nonzero vector component of the emergent light is transmitted by the analyzer. It is for this reason that such mixtures appear bright.

If one uses a polarizing microscope that is fitted with a universal stage, the angle of incidence may be varied in all three directions in space and measured as well. If a birefringent single crystal is examined in this manner, either one or two unique directions will always be found along which the crystal appears to be isotropic. These directions are the "optic axes" of the phase. Measurements of their direction, relative to the edges of the crystal, have long been used to characterize crystals. If a phase displays only one optic axis it is "uniaxial", while if it displays two it is "biaxial". There are never more than two optic axes.

---

*Snell's Law defines the relationship between the angle of refraction θ in each phase (always measured relative to the normal) to the refractive index of that phase. For two phases a and b having indices n^a and n^b, Snell's Law is $n^a \sin(\theta^a) = n^b \sin(\theta^b)$. θ is the angle between the direction of the ray and the normal to the interface.

Anisotropic phases may also differ in the sign of their anisotropy. In uniaxial materials, if the characteristic index along the optic axis is greater than the characteristic index normal to this axis the material is said to be "optically positive". If the reverse is true, it is "optically negative". It is possible by using classical microscopic techniques to ascertain whether a material is optically positive or optically negative. Qualitatively similar analyses of the sign of anisotropy may be applied to biaxial materials, but three characteristic indices exist [5].

Exactly the same principles apply to liquid crystals, but one cannot so easily control the direction of light relative to their structure because they are fluids. The optical physics of selected surfactant liquid crystal phases which can be oriented by a flat surface (such as the lamellar phase of lecithins) have been investigated [11]. Extensive investigations of the optical physics of thermotropic phases have been performed [12].

The intensity of birefringence may be observed either while a sample is at rest, or while it is being subjected to mechanical shear. When a shearing force is applied to a fluid, flow occurs. During flow the fluid's structure is momentarily altered from that which existed when the sample was at rest, and these structural changes usually enhance the optical anisotropy of the phase. Shear-dependent optical effects are described as "shear-" or "flow-birefringence" [13]. These effects represent a nonequilibrium (but characteristic) property of materials.

If the sample is isotropic at rest, shear usually causes it to become birefringent and the effect is comparatively easily observed. The ease with which shear birefringence is observed depends strongly on the rate at which the distorted structure relaxes to the equilibrium state when the shearing force is removed [13]. So far as is known, all cubic phases (both crystal and liquid crystal) display readily observed shear birefringence. The ease with which shear birefringence is observed in micellar liquid phases varies widely. Micellar solutions having long cylindrical or entangled micellar structure display readily visible shear birefringence, but this is less readily perceived as the solution is diluted and is not grossly evident in solutions that contain spherical micelles.

Shear must alter the structure of anisotropic phases just as it does that of isotropic phases, but because they are birefringent at rest the effect is far more difficult to observe.

8.3 The crystal phases

The crystal phases of surfactants are the most highly ordered and the most dense phases that surfactants form. Relative to liquids, crystals are phases of low symmetry in the sense that symmetry axes exist along only a few specific directions in crystals, while symmetry is observed along all directions in liquids. The book by Small is a particularly useful reference to information regarding the physical science of the crystal state of long chain molecules [14].

Phase structure is highly developed along all three dimensions in space within crystals. Regular periodicity, extending over many structural planes, is clearly evident when a crystal lattice is viewed from any direction. Because of this highly ordered structure, many sharp X-ray lines are observed in the powder patterns of crystals. Typically, far more lines are found with crystals than with liquid crystals.

The three-dimensional order within crystals is also responsible for the fact that, from a rheological perspective, they are solids rather than fluids [15]. Because they display high yield stresses (see Section 8.4.2), a crystal placed on a surface (which is subjected only to the weak shearing force imposed by gravity) retains its shape indefinitely at ambient gravitational field strengths. A fluid, on the other hand, spreads so as to more closely approach the center of the earth. A drop of water on a surface spreads in a matter of seconds, while the various liquid crystal phases may require minutes, days, or months to spread. Nevertheless, liquid crystals are rheologically fluids. The author has in his possession large samples of the middle and rectangular phases of soaps sealed in 2.5 cm diameter glass tubes that were prepared during the 1950s by F. B. Rosevear and O. T. Quimby. If centrifuged to the bottom of the tube, these viscous phases will spread to span the length of the tube if the tube is kept in a horizontal position for a period of many months.

It is well to keep in mind that whether or not a phase is fluid depends on the magnitude of the applied force. Even crystals are fluid if a force sufficient to exceed the yield stress is applied (as may occur within geological formations). Still, there is a vast quantitative difference between the force required to deform phases such as liquids and liquid crystals, which are normally regarded as "fluids", and that required to deform phases normally regarded as "crystals".

Crystals are highly improbable states thermodynamically; they have the lowest entropies of any of the known phase states. Under conditions such that the crystal is the equilibrium state, its low (unfavorable) entropy is more than counterbalanced by its low (favorable) enthalpy. Enthalpies are in a general way related directly to densities, and thus inversely to molar volumes.

Crystal structure is determined using X-ray studies of single crystals [6]. These are crystals within which the spatial orientation of a great many structural planes is virtually the same. Along with determination of the coordinates of each atom in the structure, uncertainties in the atomic positions (due to thermal motion) are also usually estimated. The magnitude of these uncertainties are depicted graphically using "thermal ellipsoids", which are centered at the coordinates of the time-average position of the atom and display the direction and range of thermal motion.

Single crystal structure determinations are often conducted at low temperatures (Dry Ice® or liquid nitrogen) in order to reduce this thermal motion. By so doing the crystal structure is more sharply defined and the structural analysis is facilitated. Surfactant molecules invariably exist in only one conformational structure within perfect single crystals and the thermal ellipsoids are typically small at the temperature of the study.

At the temperature of the transition of a crystal to a disordered phase by the action of heat, all crystals may be expected to differ substantially in structure from that inferred using single crystal studies. (The situation may be different when a crystal is trans-formed to a disordered state by the combined action of heat and solvation energy.) Deviations from the all-*trans* conformational structure have been shown to exist near phase transitions in straight chain hydrocarbon crystals. While such crystals display an all-*trans* conformation at low temperatures, *gauche* conformations exist near the ends of the chains in increasing numbers as the temperature approaches the melting point [16]. In the longer chain ($> C_{21}$) odd number hydrocarbons, discrete polymorphic states may also exist that differ in the number of *gauche* bonds present [17]. The important point is that crystals at ordinary temperatures are *not* perfectly structured states whose structure is precisely represented under all conditions by the single crystal structure.

The contraction in molar volume which occurs at the melting point of ice (and of a tiny number of other materials) has been long recognized to reflect unusual behavior in comparison to that of most crystal phases [18]. The molar volume of the water molecules in ice is anomalously high as a result of its open hydrogen-bonded phase structure, and decreases when ice melts because this structure partially collapses. Such behavior is never seen in fatty crystals – probably including surfactants. Fatty molecules in fluid states invariably occupy a substantially larger volume than they do in crystals [19], and because of this their melting points increase with increasing pressure, rather than decrease (as does water). Evidence also exists that the temperature of the Krafft boundary of soaps also increases with increasing pressure [20].

Most, if not all, pure surfactants exist as a crystal phase at a sufficiently low temperature, but the melting points of the crystal range from far below 0°C to greater than 350°C. The upper limit of these melting points indicates that the polar groups contribute significantly to crystal stability, since the estimated melting point of linear hydrocarbons of infinitely long chain length is about 144°C [21].

Crystal hydrates exist in a great many surfactants and will be considered in Section 8.3.4. The general properties of dry crystals and of crystal hydrates are similar, but the existence of hydrates profoundly influences the overall physical behavior of surfactants (Section 5.7). Also, the kinds of phase transitions that dry crystals and crystal hydrates undergo at their upper temperature limit of existence are usually very different.

8.3.1   The bilayer crystal structure

8.3.1.1   The molecules within bilayers

The amphiphilic molecular structure of surfactants has a pervasive influence on the phase structure of their crystals. Surfactants invariably pack within crystals so that the lipophilic groups of different molecules are associated with each other in lipophilic regions, and the hydrophilic groups are similarly collected within polar regions. The most common packing structure by far is the "bilayer". Bilayer structure is usually found in crystals of both single-chain and double-chain surfactants [22].

Stating the overall symmetry of a crystal, plus the coordinates of the atoms within the unit cell, is sufficient to fully define a crystal structure. The unit cells of surfactant crystals typically contain either a pair of molecules or a small number of pairs. If the surfactant is an ionic salt, then one or more pairs of "ion pairs" are found. (In the crystal structure of the 1/8th crystal hydrate of SDS ($X_8 \cdot W$) eight pairs would be required to accommodate the stoichiometry of the crystal, but this is an unusually large number.)

It is unique to the crystal state that such a tiny aggregate accurately reflects the structure of the phase as a whole. Since the largest dimension of the unit cells of surfactant crystals is of the order of 5 nm, a distance of 1 mm (10^6 nm) will typically span more than 10^5 unit cells. In striking contrast, it is necessary to assemble a far larger aggregate of molecules to produce a representative fragment of a liquid or liquid crystal phase structure [23]. Moreover, the number of molecules required in this fragment is indeterminate.

The conformational structure of the chain (or chains) of molecules within surfactant crystals is, in the first approximation, all-*trans* (linear). The hydrophilic group is usually positioned at one end of the lipophilic chains in both single-chain and dichain surfactants. The structure of 3-lauroyl-1,3-propanediol-1-phosphocholine monohydrate (an

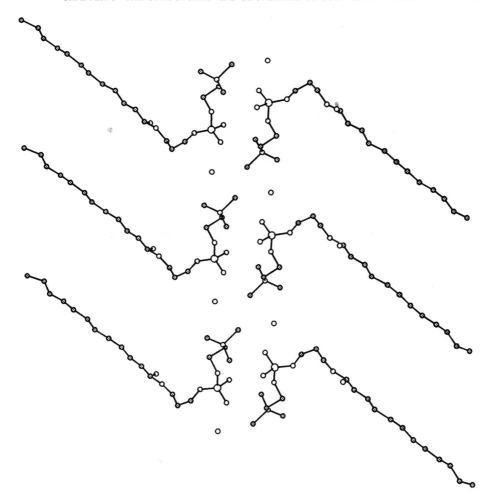

Fig. 8.1. Crystal structure of the monohydrate of 1-lauroyl-1,3-propandiol-1-phosphocholine. The head-to-head and bilayer structural features are readily visible, as is the fact that the vector of the dipole of this zwitterionic hydrophilic group lies almost within the plane of polar functional groups.

analog of a lysopolar lipid that lacks the 2-hydroxy group in the glycerol moiety, Fig. 8.1) is typical of many zwitterionic surfactant crystals [24], while the structure of the DODMAC monohydrate crystal (Fig. 8.2) is probably typical of that of ionic dichain surfactants [25].

"Dichain" surfactants are "midchain" substituted compounds in their most highly extended conformation. However, as implied by the "dichain" terminology, these molecules are usually bent near the middle (at the hydrophilic group) in most of their phase states (including crystals). This bend usually occurs in such a way that the straight parts of the chains are unequal in length, so that the shape actually resembles a lady's hairpin more closely than it does a "U". This inequality in the length of the two arms is readily accommodated in the crystal by tilting the molecule relative to the crystal planes. The hydrophilic group is at the bend – and thus at one end

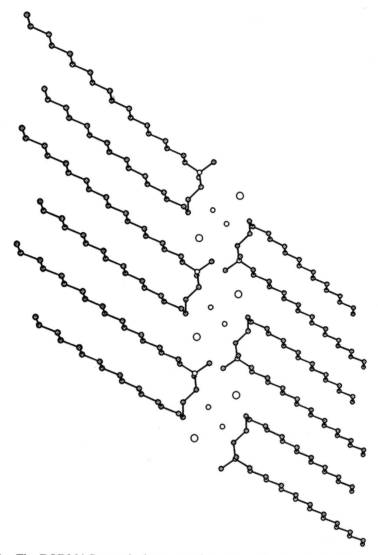

Fig. 8.2. The DODMAC monohydrate crystal structure, showing the hairpin conformational structure and the planes of water molecules (hydrogen bonded to chloride ions) between the planes of ions. The ammonium and chloride ions within each monolayer almost lie within the same plane (± 0.1 Å). Viewed from the face, each ion forms a diamond-like pattern that is interwoven with an identical pattern of oppositely charged ions. The opposing ion planes are staggered so as to provide electrostatic stabilization within the structure.

of the lipophilic region. A few specific C–C bonds must be *gauche* in such bent conformations. In the DODMAC $X \cdot W$ crystal *gauche* bonds exist at the C_1–N and the C_3–C_4 bonds of one chain; the remainder of the bonds are *trans*.

A space-filling model of a hydrocarbon chain in the all-*trans* conformation has the appearance of a bumpy bar with a rectangular cross-section. The bumps are the hydrogen atoms, and the carbon framework is buried beneath the hydrogen exterior. The fact that the cross-section of straight chain lipophilic groups is structurally anisotropic is extremely important to crystal structure and chemistry. This structural feature is principally responsible for the existence of polymorphic states and structural complexity in the crystal states of many long chain molecules [22,14]. Various patterns for arranging the *x*- and *y*-direction axes of the chains may exist within triglyceride and hydrocarbon crystal structures. In triclinic and some orthorhombic structures the long cross-wise axes of all the chains are parallel, while in other orthorhombic structures this axis in adjacent molecules is rotated by 90° about the *z*-direction (Fig. 8.3). This structural difference is perceptible in the vibrational (infrared and Raman) spectra of crystals, as well as from single crystal structure determination [26].

One should keep in mind that the conformational structure of the chain is not necessarily all-*trans* and straight in all crystals of straight-chain molecules. In lipophilic groups other than those derived from straight chain hydrocarbons, helical structures

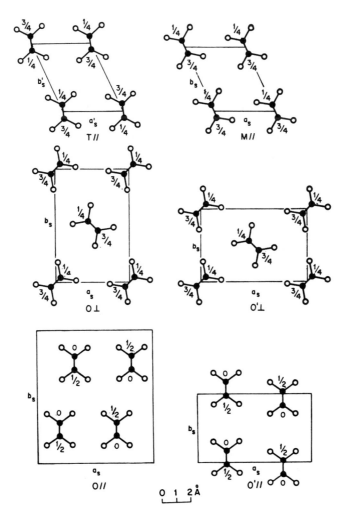

Fig. 8.3.   The arrangement of aliphatic hydrocarbon chains (as seen in cross-section) within the various triclinic (T) and orthorhombic (O) crystal structures formed by long-chain compounds. Reproduced with permission from reference [22].

are preferred over linear ones. Structural studies of small crystalline fluorocarbons [27] and also of fluorocarbon polymers [28] and of hydrocarbon polymers [29] have shown that both fluorocarbons and substituted hydrocarbon polymers crystallize in a helical rather than an extended conformation. The helices vary in the number of carbon atoms per turn and the dimensions of the helix. The existence of progressively increasing disorder as the ambient temperature approaches the temperature of crystal phase transitions must be kept in mind (Section 8.3).

It has also been suggested on the basis of Raman data, supported by the existence of polymorphism, that in AE_x surfactants a transition from an extended conformational structure in the crystal to a helical structure occurs at a value of $x = 3$ [30]. The possibility of helical structures therefore needs to be kept in mind, but it is worth noting that either an extended or a helical conformational structure may exist within the bilayer and other phase structures that surfactants form.

8.3.1.2   Assembling the bilayer

One may visualize creation of the crystal from molecules by first pairing two molecules in a collinear head-to-head or tail-to-tail arrangement, assembling many such pairs to form a bilayer, and then stacking many bilayers to form the bulk crystal. In Fig. 8.4, the individual molecules from which the pair is formed are shown not to have the extended conformational structure, while those in the pair from which the bilayer is assembled are shown to exist in an extended conformation for the sake of clarity in the drawing. (This depiction of a process by which the bilayer is formed is *not* intended to imply the actual mechanism.) If the long axis of such a pair is designated the z-direction, then many such pairs are packed together to form the bilayer so as to elaborate structure in the x- and y-directions, while keeping the polar functional groups in the same $x-y$ planes. In the bilayer assembled from the head-to-head pair (to the left in Fig. 8.4) the polar group planes exist at the center of the bilayer while the terminal methyl groups are in the outer (nonpolar) surface. In the bilayer assembled from the tail-to-tail pair (to the right in Fig. 8.4) the terminal methyl planes exist in the center and the polar groups are found on the outside (polar) surface.

It is possible to cleave a bilayer structure at either a polar or a nonpolar plane without destroying the integrity of the monolayers. The bilayer structure is uniquely "degenerate" from this perspective, in that both the polar and the nonpolar surfaces are planes of zero curvature. Neither the cylinder nor the sphere display symmetry of this kind, since the curvature of the inner surface of cylinders and spheres are opposite in sign from that of the outer surface. The symmetry of the bilayer structure is important with respect to the colloid science of dispersions of surfactant crystals (and also of the related lamellar liquid crystal phase), since both phases tend to be found at interfaces and profoundly affect the properties of the dispersion (Section 13.3.6.4). It has been shown that fatty alcohols may be prepared as crystals that have either a polar or a nonpolar outer surface by selection of the appropriate recrystallization solvent [31]. Alcohol crystals having a polar surface are readily dispersible in water but not in nonpolar solvents, while the reverse is true of crystals having a nonpolar surface.

Because the van der Waal's (contact) volume and radius of methyl groups are substantially larger than are those of methylene groups [32], the closest lateral spacing between chains is smaller than is the closest spacing between juxtaposed planes of methyl groups. The size of both must vary with temperature and pressure in all structures.

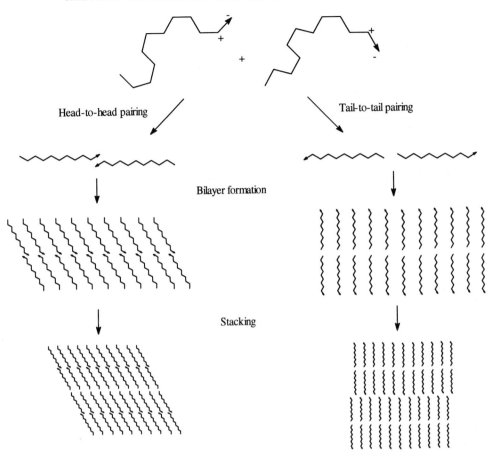

Fig. 8.4.   One way of viewing the assembly of bilayers, starting from pairs of molecules that are aligned either head-to-head (left) or tail-to-tail (right) in the z-direction. Many of these pairs are assembled within the x, y-planes to form a bilayer, and the bulk crystal results from the stacking together of many bilayers in the z-direction. Each of the monomeric molecules is shown in a bent conformation having some *gauche* bonds at the start, but in an extended all-*trans* conformation within the crystal.

Numerous quantitative variants of the bilayer structure exist in surfactant crystals. The angle of tilt of molecular pairs with respect to the x–y plane of the bilayer is one important variable, and the rotational orientation about the z-axis of adjacent molecules is another [6,14].

The bilayer structure within colloidal states. The bilayer structural element of a crystal, as described above, has macroscopic dimensions in the x- and the y-directions while having molecular dimensions in the z-direction (Section 4.19). Many colloidally structured dispersions of surfactant crystal phases may be viewed as having been constructed from fragments of a single bilayer, or from a small number of bilayers. While typically these structures are curved and closed to form spherical vesicles or liposomes [33,34], recent studies suggest that such particles may also exist as open sheets (or in three-dimensional unclosed forms) when the phase structure of the bilayer is crystal-like [35]. In these latter cases the particles do *not* enclose a volume element of the liquid phase within which they are dispersed.

The monolayer half of a bilayer is the basic structural element of surface monolayers such as exist at the air–water interface in a Langmuir trough [36]. Under these circumstances the nonpolar surface is in contact with air (which may be regarded as a "nonpolar" phase of extremely low density), while the polar surface is in contact with a water phase. The monolayer would appear to represent a more drastic perturbation of the equilibrium phase structure than exists within single bilayer particles (such as vesicles). As a result, the relationship between bulk phase behavior and monolayer behavior is likely to be weaker than it is between bulk phase and bilayer behavior. Some relationship nevertheless apparently exists, as indicated by the existence of transition temperatures in monolayers which resemble to a degree those of bulk phases [37].

8.3.1.3 Assembling the bulk phase from bilayers

One may visualize the bulk crystal phase as being formed by stacking together many bilayers in the *z*-direction (Fig. 8.4). Ordinarily the atoms in the outer planes do not penetrate significantly into the opposing plane of the adjacent bilayer during stacking (see, however, Section 8.3.2). The orientation of adjacent bilayers is rigidly defined in crystals, because adjacent layers must be exactly registered with respect to the *x*- and *y*-directions as well being parallel to one another.

Stacking within colloidal particles. Both surfaces of an isolated bilayer are true interfaces, while the surface of a bilayer that is buried within a bulk phase is *not* an interface. Surfaces within phases have been termed "structural surfaces" (Section 4.19). As a result, stacking decreases the ratio of interfacial area to volume, and thereby decreases as well the specific surface area (area/mass) and colloidal energy. Stacking is a favorable and irreversible process thermodynamically, during which the free energy of the mixture is lowered. Conversely, unstacking a bulk phase requires either the input of energy or the doing of mechanical work on the mixture.

Colloidally structured bulk phases. If a stack of bilayers of modest size is curved so as to form a closed many-layered onion-like structure, the "liposome" or "multilamellar vesicle" [33,34] results (the parallel nomenclature for a vesicle having a single bilayer is "unilamellar vesicle"). At temperatures above the Krafft eutectic liposomes are in the lamellar liquid crystal state, but when mixtures containing such particles are cooled below the Krafft discontinuity the phase structure is more highly ordered. In some cases (but not always) the structure within such particles is believed to resemble that of the bulk crystal.

Liposomes may be visible using a light microscope, and thus have macroscopic dimensions. From this perspective they may be regarded as volume elements of the equilibrium bulk phase which have been distorted by curvature so as to present an energetically favorable interface to the surrounding liquid. It thus seems likely that while liposomes possess a lower excess energy than do vesicles, they too may not be exactly equivalent, either thermodynamically or structurally, to the equilibrium phase.

Strong experimental support for this position was obtained during the DODMAC study, where it was shown that vesicular/liposomal dispersions are indeed metastable below the Krafft discontinuity. Their colloidal structure collapses (if a facile reaction path is provided), and the end products of the collapses are single crystals of the dispersed phase. No liquid phase is occluded within these single crystals, and they are not perceptibly curved [38]. From these and other data, it appears that both vesicles and liposomes (below the "chain-melting" transition) represent an energetic, non-equilibrium, colloidal form of the crystal phase which is transformed (given the opportunity) into the bulk single crystal phase. The single crystal is the lowest enthalpy form of crystal phases – and therefore of all the phase states that surfactants form.

8.3.2   Nonbilayer crystal structures

The bilayer is the basic structural element in the overwhelming majority of surfactant crystals, but it is not the only one that exists. Alternative possibilities include inter-digitated structures, and also monolayer structures [39,40]. In an interdigitated structure the head of one molecule is adjacent to the tail of the next one within the structural layer of the crystal, and the surfaces of these structural layers include both the heads and the tails of the surfactant molecules. A monolayer structure differs from both the usual bilayer and the interdigitated structure in that all the molecules within a layer are similarly oriented, but the polar surface of each layer lies against the nonpolar surface of the next layer. Finally, in one dichain cationic surfactant (DOMAC) the lipophilic chains have been shown to be extended, with the polar atoms near the middle of this extended structure and the chains themselves curved [25].

The interdigitated and monolayer crystal structures. The various crystals found within the N-alkanoyl-N-methylglucamine family of surfactants provide examples of both the interdigitated and the monolayer crystal structures. Early studies of the octanoyl (C_7 lipophilic group), nonanoyl (C_8), and undecanoyl (C_{10}) homologs showed that in the nonanoyl and undecanoyl compounds the head groups exist in an extended conformation and the molecules are interdigitated within the layers [39]. In the octanoyl homolog, in contrast, the head groups are bent (due to the presence of a *gauche* bond in the polyol chain) and the crystal has a monolayer structure [40].

In the dodecanoyl (C_{11}) homolog, two polymorphic states have been prepared and the crystal structures of both polymorphs determined [41]. In one polymorph the crystal structure is entirely analogous to the interdigitated structure of the nonanoyl and undecanoyl homologs (above). In the other polymorph the crystal structure is analogous to the monolayer structure of the octanoyl homolog (above). Phase studies have not been performed on the shorter chain homologs, but both the unary phase behavior and the binary aqueous phase behavior of the dodecanoyl homolog have been determined (Fig. 10.21) [42]. It was shown (using calorimetric and X-ray data) that the extended interdigitated structure is the equilibrium polymorph, while the bent mono-layer crystal structure is a metastable polymorph. The evidence from calorimetry is consistent with the existence of a substantial difference in densities between the two polymorphs; the equilibrium crystal ($d = 1.19 \, \text{g cm}^{-3}$) is substantially more dense than is the nonequilibrium crystal ($d = 1.15 \, \text{g cm}^{-3}$) [41]. Both crystal structures were determined at room temperature. A third polymorph (also metastable) exists, but has not been characterized.

Figure 8.5 depicts the equilibrium crystal structure of N-dodecanoyl-N-methyl-glucamine having the interdigitated bilayer structure and extended head group conformation [41]. It is evident from Fig. 8.5 that this crystal retains a layered structure. However, whereas in most surfactant crystals the functional groups of one layer do not penetrate into the space occupied by the functional groups of another, in these structures both the hydrophilic group and the lipophilic group do so; they are

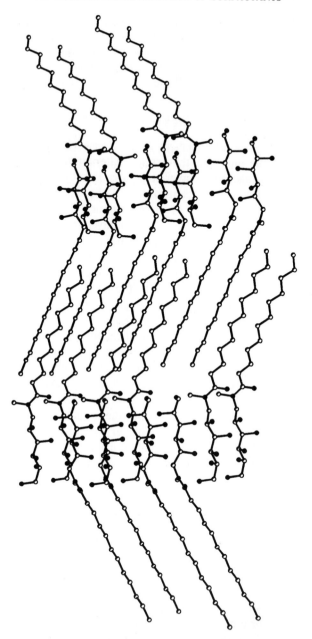

Fig. 8.5. The crystal structure of the equilibrium polymorph of *N*-dodecanoyl-*N*-methylgluca-
mine [41]. In this structure the hydrophilic groups are overlapped or interdigitated, as are the
lipophilic groups. Strong hydrogen bonding exists between adjacent molecules.

"interdigitated". The effect is to merge adjacent layers into one another. One can
imagine constructing this crystal from a molecular dimer in which two molecules lie
side-by-side with their heads overlapping, and with their tails pointing in opposite
directions (Fig. 8.5). If many such dimers were to be assembled into a bilayer (Section

8.3.1.3), the outer end of one hydrophilic group (the C_6' position of the glucosamine moiety) would lie adjacent to the inner (C_1') end of the hydrophilic group of the adjacent molecule. At the same time the C_1' end is also near the ω-methyl group of a third molecule. Such interdigitation is perhaps more likely when the hydrophilic group is so bulky that close packing within a normal monolayer cannot easily occur, and when strong side-by-side or lateral interactions may occur (such as the extensive hydrogen bonding network that exists in the above polyol structures).

The metastable polymorph of the C_{11} homolog displays the structure depicted in Fig. 8.6. In this crystal the individual molecules are all arranged in a head-to-head and tail-to-tail manner within each layer, but the polar face of one layer lies next to the nonpolar face of the adjacent layer. This comparatively rare structure is perhaps stable

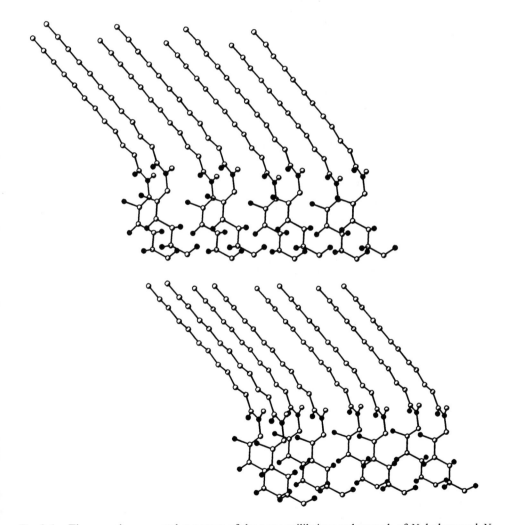

Fig. 8.6.   The monolayer crystal structure of the nonequilibrium polymorph of *N*-dodecanoyl-*N*-methylglucamine. In this structure the molecules are all arranged head-to-head within each layer, and the polar surface of one layer lies next to the nonpolar surface of the adjacent layer.

in this instance because polar interactions are satisfied by the hydrogen bonding that exists within the hydrophilic group layers, and the nonpolar face of the hydrophilic group lies next to the methyl plane of the adjacent layer.

A curved midchain monolayer crystal structure. Dioctadecylmethylammonium chloride (DOMAC) was recently discovered to possess an extraordinary crystal structure with respect to both the conformational structure of the molecules and the packing of the molecules within the crystal (Fig. 8.7) [25]. Both chains are extended away from the hydrophilic group in DOMAC crystals; they are not folded back upon one another in the usual manner. The molecules in this crystal are therefore best regarded as having a "midchain monolayer", rather than a "dichain bilayer", crystal structure.

Also, the first eight carbons next to the nitrogen of both chains (together with the nitrogen) form a smoothly curved 17-atom central chain segment. This curved segment is bracketed by the familiar all-*trans* and straight (zig-zag) outer 10-carbon segments of both chains. The curvature is three-dimensional; the carbon atoms within this segment do not lie within a plane. The curvature results from significant deviations of the dihedral angles of the carbon–carbon bonds from the idealized *trans* value of 180°. It would clearly be invalid in this crystal to restrict the analysis of conformational energies to the idealized *trans* and *gauche* forms.

The chloride ions exist at the mid-point of the amphiphilic ammonium ions. The structure of each ammonium ion is three-dimensional and chiral. Pairs of chiral molecules (having opposite chirality) fit together so as to allow space for the chloride ions. Viewed from a broad perspective, planes of chloride and ammonium nitrogen atoms exist that bisect the pairs of twisted organic cations.

The chloride ions are strongly bonded at the vertices of nitrogen tetrahedra to N^+–H groups. The extremely short N–Cl distance observed (3.05 Å) suggests that this interaction is extraordinarily strong. It may be regarded as a hybrid "hydrogen/ion pair" bond. The unusual strength of this bond may serve to compensate for the energetically costly distortion of the chain conformations away from the *trans* conformation within the curved segment.

The chloride ions also lie (within the same structure) on the face of the tetrahedron of an adjacent ammonium ion to which they are not hydrogen bonded. The N–Cl distance for these particular ion pair interactions is very long (4.25 Å); their length is almost identical to similar interactions within the DODMAC monohydrate crystal (Fig. 8.2) [25]. The strong hydrogen bond/ion pair interaction in DOMAC does not exist in DODMAC due to absence of the N^+–H functional groups.

8.3.3 Polymorphic crystal phases

The existence of both equilibrium and metastable polymorphic crystal states, and the thermodynamic relationships among them, have been considered in Section 5.5.4. Nothing need be said of the structure of equilibrium polymorphs, but the relationship between the phase structure of metastable and equilibrium crystal states deserves further consideration.

Any crystal may exist in a metastable state whose overall structure is qualitatively similar to the equilibrium structure but is quantitatively distorted. Such a state may result if the crystal is formed by freezing the melt so rapidly that sufficient time is not allowed for equilibrium of state to result. Such metastable states are *not* usually regarded as polymorphic states.

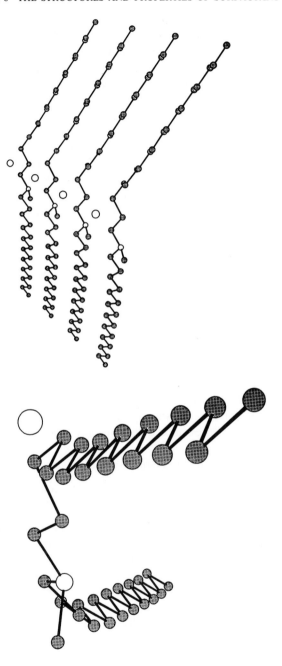

Fig. 8.7.   The crystal structure of dioctadecylmethylammonium chloride (DOMAC) [25]. The manner of packing of several ion pairs is shown at the top. The larger open circles are the chloride ions. The surfactant ion exists in a complex three-dimensional extended conformation (shown below), rather than in the more usual folded conformation (as in Fig. 8.2). The charged nitrogen and chlorine atoms are found near the center of this structure, and the planes of ion pairs bisect the layer. The readily perceptible curvature along the central fifteen atoms of the extended chain is unique among crystalline long-chain compounds.

It is entirely possible that many (if not most) crystals do not exist precisely in their equilibrium state. In studying the $X \cdot W$ crystal of DODMAC, for example, infrared data on the frequencies of the methylene stretching bands as a function of temperature suggested that the conformational structure of the chains in the sample used was, at the outset, more disordered than it would have been in the equilibrium crystal [43]. If so these crystals were necessarily metastable, yet they were large, well-formed, reflective crystals that had been produced by solvent recrystallization.

Crystal phases may also exist that are metastable and also differ qualitatively in structure from the equilibrium crystal. These are properly termed "metastable polymorphs" (Section 5.5.4). A metastable polymorph has a higher free energy than does the equilibrium crystal, and it too may exist as a disordered crystal relative to its "pseudo-equilibrium" structure. The available information on the well-studied metastable α polymorph of triglycerides is consistent with this concept (Section 5.5.4).

8.3.4   Stoichiometric crystal hydrates

Crystal hydrates exist in most aqueous surfactant systems, so it is worthwhile to consider the role that they may play. It would also be useful to be able to anticipate when crystal hydrates exist, because the historic concern with these phases has been minimal.

The most perfect and energetically stable crystals are those formed from molecules that pack densely within a structure that is space-filling. Such molecules are unlikely to form crystal hydrates. Formally charged or strongly dipolar groups within surfactant compounds strengthen crystal phases, but substituents which interfere with close packing weaken the crystal state, strongly depress melting points, and offset the stabilization conferred by polar interactions. Crystal hydrates are highly probable in crystals of strongly polar surfactants whose shape does not allow them to pack densely without water being present.

Water may serve two functions in crystal hydrates:

(1) it is space-filling and improves the crystal packing, and
(2) it may interact energetically with the polar functional groups that are present.

When both of these conditions exist, crystal hydrates are highly probable. However, it is difficult to anticipate when this is true by inspection of molecular structure. When a molecule can pack nicely in a dense crystal and strong interactions which compete with solvation exist within the dry crystal, then crystal hydrates are unlikely. Dioctadecylammonium chloride and the N-alkanoyl-N-methylglucamines, neither of which form crystal hydrates, illustrate the latter situation.

Crystal hydrate water interacts energetically with hydratable functional groups, and thus competes with liquid water for solvation of the molecule. As a result, non-hydrated crystal phases may interact more rapidly with water to form liquid and liquid crystal states (or higher crystal hydrates) than do crystal hydrates. An illustrative comparison is the DODMAC dry crystal X (which reacts with water vapor to form the monohydrate crystal in seconds) with the monohydrate $X \cdot W$ crystal itself (which reacts with water vapor to form the dihydrate $X \cdot W_2$ only after about 3 years [43]).

The water in crystal hydrates may be bound to metal cations via coordination to the oxygen, or to anions or other hydrogen bond acceptor atoms via the hydroxyl hydrogen [44]. Hydrogen bonds exist between O–H bonds and non-bonding pairs of electrons. The strongest hydrogen bond is one in which the O–H bond is collinear with a non-bonding acceptor orbital, but bifurcated hydrogen bonds (in which an O–H hydrogen is shared between two atoms) are known in crystals [44]. Bifurcated hydrogen bonds are unlikely in fluid states.

From these ideas it follows that if a surfactant molecule is bulky then crystal hydrates are more likely. Sodium dodecyl sulfate forms at least three equilibrium crystal hydrates [45], and amine oxides also form crystal hydrates [46,47]. Dodecyltrimethylammonium chloride forms one [48], dioctadecyldimethylammonium chloride (DODMAC) two [43], and dioctadecylmethylammonium chloride (DOMAC) forms at least one crystal tetrahydrate. The zwitterionic surfactant docosyldimethylammoniohexanoate forms four or more [49].

Hard evidence exists that neither dioctadecylammonium chloride (DOAC) [50] nor n-alkyldimethylphosphine oxides [51] form crystal hydrates. The better packing of ammonium ions which bear a hydrogen atom, plus the strong monopole–dipole interaction that likely exists between N^+–H groups and Cl^- anions in DOAC (Section 8.3.2) qualitatively account for its reluctance to form hydrates relative to its methyl analogs. The fact that phosphine oxides do not while amine oxides do suggests that the latter represents a case where the usually strong hydrophilicity of the functional group overcomes the tight packing expected of the small amine oxide group.

8.3.4.1   Nonstoichiometric crystal hydrates

In stoichiometric crystal hydrates the molar ratio of water to surfactant is rigidly defined. From the published work on soap–water systems [52,53], it is also necessary to consider the possibility that nonstoichiometric crystal hydrates may exist in surfactant systems. This means simply that the composition of the crystal phase is *not* stoichiometrically well defined, but varies smoothly within a finite composition range. It has been suggested that such hydrates exist in the sodium chromoglycate–water system [54].

Nonstoichiometric crystal hydrates represent a special case of the more general phenomenon of "solid solutions". Solid solutions are true solutions thermodynamically and in virtually every aspect of their physical science, but they differ from the more familiar liquid solutions in that they have a crystalline phase structure. (Liquid crystal phases are also "solutions", from this perspective.)

In an earlier phase diagram of the sodium palmitate–water system [52], the β, δ, and Ω phases are all represented as nonstoichiometric crystal hydrates over at least part of their range of existence. The δ and Ω phases are indicated to decompose in a normal peritectic manner at their upper temperature limit, but the β phase is also indicated to decompose on *cooling* (at a eutectic discontinuity). This latter behavior is most unusual for a crystal hydrate, since once formed, these phase compounds ordinarily persist at low temperatures. For a crystal hydrate to decompose on cooling is not a violation of the Phase Rule, but it would be extremely unusual.

Stoichiometric crystal hydrates are shown in some of the more recently published soap–water diagrams, but not in others; their existence is suggested by recent calorimetric investigations of soap–water phase behavior [55]. Regrettably, the question of whether or not soaps form stoichiometric hydrates must be left unanswered.

8.3.5   Plastic crystal phases

Mention has been made of disorder within crystal phases (Section 8.3), and that this disorder likely increases as the temperature approaches the melting point. It is worth noting that disorder of perhaps a rather different kind may exist in those crystals which have been termed "plastic crystals". The concept of plastic crystals is due to Timmermans [56], and is an apt term for the crystal state of molecules such as neopentane, *t*-butanol, or *t*-butyl nitrile. The characteristic property of these particular crystals is that an unusual degree of molecular motion and randomness exists which leads to fuzziness in the crystal structure. The unusual level of motion is evident from X-ray, calorimetric, and NMR data.

Ice itself has an unusual level of disorder in comparison to most crystals [18]. The water molecule is unique in that it functions as both a hydrogen bond acceptor and a hydrogen bond donor with similar ease, and moreover is symmetrical in that it may both accept two and donate two hydrogen bonds. This symmetry is responsible for many of its unique properties, among which is the disorder that exists in the crystal state. Because the directions about the oxygen tetrahedron along which the covalent and the hydrogen bonds exist are undefined and fluctuate rapidly, ice has both a disordered structure and an anomalously high entropy. In a sense, it too could be regarded as a plastic crystal.

The disorder that exists in *t*-butanol and *t*-butyl nitrile crystals is isotropic; it is not restricted to any particular direction relative to the structure. Plastic crystal states differ from liquid crystal states in this respect. Liquid crystals display well-defined crystal-like order while at the same time having liquid-like disorder. The disorder in plastic crystals is isotropic, and is generally disruptive of the structure.

8.4   The liquid crystal phases

8.4.1   General properties

The most distinctive attribute of surfactants is the capacity of their aqueous mixtures to exist as liquid crystal phases. Liquid crystals behave as fluids, and are usually highly viscous. At the same time, X-ray studies of these phases yield a small number of relatively sharp lines which resemble those produced by crystals [53]. Being fluids, it is apparent that they are less highly ordered than are crystals, but from the existence of X-ray lines and their viscosity it is also apparent that they are much more highly structured than are ordinary liquids.

Well-defined periodicity of structure always exists along one or more directions and spans many structural planes within liquid crystals, while no comparable structure exists within liquids. This long range periodicity is evident from the narrow widths of the X-ray lines, which are not much broader than are lines produced by crystal phases. This periodicity may extend in one, two, or in all three dimensions in space.

Liquids display no long-range periodicity, and as a result no sharp lines. Instead, a broad halo is observed, at short spacings, which results from the diffuse instantaneous radial distribution of molecules (about a reference molecule) that exists within liquid phases. In liquid paraffin hydrocarbons [53] this diffuse band is found at a Bragg spacing of 4.5 Å, and this same band is seen in the patterns of liquid crystal phases.

In spite of their long-range phase structure, the individual molecules within liquid crystals behave in some respects as if they existed in a liquid. The density of the lipophilic region of liquid crystals is close to that of liquid hydrocarbons [57]. NMR nuclear relaxation studies show that the mobility through space of the protons and carbons within the chain is high [58]. NMR pulsed field gradient self-diffusion coefficients demonstrate that long range molecular diffusion readily occurs [59]. While this diffusion is slower than diffusion within liquids (by two to three orders of magnitude), it is much faster than occurs within crystals. Under conditions and within time spans where the molecules within crystals are confined to the same site, those within liquid crystals are free to move about over long distances.

Hydration of the hydrophilic group partially restricts the motion of the lipophilic groups in liquid crystals [53]. This restriction is selective for the end of the chain that is attached to the hydrophilic group, and its existence is suggested by the fact that the order parameters* of methylene groups near the hydrophilic group are larger than are those of methylene groups remote from the hydrophilic group [60]. The latter closely resemble those found in the liquid state. Experimental evidence for this has been provided by NMR studies of molecules that have been deuterated in specific positions along the chain. This conclusion is further strengthened by the fact that these experimental data are consistent with molecular dynamics calculations on both micellar aggregates and bilayer structures [61].

It is extremely important always to keep in mind that these various dimensions of order and disorder exist within the same phase structure. Liquid crystals are homogeneous one-phase states of matter; they are *not* heterogeneous mixtures of liquid phases and crystal phases.

8.4.2   Rheological properties

Quantitative rheological studies of liquid crystals are supportive of the inference from casual observation that they are, in fact, fluids. They are ordinarily non-Newtonian, which means that the shear stress during flow is not a linear function of the shear strain (shear rate). They are also usually viscoelastic, which means that they display a combination of viscous and elastic responses to shear. When mechanical shear work is done on liquid crystal phases, part of the work is transformed to heat (the viscous response) while the rest is stored (the elastic response). The stored energy (but not the dissipated energy) is recovered when the shear is removed, just as a spring stores energy when it is compressed and releases it when the compressive force is removed.

A particularly important observation is that most liquid crystal phases do not display a substantial "yield stress". This means that the phases flow under stress even when the applied stress is very low. Phases which do have a large yield stress do not flow at all until the applied stress exceeds the yield stress. This is an important characteristic rheological property of solids, which distinguishes them from fluids. Recent studies suggest that cubic liquid crystals *do* display a yield [130].

Viscosity is the constant of proportionality between shear stress and shear strain. The viscosities of liquid crystals at modest shear rates range from slightly higher than that of

---

*The order parameter, S, is defined as $S = \overline{(3/2 * \cos^2 \Theta - 1/2)}$ where Θ is the angle between the director of an aligned CD_2 group and a reference direction. S can be measured by determining the splitting of the deuterium NMR signal of CD_2 groups [129].

liquid water (e.g. in dilute lamellar phases), to many orders of magnitude higher (e.g. in hexagonal and cubic phases). Nevertheless, qualitative observations suggest that the viscosities of even the cubic liquid crystal phases are smaller than that of ice, or of the surfactant crystal phases that may exist (under other conditions) in the same system. A blunt rod penetrates a cubic phase if modest force is applied, while considerable force must be applied to a sharp ice-pick in order to penetrate this phase rapidly.

Rheological studies within the lamellar phase region of the cetyltrimethylammonium bromide–hexanol–water system have been performed at various water contents [62]. This lamellar phase is actually Newtonian at extremely low shear rates ($\ll 10\,\text{s}^{-1}$), but displays shear-thickening (hardening) at higher shear rates in a continuous shear (Couette) rheometer. It is interesting to note that this hardening does not perceptibly alter its X-ray spacings, and requires months to relax.

The bicontinuous cubic liquid crystal phase in the same system (which has a body-centered structure) is viscoelastic. This phase displays a high elastic modulus, and (like the lamellar phase to which it is structurally related) displays shear-thickening.

Studies of the inverted hexagonal phase of a monoglyceride–water system using an oscillatory (rather than a continuous shear) rheometer indicated that this phase is non-Newtonian even at extremely low shear-strain amplitudes (0.0015) [63]. Also, in contrast to the lamellar and cubic phases, it exhibits shear-thinning.

It can be seen from these sketchy data that while the rheology of these phases is complex, they nevertheless behave in their response to shear as fluids rather than as solids. The rheological and acoustic properties of surfactant mixtures, including some liquid crystal phases, have been extensively investigated [64].

Acoustic properties of the cubic phases. The unusual elastic properties of cubic liquid crystal phases lead to some very interesting acoustic behavior on their part. When a test tube containing a uniform, bubble-free, sample of a cubic phase is struck with a hard tool, a characteristic musical "ringing" sound is heard [65]. This sound is distinctly different from the dull sound made by striking a test tube containing a liquid or another liquid crystal phase – a property described to the author many years ago by F. B. Rosevear. (The term "ringing gel" has been used for such phases [66], but the use of this term is unfortunate and misleading. The reason is that these phases are not "gels" at all structurally, as defined in Kruyt [67], but are single-phase liquid crystal states.)

Advantage is taken of the high viscosity of the cubic phase and acoustic effects during phase studies, because the change in the sound of the stirrer hitting the test tube is the most sensitive means that is known for detecting the initial separation of a cubic phase from solution (Appendix 4.4.3). These phases appear to precipitate preferentially at the wall of the container, rather than throughout the liquid as occurs during the separation of other liquid crystal or crystal phases.

8.4.3 Optical properties

The optical properties of liquid crystal phases vary widely. Most liquid crystals are birefringent, but cubic phases are nonbirefringent or isotropic. Concentrated liquid crystal phases are more intensely birefringent than are dilute phases. Like most optical phenomena, the intensity of the birefringence that is observed depends on sample thickness. The birefringence of weakly birefringent liquid crystals may be imperceptible in samples a few microns thick (such as exist between slide and cover slip), but may be clearly evident in samples that are about 1 cm thick. The lamellar phase of the $C_{10}E_3$–water system (Fig. 5.22) is obviously birefringent in a 15-mm diameter test tube, but no birefringence is visible between slide and cover slip (possibly because the phase is so rapidly and easily oriented by the glass surface).

The isotropic liquid crystal phases do not scatter light strongly (are "crystal clear"), but anisotropic phases are visibly turbid. It was this readily visible turbidity in the liquid crystal formed by heating cholesteryl benzoate (relative to that of the liquid that existed at higher temperatures) that led to Reinitzer's discovery of liquid crystals [68]. This turbidity is natural in that it results from the existence of structural defects within the phase [69]. A steady state level of defects probably exists in all liquid crystals at equilibrium, and (in anisotropic phases) results in perceptible scattering of light. Information regarding the nature of these defects is provided by the textures seen in polarized light, and they have been the subject of extensive investigations. Texture depends on both the phase structure itself and the nature of the defects within it; the texture is *not* synonymous with and determined entirely by phase structure. Texture is also influenced by the substrate in contact with the mixture, by the confinement (whether covered or not), by the shear history of the sample, etc. Texture is the macroscopically observable consequence of both the structure of phases and of the defects that exist within their structure.

Defects produce domains within a mixture, and boundaries between the domains. Single-crystal-like order exists within domains, but a discontinuity in refractive index exists at domain boundaries. It is these discontinuities that induce scattering – even though the mixture may be perfectly uniform in composition. While it is likely that impurities concentrate at domain boundaries, impurities are not necessary for domain structure to exist.

The magnitude of the turbidity of liquid crystals is strongly dependent on composition. A very uniform sample (13 mm thick) of the 33% lamellar phase of DODMAC that has been subjected to extensive shear of low intensity is optically clear (by casual observation), while more concentrated mixtures in the same phase region are visibly turbid [43].

Both the optical and the rheological properties of liquid crystals vary widely with composition and temperature within the same phase region. It has been shown that the lamellar phase of Aerosol OT undergoes a reversal of its optical sign as composition is varied [70,71]. At 25°C it is optically negative at <42%, it must be isotropic at 42%, and it is optically positive at >42% compositions.

8.4.4   The theoretical number of liquid crystal structures

Before considering specific liquid crystal structures, it is of interest to consider the number of liquid crystal states that may possibly exist. This question was addressed in 1931 by Hermann using symmetry analysis [72]. He predicted the existence of 18 classes that have qualitatively different symmetries. The Hermann analysis has been reviewed by Mabis, and each of these states was depicted in reciprocal space using multicolored graphics [73]. Even at the time the Mabis analysis was published (in 1962), it was claimed that many of these states had been recognized. Some were found in binary aqueous mixtures, while others had been found only in commercial mixtures.

8.4.5 Lyotropic and thermotropic liquid crystals

The first liquid crystal phases to be discovered, and the only kind mentioned in the Roozeboom treatise, were the cholesteric liquid crystals [68,74]. These are related to the nematic phases, but are modified (structurally and optically) by the existence of chirality in the molecules from which they are formed.

The liquid crystal states of most surfactants do not exist in the absence of a solvent. Solvents that induce formation of liquid crystal (or mesomorphic) states have been termed "mesogenic". Water is a particularly effective mesogenic solvent, but many polyols are also mesogenic [75].

Since solvents are required for most surfactant liquid crystals to exist, they have been termed "lyotropic" liquid crystals. (The prefix "lyo" signifies the action or influence of a solvent in other contexts, as in the "lyotropic series" of salts.) The region in composition/temperature space within which lyotropic liquid crystals exist is rigorously defined, even if rarely determined (Sections 5.6, 5.6.8).

Liquid crystal states that structurally resemble lyotropic surfactant liquid crystals are also found in molecules that do not contain operative hydrophilic groups (Section 9.5.1) [76]. These other liquid crystals do *not* depend for their existence on the presence of a solvent; they result simply from holding such compounds within an appropriate temperature range. Because their existence is dictated solely by the ambient level of thermal energy, they are commonly termed "thermotropic" liquid crystals. More than one thermotropic state may exist in the same system – each being the stable phase within a particular range of temperature (and pressure).

A minimal level of thermal energy is required for both lyotropic and thermotropic liquid crystals to exist. Also, both are transformed into a liquid at sufficiently high temperatures. The difference between them lies in the fact that *both* solvation energy ($n_W\mu_W$) and thermal energy ($3/2kT$) are required to form the lyotropic states (Section 5.6), while thermal energy alone suffices to form the thermotropic states. "Lyothermotropic" is perhaps a more apt descriptive term for the states that surfactants form than is "lyotropic".

The structural requirements that nonhydrophilic molecules must meet in order to form purely thermotropic liquid crystals have been thoroughly studied [76]. Such molecules are typically long and slender, contain a rigid central structural element (such as a steroid ring system, a biphenyl, or a stilbene moiety), possess a significant dipole moment (due to the presence of ester, nitrile, azoxy, or other dipolar but non-hydrophilic groups), and have short, flexible acyclic chains at one or both ends. The properties of thermotropic liquid crystal phases are often influenced by strong magnetic or electrical fields, and advantage is taken of this property in liquid crystal electronic displays. The physical properties of thermotropic states have been exhaustively studied because of their importance to the electronics industry [12,77].

A clean distinction cannot be made between lyotropic and thermotropic liquid crystals, because both exist in a few surfactant molecules. The presence of thermotropic states in surfactant molecules is correlated strongly with the presence of several hydroxy groups; a substantial fraction of the polyol surfactants that have been studied have been found to display thermotropic mesomorphism (Section 10.5.5.1). The thermotropic states formed by surfactants differ significantly from those formed by nonsurfactants in that the former are miscible with water. Water has a finite but small miscibility in all disordered states of low density – whether polar or not. The solubility of large molecules (other than surfactants) in water is, however, much smaller. The solubility of water in the thermotropic liquid crystal phases of surfactants is from 5 to 15%, which is far greater than the normal solubility. If a surfactant forms a thermotropic liquid crystal, the appearance of the composition-temperature space that it occupies is conventional except that it is truncated at the 100% composition boundary (Section 5.9).

The thermotropic states formed by nonsurfactant molecules are not expected to display substantial miscibility with water because the functional groups that exist in these molecules are not operative hydrophilic groups. Thus, while there may be a strong structural similarity between the phases formed by these two kinds of molecules, they differ profoundly in the miscibility of the respective phases with water. Water is miscible with the thermotropic liquid crystal phases formed by surfactants, but not with the structurally similar phases formed by nonsurfactants. Also, nematic phases are very common in thermotropic liquid crystal-forming compounds, whereas this state is comparatively rare in surfactant systems.

8.4.6   The lamellar liquid crystal phase

It is generally recognized that the lamellar phase plays a central role in the evolution of liquid crystal phase structure, and that the geometric features of the other liquid crystal phases may be viewed as perturbations which bend and reshape the lamellar phase towards either the water side or the oil side. This phase may be regarded as possessing one-dimensional long-range order. A systematic description of the evolution of liquid crystal structures, suggested by Fontell [78], is displayed in Fig. 8.8.

One may conceive of the lamellar phase as having been constructed by stacking together bilayers in exactly the same manner as one may visualize the formation of surfactant crystal phases. The assembly of bilayers from pairs of molecules that are coupled head to head (or tail to tail) has been described for crystals (Section 8.3.1.2). Once the bilayer is constructed, the bulk lamellar phase may be envisioned (also as in crystals) as resulting from the stacking of bilayers in the z-direction (Section 8.3.1.3).

The bilayer of lyotropic lamellar phases must include the proper amount of water. In most cases (but not always) more water exists in the liquid crystal than is found in crystals. Both surfactant crystal and lamellar liquid crystal phases would be described in thermotropic liquid crystal terminology as being "smectic" (layered) phases [76].

The molecules in crystals possess a unique conformational structure and are constrained in space. In the lamellar liquid crystal they are "shivering" noticeably, and are also free to move about from one position to another. That is, at any particular point in time they exist in one of an enormous number of conformational states. They are constantly and rapidly undergoing restructuring among these different conformations.

The rigid registration along the x- and y-directions which exists in crystals does not exist in the lamellar liquid crystal. The mean separation distance between bilayers, on the other hand, is well defined. The observation that narrow long spacings corresponding to this separation exist requires that the statistical variability of this separation distance be small, although fluctuations no doubt exist.

The span of compositions within which the lamellar phase exists varies widely, depending on molecular structure. For typical single-chain surfactants, the range of lamellar liquid crystal compositions is from about 60 to 90%, while for dichain surfactants it ranges from about 30 to 90%. In soap-like surfactants the lamellar phase is the

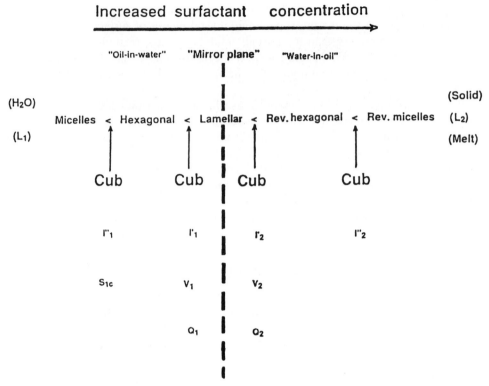

Fig. 8.8. The Fontell scheme for the dependence of surfactant liquid crystal structure on composition, illustrating the symmetry of curvature about the lamellar phase structure. Reproduced with permission from reference [78].

most concentrated liquid crystal phase, which means that this phase is the most demanding of thermal energy and the least tolerant of hydration energy of the various liquid crystal states that exist (Sections 5.6.5, 5.6.8).

In the sulfosuccinate ester class of surfactants, of which Aerosol OT (sodium bis-2-ethylhexyl sulfosuccinate) is the best known member, the lamellar phase is the most dilute liquid crystal phase that exists rather than the most concentrated [70]. In fact, this is the coexisting phase at the liquid crystal solubility boundary in these surfactants, which means that in this class of surfactants it is the phase that is most tolerant of strong hydration energies. A cubic and a hexagonal phase also exist at higher compositions (Section 11.3.2.4).

Chain structure within the lamellar phase. The properties and the state of the lipophilic groups within liquid crystals in general have been described in Section 8.4.1, and a computer simulation of the lamellar phase formed by octylammonium chloride is shown in Fig. 8.9. This figure constitutes a realistic snapshot of this phase as it might actually exist at some point in time. This representation displays not only the existence of long range order and structural regularity, but also the considerable disorder that exists in these states.

In transforming a lamellar crystal to a lamellar liquid crystal of the same composition, it is worth noting that the molecular volume increases – but the average

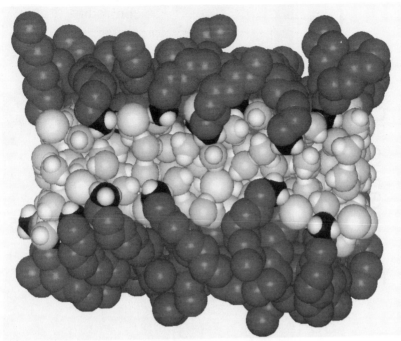

Fig. 8.9.   A computer simulation of the lamellar liquid crystal phase of octylammonium chloride, showing both the lamellar structure and the high level of disorder that exists. Courtesy of M. L. Klein.

length of the chain in the bilayers actually decreases. This shrinkage in length stems from the fact that the all-*trans* conformation represents the maximum possible extended length of the chain, and the introduction of *gauche* bonds necessarily reduces this length. The average area occupied by the molecule must increase by an amount sufficient to overcome the shrinkage in the thickness during this process, so that the expansion is anisotropic. Whether or not a net increase or decrease is observed will depend on the angle of tilt of the chains in the crystal. If they are not tilted then a decrease in spacing is to be expected, while if they are strongly tilted an increase may occur.

Changes in interlayer spacing during phase transitions are revealed with particular clarity during programmed temperature X-ray studies. Such a scan of pure DOACS (dioctadecylammonium cumenesulfonate) is shown in Fig. 8.10. At low temperatures this surfactant is obviously crystalline, as indicated by the existence of numerous diffraction lines. At 73°C it undergoes an abrupt transition to a lamellar phase, and at 121.5°C the lamellar phase melts to form an isotropic liquid. The long spacing of the crystal actually increases in passing from the crystal to the liquid crystal in this compound, but the thickness of the lamellae shrinks as the temperature is increased. Interestingly, just the opposite trends occur within the crystal phase as the temperature is varied; the dimensions of the crystal expand as the temperature increases. Also, the observation that one of the lines crosses over another suggests that the expansion is anisotropic; the coefficient of expansion is greater in some directions than in others.

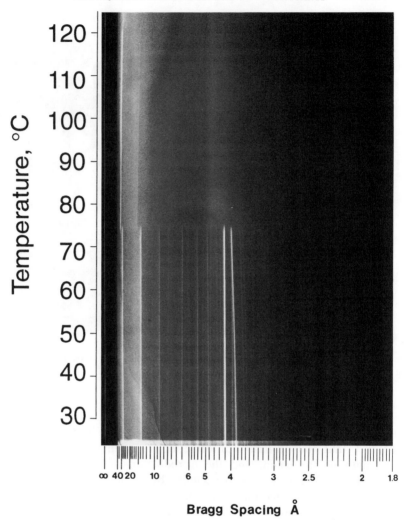

Fig. 8.10. Programmed temperature powder X-ray scan of DOACS, showing the phase transition at 73°C from the crystal to the lamellar liquid crystal phase, and at 121.5°C from the lamellar to the liquid phase. The influence of temperature on the Bragg spacings is also evident. The short spacings of the crystal phase increase with increasing temperature, and a significant increase occurs upon formation of the liquid crystal. The spacing of the liquid crystal *decreases* as the temperature increases. The long d-spacing of this liquid crystal persists in the liquid phase, but is more diffuse.

Another interesting feature of this scan is the persistence of the long d-spacing line of the lamellar phase into the isotropic liquid phase. While this line is broader and less sharply defined in the liquid, it is unmistakably present. This is extremely unusual for liquid phases (of both surfactants and nonsurfactants), and suggests that elements of the lamellar structure persist (in disordered form) within this liquid phase.

An exceptionally large amount of information regarding the state of the compound, and the dependence of state on temperature, is revealed during this one experiment. The lamellar spacings typically vary with temperature over the entire composition range, but the influence of a change of state is not easily visible unless a crystal and liquid crystal of the same composition can be compared (as above).

Areas and spacings within the lamellar phase vs. composition. Mean values for the area requirements of the molecule can be extracted from data on the long d-spacing of the lamellar phase along isotherms from plots of the d-spacing vs. the reciprocal of the volume fraction of surfactant $(1/\phi_a)$, or from the function $\log(d)$ vs. $\log(1/\phi_a)$ [79]. This is often the point of departure for analyses of the geometric aspect of phase structure, in that by making further assumptions regarding the contributions of various structural fragments the dimensions of the hydrated region of the bilayer may be estimated.

It was recognized by McBain [80] that the influence of composition on the area per molecule parameter varied widely depending on molecular structure. McBain described those systems in which the spacing between the lamellae increased with increasing water composition (while the area per head group remained constant) as "expanding" systems. Conversely, he described those systems in which the lamellar spacing remained constant as the water composition increased (which signified that the area per head group increased) as "nonexpanding" systems.

Examples of expanding systems (in which the area/molecule remains constant) include the monoglycerides and Aerosol OT. Examples of nonexpanding systems (in which the area/molecule varies with composition) included the sodium, potassium, rubidium, and cesium soaps, sodium alkyl sulfates, sodium alkanesulfonates, and ammonium salt cationic surfactants. These generalizations (in the soap systems) hold over a wide range of chain-lengths (C_8 to C_{22}). Complex behavior is very often observed in the nonexpanding systems within the water-rich compositions of the lamellar phase, in that both the area per molecule and the lamellar spacing varied with composition. Considerable information regarding this subject is cited in Ekwall's compilation of surfactant phase behavior [79].

The fine details of the structure of liquid crystal phases must surely depend upon both composition and temperature, but information regarding these variations has existed only for a short time. Physical studies of the SDS–water phases strongly suggest that defects in the idealized lamellar structure exist, and that the incidence of these defects varies with composition [81]. Near the dilute limit of the phase, evidence for the incipient formation of the next more dilute phase is found. Information of this sort was unavailable until the advent of synchrotron X-ray sources.

The viscosity of the lamellar phase increases with increasing surfactant composition [43]. Optical textures also suggest that larger domain structure exists in the dilute than in the concentrated lamellar phase [43]. This information suggests that the weak shear plane within the lamellar phase is the structural plane between the polar layers, but this has been disputed [62]. When one recognizes that the lamellar phase often lies next to the crystal, the fact that this phase has a much lower viscosity than does the considerably more dilute hexagonal phase is, at first glance, surprising. It is understandable qualitatively on the basis that weak shear planes likely exist between the lamellae in the lamellar phase while these are absent in the hexagonal. The fact that the viscosity of the lamellar phase is relatively low is critically important to the processing of surfactants (Section 13.4), because it allows this phase to be handled by engineering equipment. The more dilute hexagonal phase cannot be similarly processed, and cubic phases would be still more difficult.

8.4.7   The P'_β, L'_β, and L_α phases

X-ray and NMR phase studies of the dipalmitoylphosphatidyl choline (DPPC)–D_2O system have shown that a phase exists beneath the concentrated part of the lamellar phase region that is called the P'_β phase. The coexistence relationships between the P'_β phase and the surrounding phases have been determined using both deuterium NMR [82] and calorimetry [83] (Fig. 8.11). These two studies are generally in good agreement, except with respect to the composition range of the L_α phase at higher temperatures.

The P'_β phase has a lamellar structure, but differs from other lamellar phases in two important respects. First, the molecules are more highly ordered than in the L_α lamellar phase, as indicated by the position of the X-ray short spacings. Second, periodicity also exists within the plane of the bilayer. This periodicity has a dimension of c. 20 nm, and gives the phase a "rippled" appearance [84]. The eutectic phase reaction in which the P'_β phase is produced via hydration of the L'_β phase is probably responsible for the calorimetric "pretransition" observed in dilute mixtures of DPPC [85].

Whether or not the P'_β phase should be regarded as a liquid crystal is uncertain, because the chains in this phase are more highly ordered than is found in most liquid crystals. It will be regarded here as a liquid crystal because it appears not to be a crystal, it is clearly not a liquid, and it thus has the combination of structure and disorder generally found in liquid crystals. Also, it occupies a region within the phase diagram which resembles that of other liquid crystals.

Stripping water from the P'_β phase of DPPC (and also from the L_α phase that lies at higher temperatures) produces a different kind of lamellar phase which displays no

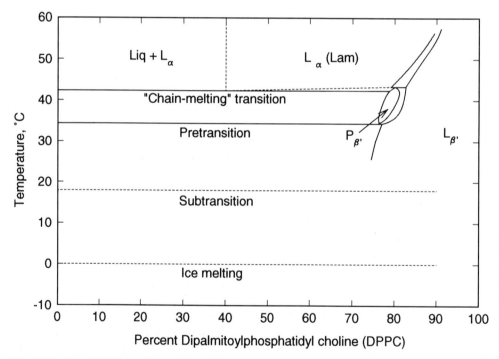

Fig. 8.11.   The P'_β, L'_β, and L_α phase regions in the dipalmitoylphosphatidyl choline (DPPC)–deuterium oxide system, determined using NMR methods [82].

evidence of "ripples", and has been termed the L'_β phase. The existence of a coexistence region between the P'_β phase and the L'_β phase has been established. The L'_β phase, like the L_α phase, spans a range of compositions. Unlike the L_α phase, however, the L'_β phase displays highly ordered chains (X-ray data). The coexistence relationships between the L'_β phase and crystal phases in the DPPC system remain unclear.

In other systems (such as diglyceride glucosides), a phase related to the L'_β phase is found which is very similar except that the molecules are not tilted relative to the structural planes. Following Luzzati, this phase has been termed the L_β phase. Its relationship to crystal phases within these systems is also unclear, but like the L'_β phase it is reported to span a range of compositions.

8.4.8 The bicontinuous cubic phase

The presence of too much water destroys the lamellar phase and produces another phase state. The next more dilute phase beyond the lamellar phase in monolong chain surfactants is often (but not always) the bicontinuous cubic phase. One may infer from its location that this phase is capable of withstanding a greater quantity of water (and of water free energy, Section 5.6.2) than is the lamellar phase. Also, the minimal level of thermal energy (temperature) that is required for the bicontinuous phase to exist is evidently less than is that required by the lamellar phase.

Cubic liquid crystal phases possess three-dimensional long-range order. The presently accepted model for one of the bicontinuous cubic phase structures was first suggested by Scriven [86]. This model has spurred intensive exploration of both the theoretical and experimental aspects of this liquid crystal subclass [87]. Moreover, it has been suggested that such cubic phase structure may be an intermediate state during biophysical processes such as cell membrane fusion, starting from states in which the lipids present are lamellar [88].

Assembly of the structure of the bicontinuous phase may be visualized by starting with a cube which has had each of its eight corners lopped off. Triangular facets would result from this process, while the faces of the cube become octahedral platelets. If many such structures are joined at the corners, a three-dimensional network results which retains cubic symmetry (Fig. 8.12). The plane surfaces of the truncated cubes may be regarded as having a structure that resembles the lamellar liquid crystal phase, and within which the lateral diffusion of surfactant and water molecules is similar. The core of these platelets contain the lipophilic groups, and the polar groups are found on the surfaces. This bicontinuous surface separates the water into two discrete and unconnected regions, one to either side of the structural elements, but both regions permeate the entire volume occupied by the phase.

Since these lamellar structural elements are constructed from molecules bound loosely together by dispersion and polar forces, the formal angular structure can be expected to relax under the influence of thermal energy to form smooth curved surfaces. Figure 8.12 presents two plausible views of the average surface within which the polar groups of the molecules exist.

While it is apparent that this surface is curved, it is also true that the curvature is at all points both positive and negative (as in a saddle), and that the net curvature is exactly zero. This particular structure is the simplest member of a large family of "periodic minimal surfaces". These can be described mathematically and are being discovered to exist not only among cubic liquid crystal states, but also in polymer

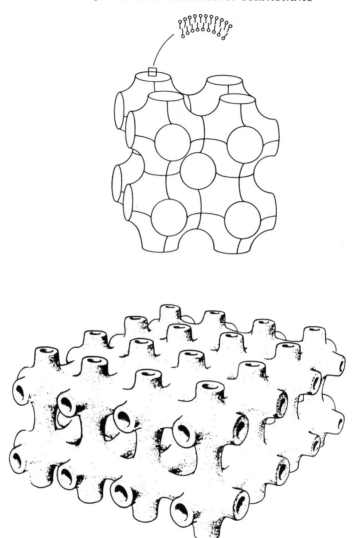

Fig. 8.12. Two views of the bicontinuous cubic phase structure. In the upper version the arrangement of the bilayers within the bicontinuous structural surface is indicated [92], while in the lower figure many of these structural elements are assembled and are smoothly curved [87]. Reproduced with permission from references [87] and [92].

systems [89]. The mathematical description can serve as a point of departure for both the theoretical treatment of the phase and the analysis of electron microscopic images of the states [90]. The relevance to biological processes of the cubic liquid crystal phases, and also their potential value as a form for pharmaceutical preparations, constitute other reasons for the serious interest that presently exists in these phase states [91].

The role of NMR in the determination of cubic phase structure. Historically X-ray methods have been relied on almost exclusively to determine phase structure, but NMR data played a key role (for the first time) in determining the structure of the bicontinuous cubic phase [92]. X-ray data establish that the structure possesses cubic symmetry and (if a sufficiently large number of lines are available) can establish the space group. X-ray data do not, however, provide the principal evidence upon which the above structure is based.

Inspection of the bicontinuous cubic phase structure reveals that it is possible for a surfactant molecule to pass smoothly from one cube to the next without ever leaving the bilayer. In fact, diffusion within this phase might be expected to closely simulate lateral diffusion within the lamellar phase itself. Similarly, it is possible for water molecules within the structure (which exist in two discrete regions to either side of the bilayers) to pass throughout the structure without ever having to cross a bilayer.

Pulsed field gradient NMR self-diffusion studies were therefore performed on a carefully selected system (monoolein–water, Fig. 5.15). Isopleths exist in this system along which one can pass directly from the lamellar phase to this cubic phase. Compositions change along this path only within the narrow miscibility gap that lies between the two-phase regions, and if the temperature span of this gap is passed quickly enough, significant disproportionation in composition should not result.

The self-diffusion coefficients of both the surfactant and the water were determined along this isopleth. The variation of temperature will necessarily change diffusion coefficients (D), but if the activation energy for diffusion remains constant then $\ln(D)$ will vary linearly with the reciprocal of temperature $(1/T)$ (as expressed by the Eyring equation [92]). Indeed, it was found that the diffusion coefficients of the surfactant along this isopleth fell on precisely the same Eyring plot within both the cubic and the lamellar phase regions. No discontinuity in the activation energy for diffusion was observed at the phase discontinuity. The same result was found for the water. (The diffusion coefficients and activation energies for the water and the surfactant are different.) This result is especially noteworthy when the vast difference in bulk viscosities that exists between these two phases is recognized.

These data eliminate those earlier models for the structure of this cubic phase in which the surfactant exists within discrete particles separated by water regions. The bicontinuous structure is consistent with X-ray data, and also with earlier NMR studies which showed that narrow high resolution NMR bands are found in the spectra of this viscous phase state [93]. The body of information that presently exists is such that the above phase structure is generally regarded as having been firmly established.

8.4.9 The "normal" hexagonal phase and the "rectangular" phase

The bicontinuous cubic phase is found only within narrow hydration energy limits which lie above that of the lamellar phase, and it too is destroyed by too high a level of hydration. The next more dilute phase that is produced is usually the "normal hexagonal" phase, or possibly its cousin the "rectangular" [94] or "monoclinic" [45] phase. The hexagonal phase may be regarded as possessing two-dimensional long-range order. Quite often the hexagonal phase is the most dilute liquid crystal phase that exists, but sometimes dilution of the hexagonal phase produces a discontinuous cubic phase before the micellar liquid phase region is reached.

The hexagonal phase may be visualized as resulting from breakup of the lamellar phase into cylindrical elements by the addition of water. The diameter of the cylinders is similar to the thickness of the bilayer, and their length is indefinitely long.

One may imagine constructing the bulk hexagonal phase by arranging these cylindrical aggregates in a parallel, hexagonally close-packed, manner (like bowling pins viewed from above). A plane passing through this structure (perpendicular to the direction of the cylinders) would reveal circular cross-sections, and the centers of adjacent cylinders lie at the vertices of equilateral triangles. Six adjacent triangles which have a common vertex form the hexagonal pattern. Two-dimensional order exists within this cross-sectional plane in this phase, one-dimensional order (only in the direction perpendicular to the bilayers) exists in the lamellar phase, and three-dimensional order exists in all the cubic phases.

Top and side views of a computer simulation of the structure of the hexagonal phase of sodium octanoate are shown in Figs 8.13 and 8.14. As in Fig. 8.9, these represent a probably realistic snapshot of the phase that depicts both the overall structure of the phase, and the disorder that exists within this structure.

Just as defects are presumed to develop in the lamellar phase as it is diluted with water, so it is as the hexagonal phase is diluted [81]. These defects may take the form of interruptions in the continuity of the cylinders. If such defects occur sufficiently often, the stubby cylindrical building block of the discontinuous cubic phase would result from this process (next section).

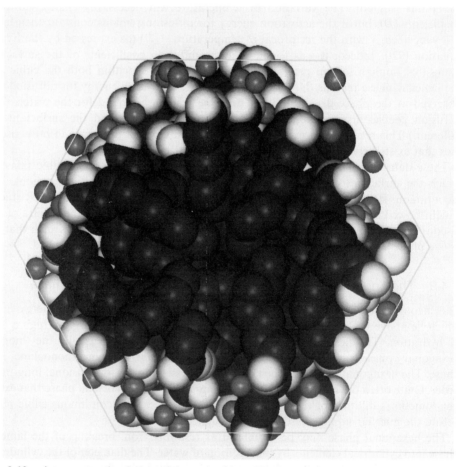

Fig. 8.13.   A computer simulation of the normal hexagonal phase formed by sodium octanoate, showing the hexagonal arrangement of the cylindrical structural elements and the disorder that exists. Courtesy of M. L. Klein.

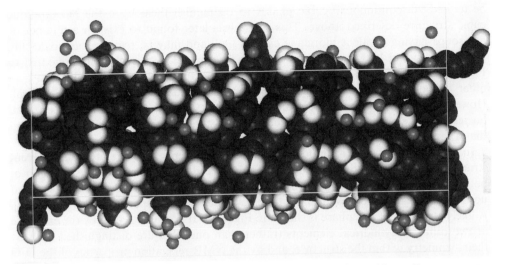

Fig. 8.14. A computer simulation of the normal hexagonal phase formed by sodium octanoate, as seen in a direction perpendicular to the axis of the cylindrical elements. Courtesy of M. L. Klein.

The rectangular phase. The rectangular (monoclinic) phase may be regarded as a distorted form of the hexagonal phase in which the cylindrical elements are flattened. If flattened cylinders were packed together in a manner qualitatively similar to that found in the hexagonal phase itself, the resulting phase structure would also be flattened and less symmetric than the hexagonal phase. In the SDS–water system, X-ray and NMR investigations suggest that the hexagonal and the rectangular phases are very closely related structurally [45]. Taken literally, this diagram indicates that the rectangular structure coexists with the hexagonal phase at low temperatures, but changes smoothly into the hexagonal structure as the phase is diluted at higher temperatures.

Both the hexagonal and the rectangular phases are birefringent. The hexagonal phase (like the lamellar phase) is expected to be optically uniaxial (Section 8.2), while the rectangular phase should be biaxial. This is very difficult to investigate, since orienting the hexagonal phase along its optic axis is highly improbable. Apparently such orientation has been accomplished, recently, in selected systems having suitable phase behavior [95]. Hexagonal phases are smectic, in thermotropic terminology, since layers of the cylindrical structural elements may be perceived in viewing the structure.

8.4.10 The discontinuous cubic phase

It has long been known that cubic phases may exist to either side of the hexagonal phase region. During the earliest studies of these phases, when it was thought that they were constructed by packing together spherical aggregates, the existence of two such phases in these locations was very puzzling.

In recent years, support for models of cubic phase structures that are based on spherical aggregates has dwindled. Instead, structures constructed from short cylindrical elements are now accepted as being more likely. These structures were first suggested for the cubic phase of dry strontium myristate at 223°C [96], and may be termed "discontinuous" cubic phases to distinguish them from the bicontinuous cubic structure described above. This phase was later found to exist in a number of dry soaps having divalent inorganic cations [96], and may be regarded as possessing three-dimensional long-range order. A view of a model of this phase is presented in Fig. 8.15. This structure is composed of two interlocking arrays of small cylinders. Two other cylinders are joined to each end of each cylinder, and each array is chiral. However, the two arrays possess opposite chirality and the structure does, in fact, possess cubic symmetry. As with bicontinuous cubic phases, it has been proposed that a family of such structures exists [98].

In a number of lysophosphatidyl choline–water systems [97], a related but different structure for the discontinuous cubic phase has been suggested from consideration of both X-ray and NMR data. As above, the basic structural elements of these discontinuous cubic phases are believed to be stubby cylinders whose aspect ratio is about 2 [97]. A likely structure of this phase is shown in Fig. 8.16. It is characterized by the existence of two different cylindrical elements (rather than one) that are distinguished both by their symmetry within the structure, and by the NMR relaxation properties of the D_2O within them.

Fig. 8.15. A photograph of a model of a discontinuous type of cubic phase such as seen in dry strontium myristate at 223°C, prepared by K. Fontell [96].

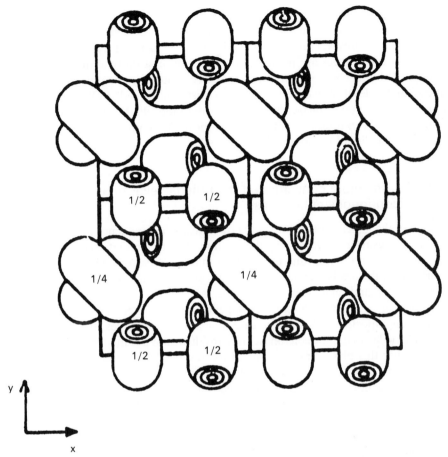

Fig. 8.16. A plausible structure of the discontinuous cubic phase of a lysophosphatidyl choline. NMR data rule out the possibility that all the cylindrical structural elements are identical, and this structure is consistent with the existing data. Reproduced with permission from reference [97b].

The qualitative impact of discontinuous structure on self-diffusion coefficient data is evident. The self-diffusion coefficients of the surfactant must be reduced, relative to that in a lamellar phase, because each of the cylindrical elements is a discrete particle. For a surfactant molecule to diffuse throughout the structure it must hop from one cylinder to the next, and the activation energy for this process should be higher than is required for diffusion within a bilayer. Both X-ray and NMR data support the existence of discontinuous structure, but details of the structures may be expected to change.

The existence of two phases of cubic symmetry that have dramatically different structures provides a plausible explanation for the existence of cubic phase regions to both the concentrated and the dilute side of the hexagonal phase region. Starting from the hexagonal phase, it is understandable that the removal of water might induce the fusion of cylindrical elements to form the bilayers of the lamellar phase. That this might occur in two discrete steps (via the bicontinuous cubic phase) is rational, even if unanticipated. Indeed, this phase is not always found.

The breakup of the long cylindrical elements of the hexagonal phase into shorter cylinders of finite length by increasing the fraction of water is also rational. Again, the actual phase structures were not anticipated before the fact, but it is reassuring that the observed phase structures are related to composition in a rational manner.

As required by its composition range, the discontinuous cubic phase (when it exists) requires a higher level of hydration energy to exist, and extends to lower temperatures, than does the hexagonal phase. No liquid crystal more dilute than the discontinuous cubic phase has been found to date; the next more dilute phase encountered is invariably the micellar liquid (in soluble surfactants).

8.4.11 Phases more concentrated than the lamellar phase: the "inverted" bicontinuous cubic phase

Returning to the lamellar phase, we may now consider the influence of changes in phase structure that result from decreasing the fraction of water present. Such changes should induce curvature in the opposite direction to that which is considered above. In soap-like surfactants this leads inevitably to crystals, but the situation is different in the case of Aerosol AOT and related surfactants [70].

As noted earlier (Section 8.4.6), the lamellar phase in Aerosol AOT occurs at far more dilute compositions than is the case in soap-like molecules. If water is removed from this lamellar phase a cubic phase may result within a finite temperature range. It is possible that this phase has an inverted (or reversed) bicontinuous cubic structure. If it does have this structure, it would qualitatively resemble the "normal" phase described in Section 8.4.8 except that the water (rather than the lipophilic tails) would occupy the core of the lamellar structural elements, and the tails would be on their surface. In such a structure the molecules to one side of the bicontinuous aqueous surface would not exist in an environment that is identical to the molecules on the other side.

8.4.12 The "inverted" hexagonal phase

If water is removed from the Aerosol AOT cubic phase, a hexagonal phase may result, again within a particular (still higher) range of temperatures [70]. The structure of this phase is geometrically similar to that of the "normal" hexagonal phase (above) – except that the core of the cylindrical elements is aqueous and the lipophilic groups are located on the perimeter. This same phase structure also exists in many ternary surfactant–oil–water systems between the lamellar liquid crystal and the microemulsion or L_2 phase regions (Section 12.4). This phase is also commonly found at low water compositions in weakly hydrophilic surfactant–water systems such as the monoglycerides. The monoglyceride hexagonal phase was originally identified as a normal phase [99], but is now believed to have an inverted structure [100].

At this time an inverted discontinuous cubic phase (which would be the structurally symmetrical counterpart of the normal analog) has not been found and identified with certainty, although one might suspect that the dry strontium and barium soap phases described above (Section 8.4.10) might fall within this category. Generally, the next more concentrated phase than the inverted hexagonal phase is the crystal.

8.4.13 The lyotropic nematic phases

A nematic phase has been observed to exist in a number of aqueous surfactant systems [101,102,103,104]. This phase is recognizable by its distinctive optical texture when viewed in the polarized light microscope. In addition, its texture changes when a strong magnetic field (of the order of 1 tesla) is applied.

The nematic phase is the least highly ordered of the liquid crystal states. The only element of symmetry that exists is the fact that the long axes of all the molecules are oriented in the same direction; disorder exists in all other respects. In smectic phases, additional constraints and less disorder exist as a result of their layered structure.

Since nematics are the least ordered liquid crystal phase, they are also usually the most stable at high temperatures in thermotropic systems. They are typically the last phase to melt, and when found usually coexist with the isotropic liquid phase [76].

Nematic textures have been seen in commercial surfactant mixtures [4], in ternary cationic surfactant–salt–water systems [102], and in binary anionic systems of surfactants containing perfluoroalkyl [103] or semi-rigid (biphenyl-containing) lipophilic groups [104]. The existence of a narrow nematic phase region lying between the lamellar and the micellar liquid phase in the cesium perfluorooctanoate–deuterium oxide system has been reported [101]. In this instance particular care was taken to establish that the phase was thermodynamically stable, since it could be found when approached from both high and low temperatures. The phase reaction that occurs at the upper temperature limit of its existence is uncertain. From the appearance of the diagram the phase appears to have decomposed via a peritectic reaction, but this is not certain.

The cesium perfluorooctanoate–deuterium oxide nematic phase is clear to the naked eye and fluid, but is birefringent between crossed polars. It displays the characteristic nematic texture and orientation in a magnetic field. The D-NMR signal is split in both the lamellar and the nematic phase in this system, but the splitting is smaller in the nematic so that both phases may be seen in biphasic mixtures. Later investigations of structurally similar perfluorinated surfactants (short chain rubidium and ammonium salts of perfluorocarboxylic acids) revealed that this phase exists in several of these systems [103]. Neutron scattering studies suggested that disk-like aggregates exist in both the nematic phase and the coexisting isotropic liquid [103].

A cautionary note must be introduced in connection with lyotropic nematic phases, especially when the evidence for their existence hinges on the alteration of optical texture by application of a strong magnetic field. Equilibrium phase behavior can be altered by such fields (Section 4.17.5), and one assumes in general that phase studies are performed at the weak ambient magnetic field strength at the surface of the earth. It is possible for nematic phases to exist as equilibrium phases only at strong magnetic fields, but not at the ambient magnetic field strength at the surface of the earth. In addition to the usual concerns with establishing phase behavior, therefore, the existence of nematic phases must be established at both normal and high magnetic field strengths before they can be regarded as equilibrium phases at normal magnetic field strengths. This appears to have been done in the above-mentioned studies, since regions of coexistence with the liquid were evident during some studies [101] and the phase was revealed by swelling (contact) methods in others [104].

8.4.14 The "intermediate" phases

It was demonstrated during a reinvestigation of the binary soap–water systems using improved calorimetric methods that two liquid crystal phases existed that had gone unrecognized for more than a half-century. Since at first the structures of these phases were unknown – only their approximate composition range and the lower temperature limit of their existence – they were termed "intermediate" phases [105]. Their existence has since been confirmed using other methods (penetration and NMR studies [106]), and it has been suggested that they exist in cationic surfactant systems as well [107]. It is possibly fair to characterize the phase states in the SDS–water system, for example, that occupy narrow composition range between the broad hexagonal and lamellar phase regions as "intermediate" phases as well. In this instance, however, the structures were firmly established using synchrotron X-ray and other studies.

One might also view the cubic phases as being "intermediate" states in that they too span narrow regions between the two principal liquid crystal phases. In fact, this phase was often overlooked during the initial phase study (as in the dodecyldimethylamine oxide–water system), and discovered only during a more detailed re-examination of the system. It is worth noting that a cubic phase does not always exist between the hexagonal and lamellar phases. The alkyldimethylphosphine oxide–water systems, for example, are reported not to have a cubic phase; the hexagonal and lamellar liquid crystal phases are separated by a liquid phase, insofar as is presently known [108].

The door has only just been opened as regards information concerning the intermediate phases. Now that their existence is known, it should be possible to determine whether or not they exist and to characterize them if they do. If this is to be done, the methods used will have to include not only the traditional isoplethal methods but the newer isothermal swelling methods as well [109]. The latter constitute the only practical approach that is virtually guaranteed to recognize the existence of such phases and provide information regarding their composition range.

8.4.15 The "K" phase

During investigations of the ternary potassium decanoate–octanol–water system, a birefringent phase was discovered to the water-poor side of the lamellar phase that is clearly different from both the lamellar (D) phase and the inverted hexagonal (F) phase, and was termed the "K" phase [110]. The X-ray reflections of the K phase are in the sequence 1 : 2 : 3 : 4, and they are spotty. Later, evidence for the existence of this phase in the hexadecyltrimethylammonium bromide–alkanol–water system, and possibly also the hexadecyltrimethylammonium sulfate–decanol–water system, was obtained. The structure of this phase, and its relationship to the neighboring phases, remains unclear.

8.5 The liquid phases

8.5.1 Binary aqueous surfactant liquids

From casual observation, from their optical texture, and in many of their properties the liquid phases formed by surfactants are no different from most liquids. However, it has been recognized since the work of Krafft [111] that considerably more structure exists in liquid surfactant solutions (over a large part of their composition range) than is found in most liquids. This structure sets them apart from the usual "unstructured" (or more properly, "normally structured") liquids.

In aqueous surfactant liquids that are rich in water the important structural feature is the micellar aggregate. The size and shape of micellar aggregates varies enormously, from a few tens of molecules (or ion pairs, in the case of short chain ionic surfactants) [112], to long cylindrical micelles of the same radius but considerably greater length [113], to enormous flexible thread-like micelles (also having a similar radius) that are cross-linked so as to create a gel-like micellar network [114]. It has been suggested that micellar aggregates may assume a hollow spherical (vesicle) structure, but this proposal cannot presently be regarded as having been firmly established. The amount of effort that has been invested in studies of the structures of micellar solutions is staggering, and its treatment in depth is beyond the scope of this book.

A computer simulation of a small "spherical" micelle of sodium octanoate is shown in Fig. 8.17. The molecules in this aggregate are indeed clustered about a point and centrosymmetric. However, it can be seen that the instantaneous structure is anything but spherical, that the surface is only sparsely covered with hydrophilic groups, that extensive contact must exist between the lipophilic groups and water, and that a complex situation exists with respect to the position of the counterions relative to the aggregated structure. Obviously, the beautifully ordered circular cross-sections of spherical micelles found in cartoon representations of this aggregate are in most details misleading.

Dilute liquid phases containing spherical micelles are Newtonian liquids whose viscosities are not much greater than that of liquid water. Liquid phases containing cylindrical (rod) micelles are considerably more viscous, and also display pronounced (often readily observable) shear-birefringence. This may often be observed in a test tube during a step-wise dilution phase study (Appendix 4.4.3) as a general brightening of the mixture (observed between crossed polars) while it is being stirred, which disappears when the stirring stops. Often the decay of the brightness requires several seconds, and is thus directly observable. It is common for micellar aggregates to be spherical in dilute solutions, but undergo a "sphere-to-rod" structural transition within a narrow composition range as the composition is increased. In many (but not all) soluble surfactants, a significant increase in viscosity occurs at concentrations within about 10% of the composition of the liquid crystal solubility boundary. Heating these viscous solutions typically causes the viscosity to drop suddenly to values suggestive of the existence of spherical micellar solutions. The existence of rod-like micelles thus appears to be not only concentration- but temperature-dependent. Rod-like micelles are favored by high concentrations and low temperatures, and the onset and disappearance of these large micelles can be rather sudden. As with micellization itself, these sudden changes are not discontinuities of state.

Viscoelastic mixtures. In a few systems (such as cationic surfactants having specific amphiphilic counterions, e.g. salicylate), the liquid phase is viscoelastic [114]. That is, it not only displays viscous flow properties, but also has perceptible elasticity. The same phenomenon has been observed in solutions of zwitterionic surfactants, such as $C_{16}APS$ [115]. The viscoelasticity is often enhanced by the presence of salts in the solutions.

Such viscoelastic liquids may display the property of "recoil". This means that when the liquid in a conical flask is set in motion by swirling the flask and the swirling is stopped, the liquid first slows to a halt and then reverses direction for a significant

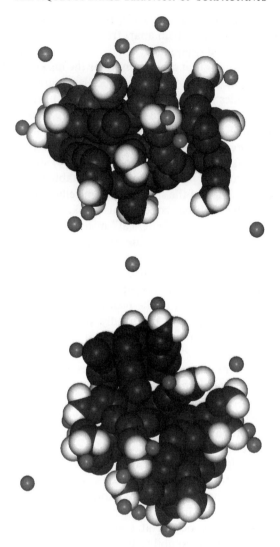

Fig. 8.17. A computer simulation of the instantaneous shape of a small micelle of sodium octanoate at two different times, showing both the severe distortion from a spherical shape and the strong dependence on time. Over long times (microseconds), the structure will appear to have an average structure that is centrosymmetric. Courtesy of M. L. Klein.

distance (perhaps half a revolution). The phenomenon is most clearly evident when a few small visible particles (such as air bubbles) are suspended in the mixture. It has recently been shown (using direct investigations by cryoelectron microscopy) that long, probably flexible, thread-like micelles exist in such viscoelastic liquids [114]. These extremely long micelles are occasionally crosslinked so as to form a gel network within the liquid phase. Such a structure provides a rational explanation for their viscoelasticity.

It was suggested in the older literature [116] that more than one sudden change, akin to the critical micellar concentration (cmc), occurs in the liquid phase of surfactants. It is likely that sudden changes in liquid structure such as those described above are responsible for these earlier observations.

Every one of the above-described variations in liquid structure occurs within a single homogeneous liquid phase. None of the above structural changes signifies the existence of discontinuities of state, regardless of how sudden or sharp they may appear to be.

Water-poor liquid phases. It has been noted (Section 3.12) that at a sufficiently high temperature the liquid phase probably spans the entire composition range in all aqueous surfactant systems. It is also clear that, at least in some surfactant systems, the very low water part of this liquid region is "unstructured" even at ordinary temperatures. This requires that, at intermediate concentrations, there must be a transition from a micellar liquid structure to these "unstructured" liquids as the concentration is increased.

The usual consequence as to phase behavior of adding water in small amounts to a crystalline surfactant is to depress the temperature of the liquidus boundary. The liquidus is the boundary at which the crystal phase disappears when mixtures of the crystal and liquid phase are heated; at temperatures just below this boundary the liquid coexists with the surfactant crystal. Inspection of this region, alone, in surfactant–water diagrams provides not the slightest hint of unusual phase phenomena elsewhere in the diagram (such as the existence of liquid crystal phases).

The influence of water on the liquidus boundary of dodecyldimethylphosphine oxide is indistinguishable from the influence of the homologous alcohols between C_1 and C_{12} (Fig. 5.3). The position of the liquidus boundary in the water system is virtually the same as in all the alkanol systems, if the compounds are compared on a mole fraction (rather than a weight fraction) basis. Since exaggerated aggregate structure (such as micelles) does not exist in such surfactant–alkanol mixtures, this result suggests that aggregate structure also does not exist in aqueous mixtures within this narrow span of low water compositions. (Comparable studies have not carried out using ionic or zwitterionic surfactants.) However, at water compositions beyond the limit of the liquidus boundary, the formation of liquid crystal phases in water sets this solvent apart from all the alkanols.

The structure of surfactant liquids must obviously vary widely across the entire composition range. At surfactant compositions below the cmc (at temperatures above the knee in the Krafft boundary (Section 5.4.1)), it is clear that such liquids are "unstructured" and that a molecular solution of the surfactant exists. At extremely high surfactant (low water) compositions it is likely that this liquid is also "unstructured", although relatively little hard evidence as to its structure exists. If so, structural complexity within liquid phases is found only at intermediate compositions. Micellar aggregates clearly exist within the water-rich region of this intermediate range of compositions, but these aggregates apparently disappear as the water composition is reduced. Intuitively one might suppose that a minimal fraction of water must be present for micellar structure to exist, but there is always the possibility of inverted (or yet-to-be-defined) structures in this transition region. The manner in which this transition from micellar to "unstructured" liquid occurs, as water composition is reduced, remains for the most part obscure.

It is likely that the structural profile in this transition region varies substantially from one surfactant to another. It has been shown, for example, that even extremely concentrated (70%) solutions of a zwitterionic surfactant (nonyldimethylammonio-propyl sulfate) display small spherical aggregates [117].

The situation as to structure in the dry liquid phase is also very unclear when the surfactant itself may exist as a thermotropic liquid crystal phase, as in Fig. 10.21.

Continuity in liquid crystal and liquid phase structure with composition. It has been widely recognized that a smooth variation in the geometric features of surfactant phases exists, and that this spans both the liquid crystal and liquid phase states. Starting from the lamellar phase, increasing the water composition (and also, necess-arily, the water activity, a_W), results first in the bicontinuous cubic phase, then the rectangular and/or hexagonal phases, and finally the discontinuous cubic phase. Cylindrical or thread-like micelles in the liquid phase bear an obvious structural resemblance to the structural elements of the hexagonal phase with which the micellar liquid phase often coexists. They also resemble, to a lesser degree, the short rod-like structural elements of the discontinuous cubic phase. It is reasonable that, as the water composition (and water activity) is increased in the liquid phase, the cylindrical aggregates should be broken into smaller (spherical) fragments having a higher hydrated area/volume ratio.

8.5.2 Ternary aqueous surfactant liquids (microemulsions)

In ternary mixtures containing water, a surfactant, and a third component that is liquid at or near the temperature of the study, another kind of liquid phase must exist (Chapter 12). In systems where the third component is a high-melting solid (as with sodium chloride) this issue is moot, but it is very important in surfactant–oil–water systems.

Two separate liquid phase regions necessarily exist in oil-containing aqueous surfac-tant systems if the oil is immiscible with water. In many such systems these regions (labeled by Ekwall as L_1 and L_2) are connected to the pure water and pure oil vertex of the phase diagram, respectively, and remain unconnected in the ternary diagram. (See, for example, the various ternary surfactant–oil–water phase diagrams in Chapter 12). If the oil component is a small amphiphilic molecule having substantial water misci-bility (such as butanol, Fig. 12.17), these two liquid regions may be fused into one if the composition of surfactant is sufficiently high. A continuous liquid phase band which spans the entire oil–water composition range exists in such cases.

The existence of mixtures that contain large fractions of both oil and water within a single phase was first recognized by Schulman in quaternary systems, although the phase behavior of these systems was at first not correctly stated. The Schulman mixtures typically included a surfactant (such as potassium oleate), water, a nonpolar (hydrocarbon) oil, and a small amphiphilic oil (such as *n*-pentanol). The latter compo-nent was termed a "cosurfactant". Mixtures that closely resemble Schulman's mixtures are common in simpler ternary systems, however, if the oil is amphiphilic rather than nonpolar.

Schulman viewed his mixtures as being emulsions of water in oil in which the size of the water "droplets" was very small. The name "microemulsion" later given to these mixtures is fully consistent with his view that they were emulsions of water in an oil phase.

It is presently generally accepted that microemulsions are more accurately described as single liquid phases that are highly structured, and may contain "aggregates" of water molecules. These water aggregates are surrounded by all the various amphiphilic molecules present. Regardless of the phase problem, Schulman's proposal as to their structure has, in fact, been substantiated within part of the composition range of the L_2 phase in many such mixtures. It has been challenged in other parts of this phase region, however. It is presently believed (largely on the basis of neutron scattering studies) that, in the more water-rich regions, such liquids may have an irregular bicontinuous structure which qualitatively resembles (to a small degree) the bicontinuous cubic liquid crystal phase structure (Section 8.4.8). This picture can be expected to evolve and become steadily more refined as research in this area proceeds.

The matter of microemulsion structure, and the physics of these phase structures, is thoroughly treated elsewhere [118,119]. It suffices for the present purpose to recognize that both water-rich and oil-rich liquid phases exist in ternary aqueous surfactant systems that also contain an oil, and that these are often structured to an unusual degree compared to most liquids.

8.6   The nomenclature systems for surfactant phase structures

The great variety of phase states that exist in surfactant systems poses a serious nomenclature problem. However, numerous uncertainties exist which preclude the development of an entirely satisfactory system at this time. Ideally such a system should be based on structural criteria, but determining the structure of fluid phases is difficult and laborious, and the fraction of the phases that have been discovered whose structure is rigorously defined is very small. This being the case, it may be preferable to hold back in naming these phases until a firm basis for structural assignments exists across a wider range of systems.

Ekwall published in 1975 a summary of the then-existing systems for naming liquid crystal phases [120]. This is updated and reproduced in Table 8.1. No new systems of nomenclature have been proposed since that time, and the issue is avoided insofar as possible in this book by the use of self-explanatory abbreviations (Lam, Cub, Hex). When more concise symbols are required, Ekwall's simple alphabetic nomenclature is useful for designating phases and identifying biphasic and triphasic regions, and these are defined within each diagram.

8.7   The absences of phase structures

The focus of this chapter to this point has been on the phase states that exist and their structures, but it is worth noting that no system is known within which all the phases described above are found. This raises the nontrivial question, "why not?".

A phase either exists or it does not, but even if it does exist it may not appear on the phase diagram. One obvious reason why a phase may not appear is that, during the phase study, the composition/temperature space where the phase exists was not explored. However, a second and more important reason is that the investigator failed to recognize its existence – even though the composition and temperature range where it is found was, in fact, explored.

Phases that exist but have been missed. Missing a phase is a general problem in phase research, but this problem has been particularly severe during studies of surfactant

TABLE 8.1

Nomenclature systems for fluid surfactant phases

Phase names	Structure	Description	Optical properties	Notation Luzzati	Notation Winsor	Notation Ekwall
One-dimensional periodicity; indefinitely extended layered structural elements; Bragg spacing ratio 1:1/2:1/3						
Neat soap "Soap boiler's neat soap"	One-dimensional lamellar Bilayer	Side-by-side molecules arranged head-to-head to form a monolayer; like surface of two monolayers juxtaposed to form a bilayer; many bilayers stacked to form the bulk phase	Birefringent	L LL L_α	G	D
Single layer neat phase	One-dimensional lamellar Monolayer	Monolayers arranged with polar surfaces next to nonpolar surfaces	Birefringent			D_s
Mucous woven phase	One-dimensional lamellar Bilayer	Probably a colloidal mixture of lamellar and liquid phases	Weakly birefringent			B
Two-dimensional periodicity; indefinitely extended linear close-packed structural elements; Bragg spacing ratio $1:1/\sqrt{3}:1/\sqrt{4}$						
Normal middle phase	Two-dimensional hexagonal "Oil-core" cylinders	Indefinitely long cylindrical elements with lipophilic groups inside. Arranged parallel and hexagonally close-packed	Birefringent	H_1	M_1	E
Inverted middle phase	Two-dimensional hexagonal "Water-core" cylinders	Indefinitely long cylindrical elements with hydrophilic groups and water inside. Arranged parallel and hexagonally close-packed	Birefringent	H_2	M_2	F
Complex normal hexagonal phase	Two-dimensional hexagonal "Oil-core" cylinders	Similar to normal hexagonal	Birefringent	H_c		H_c
Two-dimensional periodicity; close-packed linear structural elements; one or two sets of Bragg spacings; spacing ratio 1:1/2:1/3						
Normal rectangular phase	Two-dimensional orthorhombic Flattened cylinders	Similar to normal hexagonal, but with rectangular cross-section	Birefringent	R		R
Normal square ("white") phase	Two-dimensional tetragonal Cylinders	Similar to normal hexagonal, but with square instead of hexagonal cross-section. Possibly not a single phase	Birefringent			C

	Structure	Description	Birefringent			K
Inverted square phase	Two-dimensional tetragonal Cylinders	Similar to normal square phase but with water and hydrophilic groups inside cylindrical structural elements				
Three-dimensional periodicity; discrete (discontinuous) or extended (continuous) structural elements; cubic symmetry, either body-centered (space group Ia3d), face-centered, or possibly primitive (space group Pm3n) crystallographic groups						
Discontinuous normal cubic phase	Three-dimensional body-centered cubic Space group probably $Pm3n$	Stubby cylindrical elements arranged in two interwoven unconnected chiral networks	Isotropic	Q	V_1	I_1'
Discontinuous inverted cubic phase	Three-dimensional body-centered cubic Space group probably $Pm3n$	Similar to normal phase but with water and hydrophilic groups inside the cylindrical structural elements	Isotropic	Q	V_2	I_2'
Bicontinuous normal cubic phase	Three-dimensional body-centered cubic Space group probably $Ia3d$	Platelets of the lamellar (D) structure with polar outer surfaces, joined at the edges into a saddle-like (zero-mean-curvature) isotropic structure	Isotropic	F_{1b} S_{1c}	I_1''	
Bicontinuous inverted cubic phases	Three-dimensional body-centered cubic Space group probably $Ia3d$	Similar to the normal structure, except with hydrophilic groups inside the structural elements	Isotropic			I_2''
Liquid phases						
Aqueous micellar liquid, O/W microemulsion	Molecular solution, or liquid containing micellar aggregates	Normal liquid structure except for the existence of micellar aggregates	Isotropic	S_1		L_1
Oily liquid, W/O microemulsion	Molecular solution, or inverted micellar aggregates, or bicontinuous fluid structure	Liquid structure ranging from molecular solutions to inverted micellar aggregates to bicontinuous liquid structure	Isotropic	S_2		L_2

systems. For example, the intermediate phases now recognized to exist in soap–water systems (Section 8.4.14) were not found until a half-century after the initial investigations. This occurred in spite of persistent investigations of the same and of related systems by many workers during the intervening period of time.

Cubic phases have also been difficult to detect. Twenty five years ago cubic phases were considered to be rare and unusual, whereas now it is recognized that these phases are widespread and exist in many systems. Such problems become more readily understandable when one considers the physical properties of the phases that have been missed, and recognizes that they often spanned narrow ranges of composition. The cubic phase problem was exacerbated by the fact that these phases are not only viscous, but isotropic; they may easily be overlooked during examination of mixtures in a polarizing microscope for that reason.

Earlier workers were further hampered by the absence of an applicable isothermal phase study method; for studies of binary systems almost complete reliance had to be placed on isoplethal phase studies. These problems are purely experimental in nature, and will be resolved by improved awareness of the general phase behavior of surfactants, by training in phase science, and by the use of more varied and reliable methods.

Non-existent phases. A more fundamental issue arises when a phase state actually does not exist. One may never be absolutely certain that a phase really does not exist. However, when the historic pattern as to the region within which a phase is found is known, and when this region has been carefully explored using appropriate methods and the phase is not found, then one may conclude (tentatively) that the phase truly does not exist. Drawing this conclusion on the basis of such information is a positive result; it is quite different from attempting to prove a negative result.

Consideration of the (presumably bicontinuous) cubic phase that exists between the lamellar and hexagonal phases illustrates both kinds of situations. This phase does exist (although it was earlier overlooked) in the $C_{12}E_6$–water, the $C_{12}AO$–water, and the $C_{12}NMe_3^+,Cl^-$–water systems. However, it probably does not exist in the corresponding phosphine oxide–water ($C_{12}PO$–water) system [108]. It is entirely unclear why it exists in some systems, but not in others of closely related molecules.

A comprehensive study of the DODMAC–water system, using both classical methods and the isothermal DIT phase studies method, provided hard evidence that no other liquid crystal besides the lamellar phase exists in this system within the range of temperature spanned by the DIT method (25–80°C) [43]. Other data suggested (with less certainty) that other liquid crystal phases do not exist at higher temperatures. Apparently the only liquid crystal phase that exists in this system is the lamellar phase, and one may wonder why inverted structures do not appear at low water compositions.

8.8 The factors which define surfactant phase behavior

Each of the aqueous phase states that exists for a particular surfactant molecule is found only within a finite region of composition-temperature space within the phase diagram. From this it is apparent that phase behavior is strongly influenced by the system variables as well as by the molecular structure of the surfactant, so that in considering models for surfactant phase behavior both factors are equally important.

Moreover, since no surfactant displays all the phase states that are known, it is important to be able to anticipate which states exist and which do not.

8.8.1   Models for the relationship between phase structure and molecular structure

As knowledge of both molecular structure and the phase behavior of surfactants has improved, correlations between molecular structure and phase behavior and the antici-pation of phase behavior on a theoretical basis have attracted considerable attention. The focus of these investigations has been directed largely towards the structural aspect of phase behavior, and several attempts have been made during the past 50 years to develop relationships between the geometric features of surfactant molecules and the geometric structure of the phases that these molecules form. A distinction has not usually been made between the geometric structure of the aggregates that surfactants form within a liquid phase and the structure of homogeneous phases, in considering models [121].

The first serious attempt along these lines was the "R-theory" of Winsor [122]. In this approach, details of the conformational structures of surfactant molecules were neglected. They were viewed as having either a cylindrical or a wedge shape, and it was suggested that this shape dictated the geometric structures of the phases that could be formed. In 1955 Tartar proposed similar ideas that were based on analysis of (1) the surfactant lipophilic group volume, (2) the area requirements of the hydrophilic group, and (3) the extended length of the molecule [123]. In 1973 Tanford sought to bring to the fore Tartar's analysis, and extended it to the behavior of biological surfactants [124].

Most recently, this subject has been approached from the perspective of the physical theory of colloidal states [125]. In this approach the use of a packing parameter, $v/a_o l$, whose value is dictated by the molecular area requirements of the hydrophilic group (a_o), the maximum extended length of the lipophilic group (l), and the lipophilic group volume (v) of the surfactant molecule, has been suggested. This parameter offers a convenient means of considering all three parameters at once within a useful func-tion, and a theoretical basis for its use has been suggested. It is presumed that this function is influenced not only by molecular structure, but also by the amount of water present, by the presence of added salts, dissolved oils, other amphiphilic molecules, etc. The numerical value of $v/a_o l$ is thus not determined solely by the molecular structure of the surfactant; other components somehow influence its magnitude.

It is suggested that the geometry of the aggregates that will form may be anticipated from the magnitude of $v/a_o l$. The volume of a cylinder having a cross-sectional area "a" and a length "l" equals "$a \times l$". If $v/a_o l$ is < 0.5 for a given surfactant molecule, then the actual lipophilic group volume of the molecule is substantially less than its "cylindrical volume". When this is true it is supposed that aggregates of such molecules will be curved so as to enclose the lipophilic region (following the general idea proposed by Winsor). Curvature of this sort is most pronounced within spherical aggregates such as small micelles, and these structures should therefore be favored for small values of $v/a_o l$.

If $v/a_o l$ falls within the range 0.5 to 1.0 then lesser curvature should result. This situation exists in the cylindrical aggregate geometry. The cylinder displays circular curvature in the plane perpendicular to the axis of the cylinder, but no curvature at all in the direction of its axis. Finally, if $v/a_o l$ is equal to or greater than 1.0, then minimal curvature (or curvature in the opposite sense) is to be expected and it is suggested that lamellar aggregate geometry should result.

8.8.2 Theories of phase behavior

The presently existing theories for the phase behavior start essentially from this geometric analysis of molecular structure [126]. It is presumed that surfactants spontaneously organize themselves ("self-assemble") into a particular structure as a result of the balance between a solvophobic force (a reluctance to mix with the solvent, water) resulting from interactions with water, and an attractive force between the molecules. Interactions between the head groups are regarded as being generally repulsive. A crucial aspect of all these processes is that one dimension of the resulting aggregate (whether a spherical micelle, a hexagonal liquid crystal, or a lamellar liquid crystal or crystal) has molecular dimensions. Not surprisingly, an analysis of the forces involved which is sufficiently detailed to provide a reasonable prediction as to the overall phase behavior is very complex. The focus is on the liquid crystal states that exist (more than on the crystal), and in the general approach one starts from dilute aqueous mixtures. The various terms that contribute to the free energy of aggregated states may then be considered; those factors presently believed to be of greatest importance are as follows:

(1) A hydrophobic free energy gain. This is associated with the complete transfer of the apolar chains from within the solvent into the aggregate. This contribution is assumed to be a linear function of the chain length (volume) (Section 11.2). This has long been recognized to be a dominant factor with respect to the value of the cmc, but it is also important with respect to other phase states than the micellar liquid.

(2) A chain conformation free energy. This term is important in analysis of the state of liquid hydrocarbons, but its contribution in liquid crystal phases is influenced by the fact that the polar group is constrained to the structural surfaces within the phases (Section 8.4.1). Because of this constraint, the chains are more stretched out and have less freedom in liquid crystal phases than they do in liquid hydrocarbons. The numerical value of the area per head group influences the magnitude of this contribution.

(3) A surface interaction term. This term takes into account the existence of an unfavorable solvent-chain contact at this point within the structure, and the fact that the solvation of the head group is a function of composition. The terms (2) and (3) may be combined into one free energy term that is linear with head group area.

(4) A head group repulsion free energy term. This depends strongly upon the chemical nature of the head group. For ionic surfactants this can be estimated using a Poisson–Boltzman type of analysis, for polyoxyethylene surfactants an estimate can be made using polymer solution theory, and for dipolar (semipolar and zwitterionic) surfactants a short-range electrostatic dipole interaction model suffices.

(5) An entropy of mixing term. This is associated with the disorder that exists in the state at the aggregate level. For dilute phases with finite aggregates this entropy can be calculated assuming ideal mixing, while at higher concentrations estimates

based on hard sphere models for interactions within colloidal dispersions have been used. For mixtures of two or more surfactants in water an additional entropy of mixing exists that is associated with the mixing of the surfactant molecules within the aggregates.

(6) A free energy term associated with interactions between aggregates. Estimation of this term draws upon the theories of colloid stability. For charged molecules this term can also be treated using an electrostatic model, but nonionic surfactants must be treated using other models.

Such models have led to reasonable predictions as to the influence of composition on micelle geometry [127]. They also provide a basis for understanding the formation with increasing composition of the discontinuous cubic phase, the hexagonal phase, the bicontinuous cubic phase, and ultimately the lamellar phase in a cationic system. Elements of the temperature-dependent phase behavior of polyoxyethylene nonionic surfactants have been anticipated, including the increased swelling of the lamellar phase as the temperature is increased as well as its upper limit of existence [128]. It is presently felt that the theoretical analysis has effectively dealt with the complex interplay of forces that determines the phase behavior of surfactants, and that a basic understanding of the factors that influence phase behavior exists. It is also recognized that new and unexpected phenomena are continually being discovered that challenge the theory.

8.8.3   Relevant experimental observations

A critical analysis of the various models for the phase behavior of aqueous surfactant systems will not be undertaken. However, those aspects of experimental phase science that bear on the subject will be noted in this section. Some of these factors have been widely considered, while others have received less attention.

Molecular geometry as a state variable. In considering the relevance of parameters such as $v/a_o l$ to phase structure, it is crucially important to recognize that volume, v, is a state variable. (In this section the lower case "v" is used for molecular volume, while the capital "V" refers to the molar volume.) As state changes so does v, and as v changes then so must molecular area (a) and/or molecular length (l). Thus, one must always keep in mind that these parameters are not defined simply by stating the molecular structure of the surfactant – even in the absence of other components.

Also, while volume can be directly measured, the dissection of volume into area and length parameters is model-dependent. Lipophilic group volume (v) is usually calculated using algorithms derived from the densities of the homologous alkanes. Both the length, l, and the area, a, of molecules within a crystal phase may be estimated from single crystal structure data. It is worth noting that the phase geometry issue is moot in crystals, because this state (almost) invariably has a lamellar structure (Section 8.3).

Within liquid crystal phases, the dependence of the long d spacing on composition provides a measure of the area per head group within these states. For some hydrophilic groups the estimated area is independent of composition (nonexpanding groups), while for others the area does depend on composition (expanding groups) (Section 8.4.6). It is interesting that the sodium sulfonate group is classified from these data as an expanding group in the lamellar phase of the soap-like sodium alkane sulfonate surfactants, while the same functional group is nonexpanding within the same phase in the sodium sulfosuccinate diester (AOT) class of surfactants. These results suggest that the influence of composition on the area parameter in purely aqueous mixtures is dependent on other aspects of surfactant structure than just the structure of the hydrophilic group.

The length and area dimensions may also be estimated from the d-spacing of liquid crystal phases by application of models which presume that the volume of the aqueous and lipophilic regions within these phases are quantitatively related to those of liquid water and of hydrocarbons, respectively.

The area requirement of hydrophilic groups has also been estimated from analysis of the slope of $\gamma/\ln(C)$ data at the air/water interface (by application of the Gibbs model). The curvature at this interface is typically small, the phase to one side of the interface has an extremely low density, and one may legitimately wonder just how closely such data quantitatively resemble the area occupied by the same groups within condensed bulk phase states.

The important role of the system variables. It is clear from a survey of the phase behavior of surfactants that phase structure is dependent on the numerical values of the system variables. The form of surfactant–water phase diagrams, and of liquid crystal regions in particular (Section 5.6), indicates that particular structures exist only within defined regions of composition-temperature space. Those models for phase structure which predict the range of system variables within which each phase exists will be particularly useful.

Also, phase behavior is inherently thermodynamic in nature and can be approached from the perspective of the free energy of mixing function (Chapter 4). The form of this function suggested in Chapter 4 is *not* intended to represent a serious model, but rather a qualitative basis for understanding the principles of phase behavior. Ideally, a satisfactory model for phase behavior should anticipate the form of these free energy of mixing functions and the dependence of these functions on the system variables.

The data show that all manner of discontinuities of state abound in surfactant phase behavior. These too must be accounted for.

The absence of phases. The existing data clearly show that not all of the phases that might be expected from a purely geometric perspective are typically encountered, and it is presently impossible to predict when a phase will be present and when it will not. The capacity to predict this qualitative aspect of surfactant phase behavior from models would be very useful.

The experimental data problem. A particularly awkward problem for the development and testing of models is that, all too often, the existing experimental data are not sufficiently well known to provide a sound basis for evaluating models. If models could identify the errors that exist in the present database, this too would represent a very significant contribution. It is the author's personal opinion that more effort expended towards determining comprehensive equilibrium phase diagrams of high quality for a wide range of surfactant structures would be of considerable value in its own right (for example, in the context of their industrial usage), and would also provide a solid basis for model development and evaluation.

8.9 The broad picture

Aqueous surfactant mixture may exist as liquid, liquid crystal, or crystal phases. The liquid phases of surfactants are molecular solutions which display a normal liquid structure at very low concentrations (below the cmc), display micellar structure over a large fraction of the composition ranges that they occupy, but appear to be "unstructured" (in the sense of lacking micellar structure) at low water compositions at ordinary temperatures. While the cmc transition is reasonably well defined, the transition from structured micellar solutions to the concentrated unstructured liquid state is usually uncertain.

The vast majority of surfactant crystals possess a bilayer phase structure, display phase chemistry which closely resembles that of nonsurfactant compounds, and possess optical and rheological properties that resemble crystals of other long-chain organic molecules. It must be recognized that other kinds of crystal structures than the bilayer do exist, however (Section 8.3.2).

Liquid crystals are the structurally distinctive phase states that aqueous surfactant mixtures form. These usually (but not always) require the presence of water in order to exist, and are termed "lyotropic" states. Liquid crystal phases are generally fluids, but they are in some ways almost as highly structured as are crystals. Their optical and rheological properties are extremely diverse, and are influenced both by the overall structure of the phase and by its composition. Generally speaking, the one-dimensionally ordered lamellar phases are the least viscous, the two-dimensionally ordered hexagonal phases are more viscous, and the three-dimensionally ordered cubic phases are the most viscous of all. Since the lamellar phase (in soap-like surfactants) is the most concentrated, the hexagonal phase is relatively dilute, and the cubic phases are found to either side of the hexagonal phase, it can be seen that the viscosities of surfactant mixtures do *not* vary in a regular manner with composition.

It is useful to regard the lamellar liquid crystal phase as the central (mirror) phase with respect to the influence of composition on surfactant phase structure. Among those soluble, single-chain soap-like surfactants which form lamellar phases at high surfactant compositions, diluting this phase may produce the following sequence of structures: the "normal" (oil-core) bicontinuous cubic phase, the rectangular phase, the hexagonal phase, and the discontinuous cubic liquid crystal phase. Further dilution produces the micellar liquid phase. Removing water from the lamellar phase in such surfactants typically produces a crystal phase.

For those surfactants which form a dilute lamellar phase at high water composition (such as occurs in the case of Aerosol OT, Fig. 10.7), a related pattern of phase structures having inverted structure, relative to those described above, develops as water is removed from the lamellar phase. Diluting such lamellar phases yields only the micellar liquid phase, while removing water appears to produce a qualitatively similar sequence of "inverted" (water-core) liquid crystal phase structures.

The influence of system variables on the regions occupied by these phases was considered from a thermodynamic perspective in Chapter 6. By application of these concepts, and from the sequence of structures that is observed, one can infer the minimal thermal energy required for each phase structure to exist, and the relative contribution of water free energies within which each is thermodynamically stable.

The molecular geometry of surfactants is evidently important in defining the sequence of phase structures which exist, but for a particular surfactant the system variables must also be specified.

References

1. Overbeek, J. Th. G. (1952). *Colloid Science, Irreversible Systems*, Vol. I (H. R. Kruyt ed.), pp. 79, Elsevier, New York.

2. Mason, J. T., Huang, C. H. and Biltonen, R. L. (1983). *Biochemistry* **22**, 2013–2018.
3. Friedel, G. (1926). *Colloid Chemistry*, Vol. I (J. Alexander ed.), pp. 102–125, The Chemical Catalog Co., New York.
4. Rosevear, F. B. (1954). *J. Am. Oil Chemists Soc.* **31**, 628–638.
5. Hartshorne, N. H. and Stuart, A. (1970). *Crystals and the Polarizing Microscope*, 4th ed., Arnold, London.
6. Lindblom, G. and Rilfors, L. (1989). *Biochim. Biophys. Acta* **988**, 221–256; Luzzati, V. (1968). *Biological Membranes, Physical Fact and Function*, (D. Chapman and D. F. H. Wallach eds), pp. 71–124, Academic Press, New York.
7. Fontell, K. (1990). *Colloid Polym. Sci.* **268**, 264–285.
8. Lindblom, G. and Rilfors, L. (1989). *Biochim. Biophys. Acta* **988**, 221–256.
9. Azzam, R. M. A. and Bashara, N. M. (1977). *Ellipsometry and Polarized Light*, pp. 119–131, North-Holland, New York.
10. Hecht, E. (1987). *Optics*, 2nd ed., pp. 283–288, Addison-Wesley, Reading, MA.
11. Powers, L. and Pershan, P. S. (1977). *Biophys. J.* **20**, 137–152.
12. Khoo, I. C. and Simoni, F. (1991). *Physics of Liquid Crystalline Materials*, Gordon and Breach, Philadelphia; DeJeu, W. H. (1980). *Physical Properties of Liquid Crystal Materials*, Gordon and Breach, New York.
13. Hofmann, S., Rauscher, A. and Hoffmann, H. (1991). *Ber. Bunsenges. Phys. Chem.* **95**, 153–164.
14. Small, D. M. (1986). *The Physical Chemistry of Lipids. From Alkanes to Phospholipids*, Vol. 4, pp. 25–32, Plenum, New York.
15. Barnes, H. A., Hutton, J. F. and Walters, K. (1989). *An Introduction to Rheology*, Elsevier Science, Amsterdam.
16. Snyder, R. G. (1992). *J. Chem. Soc. Faraday Trans.* **88**, 1823–1833.
17. Maroncelli, M., Song, P. Q., Strauss, H. L. and Snyder, R. G. (1982). *J. Am. Chem. Soc.* **104**, 6237–6247.
18. (1972). *Water: A Comprehensive Treatise. The Physics and Chemistry of Water*, Vol. 1 (F. Franks ed.), Plenum, New York; (1973). *Water: A Comprehensive Treatise. Water in Crystal Hydrates, Aqueous Solutions of Simple Nonelectrolytes*, Vol. 2 (F. Franks ed.), Plenum, New York; (1973). *Water: A Comprehensive Treatise. Aqueous Solutions of Simple Electrolytes*, Vol. 3 (F. Franks ed.), Plenum, New York; (1975). *Water: A Comprehensive Treatise. Aqueous Solutions of Amphiphiles and Macromolecules*, Vol. 4 (F. Franks ed.), Plenum, New York; (1975). *Water: A Comprehensive Treatise. Water in Disperse Systems*, Vol. 5 (F. Franks ed.), Plenum, New York; (1979). *Water: A Comprehensive Treatise. Recent Advances*, Vol. 6 (F. Franks ed.), Plenum, New York; (1982). *Water: A Comprehensive Treatise. Water and Aqueous Solutions at Subzero Temperatures*, Vol. 7 (F. Franks ed.), Plenum, New York.
19. Swern, D. (1964). *Bailey's Industrial Oil and Fat Products*, 3rd ed., pp. 105–106, Interscience, John Wiley, New York; Bondi, A. (1968). *Physical Properties of Molecular Crystals, Liquids, and Glasses*, pp. 140–149, John Wiley, New York.
20. Wong, P. T. T., Chagwereda, T. E. and Mantsch, H. H. (1987). *J. Chem. Phys.* **87**, 4487–4497.
21. Mandelkern, L., Prasad, A., Alamo, R. G. and Stcak, G. M. (1990). *Macromolecules* **23**, 2595–3700.
22. Krog, N. J. (1990). *Food Emulsions*, 2nd ed., pp. 127–180, Marcel Dekker, New York.
23. Klein, M. L. (1992). *J. Chem. Soc. Faraday Trans.* **88**, 1701–1705.
24. Hauser, H., Pascher, I. and Sundell, S. (1980). *J. Mol. Biol.* **137**, 249–264.
25. Emge, T. J., Strickland, L. C., Oliver, J. D. and Laughlin, R. G. (1989). Presented at American Crystallographic Association Meeting, Seattle, WA, 23–29 July 1989.
26. Kodali, D. R., Atkinson, D. and Small, D. M. (1989). *J. Phys. Chem.* **93**, 4683–4691.
27. Bunn, C. W. and Howells, E. R. (1954). *Nature* **174**, 549–551.
28. Banks, R. E. (1964). *Fluorocarbons and Their Derivatives*, pp. 12–15, Oldbourne Press, London.
29. Bunn, C. W. and Holmes, D. R. (1958). *Disc. Far. Soc.* **25**, 95–103.
30. Matsuura, H. and Fukuhara, K. (1986). *J. Phys. Chem.* **90**, 3057–3059.
31. Friberg, S. and Larsson, K. (1976). *Advances in Liquid Crystals*, Vol. 2 (G. H. Brown ed.), p. 176, Academic Press, New York.

32. Bondi, A. (1964). *J. Phys. Chem.* **68**, 441–451.
33. Bangham, A. D., Standish, M. M. and Watkins, J. C. (1965). *J. Mol. Biol.* **13**, 238–252.
34. Gennis, R. B. (1980). *Biomembranes. Molecular Structure and Function*, (C. R. Cantor, Series Editor), pp. 36–84, Springer-Verlag, New York.
35. J. L. Burns, personal communication; Vinson, P. K., Talmon, Y. and Walter, A. (1989). *Biophys. J.* **56**, 669–681.
36. Adamson, A. W. (1982). *Physical Chemistry of Surfaces*, 4th ed., pp. 100–184, Interscience, John Wiley, New York.
37. Crawford, G. E., Crilly, J. F. and Earnshaw, J. C. (1981). *Faraday Symposia of the Chemical Society* **16**, 125–137.
38. Laughlin, R. G., Munyon, R. L., Burns, J. L., Coffindaffer, T. W. and Talmon, Y. (1992). *J. Phys. Chem.* **96**, 374–383.
39. Jeffrey, G. A. and Maluszynka, H. (1989). *Acta Cryst.* **B45**, 447–452.
40. Mueller-Fahrnow, A., Zabel, V., Steifa, M. and Hilgenfeld, R. (1986). *J. Chem. Soc. Chem. Commun.* 1573–1574.
41. Wireko, F. C. and Mootz, M. R. (1992). 41st Annual Denver Conference on Application of X-ray Analysis, Colorado Springs, CO, Aug. 3–7.
42. Fu, Y.-C., Glardon, A. S. and Laughlin, R. G. (1993). American Oil Chemists' Society Meeting, Anaheim, CA, April 25–29.
43. Laughlin, R. G., Munyon, R. L., Fu, Y.-C. and Fehl, A. J. (1990). *J. Phys. Chem.* **94**, 2546–2552.
44. Hamilton, W. C. and Ibers, J. A. (1968). *Hydrogen Bonding in Solids*, pp. 188–237, W. A. Benjamin, New York.
45. Kekicheff, P., Grabielle-Madelmont, C. and Ollivon, M. (1989). *J. Colloid Int. Sci.* **131**, 112–132; Kekicheff, P. and Cabane, B. (1987). *J. Phys.* **48**, 1571–1583.
46. Lutton, E. S. (1966). *J. Am. Oil Chem. Soc.* **43**, 28–30.
47. Smith, K. R., Borland, J. E., Corona, R. J. and Sauer, J. D. (1991). *J. Am. Oil Chemists' Soc.* **68**, 619–622.
48. Broome, F. K. and Harwood, H. J. (1950). *J. Am. Chem. Soc.* **72**, 3257–3260.
49. Y.-C. Fu, unreported work.
50. R. G. Laughlin, unreported work.
51. Marcott, C., Laughlin, R. G., Sommer, A. J. and Katon, J. E. (1991). *Fourier Transform Infrared Spectroscopy in Colloid and Interface Science*, Vol. 447 (D. R. Scheuing ed.), pp. 71–86, ACS Symposium Series, American Chemical Society, Washington, DC.
52. Rosevear, F. B. (1968). *J. Soc. Cosmetic Chemists* **19**, 581–594.
53. Luzzati, V. (1968). In *Biological Membranes, Physical Fact and Function*, (D. Chapman and D. F. H. Wallach eds), pp. 71–124, Academic Press, New York.
54. Cox, J. S. G., Woodward, G. D. and McCrone, W. C. (1971). *J. Pharm. Sci.* **60**, 1458–1465.
55. Madelmont, C. and Perron, R. (1974). *Bull. Soc. Chim. France* 1799–1805, 3425–3429, 3430–3434.
56. Winsor, P. A. (1974). *Liquid Crystals and Plastic Crystals*, Vol. 1 (G. W. Gray and P. A. Winsor eds), pp. 48–59, Ellis Horwood, Chichester, England.
57. Reiss-Husson, F. and Luzzati, V. (1964). *J. Phys. Chem.* **68**, 3504–3511.
58. Derome, A. E. (1987). *Modern NMR Techniques for Chemistry Research*, Vol. 6 (J. E. Baldwin ed.), pp. 7, Pergamon Press, Oxford.
59. Stilbs, P. (1987). *Prog. Nucl. Mag. Reson. Spectrosc.* **19**, 1–45.
60. Seelig, J. (1977). *Quart. Rev. Biophysics* **10**, 353–418; Seelig, J. and Browning, J. L. (1978). *FEBS Lett.* **92**, 41–44.
61. Klein, M. L. (1992). *J. Chem. Soc. Faraday Trans.* **88**, 1701–1705.
62. Bohlin, L. and Fontell, K. (1978). *J. Colloid Interface Sci.* **67**, 272–283.
63. Bohlin, L., Ljusberg-Wahren, H. and Miezis, Y. (1985). *J. Coll. Interface Sci.* **103**, 294–295.
64. Hoffmann, H. (1992). *Organized Solutions* (S. Friberg and B. Lindman, eds), Vol. 44, pp. 169–192, Marcel Dekker, New York.
65. F. B. Rosevear, personal communication.
66. Gradzielski, M., Hoffmann, H. and Oetter, G. (1990). *Colloid Polym. Sci.* **268**, 167–178.
67. Overbeek, J. Th. G. (1952). In *Colloid Science, Irreversible Systems*, Vol. I (H. R. Kruyt ed.), pp. 58, 335, Elsevier, New York.

236   THE AQUEOUS PHASE BEHAVIOR OF SURFACTANTS

68. Roozeboom, H. W. B. (1901). *Die Heterogenen Gleichgewichte vom Standpunkte der Phasenlehre. Die Phasenlehre – Systeme aus Einer Komponente*, Vol. 1, pp. 142–154, Braunschweig, F. Vieweg und Sohn; Reinitzer, F. (1888). *Monatsheft. Chem.* **9**, 421–441.
69. Kléman, M. (1983). *Points, Lines, and Walls in Liquid Crystals, Magnetic Systems and Various Ordered Media*, John Wiley, New York.
70. Rogers, J. and Winsor, P. A. (1969). *J. Colloid Interface Sci.* **30**, 247–257.
71. Franses, E. I. and Hart, T. J. (1983). *J. Colloid Interface Sci.* **94**, 1–13.
72. Hermann, C. (1931). *Z. Krist.* **79**, 186–221.
73. Mabis, A. J. (1962). *Acta Crystallogr.* **15**, 1152–1157.
74. Demus, D., Diele, S., Grande, S. and Sackmann, H. (1983). *Advances in Liquid Crystals*, Vol. 6 (G. H. Brown ed.), p. 71, Academic Press, New York.
75. Friberg, S. E., Ward, A. J. I. and Larsen, D. W. (1987). *Langmuir* **3**, 735–737; Ranavare, S. B., Ward, A. J. I., Friberg, S. E. and Larsen, D. W. (1987). *Mol. Cryst. Liq. Cryst.* **4**, 115–121; Bleasdale, T. A. and Tiddy, G. J. T. (1992). *Surfactant Sci. Ser. (Organized Solutions)* **44**, 125–141; Friberg, S. and Liang, P. (1986). *Colloid Polym. Sci.* **264**, 449–453; El-Nokaly, M., Friberg, S. E. and Larsen, D. W. (1984). *Liq. Cryst. Ordered Fluids* **4**, 441–450; Auvray, X., Perche, T., Petipas, C., Anthore, R., Marti, M. J., Rico, I. and Lattes, A. (1992). *Langmuir* **8**, 2671–2679.
76. Gray, G. W. (1962). *Molecular Structure and the Properties of Liquid Crystals*, Academic Press, New York.
77. Blinov, L. M. (1983). *Electrooptic and Magnetoopic Properties of Liquid Crystals*, John Wiley, New York.
78. Fontell, K. (1990). *Colloid Polym. Sci.* **268**, 264–285.
79. Ekwall, P. (1975). *Advances in Liquid Crystals*, Vol. 1 (G. H. Brown ed.), pp. 24–52, Academic Press, New York.
80. Reference 79, pp. 43–48.
81. Hendrixx, Y., Charvolin, J., Kekicheff, P. and Roth, M. (1987). *Liq. Cryst.* **2**, 677–687.
82. Ulmius, J., Wennerstrom, H., Lindblom, G. and Arvidson, G. (1977). *Biochemistry* **16**, 5742–5745.
83. Grabielle-Madelmont, C. and Perron, R. (1983). *J. Colloid Interface Sci.* **95**, 471–482.
84. Hentschel, M. P. and Rustichelli, F. (1991). *Phys. Rev. Lett.* **66**, 903–906.
85. Ladbrooke, B. D. and Chapman, D. (1969). *Chem. Phys. Lipids* **3**, 304–356.
86. Scriven, L. E. (1976). *Nature* **263**, 123–125.
87. Anderson, D., Wennerstrom, H. and Olsson, U. (1989). *J. Phys. Chem.* **93**, 4243–4253.
88. Siegel, D. P. (1986). *Chem. Phys. Lipids* **42**, 279–301; Lindblom, G. and Rilfors, L. (1989). *Biochim. Biophys. Acta* **988**, 221–256.
89. Thomas, E. L., Alward, D. B., Kinning, D. J., Martin, D. C., Handlin, G. L. and Fetters, L. J. (1986). *Macromolecules* **19**, 2197–2202.
90. Thomas, E. L., Anderson, D. M., Henke, C. S. and Hoffman, D. (1988). *Nature* **334**, 598–601.
91. Lindell, K., Enmgstroem, S. and Carlsson, A. (1991). *Proc. Program Int. Symp. Controlled Release Bioact. Mater., 18th*, (I. W. Kellaway ed.), pp. 265–266, Controlled Release Soc., Deerfield, IL; Carlsson, A., Olsson, H., Axell, T., Loden, M. and Bogentoft, C. (1991). *Proc. Program Int. Symp. Controlled Release Bioact. Mater., 18th*, (I. W. Kellaway ed.), pp. 267–268, Controlled Release Soc., Deerfield, IL; Carlsson, A., Bogentoft, C., Lindman, B. and Andersson, L. (1991). *Proc. Program Int. Symp. Controlled Release Bioact. Mater. 18th*, (I. W. Kellaway ed.), pp. 455–456, Controlled Release Soc., Deerfield, IL.
92. Lindblom, G., Larsson, K., Johanssen, L., Fontell, K. and Forsen, S. (1979). *J. Am. Chem. Soc.* **101**, 5465–5470.
93. Lawson, K. D. and Flautt, T. J. (1968). *J. Phys. Chem.* **72**, 2058–2065, 2066–2074.
94. Reference 79, pp. 31–32.
95. McGrath, K. M., Kekicheff, P. and Kléman, M. (1993). *J. Phys. II France* **3**, 903–926.
96. Luzzati, V., Tardieu, A., Gulik-Krzywicki, T., Rivas, E. and Reiss-Husson, F. (1968). *Nature* **220**, 485–488.
97. Lindblom, G. and Rilfors, L. (1989). *Biochim. Biophys. Acta* **988**, 221–256; Eriksson, P.-O., Lindblom, G. and Arvidson, G. (1987). *J. Phys. Chem.* **91**, 846–853.
98. Luzzati, V., Vargas, R., Mariani, P., Gulik, A. and Delacroix, H. (1993). *J. Mol. Biol.* **229**, 541–551.

99. Lutton, E. S. (1965). *J. Am. Oil Chemists Soc.* **42**, 1068–1070.
100. Krog, N. and Lauridsen, J. B. (1976). *Food Emulsions*, Vol. 5 (S. Friberg ed.), pp. 67–140, Marcel Dekker, New York.
101. Boden, N., Jackson, P. H., McMullen, K. and Holmes, M. C. (1979). *Chem. Phys. Lett.* **65**, 476–479.
102. Gault, J. D., Leite, M. A., Rizzatti, M. R. and Gallardo, H. (1988). *J. Colloid Int. Sci.* **122**, 587–590.
103. Herbst, L., Hoffmann, H., Kalus, J., Reizlein, K., Schmelzer, U. and Ibel, K. (1985). *Ber. Bunsenges. Phys. Chem.* **89**, 1050–1064.
104. Luehmann, B. and Finkelmann, H. (1966). *Colloid Polym. Sci.* **264**, 189–192.
105. Madelmont, C. and Perron, R. (1976). *Colloid Polym. Sci.* **254**, 6581–6595.
106. Leigh, I. D., McDonald, M. P., Wood, R. M., Tiddy, G. J. T. and Trevethan, M. A. (1981). *J. Chem. Soc., Faraday Trans. I* **77**, 2867–2876.
107. Henriksson, U., Blackmore, E. S., Tiddy, G. J. T. and Soderman, O. (1992). *J. Phys. Chem.* **96**, 3894–3901.
108. Chernik, G. G. (1991). *J. Colloid Interface Sci.* **141**, 400–408.
109. Laughlin, R. G. (1992). *Adv. Coll. Interface Sci.* **41**, 57–79.
110. Ekwall, P., Salonen, M., Krokfors, I. and Danielsson, I. (1956). *Acta Chem. Scand.* **10**, 1146–1159; Ekwall, P., Mandell, L. and Fontell, K. (1969). *J. Colloid Interface Sci.* **31**, 508–529, 530–539.
111. Krafft, F. and Wiglow, H. (1895). *Ber. deutsche Chem. Ges.* **28**, 2573–2582.
112. Lindman. B. and Wennerstrom, H. (1980). *Fortschritte der chemischen Forschung*, Vol. 87 (F. L. Boschke ed.), Springer, Berlin.
113. Mazer, N. A., Benedek, G. B. and Carey, M. C. (1976). *J. Phys. Chem.* **80**, 1076–1085.
114. Clausen, T. M., Vinson, P. K., Minter, J. R., Davis, H. T. and Miller, W. G. (1992). *J. Phys. Chem.* **96**, 474–484.
115. K. W. Herrmann, unreported work.
116. Reference 79, p. 6.
117. Nilsson, P.-G., Lindman, B. and Laughlin, R. G. (1984). *J. Phys. Chem.* **88**, 6357–6362.
118. Prince, L. M. (1977). *Microemulsions*, Academic Press, New York; Bourrel, M. and Schechter, R. S. (1988). *Microemulsions and Related Systems*, Marcel Dekker, New York; (1987). *Microemulsions: Structure and Dynamics*, (S. Friberg and P. Bothorel ed.), CRC Press, Boca Raton, FL.
119. (1987). *Physics of Amphiphiles. Springer Proceedings in Physics*, Vol. 21 (D. Langevin, J. Meunier, and N. Boccara eds), Springer, Heidelberg.
120. Reference 79, pp. 14–15.
121. Israelachvili, J. (1992). *Intermolecular and Surface Forces*, 2nd ed., pp. 341–365, Academic Press, London.
122. Winsor, P. A. (1968). *Chem. Rev.* **64**, 1–40.
123. Tartar, H. V. (1955). *J. Phys. Chem.* **59**, 1195–1199.
124. Tanford, C. (1972). *J. Phys. Chem.* **76**, 3020–3024; (1973). *The Hydrophobic Effect: Formation of Micelles and Biological Membranes*, pp. 74–79, John Wiley, New York.
125. Israelachvili, J., Mitchell, D. J. and Ninham, B. W. (1976). *J. Chem. Soc., Faraday Trans. 2* **72**, 1525–1568.
126. The gist of the analysis that follows was suggested by Professor Håkan Wennerstrom, Lund University.
127. Jonsson, B. and Wennerstrom, H. (1987). *J. Phys. Chem.* **91**, 338–352.
128. Strey, R., Schomaeker, R., Roux, D., Nallet, F. and Olsson, U. (1990). *J. Chem. Soc. Faraday Trans.* **86**, 2253–2261.
129. Reizlein, K. and Hoffmann, H. (1984). *Prog. Coll. Polym. Sci.* **69**, 83–93.
130. Warr, G. G., Chen, C.-M. and Prudhomme (1994). 8th International Conference on Surface and Colloid Science, 13–18 February 1994.

Part IV

MOLECULAR CORRELATIONS

Chapter 9

Surfactant and Nonsurfactant Behavior in Amphiphilic Molecules

9.1 The concept of "amphiphilic" molecular structure

Surfactants have "amphiphilic" (am-fi-fil'-ic) molecular structures [1]. "Amphi" connotes "both", and "phil" implies a love of or liking for. As applied to molecular structure, the word "amphiphilic" signifies a love of or liking for both aqueous and oily phases. Both "hydrophilic" and "lipophilic" structural fragments exist within amphiphilic molecules. Those which are hydrophilic are "water-loving", while those which are lipophilic are "lipid- or oil-loving". (The terms "hydrophobic" (water-hating) and "lipophobic" (oil-hating) are also useful on occasion, because hydrophilic is not exactly the same as lipophobic and lipophilic is not precisely hydrophobic.) Their amphiphilic structure readily explains the strong affinity that surfactants have for interfaces, since the miscibility preferences of both parts of the molecule are satisfied at this location. Either one part, or the other, will be thermodynamically "unhappy" within either an aqueous or an oily bulk phase.

An alkane (such as dodecane, $C_{12}H_{26}$) is lipophilic but is not hydrophilic. A salt (such as sodium formate, HCO_2Na) is hydrophilic but is not lipophilic. Neither the alkane nor the salt are amphiphilic. Coupling the two (by the formal abstraction of hydrogen atoms and joining the resulting radicals) results in the compound $C_{12}H_{25}CO_2Na$ (sodium tridecanoate), which is amphiphilic. The formal coupling of an alkane with water itself produces a fatty alcohol ($C_{12}H_{25}OH$), which is also amphiphilic.

Amphiphilic compounds result from substituting (almost) any polar functional group onto a wide variety of lipophilic molecules such as alkanes, alkylbenzenes, fluoro-carbons, siloxanes, etc. The substitution of halogen, sulfur, or phosphorus onto lipo-philic molecules also produces RX compounds having dipolar C–X bonds, but these are not usually regarded as being amphiphilic. Alkyl fluorides represent an interesting borderline case, because the C–F bond in these compounds is both strongly dipolar and weakly hydrogen bonding [2].

9.2 The structural features which influence phase behavior

We are here concerned broadly with the molecular structure of amphiphilic molecules, and with how their structure influences their aqueous phase behavior. The structure of these compounds has several ramifications, so it is first necessary to identify those particular features with which we will be concerned. In so doing we will take a leaf from the rules of nomenclature [3], and first define a parent amphiphilic structure.

Having done so, it is then possible to compare different parent structures (which provides one kind of information). It is also possible to decorate a particular parent structure with various substituent groups (which provides another kind). The influence on phase behavior of both the parent structure itself and of the substituent groups will be revealed by this analysis.

Ideally, only a lipophilic group (R) and a polar group (X) exist in the parent amphiphilic compound, RX. n-Alkyl groups are particularly important parent lipophilic groups, because they are the most common (and probably also the most commercially important) of all the lipophilic groups. However, alkylbenzenes or straight chain fluorocarbons also constitute logical parent lipophilic structures.

The simplest polar X groups imaginable, structurally, are the $-OH$, $-NH_2$, and $-NH_3^+$ groups. More complex functional groups (such as a carboxylic acid) differ formally from an alcohol in that a carbonyl group has been inserted between the OH of the alcohol and the carbon. If X is an ester, a second lipophilic group (which is at least as large as methyl) exists as well. It replaces the hydroxy hydrogen of the carboxylic acid.

In amine oxides, the nitrogen must bear three non-hydrogen (usually alkyl) lipophilic ligands; otherwise, tautomeric isomerism to a hydroxylamine will occur [4]:

$$RNH_2 \rightarrow O \rightleftarrows RNHOH$$

The hydroxylamine tautomer lacks the strongly dipolar structure of the amine oxide, and is entirely different in its physical behavior. (This is not true of oxides of primary and secondary phosphines, which exist in the dipolar form.) This is a characteristic feature of most semipolar functional groups; they are chemically stable only if no hydrogen substituents exist on the positive atom, and this requires that lipophilic substituents exist on this atom. An important consequence of this fact is that, in reality, one must always consider the influence of these substituents on physical behavior in addition to considering the intrinsic polarity of the polar bonds that exist. Substitution is also possible on the lipophilic group R, but may be treated separately.

The essential structural features that must be considered in analyzing the phase behavior of amphiphilic compounds therefore include:

(1) the structure of the polar functional group (X),
(2) the structure of the parent lipophilic group (R),
(3) the structure and position of substituent groups on the polar functional group, and
(4) the structure and position of substituent groups on the lipophilic group.

The dimensions of "structure", in each of these elements, include (besides the connectivity of the atoms involved) the volume of the group, its conformational structure(s) (or shape), and the electron distribution within the element.

9.3 The concept of hydrophilicity

It is useful to start with consideration of the structure of the polar functional group X, which exerts a profound influence on aqueous phase behavior. Two very different kinds of phase behavior have been observed, one of which is exemplified by that of the soaps. Soaps are commonly regarded as being strongly hydrophilic compounds, and we may attribute to the polar functional group of soaps an intrinsic property which will be termed "hydrophilicity". A soap will be said to possess an "operative hydrophilic group" – that is, a functional group that is capable of conferring hydrophilic behavior onto soap molecules.

Fatty acid methyl esters display an altogether different kind of behavior with water which is suggestive of weak interactions with this solvent [1]. Methyl esters will be said to lack "hydrophilicity", or more precisely, to be "weakly hydrophilic". The methyl ester group will be regarded as being "inoperative" as a hydrophilic group.

Two important questions may be posed with respect to hydrophilicity: one is "how may one recognize it?", and the other is "what is it?". To answer these questions, it is first necessary to comprehend the range of polarity that may exist within polar functional groups. We will start with the uncharged dipolar groups, and then consider the formally charged (monopolar) groups.

9.4 The range of polarity within dipolar functional groups

The relative polarities of dipolar bonds are most clearly revealed by comparing their "reduced bond moments", a concept introduced by Fajans in 1928 [5]. The reduced moment, μ_r, is defined as

$$\mu_r = \frac{\mu_{obsvd}}{d \times 4.80} \tag{9.1}$$

where μ_{obsvd} is the experimentally measured dipole moment of the bond (in Debye units, esu-m $\times 10^{20}$). Bond moments are derived from the molecular moments of conformationally simple model compounds – usually methyl derivatives [6]. The bond moment is calculated (using vector analysis) from the molecular moment and bond angles, assuming that the moments of weakly dipolar bonds (such as $X-CH_3$) are equal to values found (using similar methods) in related molecules. In calculating the moment of the $N \rightarrow O$ bond in trimethylamine oxide, for example, the $C-N$ bond moment in the oxide is taken to be equal to that found in trimethylamine [6].

The constant 4.80 (in eq. 9.1) is dictated by the numerical value of the electrostatic unit of charge (4.80 $\times 10^{-10}$ esu), and the parameter d is the measured internuclear distance within the bond (in Å, or m $\times 10^{-10}$). The denominator in eq. 9.1 thus equals the calculated moment for a hypothetical bond in which unit charges are separated by a distance equal to the bond length. Dividing the observed moment by this hypothetical moment normalizes moment data with respect to bond length, and provides a better measure of the fractional excess charge on each atom than does the experimental bond moment. This is the quantity of interest.

The reduced bond moments of various dipolar groups are displayed in Fig. 9.1 [1]. Electronically similar types of functional groups are clustered together, and their reduced moments are scaled along the y-axis. To the right are singly-bonded groups, within which no substantial multiple bonding exists. Charge separation within these bonds exists mainly because of the differences in the electronegativities of the bonded atoms.

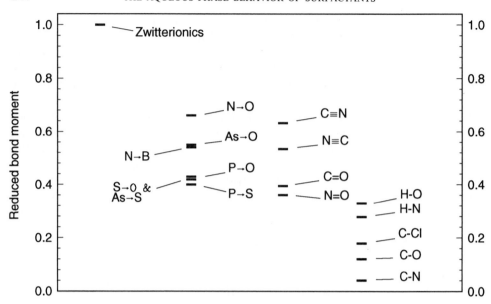

Fig. 9.1. Reduced bond moments of the common dipolar functional groups. Singly bonded groups are to the right, carbonyl and related groups next to these, and semipolar groups are to the left. The uniquely dipolar zwitterionic groups have, by definition, a reduced bond moment of one.

Those functional groups included in the second group from the right have double or triple bonds between second row elements. The second most strongly dipolar group in chemistry (the nitrile) falls within this group. Substantial polarity also exists among carbonyl functional groups. From the fact that symmetrical double and triple bonds (as in acetylene, ethylene, dinitrogen, and azo compounds) possess zero dipole moments, it is evident that (just as in single bonds) the electronic asymmetry of multiple bonds also arises from the difference in electronegativities of the bonded atoms. The degree of polarization in multiple bonds is significantly higher than is that in single bonds, however.

The third group from the right in Fig. 9.1 includes the "semipolar" functional groups. This is a convenient catch-all term for all those dipolar functional groups that do not have $2p\pi$ multiple bonds. It includes aliphatic amine oxides (which contain the most strongly dipolar group in chemistry), boron–nitrogen Lewis acid–base adducts, and phosphorus, sulfur, and arsenic bonds to oxygen or nitrogen.

The "group" of one, to the extreme left, represents the various zwitterionic dipolar groups. Formally charged substituents exist within zwitterionic groups, and they require special consideration (Section 9.4.2).

9.4.1   The electronic structure of dipolar bonds

Carbonyl and nitrile groups. The electronic structure of carbonyl groups results from the overlap of filled 2p orbitals on oxygen with empty 2p orbitals on carbon within molecular orbitals having π symmetry [7]. The bond order of this π bond (which is superimposed on another bond between the same atoms having σ symmetry) is inversely related (approximately) to its reduced bond moment. The weaker the π bond, the more strongly polar is the group. Such groups are often described as having "2pπ" multiple bonds.

Both the carbon and the oxygen are nominally sp^2 hybridized in a carbonyl group. As a result, the two filled nonbonding orbitals on oxygen (together with the C–O sigma bond) assume a planar, pseudotrigonal geometry. These orbitals lie within the plane defined by the carbonyl group itself plus the two atoms bonded to the carbonyl carbon. The atomic dipole moment of these nonbonding orbitals is very large. They serve as hydrogen bond acceptor sites and are very important with respect to hydrophilicity.

In the nitrile group, the carbon and the nitrogen are both sp (instead of sp^2) hybridized, and overlap occurs between *two* sets of filled and unfilled orbitals (rather than one). Only one non-bonding orbital exists (on nitrogen), and its axis is colinear with the C–N nuclear axis. The isonitrile is qualitatively similar, but the two-coordinate atom is the nitrogen and the nonbonding orbital is on the terminal one-coordinate carbon.

Hydrogen bonds may form to the π electrons of all these groups (particularly within crystal phases), but such hydrogen bonds are very weak in comparison to those formed to the nonbonding orbitals [8].

Semipolar bonds. The electronic structures of semipolar bonds are extremely diverse. A formal charge may be assigned to each atom by examining the σ bond structure, assuming that a symmetrical distribution of electrons exists within each bond. (This limiting structure, if it existed, would have the theoretical bond moment used above to define reduced bond moments.) Semipolar bonds are conventionally represented by an arrow which points towards the negative end of the bond. Trimethylamine oxide serves as an example:

$$
\begin{array}{ccc}
\text{CH}_3 & & \text{CH}_3 \\
| & & | \\
\text{CH}_3-\text{N}^+-\text{O}^+ \rightarrow \text{CH}_3-\text{N} & \rightarrow \text{O} & \\
| & & | \\
\text{CH}_3 & & \text{CH}_3
\end{array}
\qquad (9.2)
$$

The bond in the actual molecule differs from the formal dipolar structure in two important ways:

(1) the bonding electrons are not symmetrically distributed (owing to differences in electonegativities), and
(2) back-bonding may exist.

Back-bonding is the overlap of filled ground state nonbonding 2p orbitals on oxygen with unfilled ground state d orbitals on the positive atom [9]. When it exists, back-bonding causes a flow of electron density from the negative atom back towards the positive atom. This effect opposes the formal charge separation found in the formal purely sigma bond structure, and significantly reduces the bond polarity. The qualitative effects of backbonding are identical to those of π bonding in carbonyl groups. However, back-bonding differs importantly from π bonding because 3d (or 4d) orbitals are involved (on the positive atom), instead of 2p orbitals. Because of this difference, the nonbonding orbitals are probably not severely constrained within a plane (as in carbonyl groups). In oxides filled 2p orbitals on oxygen are involved, while in sulfides (P $\rightarrow$ S and As $\rightarrow$ S) filled 3p orbitals on sulfur are involved.

The degree of back-bonding varies widely among semipolar groups. In amine oxide and Lewis B–N compounds the outer shells of both atoms are filled and may be regarded as "saturated". Back-bonding does not exist within bonds between saturated atoms, but is very important if the positive atom is in the third or fourth row of the periodic table. For these atoms unfilled ground state orbitals exist, and such atoms are to be regarded as "unsaturated" in the above usage of the word.

The degree of back-bonding is much stronger if the formal charge on the positive atom is +2 (as in sulfones, imines of sulfones, and sulfate diesters) than if it is +1 (as in sulfoxides and phosphoryl compounds). Back-bonding is relatively weak if the formal charge is zero (as in Si–O compounds). It is strengthened if electronegative oxygen or nitrogen substituent groups replace carbon substituents on the positive atom. Back-bonding is also weaker if the unfilled d orbitals are in the fourth electron shell (as in arsenic or selenium) than if they are in the third electron shell (as in silicon, phosphorus, and sulfur).

For semipolar bonds derived from elements in the fifth column of the periodic table (Fig. 9.1), it would be expected from the above generalizations that the $P \rightarrow O$ bond of phosphate esters $((RO)_3P \rightarrow O)$ should be the most weakly dipolar bond, and that the $As \rightarrow O$ bond in trialkylarsine oxides $(R_3As \rightarrow O)$ should be the most strongly dipolar. The difference in electronegativity is greater between As and O than between P and O, 4d orbitals are invoked in the $As \rightarrow O$ bond while 3d orbitals are invoked in the $P \rightarrow O$ bond, and the substituents on P in the phosphate ester are oxygen while those on As are carbon. These expectations are fully consistent with the available data.

In aliphatic amine oxides the difference in electronegativity between nitrogen and oxygen in the $N \rightarrow O$ bond is small, but no back-bonding exists. Experimentally this bond is found to be the most strongly dipolar group in chemistry, and this fact serves to emphasize the considerable importance of back-bonding to bond polarity. N–O bonds having $2p\pi$ bonds (such as in nitroso, nitro, and pyridine N-oxide groups) are carbonyl-like, and as a result are vastly different electronically (and also in their physical behavior) from the $N \rightarrow O$ bond in aliphatic amine oxides.

In sulfones the formal charge on sulfur is +2 while in sulfoxides it is +1. A sulfone is therefore expected to be weakly polar in comparison with a sulfoxide. This expected difference is actually not evident from the measured bond moments, which are reported to be similar [10]. However, it is entirely consistent with the dramatic differences in relative hydrophilicities between these two groups (Section 10.5.4.1).

9.4.2   The polarity of ionic and zwitterionic groups

A relatively strong electric field exists in the vicinity of dipolar functional groups because of their high dipole moments – even though they bear no net charge. However, a far stronger electric field exists about ions. Ions may only be prepared and studied as electrically neutral sets (salts) (except in a mass spectrometer or gas discharge). "Ionic surfactants" are salts in which at least one ion is amphiphilic. When the cation is amphiphilic and the anion hydrophilic, one speaks of "cationic surfactants"*. When

---

*The term "cationic surfactant" is less precise than is "cationic surfactant salt", but in the interest of conserving space and to avoid confusion such as results when an additional salt is present, the shorter term will be used. The same problem exists in "anionic surfactants", and this term will also be used for the same reasons.

the anion is amphiphilic and the cation hydrophilic, one speaks of "anionic surfactants". Salts in which both ions are amphiphilic have been termed "catanionic" surfactants [11]. Salts in which neither ion is amphiphilic are not surfactants.

The electronic bond structure between atoms within ionic groups does not differ qualitatively from that found in neutral functional groups, but it is quantitatively distorted by the formal charge. Within aliphatic ammonium ions (R_4N^+) single bonds which superficially resemble those found in amines exist, but there are no nonbonding orbitals. Also, these groups are neither hydrogen bond acceptors nor hydrogen bond donors. If an N^+-H bond exists in an ammonium ion (in salts of amines and strong acids), it may function as a strong hydrogen bond donor [12].

Within carboxylate ($-CO_2^-$) groups, $2p\pi$ bonds exist just as in neutral carbonyl groups. However, the π bond order is reduced and a fractional formal negative charge exists on each oxygen. This group is, as a result, a very strong hydrogen bond acceptor [13]. Within sulfonate ($-SO_3^-$) and sulfate ($-OSO_3^-$) groups, the $S \rightarrow O$ bonds qualitatively resemble those of sulfones but a fractional formal negative charge again exists on each oxygen. Since three such bonds exist, the fractional excess charge on each oxygen in sulfonates and sulfates is probably smaller than on the carboxylate oxygens. These groups are probably much weaker hydrogen bond acceptors than is the carboxylate anion, but the evidence on this point is weak.

Zwitterionic functional groups. Zwitterionic groups differ structurally from all those mentioned above [14]. Formally charged substituent groups exist in zwitterionic compounds, but positive and negative charges exist in equal numbers and the molecule as a whole is neutral. The charged substitutents encountered are exactly the same ones found in ionic surfactants (above).

The substituents in zwitterionic groups are electronically insulated from each other, usually by saturated divalent radicals. As a result, orbital overlap (conjugation) between the different functional groups does not exist. An example is betaine (trimethylammonioacetate):

$$(CH_3)_3N^+CH_2CO_2^-$$

The methylene group insulates the ammonio group from the carboxylate group in betaine with respect to orbital overlap, but it does *not* preclude inductive electrical interactions between them through space.

Because of their formally charged substituents the electric field in the vicinity of zwitterionic groups is presumably much stronger than that which exists around dipolar bonds, but it is probably smaller than that which surrounds ions [15]. The reduced dipole moment concept is useless for characterizing zwitterionic compounds, because their reduced bond moments are all necessarily 1 (Fig. 9.1). This does *not* mean that they all have the same polarity.

From a nomenclature perspective, zwitterionic compounds are polyfunctional molecules [14,16]. Generally, an anionic fragment (acetate, or $-CH_2CO_2^-$, in betaine) is taken as the parent compound, and the positive substituent is named by changing the "ium" suffix in the ion (Me_3N^+H) to "io" in the radical (as in trimethylammonio, Me_3N^+-). In the context of surfactant science, however, the zwitterionic group will be regarded as a single functional group. Three elements of zwitterionic structure must be considered: the positive substituent, the negative substituent, and the divalent radical which "tethers" them together [17].

Since zwitterionic molecules are electrically neutral, they are in every sense of the word nonionic compounds [14]. In striking contrast to true salts (such as sodium chloride), zwitterionic compounds neither undergo ion exchange chemical reactions nor enhance the conductivity of their solutions. They *do* significantly enhance the dielectric properties of their solutions.

It is proper to regard zwitterionic molecules as "inner salts" – provided the fundamental differences between "inner salts" and "true salts" are recognized. The formulae of "inner salts" are often depicted including species such as hydroxide ion, but zwitterionic compounds are often prepared as perfectly dry materials. The "inner salt" description of zwitterionic compounds is archaic.

9.5 A process for recognizing hydrophilicity

We now return to the first question: "how may one recognize the existence of hydrophilicity?". Deciding how to answer this question properly is a non-trivial matter. Suppose, for example, that the capacity of a functional group to confer surface activity onto an amphiphilic molecule containing that group was taken to be the criterion of hydrophilicity. The name "surfactant" itself suggests this possibility.

If surface activity were the criterion of hydrophilicity, however, one would find that virtually all of the known dipolar functional groups should be regarded as hydrophilic groups. Most (if not all) long chain molecules which have polar functionality at one end spread on water (whether soluble or not) and, as a result, reduce the tension at the air–water interface [18]. The soaps and the fatty acid methyl esters do not differ in this property, for example. Also, those hydrocarbons that spread on water also reduce this interfacial tension [18]; wherein lies the "hydrophilicity" of these compounds?

Even a casual comparison of a soap (on the one hand) with a fatty acid methyl ester (on the other) indicates that large differences in physical behavior exist between them: The soap is viewed as a "surfactant", while the methyl ester is not. The presence of "hydrophilicity" in a functional group may therefore be defined as *the capacity of a functional group to confer surfactant behavior onto amphiphilic molecules which contain that group*. A criterion that is capable of recognizing the existence of surfactant behavior, and also its absence, is needed. Phase behavior serves well as that criterion [1].

During the period from 1958 to about 1970 the available information from the literature on the phase behavior of surfactants was collected, and this information was supplemented by the results of a systematic and wide-ranging survey of the phase behavior of amhiphilic molecules. During this survey a formal experimental procedure evolved which permitted the presence or absence of surfactant behavior to be determined. By following this procedure, a "yes" or "no" answer to the question of whether or not a given compound displayed surfactant behavior was obtained in virtually every case. (Only one instance was encountered, the N,N-dimethylamide group in which some criteria for surfactant behavior were obeyed while others were not. This lone exception will be considered later.)

This procedure included the following steps:

(1) A pure compound was synthesized in which a functional group, $-X$, is attached to a proven lipophilic group, $R-$, to form the compound RX.

(2) The thermal stability of *RX* and its chemical stability in the presence of water under the conditions of the study were determined (if not already known). Only compounds that are thermally and hydrolytically stable under the screening conditions were pursued.

(3) The phase behavior of *RX* in the binary *RX*–water system was screened.

(4) The observed phase behavior was compared against defined criteria for surfactant behavior. It could be decided, from this comparison, whether or not the compound *RX* displayed surfactant behavior.

(5) If the compound *RX* displayed surfactant behavior then the functional group -*X* was classified as an operative hydrophilic group, and vice versa.

Selection of the lipophilic group, R. While a number of *R* groups might have been selected for screening purposes, *n*-alkyl groups were most commonly used. In sodium laurate, which is a C_{12} soap, the lipophilic group is actually C_{11}, but the even-chain synthetic intermediates are far more readily available than are the odd-, and qualitative behavior is not altered by adding or deleting one methylene group. Therefore, most derivatives were even-chain compounds.

Dodecyl derivatives were particularly useful for several reasons. First, they are readily accessible synthetically and information already existed on many dodecyl surfactants. In addition, the Krafft discontinuities of dodecyl surfactants rarely exceed 100°C and their liquid crystal regions rarely lie below 0°C. Because of these characteristics, this homolog is relatively easy to study and false negative results are unlikely.

Purity criteria. The screening must be performed on a sample of sufficient purity that its phase behavior is characteristic of the assigned structure (is not dictated by the impurities present) (Appendix 4.2). Although accurate assay data were rarely available, the samples investigated typically displayed no impurities (other than homologs) using thin layer chromatography. They were probably of at least 97% purity, and often greater. The presence of homologs would not have qualitatively altered the result.

Stability criteria. A compound has to be sufficiently stable thermally to survive the temperatures encountered (up to 100°C) to yield meaningful physical data. Moreover, it has to be unreactive towards water within this temperature range. These stability criteria eliminated from serious consideration a number of potentially interesting functional groups such as oxonium salts (R_3O^+,X^-), sulfonium salts (R_3S^+,X^-), imines of carbonyl compounds ($R_2C=NH$), sulfimines ($R_2S \rightarrow NH$), and phosphoryl imines ($R_3P \rightarrow NH$). Imines of four-coordinate sulfur (sulfoximines ($R_2S(\rightarrow O) \rightarrow NH$) and sulfone diimines ($R_2S(\rightarrow NH)_2$)) are both stable thermally and sufficiently resistant to hydrolysis to survive the screening. These were the only *N*-terminal functional groups that could be investigated other than the amines themselves. Reactivity towards oxygen (as for example in alkylphosphines) poses severe experimental problems because air must be excluded, but does not otherwise preclude the evaluation of hydrophilicity.

Screening process. The screening process had to provide data that could be evaluated against defined criteria for hydrophilic behavior. In the simplest test, this involved systematically scanning the entire composition range in intervals no larger than 10% and within the temperature range of 0 to 100°C [19]. The principal objectives were to:

(1) define the boundaries of the liquid phase,

(2) search for the existence of liquid crystal phases, and
(3) determine the melting point of the dry crystal.

If the compound decomposed at a temperature lower than its melting point, the decomposition temperature was recorded.

As much information as possible was collected about the liquid crystal regions, but this additional information was not required merely to decide whether or not surfactant phase behavior existed. (Such information *was* useful in ranking groups as to their relative hydrophilicity, however.) Detailed phase studies were sometimes available from the literature, or were performed (for other reasons) on a few compounds. Detailed phase studies provide unimpeachable information as to whether or not hydrophilicity exists, but are not absolutely required for screening purposes.

9.5.1   The phase criteria of surfactant behavior

The criteria used to recognize surfactant behavior are:

(1) *the liquid crystal criterion:* the existence of lyotropic liquid crystal states,
(2) *the ΔT criterion:* the existence of a Krafft discontinuity whose temperature is significantly lower than the melting point, and
(3) *the miscibility gap criterion:* distortion of the shape or location of liquid–liquid miscibility gaps (if they exist), in comparison to their usual shape and location in diagrams of nonsurfactant systems.

The liquid crystal and ΔT criteria are more or less universally applicable, and thus especially important. While quantitatively defining the range of compositions and temperatures within which liquid crystals exist is very difficult, merely recognizing whether or not they exist is not. Also, recognizing the existence of a Krafft discontinuity, determining its temperature, and establishing that a difference exists between it and the melting point are also relatively straightforward experiments even if determining the numerical value of ΔT is not. Miscibility gaps are not found in many aqueous surfactant systems, but when this phenomenon does exist this boundary is easily determined and the criterion is useful.

ΔT is defined as

$$\Delta T = T_{\mathrm{mp}} - T_{\mathrm{Krafft}} \qquad (9.3)$$

The basis for using this parameter lies in the fact that the formation of the liquid state by melting the dry compound occurs solely by the action of thermal energy, whereas the formation of aqueous liquid or liquid crystal phases is assisted by hydration (Section 5.6.2) [20]. Because of the contribution from hydration energy, less thermal energy is required to form these fluid states in aqueous mixtures than is required in the dry compound; as a result, the temperature at which they first form is lowered. If a functional group is not hydrophilic, lowering of the crystal-melting temperature is non-existent or extremely small.

The determination of Krafft discontinuity temperatures should be performed at compositions which are greater than the solubility at the Krafft discontinuity (which for most soluble surfactants is between 25 and 60%). If a slurry of crystals of the appropriate concentration is heated, it remains qualitatively unchanged until the Krafft temperature is reached. (An increase in the extent of dissolution of the crystals will commence at the "Krafft points", however (Section 5.4.5).) At the Krafft discontinuity, profound changes occur, which typically include a dramatic increase in viscosity, a clearing of the opaque slurry to form a translucent and viscous mixture, and (usually) the appearance of birefringence. The birefringence is most reliably determined by direct observation of a thick (1 cm) sample between crossed polars (Section 8.2). The supplementation of such observations by later investigations using the polarized light microscope, on selected mixtures of known composition, provides additional proof of the existence of liquid crystals and often a probable assignment of their structure.

If an R–X compound is *not* a surfactant, qualitatively different phase behavior is observed at the temperature where the crystal "melts" in aqueous mixtures. This behavior often resembles that shown in Fig. 9.2. In this system, a depression of the melting point occurs due to the solubility of water in the compound. However, this melting point depression boundary quickly encounters a liquid/liquid miscibility gap, which spans almost the entire composition range. A eutectic results at the temperature at which these two boundaries cross; the dilute liquid, the concentrated liquid, and the crystal usually coexist at this eutectic. At temperatures below this eutectic a slurry of crystals exists which, over most of the composition span, is transformed on heating at the eutectic temperature into two coexisting liquids. The temperature of this transformation is independent of composition within the liquid/liquid miscibility gap at the eutectic temperature.

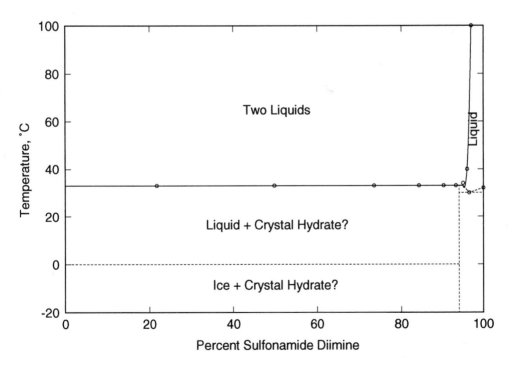

Fig. 9.2. The phase diagram of *S*-dodecyl-*N*-methylsulfonamide-bis(*N*-methylimine)–water, illustrating nonsurfactant phase behavior in an amphiphilic compound.

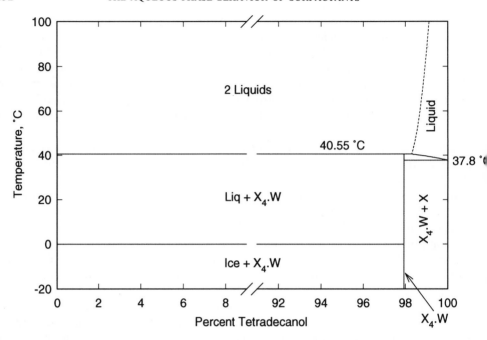

Fig. 9.3. A tentative phase diagram of the tetradecanol–water system, based on literature data and information from DIT studies [21].

Other kinds of phase behavior than this are possible in aqueous nonsurfactant systems. Fatty alcohols, for example, react with water to form a crystal hydrate whose peritectic temperature lies above their melting points (Fig. 9.3). As a result, dilute mixtures of crystals and liquid in these systems are transformed into two liquids at a temperature that is actually *higher* than the melting point of the dry alcohol. (The existence of a hydrate whose peritectic lies below the melting point would not alter the behavior shown in Fig. 9.2.) In none of these systems do liquid crystals exist, ΔT is very small (or negative), and neither of these kinds of phase behavior meets either the liquid crystal or the ΔT criterion of surfactant behavior.

9.6   The existence of hydrophilicity vs. molecular structure

9.6.1   Hydrophilicity in monofunctional amphiphilic molecules

In evaluating hydrophilicity vs. structure, it is useful first to consider the "simple" question of whether or not a given functional group is, by itself, an operative hydrophilic group. To address this question, only data on monofunctional (RX) molecules are utilized; consideration of polyfunctional (RX_n) compounds is deferred for the moment.

Formally charged compounds and substituent groups. Two important results emerged from this survey of monofunctional amphiphilic compounds. First, it was found that all the amphiphilic salts that were studied displayed surfactant behavior; no instances were discovered in which salts did not display surfactant behavior. This result indicates that all the sets of formally charged substituent groups that were examined ($-CO_2^-$,Na^+, $-N(Me)_3^+$,Cl^-, etc.) are operative as hydrophilic groups.

The same result was found among zwitterionic structures. All of the zwitterionic compounds examined displayed surfactant behavior, and no instances of non-surfactant behavior were found within this class of compounds as well.

Thus, all of the known structural arrangements of neutral sets of formally charged functional groups are operative as hydrophilic groups, regardless of whether they exist in separate molecules (ionic salts) or in the same molecule (zwitterionic compounds). Since ions cannot be examined as such, the role that each ionic molecule (in salts) or each charged substituent (in zwitterionics) plays with respect to hydrophilicity is a matter to be considered separately.

Dipolar functional groups. The second major result of this survey was that only a minor fraction of those dipolar groups investigated are operative as hydrophilic groups. Further, it was impossible to anticipate *a priori* whether or not a particular functional group is operative using, for example, data on reduced bond moments. Both strongly and weakly dipolar groups were found among both operative and nonoperative functional groups.

Two lists of functional groups may be constructed which summarize the results of these screening experiments. In Table 9.1 those groups are listed which were shown *not* to be operative as hydrophilic groups in monofunctional RX molecules. In Table 9.2 those groups are listed which *were* shown to be operative. Specific examples of the molecules examined are also shown, but in most cases other homologs were examined as well.

9.6.2   Hydrophilicity in polyfunctional amphiphilic molecules

Polyfunctional molecules containing operative functional groups. If one selects from Table 9.2 any operative functional group and creates (using this group) a new poly-functional molecule (RX_n) having a proven lipophilic group, such polyfunctional compounds invariably display surfactant behavior. This has been shown to be true in a great many instances and within all classes and subclasses. Varying hydrophilicity in this manner may be regarded as an "extensive" variation of hydrophilicity, while varying hydrophilicity by changing the group X to a different group X' (comparing RX and RX') may be regarded as an "intensive" variation of hydrophilicity. Alternatively, one may regard each group X and X' as having an "intrinsic" hydrophilicity which in the case of X is different from the intrinsic hydrophilicity of X'.

Polyfunctional molecules containing non-*operative functional groups.* If one creates polyfunctional amphiphilic molecules by combining in the same molecule two or more groups selected from the list of nonoperative functional groups, the results are more complex. For most of the functional groups in Table 9.1, polyfunctional analogs (RX_n) behave in a manner that is qualitatively similar to the monofunctional parent compound. For example, increasing the number of ester or nitrile groups in an amphiphilic molecule does *not* qualitatively alter the physical behavior; the resulting compound still displays nonsurfactant phase behavior.

A few cases exist, however, in which this is not true. For these select groups, poly-functional compounds (RX_n) *do* display surfactant behavior even though their

TABLE 9.1

Functional groups that are *not* operative as hydrophilic groups in monofunctional *RX* molecules

Functional group name		Compound studied
Carbonyl		
$RCH=O$	Aldehyde	$C_{11}H_{23}CH=O$
$R_2C=O$	Ketone	$C_{11}H_{23}(CH_3)C=O$
$R(RO)C=O$	Ester	$C_{11}H_{23}(CH_3O)C=O$
$R(HO)C=O$	Carboxylic acid	$C_{11}H_{23}(HO)C=O$
$RN=C=O$	Isocyanate	
		$\overset{\displaystyle CH_2}{\overbrace{C_{12}H_{25}CH-CH-N=C=O}}$
$RC_2N_2O_2H$	Sydnone	$N-C_{12}H_{25}C_2N_2O_2H$
$R(H_2N)C=O$	Amide	$C_{11}H_{23}(H_2N)C=O$
$RC_3H_3N_4O$	Cyanoguanidine	$C_{11}H_{23}C(O)NHC(NH)NHC\equiv N$
$RC_3H_5N_4O_2$	Guanylurea	$C_{11}H_{23}C(O)NHC(NH)NHC(O)NH_2$
$RC_3H_4N_3O_3$	Biuret	$C_{11}H_{23}C(O)NHC(O)NHC(O)NH_2$
Nitronyl		
$RN=O$	Nitrosoalkane	$C_{12}H_{25}N=O$
$RN(=O)\rightarrow O$	Nitroalkane	$C_{12}H_{25}N(=O)\rightarrow O$
$RON=O$	Nitrite ester	$C_{12}H_{25}ON=O$
$RON(=O)\rightarrow O$	Nitrate ester	$C_{12}H_{25}ON(=O)\rightarrow O$
$R_2C=N(R)\rightarrow O$	Nitrone	$C_{11}H_{23}C(C_6H_5)C=N(i\text{-}C_3H_7)\rightarrow O$
$R_2NN=O$	Nitrosamine	$C_{12}H_{25}(CH_3)NN=O$
$RC_5H_4N=O$	Pyridine N-oxide	$4\text{-}C_{12}H_{25}C_5H_4N=O$
Nitrile/isonitrile		
$RC\equiv N$	Nitrile	$C_{11}H_{23}C\equiv N$
$RN\equiv C$	Isonitrile	$C_{11}H_{23}N\equiv C$
Lewis acid/base complexes		
$R_3N^+B^-H_3$	Amine borane	$C_{12}H_{25}N^+(CH_3)_2B^-H_3$
$RN^+H_2B^-X_3$	Amine haloboranes	$C_{12}H_{25}N^+H_2B^-X_3$ (X=F,Cl)
$RN^+(CH_3)_2B^-X_3$	Amine haloboranes	$C_{12}H_{25}N^+(CH_3)_2B^-X_3$ (X=F,Cl)
Semipolar		
RSO_2CH_3	Sulfone	$C_{12}H_{25}SO_2CH_3$
$R(CH_3)_2P\rightarrow S$	Phosphine sulfide	$C_{12}H_{25}(CH_3)_2P\rightarrow S$
$R(CH_3)_2As\rightarrow S$	Arsine sulfide	$C_{12}H_{25}(CH_3)_2As\rightarrow S$
$RSO_2N(CH_3)_2$	Sulfonamide	$C_{12}H_{25}SO_2N(CH_3)_2$
$RSO(\rightarrow NCH_3)N(CH_3)_2$	Sulfonamide methylimine	$C_{12}H_{25}SO(\rightarrow NCH_3)N(CH_3)_2$
$RS(\rightarrow NCH_3)_2N(CH_3)_2$	Sulfonamide bis(methylimine)	$C_{12}H_{25}S(\rightarrow NCH_3)_2N(CH_3)_2$
Single bond		
ROH	Alcohols	$C_{12}H_{25}OH$
$ROCH_3$	Ethers	$C_{12}H_{25}OCH_3$
$RNHCH_3$	Methylamines	$C_{12}H_{25}NHCH_3$
$RN(CH_3)_2$	Dimethylamines	$C_{12}H_{25}N(CH_3)_2$
RSH	Thiols	$C_{12}H_{25}SH$
$RSCH_3$	Thioethers	$C_{12}H_{25}SCH_3$
RPH_2	1° Phosphines	$C_{12}H_{25}PH_2$
$RPHCH_3$	2° Phosphines	$C_{12}H_{25}PHCH_3$
$RP(CH_3)_2$	3° Phosphines	$C_{12}H_{25}P(CH_3)_2$

TABLE 9.2

Functional groups that *are* operative as hydrophilic groups in monofunctional *RX* molecules

Functional group name	General formula
Ionic class, anionic subclass	
Alkyl carboxylate (alkanoate) salts (soaps)	RCO_2^-, M^+
Alkanesulfonate salts	RSO_3^-, M^+
Alkyl sulfate salts	$ROSO_3^-, M^+$
N-Alkylsulfamate salts	$RNHSO_3^-, M^+$
Alkylsulfinate salts	RSO_2^-, M^+
S-Alkylthiosulfate salts	$RSSO_3^-, M^+$
Phosphonate salts	$RPO_3^=, 2M^+$
Phosphate monoester salts	$ROPO_4^=, 2M^+$
Phosphinate salts	$R(R')PO_2^-, M^+$
Nitroamide salts	RN^-NO_2, M^+
Trisulfonylmethide salts	$RSO_2(CH_3SO_2)_2C^-, M^+$
Xanthate salts	$RSCS_2^-, M^+$
Ionic class, cationic subclass	
Quaternary ammonium salts	$RN^+(CH_3)_3, X^-$
Primary, secondary, tertiary ammonium salts	$RN^+H_n(CH_3)_{3-n}, X^-$
N-Alkylpyridinium salts	$RNC_5H_5^+, X^-$
Quaternary phosphonium salts	$RP^+(CH_3)_3, X^-$
Ternary sulfonium salts	$RS^+(CH_3)_2, X^-$
Ternary sulfoxonium salts	$RS^+(\rightarrow O)(CH_3)_2, X^-$
Bis(phosphoranylidyl)ammonium salts	$[R(CH_3)_3P \rightarrow N \leftarrow P(CH_3)_3R]^+, X^-$
Nonionic class, zwitterionic subclass	
Ammonioacetates (betaines)	$R(CH_3)_2N^+CH_2CO_2^-$
Ammoniohexanoates	$R(CH_3)_2N^+(CH_2)_5CO_2^-$
Ammonioalkanesulfonates (sulfobetaines)	$R(CH_3)_2N^+(CH_2)_3SO_3^-$
Ammonioalkyl sulfates	$R(CH_3)_2N^+(CH_2)_nOSO_3^-$
Trimethylammonioethyl alkylphosphonates	$RPO_2^-OCH_2CH_2N^+(CH_3)_3$
Trimethylammonioethylphosphate acylglyceryl esters (lysolecithin and lecithins)	$RCO_2CH_2CH(OH)CH_2OPO_2^-O-$ $(CH_2)_2N^+(CH_3)_3$
Nonionic class, dipolar subclass	
Aliphatic amine oxides	$R(CH_3)_2N \rightarrow O$
Phosphine oxides	$R(CH_3)_2P \rightarrow O$
Phosphonate esters	$R(CH_3O)_2P \rightarrow O$
Phosphate esters	$RO(CH_3O)_2P \rightarrow O$
Arsine oxides	$R(CH_3)_2As \rightarrow O$
Sulfoxides	$R(CH_3)S \rightarrow O$
Sulfoximines (sulfone imines)	$R(CH_3)S(\rightarrow O) \rightarrow NH$
Sulfone diimines	$R(CH_3)S(\rightarrow NH)_2$
Ammonioamidates	$RC(O)N^-N^+(CH_3)_3$
*Amides	$RC(O)N(CH_3)_2$
Nonionic class, single bond subclass	
Primary amines	RNH_2

*Borderline – see text and Table 9.3.

monofunctional analogs (RX) do not. The two particularly important examples of this behavior are the hydroxy and the ether groups. The amide group might well also be included, but the documentation to support this listing is very sparse. Because this characteristic sets these particular "nonoperative" functional groups apart from the rest of those in Table 9.1, it is useful to form two subgroups from the list in Table 9.1. One subgroup contains those functional groups which, in polyfunctional analogs, remain inactive as hydrophilic groups. The other subgroup (which is listed separately in Table 9.3) includes those groups which do display surfactant behavior in polyfunctional molecules. Specific examples of molecules containing these interesting groups are also listed in Table 9.3.

A complete listing of operative hydrophilic groups must therefore include both those listed in Table 9.2, and those in Table 9.3. The balance of the functional groups in Table 9.1 are inoperative in both mono- and polyfunctional amphiphilic structures.

This three-group classification, represented by Tables 9.1, 9.2 and 9.3, better describes the physical behavior of functional groups over a wide range of structures than does the earlier two-group classification based on Tables 9.1 and 9.2 [1]. Except for the amide, the screening data clearly require that the groups in Table 9.3 also belong in Table 9.1. It is also true that those groups listed in Table 9.3 differ fundamentally (as regards their hydrophilicity) from the balance of those listed in Table 9.1 in possessing greater intrinsic hydrophilicity.

This survey of monofunctional compounds was comprehensive; it included all of the thermally and hydrolytically stable functional groups that are commonly encountered, plus some unfamiliar ones as well. A similarly comprehensive survey of polyfunctional compounds is inconceivable because the number of these compounds that can be imagined to exist is so enormous. The data that exist on polyfunctional compounds hinge principally on the availability of the compound (Section 14.9).

TABLE 9.3

Functional groups that are non-operative as hydrophilic groups in monofunctional RX molecules, but operative in polyfunctional molecules

Functional group name	Compounds studied
Hydroxyl	$C_nH_{2n+1}OH$
Ether	$C_nH_{2n+1}OCH_3$
Amide*	$RC(O)N(CH_3)_2$
Examples	
Alkane diols	$RCH(OH)CH_2OH$
Monoglycerides	$RCO_2CH_2CH(OH)CH_2OH$
Sugar glycosides	$RO(C_6H_{10}O_5)$
N-Acyl-N-methylglucamines	$RC(O)N(CH_3)CH_2(CHOH)_4CH_2OH$
Polyoxyethylene glycol alkyl ethers	$R(OCH_2CH_2)_nOH$
Polyoxyethylene glycol alkylphenyl ethers	$RC_6H_4(OCH_2CH_2)_nOH$
Polyoxyethylene glycol fatty acid esters	$RC(O)(OCH_2CH_2)_nOH$
Polyoxyethylene glycol methyl alkyl ethers	$R(OCH_2CH_2)_nOCH_3$

* Borderline data; weakly stable equilibrium or metastable liquid crystals were observed, but the miscibility gap is broad and the magnitude of ΔT is very small.

9.7 A structural classification of hydrophilic groups

It is both convenient and useful to classify hydrophilic groups on a structural basis. All the functional groups which are operative, including both those in Table 9.2 and those in Table 9.3, are encompassed by the following scheme:

Ionic class
    Anionic subclass
    Cationic subclass
Nonionic Class
    Zwitterionic subclass
    Dipolar subclass
    Single bond subclass

This scheme was followed in organizing Table 9.2 and Appendix 2.

Having determined which functional groups are operative as hydrophilic groups, we now turn to the quantitative aspect of hydrophilicity and the interesting question, "what is it?".

References

1. Laughlin, R. G. (1978). *Advances in Liquid Crystals*, Vol. 3 (G. H. Brown ed.), pp. 41–98, Academic Press, New York.
2. Shibakami, Motonari and Sekiya, Akira (1993). *Bull. Chem. Soc. Jpn.* **66**, 315–316; Sinitsyna, T. A., Saloutina, L. V. and Zapevalov, A. Ya. (1990). *Zh. Prik. Spektrosk.* **53**, 939–944.
3. (1969). *Nomenclature of Organic Chemistry. Sections A, B, and C*, pp. C-87, 142–143, Butterworth, London; (1991). *Handbook of Chemistry and Physics* (D. R. Lide ed.-in-chief), 72nd ed., pp. **2**-45-2-74, CRC Press, Boca Raton, FL.
4. Hudson, R. F. (1965). *Structure and Mechanism in Organo-Phosphorus Chemistry* (A. T. Blomquist ed.), pp. 83, 119–123, Academic Press, London.
5. Fajans, K. (1967). *Structure and Bonding* **3**, 88–105.
6. Smythe, C. P. (1955). *Dielectric Behavior and Structure*, pp. 228–259, McGraw-Hill, New York.
7. Pauling, L. (1960). *The Nature of the Chemical Bond*, 3rd ed., pp. 136–142, Cornell University Press, Ithaca, New York.
8. Joesten, M. D. and Schaad, L. J. (1974). *Hydrogen Bonding*, pp. 274–275, Marcel Dekker, New York.
9. Cotton, F. A. and Wilkinson, G. (1988). *Advanced Inorganic Chemistry*, 5th ed., pp. 58–76, John Wiley, New York.
10. Carlson, E. R. and Meek, D. W. (1974). *Inorg. Chem.* **13**, 1741–1747.
11. Jonsson, B. and Kahn, A. (1987). *J. Phys. Chem.* **91**, 3291–3298; Jokela, P. and Jonsson, B. (1988). *J. Phys. Chem.* **91**, 1923–1927; Friberg, S. E., Sun, W. M., Yang, Y. and Ward, A. J. I. (1990). *J. Colloid Interface Sci.* **139**, 160–168.
12. Colthup, N. B., Daly, L. H. and Wilberly, S. E. (1990). *Introduction to Infrared and Raman Spectroscopy*, 3rd ed., pp. 343–344, Academic Press, Boston.
13. Verrall, R. E. (1973). *Water: A Comprehensive Treatise. Aqueous Solutions of Simple Electrolytes*, Vol. 3 (F. Franks ed.), pp. 236–238, Plenum, New York.
14. Laughlin, R. G. (1991). *Langmuir* **7**, 842–847.
15. Cohn, E. J. and Edsall, J. T. (1943). *Proteins, Amino Acids, and Peptides as Ions and Dipolar Ions*, p. 11, Hafner, New York.
16. Use ref. 3.
17. Chevalier, Y., Germanaud, L. and Le Perchec, P. (1988). *Colloid Polym. Sci.* **266**, 441–448.

18. Adamson, A. W. (1982). *Physical Chemistry of Surfaces*, 4th ed., pp. 103–111, Interscience, John Wiley, New York.
19. Laughlin, R. G. (1976). *J. Colloid Interface Sci.* **55**, 239–241.
20. Shinoda, K. (1967). *Solvent Properties of Surfactant Solutions*, Vol. 2 (K. Shinoda ed.), pp. 12–16, Marcel Dekker, New York.
21. Small, D. M. (1986). *The Physical Chemistry of Lipids. From Alkanes to Phospholipids*, Vol. 4, pp. 247–253, Plenum, New York.

Chapter 10

Hydrophilicity and Proximate Substituent Effects on Phase Behavior

10.1 Introduction

In Chapter 9 it was suggested that the capacity to display surfactant behavior hinged upon the existence of "intrinsic hydrophilicity" in the polar functional group. The present chapter is concerned with the scaling of intrinsic hydrophilicity relative to molecular structure [1]. In so doing, it is necessary to consider also the influence on phase behavior of those substituents that are on or near the hydrophilic group.

The information in this chapter, together with that in Chapter 9, suggests a reasonable model for hydrophilicity. The structural evolution of hydrophilic groups starting from the simplest surfactant – the fatty alcohol – can be perceived by using this model in combination with the correlations of phase behavior with structure.

10.2 The intensive variation of hydrophilicity

As has been noted (Section 9.6.2), hydrophilicity may be varied in either an "intensive" or an "extensive" manner. We will first consider intensive variations by inquiring into the relative intrinsic hydrophilicities of those hydrophilic groups that are operative in monofunctional amphiphilic molecules (Table 10.2). Once that is accomplished, the influence of varying hydrophilicity in an extensive manner (in polyfunctional surfactants) may be considered.

10.2.1 Extracting intrinsic hydrophilicity from phase information

Varying the magnitude of hydrophilicity is expected to influence the phase criteria of surfactant behavior (Section 9.5.1) as follows.

The liquid crystal criterion. The thermal stability of liquid crystal phases (the highest temperature at which they exist) is most clearly evident at upper azeotropic points, where the phase is reversibly transformed to a liquid of the same composition (Section 5.6.1). Increasing the hydrophilicity should increase the temperature of this phase transition. Obviously, only data on phases of corresponding structure should be compared.

Data on the upper temperature limit of liquid crystal phases that are squeezed between other liquid crystal regions (Section 5.8) are not useful for this purpose, since the peritectic reaction (which usually occurs at this discontinuity) is a disproportion reaction during which two other phases are formed. Since this reaction is obviously influenced by the thermodynamic parameters of another phase besides the liquid crystal and the liquid, its temperature is not a good indicator of thermal stability.

The composition range of the liquid crystal region may be useful in some comparisons between closely related molecules, but the lipophilicity of a surfactant has a greater influence on this parameter than does its hydrophilicity (Section 11.3.1).

The ΔT criterion. ΔT (the melting point temperature minus the Krafft discontinuity temperature) reflects the difference in thermal energies that exists at these two important thermodynamic discontinuities. Increasing the hydrophilicity should increase the magnitude of ΔT (Section 9.5.1).

The principal barrier to the widespread use of ΔT is chemical instability of the surfactant at its melting point. When this situation exists the melting point cannot be determined and, as a result, ΔT cannot be quantitatively measured. Often Krafft discontinuity data exist, however, and the difference between these data and the crystal decomposition temperature places a lower limit on the value of ΔT. Sometimes this is useful information.

If the crystal phase at the melting point is qualitatively the same phase that coexists with the saturated liquid below the Krafft discontinuity, ΔT is a straightforward criterion of hydrophilicity. Unfortunately, this ideal situation exists for only a few surfactants such as the phosphine oxides (Section 5.5). The straightforward interpretation of ΔT may also be complicated by the existence of polymorphism in the dry crystal, by transformation of the dry crystal into a liquid crystal (or another disordered phase) before it melts, or by the fact that a crystal hydrate (rather than the dry crystal) is the coexisting phase below the Krafft discontinuity. The existence of these phenomena will complicate attempts to quantify hydrophilicity using the ΔT parameter, but they do not detract from its value as a means of inferring relative hydrophilicities. In fact, ΔT is useful both as a means of ranking compounds within subclasses, and of classes and subclasses relative to each other.

The miscibility gap criterion. If miscibility gaps exist, increasing the hydrophilicity reduces the area spanned by these gaps. Two different kinds of gaps are found (Section 5.10.4), and the influence of varying hydrophilicity on the boundary position (and critical temperature) depends on which gap one is dealing with. If the gap is defined by a lower consolute boundary (phase separation occurs on heating), increasing the intrinsic hydrophilicity *raises* the boundary. If the gap is defined by an upper consolute boundary (phase separation occurs on cooling), increasing the intrinsic hydrophilicity *lowers* the boundary.

If the intrinsic hydrophilicity is sufficiently large, liquid/liquid miscibility gaps disappear altogether. The mere presence of a miscibility gap is therefore significant, because it indicates that the hydrophilicity is *relatively* low. Conversely, the absence of this phenomenon signifies that hydrophilicity is *relatively* high. The disappearance of either gap, as hydrophilic group structure is varied, signifies an increase in hydrophilicity and the appearance of either gap signifies a decrease in hydrophilicity. Other structural features besides hydrophilicity that influence phase behavior must be taken into account in selecting molecules for comparison, of course.

The presence or position of a miscibility gap is of particular value in comparing the relative hydrophilicity of particular hydrophilic groups within a family of related surfactants, but does not necessarily reflect the relative hydrophilicities of different subclasses. For example, it is very useful in comparing the individual members of the semipolar, or the polyoxyethylene, or the zwitterionic subclasses. The existence of the upper consolute boundary in the case of the zwitterionic subclass does *not*, however, signify that this subclass is weakly hydrophilic relative to semipolar (Section 10.5.3) or polyoxyethylene surfactants (Section 10.5.4). Rather, the body of evidence taken as a whole indicates that the reverse is true.

In real diagrams, miscibility gaps often interfere with other phase phenomena (such as liquid crystal regions or the Krafft boundary). These interferences not only complicate the phase behavior, but compromise the extraction of relative hydrophilicities. Nevertheless, if care is used in selecting molecules for comparison the forms of boundaries are similar, and those fragments of the boundary that remain visible are useful.

The influence of substituents on phase behavior will now be considered.

10.2.2   The influence of substituents on phase behavior

It is necessary to address the influence of substituents on phase behavior before extracting hydrophilicity from phase information, for it is important to eliminate (insofar as possible) the influence of substituents on this analysis.

In Section 10.2 a distinction was made between substituents "on or near" the hydrophilic group and substituents far from this group. This distinction is based on data regarding the influence of substituent position on phase behavior. Substituents near ("proximate" to) the hydrophilic group often influence phase behavior not only to a different degree, but often in a different manner, than do substituents "remote" from the hydrophilic group (Section 10.5.3.2).

The position of hydroxy substitution. To illustrate, hydroxy substitution in the middle of a lipophilic chain dramatically lowers the Krafft discontinuity temperature, while hydroxy substitution next to the hydrophilic group strongly *increases* this temperature. This can be seen from the dependence of the temperature of the Krafft boundary of sodium hydroxypalmitates on hydroxy position (Fig. 10.1) [2].

Substitution of a hydroxy group in the β-position has been found to increase the temperature of the Krafft point (at 1%) in sodium dodecyl sulfate, sodium dodecanesulfonate, hexadecyldimethylammoniopropanesulfonate, and dimethyldodecylamine oxide:

Compound	Krafft point (°C)		Krafft point (°C)	Difference
$C_{12}H_{25}OSO_3^-,Na^+$	17	β-hydroxy	31	14
$C_{12}H_{25}SO_3^-,Na^+$	36	β-hydroxy	74	38
$C_{16}H_{33}(CH_3)_2N^+(CH_2)_3SO_3^-$	29	β-hydroxy	85	56
$C_{12}H_{25}(CH_3)_2N \rightarrow O$	<0	β-hydroxy	15	>15

It is noteworthy that the β-hydroxy substituent in the alkyl sulfate is in the γ-position relative to the sulfur atom. This may be responsible for the smaller difference in Krafft points for the alkyl sulfate than for the alkane and zwitterionic sulfonates.

The influence of hydroxy substitution and of hydroxy position on the lower consolute boundary can also be seen by comparing the α-, β-, and γ-hydroxydodecyldimethylphosphine oxides with the unsubstituted parent compound. The diagrams of the parent phosphine oxide and of the γ-substituted compound are shown in Fig. 10.2, and phase information (but no diagrams) is available on the other two. It can be seen from Fig. 10.2 that dodecyldimethylphosphine oxide displays a high degree of miscibility with

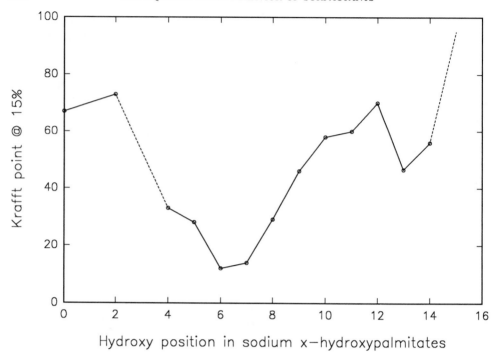

Fig. 10.1. The Krafft boundary temperature at 15% composition of the isomeric sodium *x*-hydroxypalmitates vs. position of substitution (*x*).

water (both as liquid and as liquid crystal phases), but that a closed loop of miscibility intrudes in water-rich mixtures at higher temperatures.

An α-hydroxy substituent dramatically reduces the miscibility of this compound with water below 100°C. Liquid crystal regions were found to exist, but the large liquid region of the unsubstituted compound is erased by immiscibility of the liquid crystal phases at low temperatures, and by the presence of a broad liquid/liquid miscibility gap at high temperatures. (The data suggest that the phase behavior of the α-hydroxyphosphine oxide qualitatively resembles that of the β-hydroxy sulfoxides shown in Fig. 10.3.) The β-hydroxy isomer is substantially more miscible with water, since its phase behavior is qualitatively similar to that of the parent compound except that the miscibility gap lies at a somewhat higher temperature. The γ-hydroxy substituent further increases the miscibility with water. Liquid crystal regions of roughly comparable stability exist, but the miscibility gap is erased below 100°C (Fig. 10.2). The closer the hydroxy substituent is to the P → O hydrophilic group, the more strongly it impairs the net hydrophilicity of the molecule.

The stereochemical configuration of hydrophilic groups. All of the above hydroxy compounds contain only one dissymetric center and therefore exist as a racemic pair, but a hydroxysulfoxide is different. Because both the carbon bearing the hydroxy group and the sulfoxide sulfur are dissymetric atoms, two pairs of diastereomeric isomers exist. Diastereomeric isomers (such as glucose and galactose) are both chemically and physically different, and this difference may extend to their phase properties.

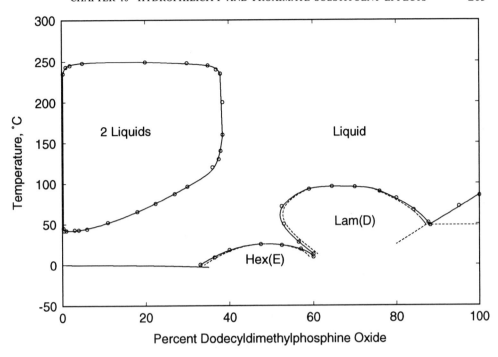

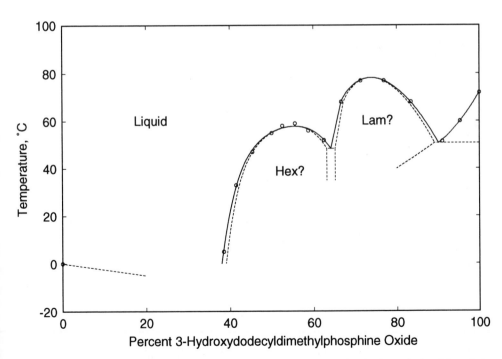

Fig. 10.2. The dodecyldimethylphosphine oxide–water (upper) and γ-hydroxydodecyldimethyl-phosphine oxide–water (lower) systems.

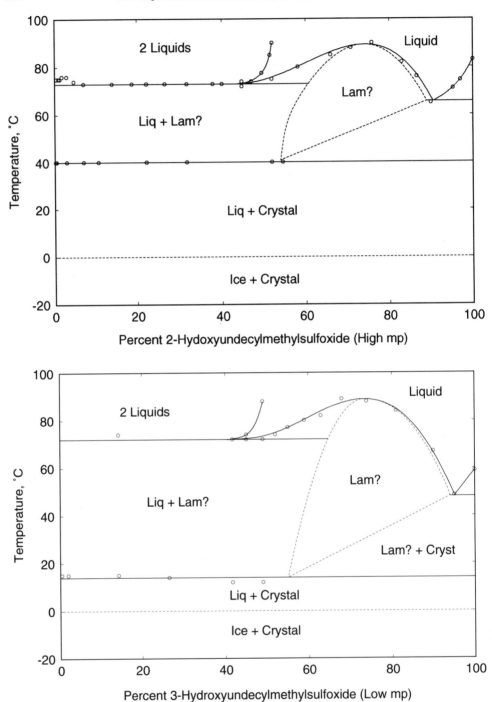

Fig. 10.3. The two stereoisomeric 2-hydroxyundecyl methyl sulfoxide–water systems. Upper diagram, high melting isomer; lower diagram, a fraction rich in the low melting isomer. The melting points and Krafft discontinuities differ, but the liquid crystal region and miscibility gap are unaffected.

The aqueous phase behavior of the two diasteromeric isomers of 2-hydroxyundecyl methyl sulfoxide have been explored [3]. The stereochemical configurations of these two compounds are unknown, but their melting points have been determined. The phase diagrams (shown in Fig. 10.3) reveal that while the higher melting compound also has the higher Krafft eutectic temperature, the position of the liquid crystal region is virtually identical in both compounds. The stereochemical difference between them does influence the thermodynamics of the crystal phase and of phase equilibria in which crystal phases are involved, but has no perceptible influence on transitions involving liquid or liquid crystal phases.

Stereochemistry has also been shown not to influence critical micelle concentrations (cmc). In a series of N-alkylglyconic acid amides, the various homologs of the mono- and disaccharide compounds were found to fall on the same log(cmc) vs. n curve, irrespective of the stereochemistry of the sugar group [4].

Position of methyl substitution. Methyl substitution in sodium octadecyl sulfate results in a progressively lower Krafft boundary as the substituent is moved away from the hydrophilic group (Section 11.3.2.6, Fig. 12.9). Comparison of these data with those in Fig. 10.1 indicates that methyl and hydroxy groups have opposite effects as proximate substituents on the Krafft boundary, but similar effects on this boundary as remote substituents. α,α-Diethyl substitution in soaps is reported to strongly affect their physical properties, but no phase information has been reported.

The position at which the transition from proximate to remote substituents occurs is probably either the β- or the γ-position. In amine oxides the α-hydroxy compound is unstable and cannot be examined, the β-hydroxy substituent raises the Krafft boundary (above), and the γ-hydroxy substituent lowers it.

Methylene insertion (homologation) into proximate substituents. Homologation in the α-position of sodium decanoate, to form sodium α,α-dibutyl decanoate, has a drastic influence on its physical behavior. Not only is the Krafft boundary low, but the normal hexagonal phase is no longer found. Also, the lamellar liquid crystal phase is found at exceptionally low surfactant compositions (Fig. 10.4). The phase behavior resembles in some degree that found in sodium AOT (Fig. 10.7).

Methylene insertions into proximate substituents on the nitrogen (or phosphorus) of dimethyldodecylamine (or phosphine) oxides amounts to homologation of the proximate substituent groups. Methylene insertions into the N–H bond of alkylammonium salts represents a similar structural perturbation. The homologation of proximate substituents influences phase behavior in a different manner than does the homologation of the lipophilic chain. The differences are revealed by comparing C_{12}diethylphosphine oxide and C_{14}dimethylphosphine oxide with C_{12}dimethylphosphine oxide (Fig. 10.5).

The C_{12}dimethyl compound is highly soluble because the miscibility gap and liquid crystal regions are widely separated. Substituting ethyl for methyl groups increases the lipophilic volume without changing the length of the lipophilic group. This structural modification enlarges the miscibility gap, but weakens the thermal stability of liquid crystals. Since both boundaries are moved in the same direction, qualitatively similar phase behavior is observed in the dimethyl and diethyl compounds. Inserting two methylene groups into the long chain, to form the isomeric C_{14}dimethyl analog, has an entirely different impact on phase behavior. This structural change also enlarges the miscibility gap – but it does *not* reduce the thermal stability of the liquid crystals. The result is "pivotal" phase behavior which closely resembles that found in the $C_{10}E_4$– water system (Section 5.10.7).

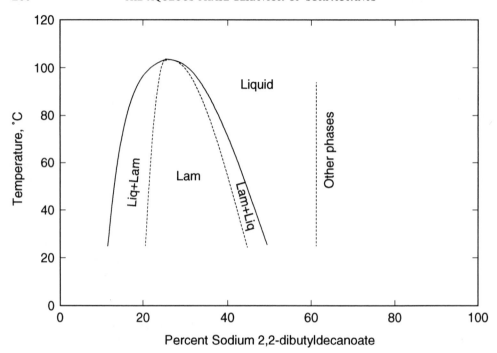

Fig. 10.4. The sodium α,α-dibutyl decanoate–water system. A dramatic difference in phase behavior exists relative to that of sodium dodecanoate.

In cationic and zwitterionic surfactants, the positive ammonio substituents may bear either N^+–H or N^+–R groups. If N^+–H groups exist then strong ion pair/hydrogen bonding is expected to occur to anionic groups in the crystal state. In the $R_2N^+HCH_3,Cl^-$ (DOMAC) crystal, for example, the chloride anion is located at the vertex of the nitrogen tetrahedron that bears the proton, and is engaged in a very strong hydrogen bond–ion pair interaction at a very short N–Cl distance (3.05 Å). In the related quaternary ammonium chloride salt (DODMAC), the chloride anion sits on the face of the nitrogen tetrahedron, the N–Cl distance is much greater (4.25 Å), and the strength of the interaction therefore presumably much weaker. Both kinds of interactions exist in the DOMAC crystal. The insertion of a methylene group in the N^+–H group, to form the N^+–CH_3 group, thus profoundly alters both the nature and the strength of polar interactions between the ions in the crystal (Section 8.3.2).

N-Methyl substitution in dodecylammonium chloride results in progressive lowering of the Krafft boundary as the number of methyl groups is increased. This boundary is an equilibrium boundary in the $-NH_3^+$ and the $-NH_2CH_3^+$ compounds, but is metastable in the di- and trimethyl compounds [5]. Estimates of ΔT in this series (Section 5.4.4) illustrate this effect:

Compound	T_{Krafft}	T_{mp}	ΔT
$C_{12}H_{25}NH_3^+,Cl^-$	32.5	225	98
$C_{12}H_{25}NH_2CH_3^+,Cl^-$	23	177	154
$C_{12}H_{25}NH(CH_3)_2^+,Cl^-$	−5	190	95
$C_{12}H_{25}N(CH_3)_3^+,Cl^-$	−16	—	—

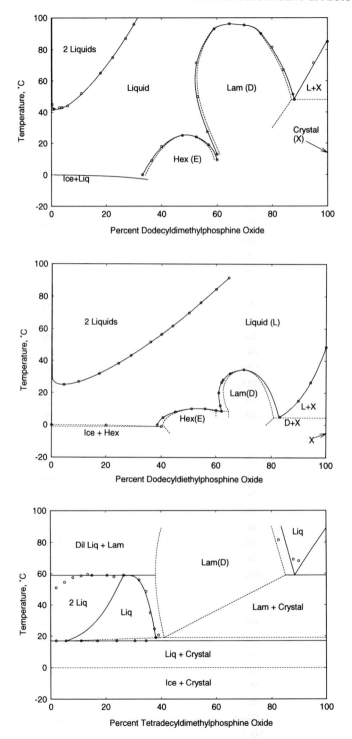

Fig. 10.5. The (upper) dodecyldimethyl-, (middle) dodecyldiethyl-, and (lower) tetradecyldimethylphosphine oxide–water systems. The tetradecyl homolog displays pivotal behavior.

Methylation of the positive nitrogen also tends to promote the existence of cubic phases. Two such phases are found in the $C_{12}H_{25}N(CH_3)_3^+,Cl^-$–water system, for example, while none was found in the $C_{12}H_{25}NH_3^+,Cl^-$–water system [5].

A similar effect on the Krafft boundary is observed from substituting a methyl group for a proton on the positive nitrogen atom of zwitterionic surfactants. Phase data exist which show that zwitterionics having a quaternary alkyldimethylammonio substituent ($R(CH_3)_2N^+-$) have much lower Krafft boundaries than do the corresponding alkyl-ammonio (RH_2N^+-) compounds, but the actual diagrams of related compounds illustrating this relationship have not been determined.

The further homologation of methyl groups in zwitterionic surfactants to form ethyl groups usually (but not always) lowers the Krafft boundary.

Compound	Proximate substituents	Krafft eutectic (°C)
$C_{12}H_{25}N^+(C)_2C-C-C-OSO_3^-$	dimethyl	58
$C_{12}H_{25}N^+(C_2)_2C-C-C-OSO_3^-$	diethyl	43
$C_{12}H_{25}N^+(C)_2C-C-SO_3^-$	dimethyl	71
$C_{12}H_{25}N^+(C)(C_2)C-C-SO_3^-$	methyl ethyl	47
$C_{12}H_{25}N^+(C)_2C-C-C-SO_3^-$	dimethyl	<0
$C_{12}H_{25}N^+(C_2)_2C-C-C-SO_3^-$	diethyl	10

These data show that the influence on the Krafft eutectic of replacing methyl proximate substituents by ethyl depends on the specific compound involved.

Effects of unsaturation. The introduction of unsaturation may be viewed as either insertion of a vinylene ($-CH=CH-$) linkage, or removal of hydrogens from adjacent methylene groups (desaturation). No information exists as to the influence of unsaturation proximate to the hydrophilic group on phase behavior. The influence of remote unsaturation will be considered in Chapter 12.

10.3 Choosing molecules for the evaluation of intrinsic hydrophilicity

It is apparent from the above that substituents exert a strong influence on phase behavior, and that this structural feature must be well controlled in evaluating hydrophilicity. Otherwise the substituent (rather than hydrophilicity) will dictate the result.

"Isoelectronic" sets of molecules are highly preferred for extracting hydrophilicity from phase data. Just as the word implies, the number of electrons (and consequently of protons) is precisely the same among members of isoelectronic sets – but the "spatial distribution" of protons differs. A particularly important isoelectronic series, insofar as surfactants are concerned, is methane, ammonia, water, hydrogen fluoride, and neon:

$$CH_4 \quad NH_3 \quad OH_2 \quad FH \quad Ne$$

The removal of a proton from methane (to form the methide anion) leaves a nonbonding pair in the place of one C–H bond (CH_3^-). If one could, by some magical process, insert this proton (and a neutron as well) into the nucleus of the methide carbon, this carbon would become a nitrogen and ammonia would result. By similarly "pushing protons" into nuclei (and adding the appropriate number of neutrons) ammonia may be "converted" to water, water to hydrogen fluoride, and hydrogen fluoride to neon. Related molecules in the same column of the Periodic Table but different rows (e.g. methane and silane) are *not* isoelectronic; nevertheless, such comparisons may also be interesting because factors such as coordination numbers to the heteroatom will be similar.

Comparing isoelectronic molecules has very important advantages over other kinds of molecular comparisons because both molar volume and covalent bond geometry remain nearly constant. Obviously neither of these parameters is precisely the same in the various members of isoelectronic sets, but in no other comparisons are the quantitative differences so small.

10.4 Supporting evidence for relative hydrophilicity data

It would be unwise to base an assessment of intrinsic hydrophilicity solely on phase information; such inferences will be far more convincing if they are consistent with other relevant kinds of physical data. The three most important kinds of supporting data for hydrophilicity are polarity, the thermodynamics of hydrogen bonding in non-aqueous solvents, and basicity.

Polarity. It has been repeatedly emphasized, to this point, that polarity (as defined by reduced dipole moments) *cannot* be used to distinguish hydrophilic from nonhydrophilic groups. The suggestion that this parameter is useful for scaling differences in intrinsic hydrophilicity therefore warrants an explanation.

It will be seen (Section 10.6) that hydrophilicity has several qualitatively different attributes, one of which is the thermodynamic strength of hydration. For a functional group to operate as a hydrophilic group, it must satisfy (at some minimal level) *all* of these attributes. Once the minimal requirements for hydrophilicity are met, however, the further quantitative variation of the thermodynamics of hydration is important because it does influence the magnitude of intrinsic hydrophilicity. Since the thermodynamics of hydration *are* related to polarity, it is reasonable to assume that polarity may serve to scale relative hydrophilicity – provided one stays within a series of proven hydrophilic groups.

Hydrogen bonding thermodynamics. As the bond dipole moment is varied, so are the atomic moments of the nonbonding electron pairs [6]. The strength of hydrogen bonds formed to these electron pairs is strongly influenced by their atomic moment, and hydrogen bonding directly influences hydrophilicity.

Within the single bond and semipolar subclasses, it is possible to determine the thermodynamics of hydrogen bonding between a surfactant and phenol in carbon tetrachloride solutions [1]. Both phenol and these surfactants, separately, behave nearly ideally in this solvent. The relevant thermodynamic parameters may therefore be determined from the temperature dependence of association constants.

During these solution studies, both the strength *and* the number of hydrogen bonds formed to an acceptor molecule can be determined. It has been shown by such data that an amine oxide is capable of accepting from phenol three hydrogen bonds of significant strength. A phosphine oxide, on the other hand, accepts only two hydrogen bonds, and a phosphine sulfide only one. A tertiary amine forms (as expected) only one hydrogen bond, but this bond is of exceptional strength. The number (one) and the strength of the hydrogen bond formed by a tertiary amine are consistent with its electronic structure and basicity.

The enthalpy of hydrogen bonds can be inferred, independently, from data on the frequency of the O−H infrared stretch band in the hydrogen bonded state relative to that in the unassociated state [7]. (The entropy of the interaction cannot be determined using infrared data.) Such infrared studies with phenol have been performed on a great many dipolar functional groups [7].

A useful correlation exists between hydrogen bonding data and hydrophilicity. If the enthalpy of the hydrogen bond of the 1 : 1 base : phenol complex is less than $5 \, \text{kcal mol}^{-1}$ ($21 \, \text{kJ mol}^{-1}$), the functional group is invariably found to be inoperative as a hydrophilic group. Further, the magnitudes of hydrogen bond enthalpies correlate reasonably well with the rank order of intrinsic hydrophilicity, as inferred from phase behavior.

Basicity. Varying the atomic dipoles of nonbonding electron pairs influences the basicities of molecules as well as their hydrogen bond acceptor properties. During the chemical reaction of a molecule (as a base) with water (as an acid), a new chemical bond is formed to the proton via the nonbonding pair. This is a very different process from physical association via hydrogen bonding of a hydroxy group with the electron pair. Hydrogen bonding data should be more relevant to hydrophilicity than basicity data, but the latter are nevertheless useful when hydrogen bonding data are unavailable. A vast amount of information exists regarding the basicity of all manner of functional groups.

If a base is titratable in water, the pKa of its conjugate acid can be precisely and accurately determined. A great many important hydrophilic groups are not basic in water, but a quantitative measure of their basicity is often provided in these cases by studies of proton transfer in strong acid media. The existence of proton transfer is evident from data such as NMR chemical shifts, and these data may be evaluated using the H_0 acidity function (or related functions) [8,9,10]. H_0 provides a measure of medium acidity in strong acid solutions, and merges with the pH scale in the limit of infinite dilution with water. Studies of this area have long been a major concern of physical organic chemists because of its relevance to the kinetics of acid catalysis.

The H_0 function allows the free energy of proton transfer equilibria to be expressed in the familiar terms of an equivalent pKa. While data based on H_0 are far less accurate than are pKas measured directly by potentiometric titration, they are nevertheless useful if the observed differences are sufficiently large. To illustrate, the value of H_0 at the point of half neutralization of a phosphine oxide is -1.5 [10], while the pH of a molecular solution of an amine oxide under the same circumstances is 4.65 [11]. Taken literally, these data would indicate that the amine oxide is half-neutralized in $2.2 \times 10^{-5} \, \text{M}$ acid (which is true), while the phosphine oxide is half-protonated in $31 \, \text{M}$ acid (which is nonsensical). The actual concentration of sulfuric acid at which an H_0 of -1.5 is attained is 26 wt per cent ($3.2 \, \text{M}$).

The significance of such differences in "pKa" values is more apparent when they are transformed into the free energies of proton transfer for the reaction

$$BH^+ \rightarrow B + H^+$$

From the pKa data, it follows that for the amine oxide this free energy change is $6.38 \, \text{kcal}$ ($26.7 \, \text{kJ}$) and for the phosphine oxide it is $-2.1 \, \text{kcal}$ ($-8.8 \, \text{kJ}$). The difference in pKas thus suggests that a difference in the free energies of proton transfer exists, and that this is approximately 8.5 kcal (35 kJ).

Relevant hydrogen bonding data are not available for ionic and zwitterionic surfactants, because these compounds are insoluble in solvents that are suitable for hydrogen bonding studies. As a result, basicity data are particularly useful for these functional groups.

10.5 The relative intrinsic hydrophilicities of hydrophilic groups

10.5.1 Ionic class, anionic subclass

Data on phase behavior suggest that the relative hydrophilicities of the sodium or potassium salts of the three major anionic hydrophilic groups are:

$$-CO_2^-,M^+ \gg -SO_3^-,M^+ > -OSO_3^-,M^+$$

The strong hydrophilicity of carboxylate salts is clearly evident from the phase behavior of the soaps, which are the only anionic surfactants that can be systematically studied over the full temperature range of their liquid crystal states [12]. Some important phase parameters are collected in Table 10.1. The mere existence of these data is a tribute to the remarkable thermal stability of these compounds.

The most thermally stable liquid crystal phase presently known is the aqueous Lam phase of potassium laurate (339°C). The largest value of ΔT that is known is found in the potassium myristate–water system (383°C). In Fig. 11.3 the thermal stabilities of the hexagonal and lamellar liquid crystal phases are displayed graphically. Sodium stearate forms the least stable liquid crystals, and has the smallest ΔT, among the common soaps. These parameters are still numerically large (279 and 215°C, respectively) and suggest that very strong hydrophilicity exists in comparison with that of most other surfactants. The value of ΔT is in all cases larger in potassium than in sodium soaps, for reasons that are unclear. No liquid/liquid miscibility gaps exist in any binary aqueous univalent metallic soap system. A liquid/liquid miscibility gap does exist in the sodium palmitate–water–sodium chloride system at very high salt compositions; it is called the

TABLE 10.1
Phase parameters of the sodium and potassium soaps

Soap	T_m, Hex (°C)	C_m, Hex (%)	T_m, Lam (°C)	C_m, Lam (%)	T_u (°C)	T_{mp} (°C)	ΔT (°C)
$C_{11}CO_2^-,Na^+$	150	46	294	78	41	310	269
$C_{13}CO_2^-,Na^+$	176	45	292	81	62	317	255
$C_{15}CO_2^-,Na^+$	166	44	279	76	69	288	219
$C_{17}CO_2^-,Na^+$	168	41	279	75	75	290	215
$C_{11}CO_2^-,K^+$	180	45	339	70	<0	375	>375
$C_{13}CO_2^-,K^+$	201	50	336	74	10	393	383
$C_{15}CO_2^-,K^+$	209	49	328	73	31	380	349
$C_{17}CO_2^-,K^+$	211	52	320	73	49	345	296

"lye-nigre bay" (Chapter 12). Cubic phases are found in potassium soap–water systems, but not in sodium [13].

These data indicate that the alkali metal carboxylate salts are probably the most strongly hydrophilic groups that are ordinarily encountered. In fact, the interaction of the carboxylate group with water is sufficiently strong that it reacts chemically by proton transfer to a significant degree, to form the carboxylic acid (Section 2.2.1). Both the extent of this reaction and the nature of the reaction product vary widely depending on conditions, but this result too is consistent with the conclusion that the carboxylate group is exceedingly strongly hydrophilic.

The phase information available on metal salts of alkanesulfonate or alkyl sulfate surfactants is far more limited, because these compounds are not sufficiently stable thermally to survive phase studies. A partial study of sodium decanesulfonate has been reported (Fig. 5.6) [14]. Dodecanesulfonic acid reacts quantitatively with water to form hydronium dodecanesulfonate ($C_{12}H_{25}SO_3^-,H_3O^+$), which is a relatively stable surfactant thermally. Its diagram (Fig. 10.6) indicates that this compound also is strongly hydrophilic. A phase study of a hydronium alkylsulfate is impossible, for reasons of instability by thermal or hydrolytic cleavage of the sulfuric acid ester link.

Strong support for the above ranking of these anionic groups is also provided by the phase behavior of zwitterionic surfactants containing them (Section 10.5.3.1).

Counterion effects. The importance of recognizing the contributions of both ions within a salt has been repeatedly emphasized, and nowhere is this principle more important than in treating the phase behavior of ionic surfactants. It has been

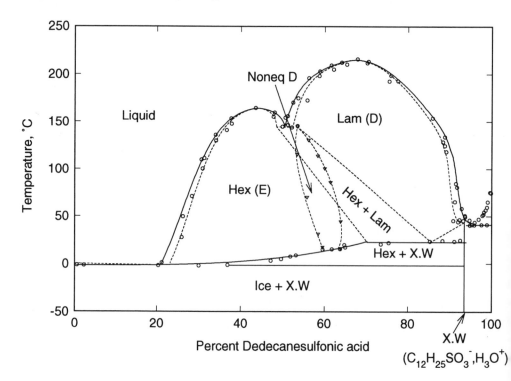

Fig. 10.6. The dodecanesulfonic acid–water system [14]. A monohydrate phase compound exists whose molecular structure is likely hydronium dodecanesulfonate ($C_{12}H_{25}SO_3^-,H_3O^+$).

known since before McBain's time that the solubility of calcium and magnesium salts of anionic surfactants is less than the solubility of the alkali metal salts, but (amazingly) no definitive phase study of a magnesium or calcium soap has been reported. Aluminum soaps are used in commerce, but even the molecular composition of these compounds is open to question.

The aqueous phase diagrams of a series of divalent ion salts of the "AOT" (bis(2-ethylhexylsulfosuccinate) class of surfactants have been reported [15]. The effect of replacing sodium with magnesium or calcium is to shrink the range of compositions spanned by the lamellar phase without dramatically altering other aspects of the phase behavior (Fig. 10.7). Substituting barium for these counterions has a profound effect on phase behavior, however. A large liquid/liquid (upper consolute?) miscibility gap exists, but the lamellar phase vanishes, and only the inverted hexagonal liquid crystal phase remains [16].

Ammonium salts of alkyl sulfates and alkyl ethoxy sulfate surfactants are widely used in personal care products (shampoos, etc.). No definitive phase information exists on these compounds, but surface tension data on solutions of pure ammonium soaps (which display anomalously low tensions) suggest that complex physical behavior is to be expected [17].

Variations of proximate substituent groups on the ammonium counterions of anionic surfactants (1-carboxyalkane-1-sulfonates) have been studied [18]. It has been found that unsubstituted ammonium surfactants display the highest Krafft points, and that substitution of ethyl or 2-hydroxyethyl groups on the ammonium ion lowers the Krafft point. A single 2-hydroxyethyl substituent in the counterion has an effect similar to that of an ethyl substituent.

Tetramethylammonium laurate (Fig. 10.8) is a highly soluble surfactant whose Krafft boundary lies below 10°C (and is probably metastable), but the liquid crystal regions of this soap are roughly similar to those of sodium laurate. Tetrabutylammonium laurate, however, displays profoundly different phase behavior from the familiar soaps (Fig. 10.8) [19]. While only primitive phase data exist, they are sufficient to reveal that this compound is soluble to the extent of 80% in water over a wide range of temperatures, and that liquid crystal regions are almost non-existent. Only a small hexagonal phase was identified and it is unstable above 15°C. Unidentified phases exist at compositions >80%. (Tetrabutylammonium hydroxide is an excellent titrant for acids in dipolar aprotic solvents because precipitation rarely interferes with the analysis. The data in Fig. 10.8 document the unusual solubility in water of this salt of laurate.) No liquid/liquid miscibility gaps were observed in any of these systems.

Supporting evidence. The rank order of hydrophilicity of anionic groups derived from phase data is supported by basicity data. The pKa of carboxylic acids in dilute aqueous molecular solutions is about +5, while that of alkylsulfuric acids is about −12 (H_0 scale). Alkanesulfonic acids are known to be weaker acids than is sulfuric acid, and a reasonable estimate of their pKa would be −9. The existence of a difference between sulfuric and sulfonic acids is well documented, if not quantitatively scaled, by the dependence on composition of the H_{GF} acidity function of aqueous methanesulfonic acid and sulfuric acid mixtures (H_{GF} is an electrochemically-defined acidity function which closely resembles the indicator-based H_0 function) [20]. Methanesulfonic acid and sulfuric acid are isoelectronic, and the observation that the sulfonic acid is the weaker acid is consistent with the fact that carbon is less electronegative than is oxygen.

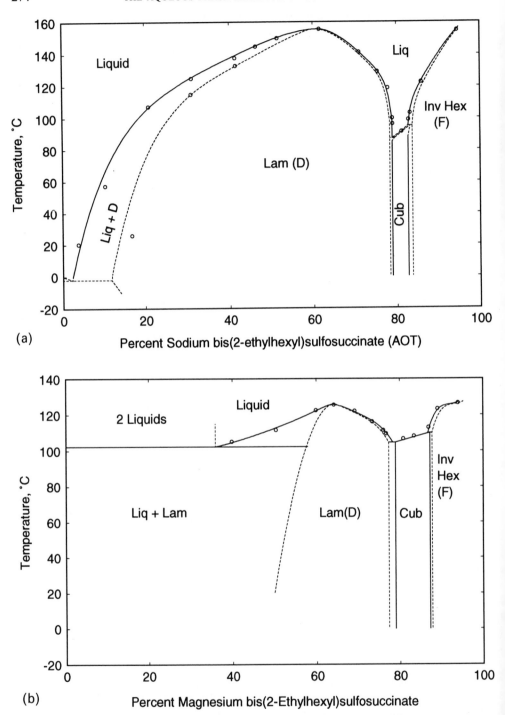

Fig. 10.7. The sodium, magnesium, calcium, and barium bis(2-ethylhexyl) sulfosuccinate–water systems. The composition span (swelling) of the liquid crystal is dramatically reduced in the divalent ion salts in comparison to sodium, and the liquid/liquid miscibility gap is greatly enlarged in the barium salt system.

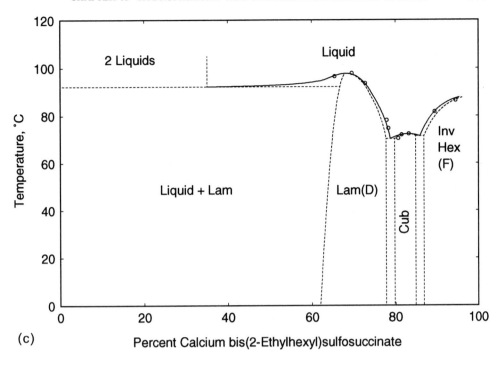

(c)

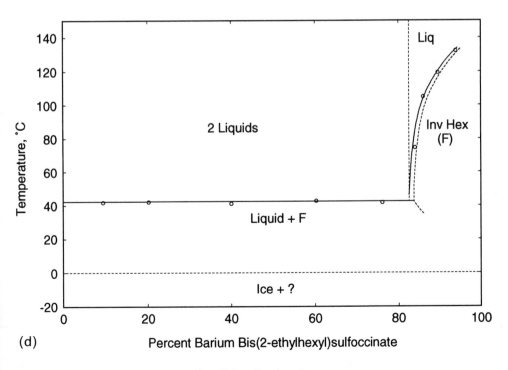

(d)

Fig. 10.7. Continued.

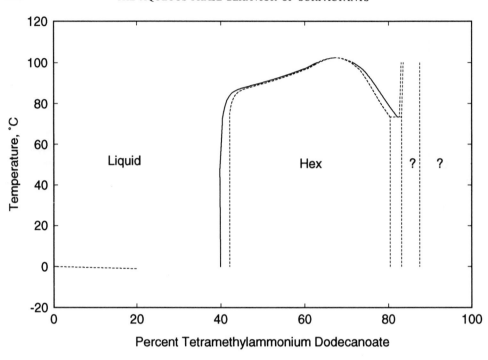

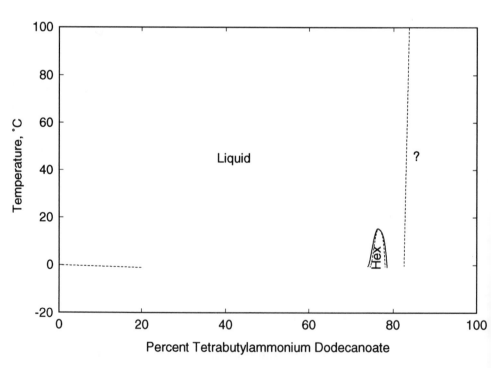

Fig. 10.8. The tetramethylammonium (upper) and tetrabutylammonium (lower) dodecanoate–water systems.

An arylsulfonic acid should be a somewhat stronger acid than an alkylsulfonic acid, because an sp^2 carbon is more electronegative than is an sp^3 carbon [21]. An arylsulfonate salt should be slightly less hydrophilic than an alkane sulfonate salt, but experimental support for this premise does not exist.

Taking the above pKa estimates at face value, a carboxylate anion is seen to be *17 pKa units* more basic than an alkyl sulfate and 14 units more basic than is an alkane-sulfonate. These are enormous differences in basicity.

Other anionic hydrophilic groups. No phase data exist upon which an evaluation of the relative hydrophilicity of other anionic groups can be based. Basicity data are therefore the only information available upon which a probable ranking of their hydrophilicities can be based. The pKa values of the conjugate acids of several anionic groups are shown in descending order in Table 10.2 [22,23]. The higher the pKa value, the stronger is the basicity. Therefore, insofar as hydrophilicities correlate with basicity, the sodium or potassium salts of the above groups are ranked in order of decreasing hydrophilicity.

The only common anionic groups more basic than the carboxylate are the various phosphorus acid dianions. The phosphinate anion is slightly less basic than the carboxylate anion. Phosphinates have been virtually unexplored, but (if no hydrogens exist on the phosphorus) they are thermally stable and resistant to oxidation. They should be slightly less hydrophilic than carboxylate salts, and display less severe hydrolysis chemistry. This group could be of interest in both anionic and zwitterionic surfactants, but a barrier to its use is the lack of a good synthesis.

A variety of phosphorus monoanions exist that are less basic than phosphinates but more basic than sulfonates. These groups will titrate as neutral salts in water, and surfactants based on them should not display significant hydrolysis chemistry (Section 10.7.2.1). Nevertheless, they should be substantially more hydrophilic than are sulfate and sulfonate salts. It is apparent from this analysis that a wide range of hydrophilicity exists among anionic groups which is still unexplored (in the phase behavior context).

It is likely that the above data are also relevant to the intrinsic hydrophilicity of cationic surfactants which contain these molecules as counterions. These data should also be relevant to the hydrophilicity of zwitterionic compounds which contain these functional groups as the anionic substituent group (Section 10.5.3.1). It is noteworthy that the basicities of all the above groups are probably greater than that of the halide anions (except fluoride).

TABLE 10.2
Relative basicities of various anionic functional groups

Molecular structure	pKa of conjugate acid	Chemical name
$CH_3PO_3^=$	7.85	Phosphonate dianion
$HOPO_3^=$	7.21	Phosphate dianion
$CH_3OPO_3^=$	6.35	Phosphate ester anion
$CH_3CO_2^-$	4.76	Carboxylate anion
$(CH_3)_2PO_2^-$	2.93	Phosphinate anion
$CH_3SO_2^-$	2.3	Sulfinate anion
$(HO)_2PO_2^-$	2.2	Phosphoric acid monoanion
$CH_3(CH_3O)PO_2^-$	2.1	Phosphonate ester anion
$(CH_3O)_2PO_2^-$	1.3	Phosphate diester anion

10.5.2   Ionic class, cationic subclass

The phase behavior of cationic surfactants suggests that they, too, are strongly hydrophilic [5]. In dodecyltrimethylammonium chloride, for example, the maximum temperature (T_m) of the cubic phase is 85°C, T_m of the hexagonal phase is 169°C, and T_m of the lamellar phase is 210°C. T_u is approximately -16°C (metastable Krafft boundary) and the melting point (with decomposition) is probably >150°C. ΔT cannot be actually defined, but is probably >165°C. The general phase behavior of this compound resembles that of potassium soaps.

Salts of long chain amines and a strong acid constitute a very important kind of cationic surfactant (Section 10.7.1). They are easier to study than are quaternary ammonium salts because they are far more stable thermally. They are also acidic, but the consequences of dissociation are minimal in comparison to those observed for soaps.

Primary alkylammonium salts may hydrolyze in water as follows:

$$RNH_3^+,X^- \rightleftharpoons RNH_2 + H^+,X^-$$

The amphiphilic product of this hydrolysis (RNH_2) is a nonionic surfactant. The hydrolysis of soaps, in contrast, produces an amphiphilic compound (a carboxylic acid) which is *not* a surfactant. The interaction between an alkylammonium salt and its hydrolysis product involves an interaction between two surfactants and water, and the ternary system octylammonium chloride–octylamine–water has been investigated (Fig. 12.24). The interactions between a soap and its hydrolysis product involve an interaction between a surfactant, an amphiphilic oil, and water. The consequences of hydrolysis with respect to phase behavior are for this reason profoundly different in these two systems (Sections 12.4, 12.6).

Salts of primary amines are unique among the various ammonium salt surfactants with respect to the physical consequences of their hydrolysis chemistry. This is because the hydrolysis of salts of secondary and tertiary amines produces (as with the soaps) amphiphilic products which are not surfactants (e.g. $RNHCH_3$ and $RN(CH_3)_2$). The physical science of these systems have not been studied.

One would expect, from the dissociation constants of ammonium salts, that an amine salt would display hydrolysis to a degree similar to that found in soaps. The pKb of a carboxylate (14-pKa) is about 9, and the pKa of a molecular solution of an alkylammonium salt is about 10. Nevertheless, amine salts and water behave as simple binary systems – and thus differ profoundly from the soaps as regards acid-soap phenomena. The reason for this difference is not known.

The phase diagram of dodecylammonium chloride (Fig. 5.11) is almost unique among surfactant diagrams, because the hexagonal liquid crystal region is surrounded by the liquid region [24]. (It has been found by Fontell that octyl- and decylammonium chloride–water systems also display a hexagonal phase island.) T_m of the hexagonal phase is 108°C, and the (presumed) lower azeotropic temperature limit of this phase is 45°C. The Krafft discontinuity lies at 32°C (13°C lower), and the coexisting crystal phase is a hemihydrate ($X_2 \cdot W$).

This system provides a well-documented example of a metastable crystal phase that separates on cooling aqueous solutions. The temperature at which this phase redissolves is indicated by the dashed Krafft boundary just below the equilibrium boundary.

Solutions held between the two boundaries will, in time, undergo phase separation to form the equilibrium crystal. This crystal then dissolves on heating at the equilibrium boundary temperature. The T_m of the lamellar phase is 267°C.

The dry crystal is reported to undergo a transformation to a thermotropic liquid crystal at 200°C, and this phase melts to an isotropic liquid at 225°C. ΔT may be taken to be 188°C, recognizing that its significance is compromised by intervention of the thermotropic liquid crystal phase at the melting point (Section 10.2.1). A liquid/liquid miscibility gap does not exist.

These phase properties suggest that this compound is strongly hydrophilic. The range of temperatures within which liquid crystals exist (32–267°C, or 235°C) is somewhat smaller than is found in a sodium soap having a similar (C_{11}) lipophilic group (41–294°C), but the span of temperatures (235°C) is about the same as for a sodium soap having a C_{13} lipophilic group (62–292°C, or 230°C).

Counterion effects. The anionic molecule in cationic surfactant salts exerts a profound influence on their phase behavior. In many instances the positive nitrogen of the cation is fully alkylated and no nonbonding electron pairs exist; the hydration of such an ion must be relatively weak. In fact, the correlations suggest that the anionic counterion may be the principle (if not the only) site of hydration in liquid and liquid crystal phases formed by quaternary ammonium salts. This conclusion is consistent with single crystal X-ray studies of quaternary ammonium halide crystal hydrates, in which the water molecules are strongly hydrogen bonded to halide anions but (as expected) show no evidence of strong interactions with quaternary ammonium ions [25,26].

Hydration of the counterion may also be important in amine salts such as dodecylammonium chloride. Strongly hydrogen bonding N^+-H groups exist in the amphiphilic ion in these compounds, and these N^+-H groups could donate hydrogen bonds to water. The relative importance to hydrophilicity of donor interactions by the cation, and acceptor interactions by the anion, is hard to assess.

Born cycle analyses have shown that the strength of hydration of the halide anions decreases in the order [27]:

$$F^- > Cl^- > Br^- > I^-$$

Relative basicities fall in the same order; F^- is the strongest base (it is titratable in water, $pKa = 3.1$), so that the acidity of concentrated aqueous hydrohalic acids decreases in the order [9]:

$$HI > HBr > HCl > HF$$

Non-basic anions (such as perchlorate or tetrafluoroborate) are also probably very weakly hydrated, and should therefore be weakly hydrophilic. The sulfate dianion is comparatively basic [28], and should be strongly hydrophilic compared to, for example, the methylsulfate ($CH_3OSO_3^-$) or methanesulfonate ($CH_3SO_3^-$) anions. Indirect evidence for this exists in the fact that particles of ammonium sulfate in the atmosphere are unstable and decompose to ammonia and ammonium bisulfate, but not further to ammonia and sulfuric acid [29].

The same functional groups that are utilized as hydrophilic groups in anionic surfactants may also serve as the anionic counterion in cationic surfactant salts. Their relative intrinsic hydrophilicities should apply in this role as well as when they reside in the amphiphilic molecule (Section 10.5.1). One may therefore expect the following order of hydrophilicity to exist within a series of cationic surfactant salts having the same amphiphilic cationic molecule and the following anions:

$$OH^- > CO_3^= > F^- > CH_3CO_2^- \ggg SO_4^= \gg CH_3SO_3^- > CH_3OSO_3^-$$

Quaternary ammonium hydroxide surfactants have recently received considerable attention. These compounds have a serious chemical stability problem, since they are the intermediates in the Hoffman degradation of ammonium salts to olefins (or alcohols) and amines. Fortunately, this reaction is retarded sufficiently by water that the compounds can be investigated in dilute aqueous mixtures at moderate temperatures.

Monolong chain hydroxide surfactants display unusually high counterion dissociation in solution [30], and hydroxide salts that have unusually large lipophilic groups remain water-soluble [31]. Acetate salts resemble hydroxides in these physical properties and are considerably more stable chemically [31].

In most ionic surfactants, one ion is amphiphilic while the other is predominantly hydrophilic. However, ionic surfactants also exist in which both ions are amphiphilic (Section 9.4.2), and this profoundly changes the phase behavior. Small lipophilic groups (such as the methyl group in methylsulfate salts) have minimal effect on phase behavior, but larger groups (such as in toluene- or cumenesulfonates) introduce significant changes. For example, a liquid/liquid miscibility gap was observed in the ditosylate salt of a dicationic surfactant in having two ammonio substituents separated by a polyoxyethylene linkage. This phenomenon, which is extremely unusual in ionic surfactants, does not exist in the dichloride salt (Fig. 10.9) [5].

Other evidence that the anion strongly influences phase behavior is found in comparing the DODMAC–water system (in which the anionic molecule is chloride) [32] with the DOACS–water system (in which the anionic molecule is cumenesulfonate) [33]. The composition range of the lamellar liquid crystal region is vastly different in these two diagrams; in the chloride salt this phase spans the composition range of 31–95%, while in the cumenesulfonate salt it spans the range of 94–100%. DODMAC has a quaternary ammonium cation while DOACS has a dialkylammonium cation, and the influence of this structural feature is unknown. This difference is not likely to be critically important, since the composition range of liquid crystal phases is not very different in alkyltrimethyl- and alkylmethylammonium chlorides.

10.5.3   Nonionic class, zwitterionic subclass

A characteristic feature of monolong chain zwitterionic surfactants is that the coexisting phase at the liquid crystal solubility boundary just above the Krafft discontinuity is almost invariably a cubic phase [1]. This cubic phase likely has the discontinuous structure (Section 8.4.10). A hexagonal phase often lies next to the cubic phase at higher compositions, and replaces the cubic phase as the saturating phase at higher temperatures. In ionic surfactants this cubic phase is often absent and the hexagonal phase itself is the saturating phase.

The liquid crystal regions of zwitterionic surfactants are stable to >150°C, but their upper limit cannot usually be determined because of chemical instability. Since crystal decomposition temperatures are typically >150°C and the Krafft discontinuity is often below room temperature, it is clear that ΔT within this subclass often exceeds $150°C$. This too suggests that zwitterionic surfactants are strongly hydrophilic.

Some zwitterionic surfactants display an upper consolute boundary [34], whose general features have been described in Sections 5.10.2 and 5.10.4. The form of this boundary is distinctive and there is no indication that it is part of a closed loop. Its existence does *not* signify that zwitterionic surfactants are weakly hydrophilic (relative to the other hydrophilic group subclasses), since zwitterionic surfactants by the other two phase criteria of hydrophilicity are strongly hydrophilic. Nevertheless, the position of the upper consolute boundary is influenced by the *relative* hydrophilicity of zwitterionic compounds in much the same way as is the lower consolute (closed loop) boundary of more weakly hydrophilic compounds.

Miscibility gaps typically become larger with increasing lipophilicity, but an exception to this correlation exists with respect to the effect that nonpolar proximate substituents have on the upper consolute boundary. The upper consolute boundary is prominent in zwitterionic surfactants having proximate methyl substituents, but *does not exist in related surfactants having proximate ethyl substituents* [35]. This remarkable correlation is found in both ammonio sulfates and ammonio sulfonates, and dramatically illustrates the principle (Section 10.2.2) that proximate substituents influence phase behavior differently than do remote substituents. It also suggests that the factors which dictate the position of the upper consolute boundary of zwitterionics are different from those which influence the position of the lower consolute (closed loop) boundary of weakly hydrophilic surfactants.

The influence of proximate substituent variation on the lower consolute boundary can readily be seen in phosphine oxides. Enlarging proximate substituents and increasing the length of the lipophilic group in these molecules moves both boundaries in the same direction (Fig. 10.5), but to a different extent (Section 10.5.4.1). In striking contrast, enlarging the long chain in zwitterionic surfactants always increases the area of the upper consolute boundary while enlarging the proximate substituents shrinks this gap to the point that it vanishes (above).

The structural features that influence zwitterionic hydrophilicity. The fact that zwitterionic molecules are, in reality, polyfunctional compounds must be confronted in analyzing their hydrophilicity. Two structural features appear to be particularly important:

(1) the nature of the anionic substituent, and
(2) the length (lipophilicity) and hydrophilicity of the tether.

The structure of the cationic substituent group appears not to strongly influence hydrophilicity.* One of the few comparisons possible is that between dodecyldimethyl-ammoniopropanesulfonate and its phosphonio analog [35]. The upper consolute boundary of the phosphonio compound is $9°C$ higher than is that of the ammonio analog, but otherwise no substantial differences in phase behavior were discovered. (The data on the phosphonio compound are limited.) The adsorption properties of these two compounds are more strongly influenced by the exchange of phosphonio for ammonio groups than are the phase properties (Table 10.3). The structures

---

*In many zwitterionic surfactants the positive atom is within the longest chain of atoms in the molecule, and the term "substituent" might appear strange. Nevertheless, this atom is treated as a substituent group in naming the compounds. If the lipophilic group is attached to this "onio" atom then an "in-line" structure results. If it is attached to the tether, the "onio" atom is pendant and the structure is branched.

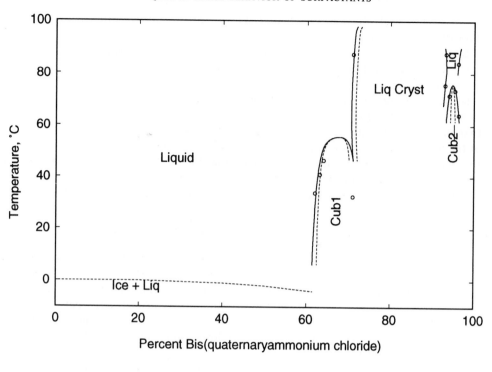

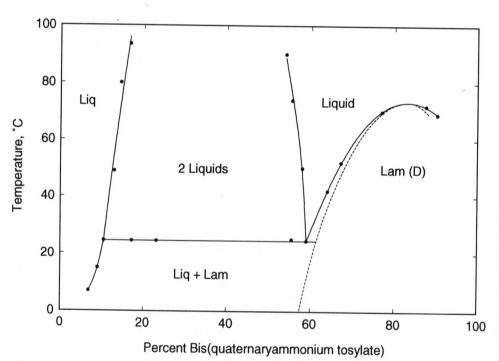

Fig. 10.9. The ethoxylated bis-quaternaryammonium salts, $C_{12}(Me)_2N^+-E_9-N^+(Me)_2C_{12}$ dichloride– (upper) and ditosylate–water (lower) systems.

of proximate substituents on the cationic group are also extremely important (above).

10.5.3.1 Influence of the anionic substituent

The anionic group of zwitterionic surfactants has a profound influence on their phase behavior and hydrophilicity. In fact, the differences in hydrophilicity among sulfate, sulfonate, and carboxylate groups are more clearly revealed by the phase behavior of zwitterionic surfactants than by the behavior of ionic surfactants.

Data on miscibility gaps are particularly useful in evaluating the hydrophilicity of zwitterionics, since neither the liquid crystal nor the ΔT criteria can be used. It is evident from a large number of comparisons that the area of this gap increases in the order:

$$\text{onio-SO}_3^- < \text{onio-OSO}_3^-$$

It is also significant that the *upper consolute boundary apparently does not exist in ammonio carboxylate zwitterionic surfactants.* From these correlations, it is concluded (in agreement with the inference from data on anionic surfactants, Section 10.5.1) that the carboxylate anion is far more hydrophilic than is the sulfonate, and the sulfonate is slightly more hydrophilic than is the sulfate:

$$-\text{CO}_2^- \gg -\text{SO}_3^- > -\text{OSO}_3^-$$

It is difficult to compare directly zwitterionic analogs of the same chain-length. The area spanned by the upper consolute boundary is increased by lengthening the lipophilic group, but so is the temperature of the Krafft discontinuity. It is often necessary to compare different homologs to bring the miscibility gap into view, and then seek to compensate for their differences in chain-length.

The liquid/liquid miscibility gap in ammoniopropyl sulfate–water systems persists even to chain-lengths as short as C_8. The upper critical temperature of the C_8dimethylammoniopropyl sulfate is 32°C, that of the C_9 homolog is 69°C (an increase of 37°C), and that of the C_{10} homolog is 90°C (a further increase of 31°C). The critical temperature of C_{12}dimethylammonioethyl sulfate is >120°C and its Krafft discontinuity lies at 86°C.

In ammoniopropanesulfonates (APS) the critical temperatures of the C_8, C_9, and C_{10} homologs must be significantly lower than those of the sulfates, since the critical temperatures of C_{12}APS, C_{14}APS, and C_{16}APS homologs are −0.5, 12, and 20°C, respectively (Fig. 10.10).

In C_{16}APS the upper consolute boundary is metastable since the Krafft discontinuity lies at 29°C. This upper consolute boundary is nevertheless easily determined, because separation of the second liquid phase on cooling is much faster than is separation of the crystal phase. Solutions of this surfactant can be cooled below the Krafft boundary for days or weeks without crystal formation, provided seed crystals are absent.

Absence of the upper consolute boundary in ammonio carboxylate surfactants. The upper consolute boundary probably does not exist in ammonio carboxylates. Whether or not it exists in ammonioacetates cannot be clearly discerned because of their relatively high Krafft boundaries, but its absence can easily be demonstrated in ammoniohexanoates.

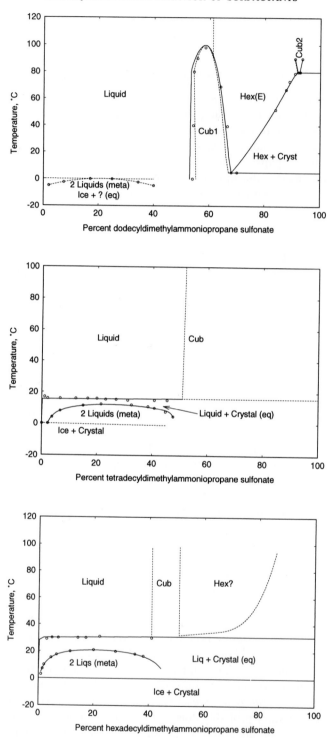

Fig. 10.10. The dodecyl- (upper), tetradecyl- (middle), and hexadecyldimethylammoniopropanesulfonate (APS)–water (lower) systems.

During explorations of ammoniohexanoate surfactants [36,37] it was found that the upper consolute boundary did not exist in the C_{12} to C_{16} homologs, although this phenomenon had earlier been found in ammonio sulfonates and sulfates having much shorter lipophilic groups. The chain-length of the lipophilic group was therefore extended in an attempt to introduce this boundary – but without success. Ammoniohexanoates having a chain-length as long as C_{42} were synthesized, and the phase behavior of homologs up to C_{30} were scanned. (The Krafft boundary of the C_{42} homolog lies >100°C and its phase behavior at higher temperatures was not investigated.) In no case was the upper consolute boundary observed.

The difference in lipophilicity between a C_{30} surfactant and a C_8 surfactant is enormous (Section 11.2). Had the upper consolute boundary existed, it should have been apparent at the C_{30} chain-length. Since it was not found, and also does not exist in ammonioacetates of conventional chain-lengths, it seems likely that it does not exist at all in ammonio carboxylates.

Ultralong-chain surfactants. This unexpected result led to recognition of the existence of a distinctive group of surfactants which have been termed "ultralong-chain" surfactants. An "ultralong-chain" surfactant may be defined as a surfactant that has a lipophilic group chain-length of C_{20} or greater and is soluble at ambient temperature. Such compounds are unusual because the C_{20} homolog of practically all types of surfactants is insoluble near room temperature.

An important characteristic of ultralong-chain surfactants is that their cmcs are vanishingly small. The monomer concentrations in solution are essentially zero, so that transport (and other phenomena) governed by monomer concentrations are suppressed relative to normal surfactants.

To be soluble at a chain-length of C_{20} or greater, a surfactant either must possess an unusually strong hydrophilic group, or a moderately strong hydrophilic group and be low-melting. Of the former type, the simplest (and most synthetically accessible) ultralong-chain hydrophilic group is the ammoniohexanoate or "AH" group ($-N^+(CH_3)_2CH_2CH_2CH_2CH_2CH_2CO_2^-$), while polyoxyethylene derivatives of erucic acid are representative of the latter type [38]. The ammonioacetate (betaine) group is *not* an ULC hydrophilic group, as the Krafft discontinuities of these compounds are high and the solubility at ambient temperatures is poor (for example, the Krafft discontinuity of C_{22} betaine is >100°C, while that of the C_{22}AH is 33°C). C_{22}AH is soluble at body temperature and its solutions readily supercool, so that this compound may easily be studied in solution at room temperature.

Simple ammonio sulfates and sulfonates are not ultralong-chain hydrophilic groups, but polyfunctional analogs of these molecules are (next section).

10.5.3.2 Influence of the tether

Another structural feature of zwitterionic surfactants which is important with respect to their phase behavior is the tether [39]. In most compounds this is a saturated aliphatic polymethylene diradical. Rigid tethers such as ethylenic or aromatic diradicals can be imagined, but the phase behavior of compounds having such tethers has not been reported.

The length of the tether is best regarded as the number of atoms separating the centers of charge, rather than simply the number of methylene groups. On this basis three atoms exist in the tether of ammoniopropanesulfonates,

$$N^+CH_2CH_2CH_2SO_3^-$$

while four exist in ammoniopropyl sulfates

$$N^+CH_2CH_2CH_2OSO_3^-$$

This is because the divalent oxygen in the sulfate ester is approximately equivalent to a methylene group in so far as bond lengths and geometries are concerned. For this reason it is better to compare ammonioethyl sulfates ($-N^+CH_2CH_2OSO_3^-$) with ammoniopropanesulfonates ($-N^+CH_2CH_2CH_2SO_3^-$) than to compare ammonioethyl sulfates with ammonioethanesulfonates ($-N^+CH_2CH_2SO_3^-$) in extracting information regarding hydrophilicity.

Considerable evidence suggests that the *hydrophilicity of zwitterionic groups increases smoothly as the length of the tether increases, and attains a plateau at about four atoms.* The largest increase occurs in the interval between one and two atoms, a smaller increase occurs between two and three, and a still smaller increase between three and four. Beyond four atoms, changes in hydrophilicity appear to be very small. No irregularity exists at any specific tether length to suggest that particularly strong interactions exist within the group at that length.

If the tether is an aliphatic chain, inserting methylene groups into it increases *both* the hydrophilicity and the lipophilicity. In ammonio sulfonates having differing tether lengths, the area of the upper consolute boundary passes through a minimum at C_3 [40]. The higher upper critical temperature of the butane sulfonate than of the propane sulfonate suggests that while hydrophilicity does not increase greatly in passing from a trimethylene to a tetramethylene tether, the lipophilicity does continue to increase. This conclusion is supported by the fact that the cmcs of ammoniophosphonate surfactants pass through a maximum as the tether length is increased [39].

It is of interest to compare the phase behavior of dodecyldimethylammonio-undecanoate with that of docosyldimethylammonioacetate.

$$C_{12}H_{25}N^+(CH_3)_2CH_2CH_2CH_2CH_2CH_2CH_2CH_2CH_2CH_2CH_2CO_2^- \text{ vs.}$$
$$C_{22}H_{45}N^+(CH_3)_2CH_2CO_2^-$$

These compounds differ by only one methylene group, but their phase behavior is dramatically different. The phase diagram of the C_{22} ammonioundecanoate (Fig. 5.5) shows that it is a highly soluble surfactant so that the Krafft boundary of C_{12} ammonioundecanoate will certainly be metastable (lie $<0°C$). This boundary in the C_{22} ammonioacetate lies above 100°C, in striking contrast.

This comparison clearly indicates that methylenes inserted in the tether do not exert the same influence on phase behavior as do methylenes inserted in the lipophilic chain. These two compounds likely differ in the stability of their crystal lattices as well as in their hydrophilicity, so this observation does not invalidate the above concept that the tether influences the hydrophilicity of the compound.

pKa and Rf data vs. hydrophilicity. The indication from phase behavior that the hydrophilicity of zwitterionic surfactants increases smoothly with tether length and reaches a plateau at about four atoms is supported by pKa data on ammoniocarboxylic acids, and also by R_f data on ammonio carboxylate surfactants. The pKas of amino acids having a wide range of structures have been accurately determined [41], and it has been noted by Weers *et al.* that the pKa of the carboxylic acid in the series $H_3N^+(CH_2)_nCO_2H,X^-$ varies linearly with $1/n$ over a wide range of n [42].

$$pK_a = k \times \left(\frac{1}{n}\right) + C \qquad (10.1)$$

It is particularly satisfying that the value of the constant, C, equals the pKa of soluble and unassociated unsubstituted alkanoic acids (which are independent of chain-length above acetic acid). These pKa data are fully consistent with the correlation of tether length with hydrophilicity suggested above.

R_f values found during thin-layer chromatographic studies (in methanol on silica gel) of zwitterionic ammonio carboxylate surfactants having various tether lengths also parallel hydrophilicity (Table 10.3). The ammonioacetate ($n = 1$) has the highest R_f, the propionate ($n = 2$) a significantly smaller value, the butyrate ($n = 3$) a slightly smaller value, and thereafter the R_f is nearly independent of n. At a value of n equal to 10, a slight rise in R_f is observed. This is consistent with the notion that insertion of methylene groups into the tether has a small influence on lipophilicity (see above).

The validity of R_f data as an indicator of hydrophilicity is suggested by several observations. First, these data are essentially independent of lipophilic chain-length (within the C_{10} to C_{16} range) for alkyldimethylammonioethyl sulfates (0.43), alkyldimethylammoniopropyl sulfates (0.37), alkyldimethylammoniopropanesulfonates (0.31), and alkyldiethylammoniopropanesulfonates (0.45). The numerical value of this parameter is dominated by the structure of the hydrophilic group.

Further, comparisons of alkyldimethylammoniopropanesulfonates and alkyldiethylammoniopropanesulfonates indicate that R_f is sensitive to proximate substituent effects. These are to be regarded as part of the hydrophilic group and thus directly influence hydrophilicity.

Finally, R_f values are higher in 90 : 10 v : v methanol : water than in pure methanol. This is a rational relationship for a parameter that is presumed to be sensitive to polar group properties. R_f data in both of these developing solvents are given in Table 10.3.

These data support the correlations that have been inferred from phase data (above) regarding the influence of anionic group structure, tether length, phosphonio substitution, and proximate substituent group effects on hydrophilicity. The large influence of phosphonio substitution on R_f suggests that this structural modification significantly influences adsorption properties, even though its effect on phase behavior is small (Section 10.5.3). The direction in which replacing nitrogen by phosphorus influences both properties is the same.

Polyoxyethylene tethers. Ammonio sulfate and sulfonate surfactants having polyoxyethylene tethers have been reported in the patent literature [43]. These compounds interact unusually strongly with calcium ion in solution, and also display interesting phase behavior. These complex polyfunctional groups may serve as ultralong-chain hydrophilic groups in spite of the relatively weak hydrophilicity of their anionic substituent groups.

Partial phase diagrams of the $C_{22}AE_9SO_3-$ and $C_{22}AE_9SO_4-$water systems illustrate their phase behavior (Fig. 10.11). As found elsewhere (Section 10.5.3.1), the sulfonate

TABLE 10.3
R_f parameters of various zwitterionic compounds

Hydrophilic group structure	R_f in methanol	R_f in 90 : 10 methanol : water
Anionic group variation		
$-N^+(C)_2C-C-O-SO_3-$	0.43	0.55
$-N^+(C)_2C-C-C-SO_3-$	0.29	0.43
$-N^+(C)_2C-C-C-CO_2-$	0.09	0.19
Charge separation		
$-N^+(C)_2C-C-SO_3-$	0.38	0.53
$-N^+(C)_2C-C-C-SO_3-$	0.29	0.43
$-N^+(C)_2C-C-O-SO_3-$	0.43	0.55
$-N^+(C)_2C-C-C-O-SO_3-$	0.37	0.51
$-N^+(C)_2C-CO_2-$	0.29	0.43
$-N^+(C)_2C-C-CO_2-$	0.11	0.24
$-N^+(C)_2C-C-C-CO_2-$	0.09	0.19
$-N^+(C)_2C-C-C-C-CO_2-$	0.09	0.18
$-N^+(C)_2-(C)_{10}-CO_2-$	0.11	0.24
Hydroxy substitution		
$-N^+(C)_2C-C(OH)-C-SO_3-$	0.49	0.60
$-N^+(C)_2C-C-C-SO_3-$	0.29	0.43
Phosphonio substitution		
$-P^+(C)_2C-C-C-SO_3-$	0.41	0.51
$-N^+(C)_2C-C-C-SO_3-$	0.29	0.43
Proximate substituents		
$-N^+(C)_2C-C-C-SO_3-$	0.29	0.43
$-N^+(C_2)_2C-C-C-SO_3-$	0.44	0.53
$-N^+(C_3)_2C-C-C-SO_3-$	0.70	—
$-N^+(C_6)_2C-C-C-SO_3-$	0.79	0.81
$-N^+C(C_2)C-C-C-SO_3-$	0.36	0.48
$-N^+(C_2)_2C-C-C-O-SO_3-$	0.50	0.59
$-N^+(C)_2C-C-C-O-SO_3-$	0.37	0.51

compound appears to be more strongly hydrophilic than the sulfate. The Krafft boundary of the sulfonate is metastable, so that the C_{22} and higher homologs should remain soluble at room temperature. The Krafft discontinuity of the sulfate (46°C) is relatively low for a C_{22} surfactant, and a lower consolute boundary having a most unusual shape is seen. A cubic phase is prominent in both systems. The oxyethylene groups clearly modify the phase behavior and appear to enhance the hydrophilicity of these compounds.

Hydroxy substitution in the tether. In 2-hydroxy-3-ammoniopropane-1-sulfonates, $R-N^+(C)_2C-C(OH)-C-SO_3^-$, the hydroxy substituent in the tether is in the beta position with respect to both the sulfonate and the ammonio groups. Not surprisingly, the diagram of the C_{12} homolog shows a high Krafft boundary. The R_f data in Table 10.3 suggest that this compound is less hydrophilic than is the unsubstituted surfactant. Again, the substitution of a polar (hydroxy) functional group in the

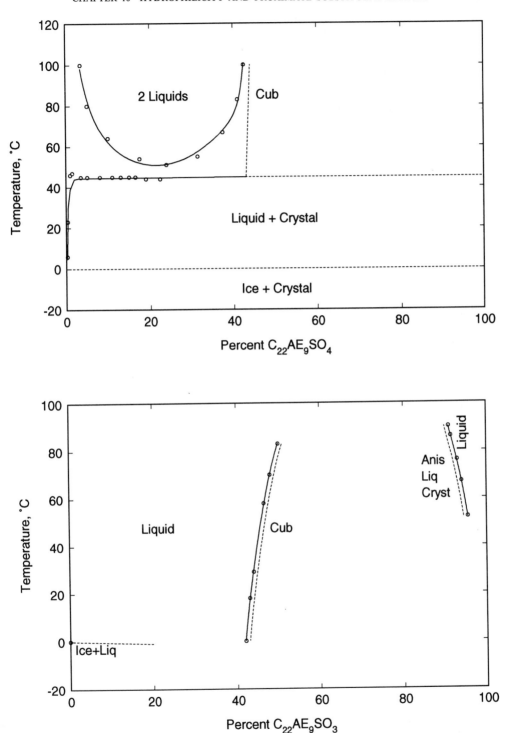

Fig. 10.11.   Phase diagrams of ethoxylated ammonio sulfate and sulfonate zwitterionic surfactants: $C_{22}AE_9SO_4-$ (upper) and $C_{22}AE_9SO_3$–water (lower) systems.

proximity of a stronger hydrophilic group *diminishes*, rather than increases, the net hydrophilicity of the molecule (Section 10.2.2).

10.5.4   Nonionic class, semipolar subclass

10.5.4.1   Group V and VI oxides

Semipolar hydrophilic groups that are Group V and VI oxides will now be considered. The phase diagrams of aliphatic amine oxides (Fig. 10.12) and arsine oxides [60] are distinguished by the presence of liquid crystal phases which extend to >100°C and the absence of a liquid/liquid miscibility gap. This gap is absent in hexadecyldimethylamine oxide (Fig. 5.7) as well as in the dodecyl homolog (Fig. 10.12), but the position of the hexagonal phase solubility boundary in the hexadecyl homolog is quite unusual. This boundary is ordinarily found at concentrations above 25%, and only rarely (as in this compound) at about 1%.

The liquid crystal phases of dodecyldimethylphosphine oxide and dodecylmethyl sulfoxide extend to similar temperatures (97 and 93°C, respectively), but the liquid/liquid miscibility gap is much more prominent in the sulfoxide. At 100°C, where this gap is visible in both compounds, it spans the composition range from 0 to 30% in the phosphine oxide but from 0 to 70% in the sulfoxide. Interaction between the miscibility gap and the liquid crystal region in the sulfoxide results in "monoglyceride-like" phase behavior (Section 5.6.8), so that solubility in the dilute liquid phase is small. Pivotal phase behavior (Section 5.10.7), which may also result from the interference of miscibility gaps and liquid crystal regions, is not observed.

In decyl methyl sulfoxide the miscibility gap and the liquid crystal regions are both expected to shrink, and the phase behavior should qualitatively resemble that of the C_{12} phosphine oxide. It is known that the C_{10} sulfoxide is highly water soluble, but its diagram has not been determined.

Taken as a whole, the above information suggests that the relative hydrophilicities of these groups fall in the order:

$$N \rightarrow O, As \rightarrow O > P \rightarrow O > S \rightarrow O$$

The phase information by itself does not allow ranking $N \rightarrow O$ relative to $As \rightarrow O$.

Supplemental information on Group V and VI oxides.   Reduced bond moment and pKa data on the above four groups are as follows:

Functional group	μ_r	pKa
S → O	0.42	−2.8
P → O	0.43	−1.5
As → O	0.55	3.75
N → O	0.65	4.65

These data both rank these groups in the same order as does the phase information, and also suggest that the arsine oxide falls between the phosphine and the amine oxide:

$$N \rightarrow O > As \rightarrow O > P \rightarrow O > S \rightarrow O$$

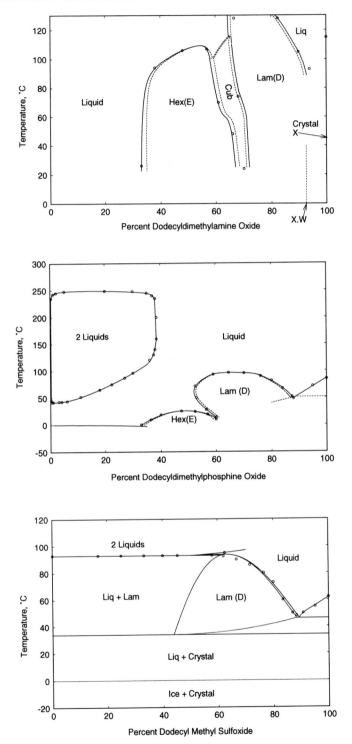

Fig. 10.12. Phase diagrams of the dodecyldimethylamine oxide– (upper), dodecyldimethylphosphine oxide–(middle), and dodecyl methyl sulfoxide–water (lower) systems.

The difference between the phosphine oxide and the sulfoxide is smaller than that between the phosphine oxide and the amine oxide with respect to both bond moments and basicities, which is consistent with the observed phase behavior.

Hydrogen bonding studies with phenol provide additional support for this ranking. The thermodynamic parameters for these association reactions are tabulated below, along with information regarding the number of hydrogen bonds observed.

Compound	ΔH kcal(kJ)	$-T\Delta S$ kcal(kJ)	ΔG	No. of hydrogen bonds
$C_{12}H_{25}N(O)(CH_3)_2$	−8.0(33)	<2.6(<11)	>−5.4(23)	3
$C_{12}H_{25}P(O)(CH_3)_2$	−7.8(33)	3.3(14)	4.5(19)	2
$C_{12}H_{25}S(O)CH_3$	−6.3(26)	3.0(13)	3.3(14)	2

The entropic free energy contributions to the association thermodynamics are (as expected) unfavorable, and are all similar in magnitude. The enthalpic free energies are favorable and decrease in the order $N \rightarrow O > P \rightarrow O > S \rightarrow O$. As noted elsewhere (Section 10.4), the amine and the phosphine oxide differ is the number of hydrogen bonds formed; the amine oxide forms three strong hydrogen bonds, while the phosphine oxide and sulfoxide both form only two.

Hydrogen bond enthalpy data do not quantitatively scale these groups in the same way as do reduced bond moment and basicity data, but the order is the same. The qualitative difference in the multiplicity of hydrogen bonding to the amine oxide, in comparison to that of the other two groups, is noteworthy.

Polyfunctionality in dipolar surfactants. The accumulation of dipolar hydrophilic groups generally increases the hydrophilicity of a molecule, but the position of the two groups relative to each other is of considerable importance. Addition of a second dimethylamine oxide group on the 2-carbon yields a strongly water-soluble surfactant, but also reduces chemical stability due to facilitation of the Cope elimination reaction [44]. Inserting a second group into the 3-position also yields a strongly hydrophilic molecule, but without reducing the chemical stability.

Introducing a second methylsulfinyl group in the 2-position of dodecyl methyl sulfoxide visibly enhances its hydrophilicity, as the disulfoxide is water-soluble. However, introducing a second methylsulfinyl group in the 1-position (as in 1,1-bis(methylsulfinyl)dodecane) yields a compound that is highly crystalline and immiscible with water below 100°C.

10.5.4.2   Phosphinyl (P → O) esters and amides

An ethyl phosphine oxide ($-P(\rightarrow O)CH_2CH_3$), a phosphonic *N*-methylamide ($-P(\rightarrow O)NHCH_3$), and a phosphonate methyl ester ($-P(\rightarrow O)OCH_3$) constitute an isoelectronic series within which the electronegativity of the proximate substituent groups on phosphorus is varied. If only one ethyl group on an alkyldiethylphosphine oxide is replaced with a methylamino or methoxy group, phosphinate amides and esters result. These are too unstable to hydrolytic cleavage to be examined in water, but if both ethyl groups are replaced with methylamino or methoxy groups the resulting phosphonic acid derivatives are more stable. The amide has borderline stability in water (reacts, but slowly), while the ester is stable. The phosphate ester, in which all three of the ligands to the phosphinyl (P → O) phosphorus are alkoxy groups, is highly resistant to hydrolysis.

The phase diagrams of diethyldodecylphosphine oxide, dimethyl dodecylphosphonate, and dimethyldodecyl phosphate are shown in Fig. 10.13. (This particular phosphate is *not* isoelectronic with the other two, as it contains an extra methylene group.) The diagram of the phosphine oxide differs from that of both esters in that a wide separation exists between the liquid crystal regions and the miscibility gap. Also, the liquid crystal phases of the diethyl phosphine oxide are anomalously unstable thermally relative to those of the dimethyl analog. The inference as to relative hydrophilicity based on the liquid/liquid miscibility gap criterion thus differs from the inference based on the liquid crystal criterion.

Compound	T_m, Lam	T_m, Hex	T_{mp}	ΔT
$C_{12}H_{25}P(O)(CH_2CH_3)_2$	33	10	48	>48
$C_{12}H_{25}P(O)(OCH_3)_2$	56	none	21	>21
$C_{12}H_{25}OP(O)(OCH_3)_2$	24	none	—	—

These data suggest the possibility that some stabilization of liquid crystal phases in the ester may result from hydration of the methoxy groups. Hydration of the methoxy oxygen should be much weaker than is that of the phosphinyl oxygen, however, and no supporting evidence for such interactions exists.

The liquid crystal phase is more thermally stable, and the miscibility gap substantially larger, in the phosphate ester than in either phosphonate ester. Thus, both the liquid crystal and miscibility gap criteria suggest that the phosphate ester is less strongly hydrophilic than is the phosphonate ester.

Dodecyl methyl methylphosphonate was prepared and its phase behavior characterized. This enables a comparison to be made between isomeric structures having different molecular shapes, but nearly identical charge distribution. Virtually no difference in phase behavior was observed (Figs 10.13, 10.14). Just as with stereoisomerism (Section 10.2.2), shape *per se* has little influence on the phase behavior of fluid phases.

The phase behavior of these phosphorus esters differs qualitatively from that of dodecyl methyl sulfoxide. Whereas in the sulfoxide the interaction of the miscibility gap and the liquid crystal regions produced monoglyceride-like phase behavior, the same interaction in phosphoryl esters results in a liquid phase which lies parallel to and just above the liquid crystal solubility boundary. The same phase phenomenon has been reported to exist in the $C_{10}E_3$–water system [45]. It is possible that this narrow liquid phase, which has a bluish color, has the same structure as the anomalous (L_3) phase of the $C_{10}E_4$–water system. If so, this suggests that these compounds display pivotal phase behavior (Section 5.10.7).

Alternatively, these mixtures may be biphasic and possess colloidally structure. Phase dispersions which were brightly colored (red, green, blue) were observed in dilute mixtures of phosphoryl esters with water, and similar observations have been made in aqueous *N*-acylglutamate surfactant mixtures [46]. None of these compounds is intrinsically colored, so that such phenomena must result from interference phenomena stemming from either the existence of well-defined colloidal structure [47] or the existence of structural features having unusually large dimensions [48]. Details of the phase behavior of the phosphinyl esters remain obscure, and they should be re-examined using modern swelling methods.

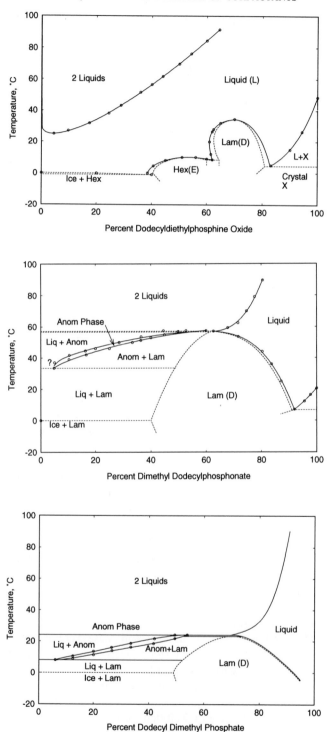

Fig. 10.13.   The diethyldodecylphosphine oxide– (upper), dimethyl dodecylphosphonate– (middle), and dimethyl dodecyl phosphate–water (lower) systems.

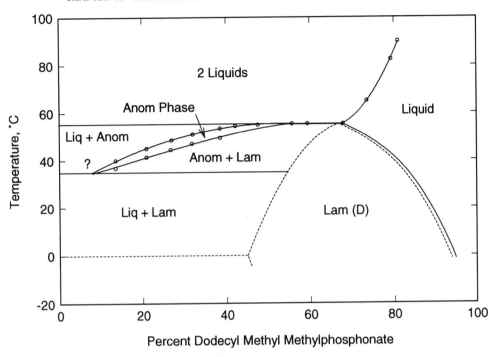

Fig. 10.14. The methyl dodecyl methylphosphonate–water system.

Supplemental information on phosphinyl esters. Estimates of hydrogen bonding enthalpies from frequency shift data, the frequency of the $P \rightarrow O$ stretching frequency itself, and ^{31}P NMR data are consistent with inferences as to relative hydrophilicity based on phase behavior.

Compound	$\Delta v(OH)$	$v(P \rightarrow O)$	$\tau(^{31}P, ppm \ vs. \ H_3PO_4)$
$C_{12}H_{25}(CH_3)_2P \rightarrow O$	7.83	1163	-42.4
$C_{12}H_{25}(CH_3O)_2P \rightarrow O$	6.31	1231	-32.4
$C_{12}H_{25}O(CH_3O)_2P \rightarrow O$	—	1282	$+0.9$

The phosphonate ester forms a weaker hydrogen bond to water than does the phosphine oxide. The frequencies of the $P \rightarrow O$ bond increase in the order phosphine oxide < phosphonate ester < phosphate ester, suggesting that the extent of back-bonding increases in the same order. Finally, the order of the ^{31}P NMR chemical shift data suggests that the electronegativity of phosphorus is greatest in the phosphate, least in the phosphine oxide, and intermediate in the phosphonate. Phase behavior, hydrogen bonding, infrared, and NMR data are in good agreement with respect to the rank order of hydrophilicity within this series of hydrophilic groups.

10.5.4.3 *N*-Terminal hydrophilic groups

Very few *N*-terminal hydrophilic groups are hydrolytically stable. Those having sufficient stability in water to withstand investigation include only sulfoximines and sulfone diimines [49], both of which are imines of four-coordinate sulfur compounds (Section

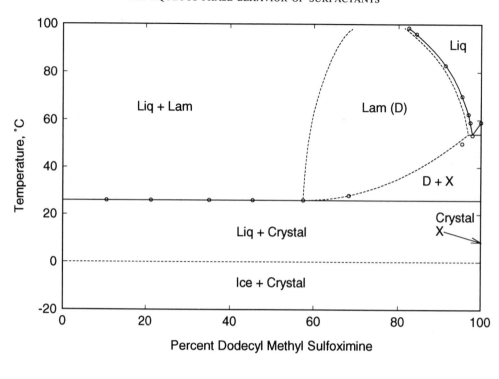

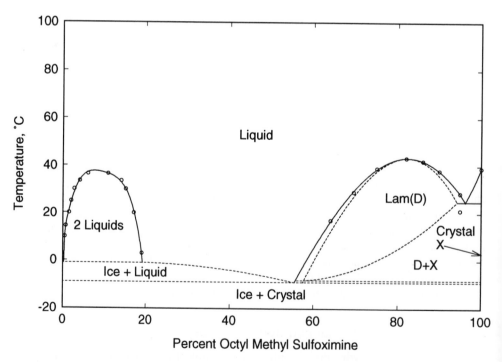

Fig. 10.15. The dodecyl methyl sulfoximine (sulfone imine)–water (upper) and octyl methyl sulfoximine–water (lower) systems.

9.4.1). The phase diagrams of the dodecyl methyl homologs of these functional groups are shown in Figs 5.18 and 10.15. It can be seen, by comparison with Fig. 10.12, that the phase behavior of the sulfoximine closely resembles that of dodecyl methyl sulfoxide while that of the sulfone diimine resembles that of the isoelectronic phosphine oxide (Fig. 10.12). These results suggest that the hydrophilicities of these pairs of molecules are similar as well.

The narrowness of the liquid/liquid miscibility gap of dodecyl methyl sulfone diimine is quite remarkable. It signifies that even several degrees above the critical temperature, the compositions of coexisting phases are remarkably similar. For example, at 30°C (3°C above the critical temperature) a *c.* 0.1% liquid (99.9% water) coexists with a 3% liquid (97% water). The distortion of the free energy of mixing curve that is imposed by this functional group must be quite severe.

Steric impairment of hydrophilicity. It will be recalled that the $-NH_2$ group is an operative hydrophilic group while the $-NHCH_3$ group is not. A qualitatively similar result is found in comparing the $-OH$ group with the $-OCH_3$ group in polyoxyethylene surfactants (Section 10.5.5.2). These observations suggest that methyl substitution on a hydrogen bond acceptor atom impairs hydrophilicity.

The effect of *N*-methyl substitution was also evaluated in sulfoximines, which are far more hydrophilic than is a lone hydroxy group (Fig. 10.16). The result is to reduce the thermal stability of the lamellar liquid crystal phase and the temperature of the Krafft discontinuity, but qualitatively similar surfactant behavior is retained.

Upper consolute boundary. It was observed, unexpectedly, that octyl methyl sulfoximine displays an *upper* consolute boundary (Fig. 10.15). The location and symmetry of

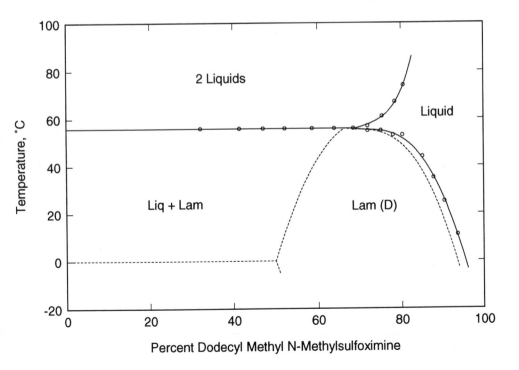

Fig. 10.16. The *S*-dodecyl-*N*,*S*-dimethyl sulfoximine–water system.

TABLE 10.4
Basicity data on *N*-terminal hydrophilic groups and related compounds

Compound	pKa	Compound	pKa	ΔpKa
$(CH_3)_2S(\rightarrow NH)_2$	+5.6	$(CH_3)_3P \rightarrow O$	−1.5	−7.1
$(CH_3)_2S(\rightarrow NH)(\rightarrow O)$	+3.2	$(CH_3)_2S \rightarrow O$	−2.8	−6.0
ΔpKa	−2.4		−1.3	

this boundary is reminiscent of those observed in selected zwitterionic surfactant–water systems.

Supplemental information on N-*terminal groups.* These groups have been studied comparatively little, but two bits of supplemental information exist. One is that the vibrational spectrum of crystals indicates that the N–H hydrogen of the sulfone diimine group is engaged in very strong hydrogen bond donor interactions with the nearby S → NH nitrogen atom [49]. This is unexpected, since the N–H bonds of amines are very weak hydrogen bonding donors, and the partial negative charge on the nitrogen should not favor such an interaction. Nevertheless, the spectral data indicate that such hydrogen bonding exists.

The basicity of these two compounds relative to each other, and to the isoelectronic P → O compounds, are also of interest (Table 10.4). Vertical comparisons within Table 10.4 provide data on structurally similar groups which differ in hydrophilicity, and in each case the more strongly hydrophilic group is also the more basic. Horizontal comparisons allow one to compare terminal nitrogen and terminal oxygen groups that possess similar hydrophilicity. From these comparisons, it can be seen that an *N*-terminal group must be 6–7 pKa units more strongly basic than an analogous *O*-terminal group in order for the two groups to display comparable hydrophilicity.

10.5.4.4 Ammonioamidates

Ammonioamidates are a particularly interesting hydrophilic group because they are the only operative hydrophilic group to possess a carbonyl-like structure. The phase diagram of the *N,N,N*-trimethylammoniododecanamidate–water system is shown in Fig. 10.17. The compound is highly water-soluble and forms liquid crystal phases that are moderately stable (the lamellar phase extends to 93°C). ΔT is greater than 40°C, and the liquid/liquid miscibility gap is absent below 100°C. From these data, it is concluded that this functional group is moderately strongly hydrophilic.

These compounds are not indefinitely stable in water; hydrolytic cleavage to a hydrazinium salt of the carboxylic acid may occur, especially in ammonioacetamidates. Compounds having the long chain on the carbonyl group and a trimethylammonio substituent, as in Fig. 10.17, are more resistant to hydrolytic cleavage.

Supplemental information. Two kinds of supplemental information are consistent with the relatively strong hydrophilicity suggested by these phase data. One is the basicity of the compound; the pKa of the conjugate acid of $C_{11}CON^-N^+(CH_3)_3$ (as the monomer) is 5.3, which makes this compound almost 1 pKa unit *more* strongly basic than a carboxylate salt. This is a very high pKa for a carbonyl compound, but a very low pKa for a derivative of an amide anion. The position of the proton in the conjugate acid of an ammonioamidate (whether on nitrogen or oxygen) has not been determined.

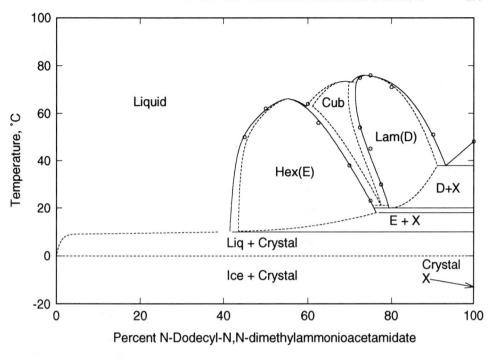

Fig. 10.17. The *N*-dodecyl-*N,N*-dimethylammonioacetamidate–water system.

Second, the frequency of the C=O stretching band in the infrared spectra of
ammonioamidates indicates that severe distortion of the usual carbonyl bond structure
exists. The vibrational frequency of the C=O bond in a variety of carbonyl compounds
is given in Table 10.5, and the frequency of the ammonioamidate group is seen to be
extremely low. This signifies the existence of a low π bond order, and is consistent
with the unique ability of this functional group to serve as a hydrophilic group. The
ammonioamidate remains a special case among operative hydrophilic groups.

An overall pictorial view of the relative hydrophilicity of the various semipolar
hydrophilic groups is presented in Fig. 10.18.

TABLE 10.5
C=O stretch frequencies among various carbonyl groups

Name	Structure	v(C=O)
Carbonate ester	ROC(=O)OR	1760 ± 20
Carboxylate ester	RC(=O)OR	1742 ± 8
Aldehyde	RCH=O	1730 ± 10
Ketone	R_2C=O	1715 ± 10
Amide	RC(=O)NR_2	1655 ± 25
Ammonioamidate	RC(=O)$N^-N^+R_3$	1580
Carboxylate	RCO_2^-,M^+	1595 ± 55 (asym)
Carboxylate	RCO_2^-,M^+	1405 ± 45 (sym)
Carboxylate	RCO_2^-,M^+	1500 (mean)

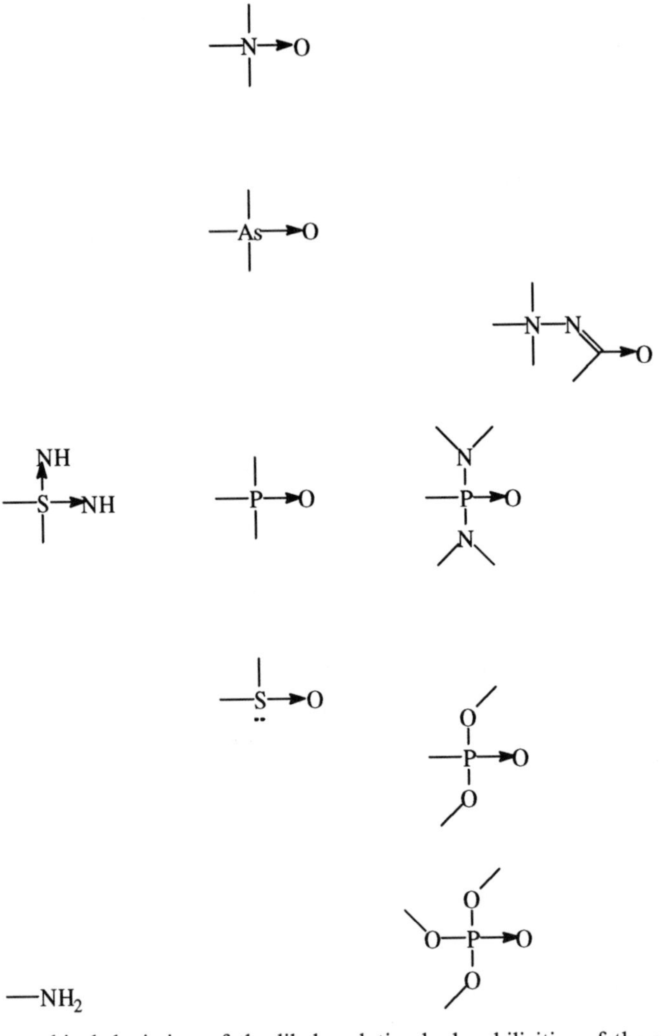

Fig. 10.18. A graphical depiction of the likely relative hydrophilicities of the various dipolar hydrophilic groups.

10.5.5   Nonionic class, single bond subclass

This subclass contains only one member that is operative in monofunctional compounds – the primary amino ($-NH_2$) group (Table 10.2). The phase behavior of various primary amines was explored in the 1940s [50], and information on other homologs has been provided more recently [51]. Liquid crystal phases do not exist in the C_6 homolog at 25°C, but appear in the C_7 and higher homologs. The aqueous diagram of the C_8 homolog is displayed in Fig. 10.19. The dodecyl homolog displays monoglyceride-like phase behavior; it is poorly soluble in water, so that crystals or liquid crystals separate in dilute mixtures and only the concentrated liquid phase region exists. The data suggest that two liquid crystal phases and three crystal hydrates exist, and that the value of ΔT is 14°C. The work bears repeating.

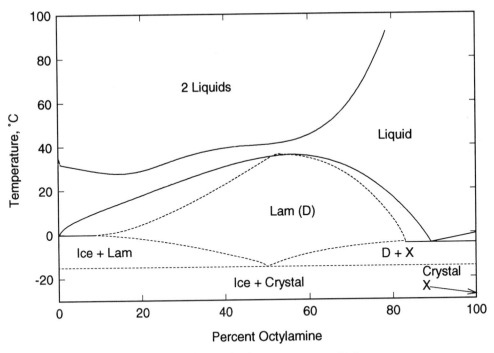

Fig. 10.19.   The octylamine–water system [50].

It is readily apparent from the above information that the amino group is weakly hydrophilic. The lower consolute boundary is prominent even with a short C_8 lipophilic group, and monoglyceride-like behavior exists in higher homologs. The liquid crystal phases formed by the C_{12} homolog are weakly stable thermally, and the value of ΔT in the $C_{12}NH_2$–water system is very small. The fact that only a single methyl substituent on nitrogen destroys the ability of the primary amino group to operate as a hydrophilic group (Table 10.1) is also consistent with this appraisal. Methylation of the more strongly hydrophilic sulfoximine nitrogen weakens its hydrophilicity, but does not eliminate it (Figs 10.15, 10.16).

10.5.5.1   Polyhydroxy surfactants

A single hydroxy group is inoperative as a hydrophilic group, but the presence of only two hydroxy groups suffice to introduce surfactant phase behavior. Alkane-1,2-diols and monoglycerides constitute important examples of such surfactants [52,53]. Monoglycerides possess an ester group plus a 1,2-diol group, but the ester group contributes little to their hydrophilicity (Fig. 5.15). The exact role of the ester group in the hydration of monoglycerides remains uncertain. (Consideration of ternary phase equilibria suggest that groups such as nitriles, esters, and aldehydes are not entirely devoid of hydrophilicity (Section 13.3.1).) As the number of hydroxy groups increases, so does the hydrophilicity.

The preparation of "simple" analogs of 1,2-diols having larger numbers of OH groups is not straightforward synthetically, so that most of the available information

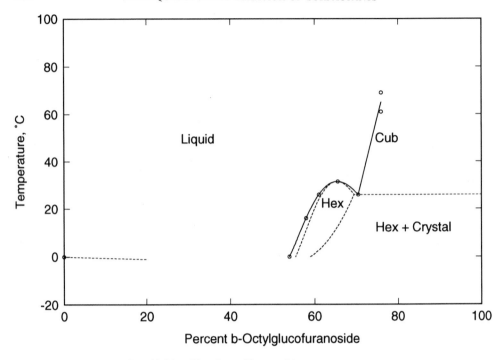

Fig. 10.20.   The β-octylfuranoside–water system.

on polyol surfactants is derived from studies of sugar derivatives. The diagram of a glycoside is shown in Fig. 10.20 [54]. Sugar esters have also been prepared [55]. The functional groups in these molecules include a cyclic ether and various numbers of hydroxy groups, and the analysis is messy. Nevertheless, it is clear that an alkyl glycoside (which has two ether and four hydroxy groups) is strongly hydrophilic in comparison to, for example, an amino group (Fig. 10.19). Phase studies of various sucrose monoesters (seven hydroxy groups, three ether oxygens, one ester group), reveal that these compounds too are highly water-soluble and do not display a lower consolute boundary below 100°C [52]. Being water-soluble such esters will not be indefinitely stable in water, but neither do they hydrolyze quickly.

Phase information on *N*-acyl derivatives of *N*-methylglucamine has recently been reported (Fig. 10.21 [56]). These molecules are analogs of the sugar glycosides that contain five hydroxy groups, plus an *N*-methylamide group. They are more closely related structurally to sorbitol than to sugars, since the 1-carbon is in the alcohol (rather than the aldehyde) oxidation state. Also, they are acyclic.

This diagram suggests that this functional group is moderately strongly hydrophilic, since the liquid crystal phases are rather stable thermally. Chemical decomposition (cyclization via dehydration to a tetrahydrofuran) occurs, but only above 120°C. An interesting feature of the diagram is that the boundary of the hexagonal liquid crystal region extends to temperatures well below the Krafft discontinuity. It is also of interest that the material exists as an equilibrium thermotropic liquid crystalline phase. (This is also true of alkyl glycosides [53].)

Two metastable polymorphic crystals of this compound were discovered, the crystal structure of one of the metastable phase and of the equilibrium phase determined, and

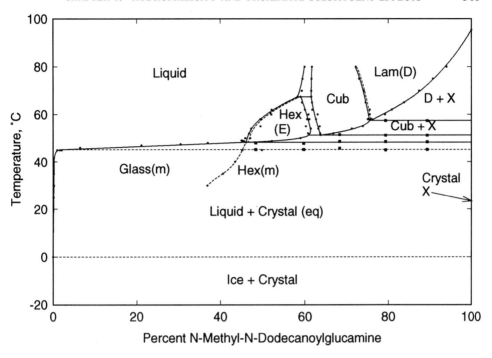

Fig. 10.21. The *N*-dodecanoyl-*N*-methylglucamine–water system.

the structural differences between the metastable and the equilibrium crystal were revealed (Section 8.3.2). Since this diagram was determined using swelling methods, data on the lower boundaries of the liquid crystal phases were obtained. These data clearly reveal the smooth decrease in the miscibility of the compound with water as the temperature is lowered – irrespective of the structure of the fluid state that exists above this miscibility boundary.

Thermotropic liquid crystal states. Considerable evidence exists which suggests that if three or more hydroxy groups exist in a surfactant molecule, the compound tends to exist as a thermotropic liquid crystal. While not seen in monoglycerides, it has been reported that monoesters of polyols having more than two hydroxy groups exist as dry liquid crystals [57]. Later binary phase studies of several types of polyol surfactants including alkyl glycosides [54], *N*-alkyl glyconamides [58], *N*-acyl-*N*-methylglucamines (Fig. 10.21) [56], and polyglycerol monoesters [59] all show this feature in their phase behavior.

10.5.5.2 Polyoxyethylene surfactants

The most important nonionic surfactants presently in commercial use are the poly-oxyethylene (POE) compounds. A long polyoxyethylene chain exists in the middle of these molecules, a lipophilic group is attached at one end, and a hydroxy group exists at the other end. In most POE surfactants the lipophilic group is linked to the POE moiety by an ether bond, but in some it is linked by an ester [55] and in a third type the lipophilic group is an alkylphenyl group attached by an ether link [60] (Fig. 10.22).

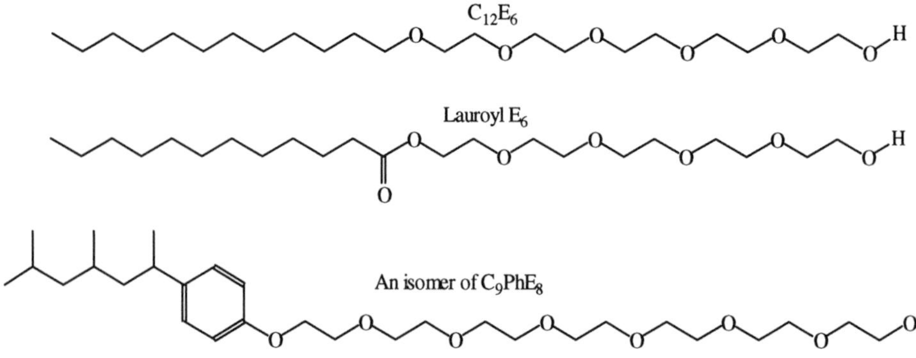

$C_{12}E_6$

Lauroyl E_6

An isomer of C_9PhE_8

Fig. 10.22. Typical examples of polyoxyethylene surfactant structures in which the hydrophilic group is linked to the lipophilic group by an alkyl ether link (upper structure), an ester link (middle structure), and a phenyl ether link (lower structure).

These compounds are easily and inexpensively prepared on a commercial scale by the addition of ethylene oxide to a fatty alcohol, a fatty acid, or an alkylphenol. Individual POE surfactants having a defined structure can be synthesized in the laboratory with reasonable effort (Figure 10.23). Commercial surfactants invariably display a distribution in the number of oxyethylene groups [61].

POE surfactants epitomize the variation of hydrophilicity in an "extensive" manner. Neither the hydroxy nor the ether functional groups are by themselves operative hydrophilic groups, but the accumulation of ether groups (with retention of the hydroxy group) yields polyfunctional molecules whose hydrophilicity varies over a wide range. Moreover, if the lipophilic group is kept constant, both the size of the hydrophilic group relative to the lipophilic group, and also the balance between hydrophilicity and lipophilicity (the "HLB"), may easily be varied. HLB has been discussed in connection with the classes of phase behavior (Section 5.10.7), and more will be said of HLB later (Section 11.4).

Individual POE surfactants vary widely in their phase behavior [63], and considerable complexity might have been expected to exist in commercial mixtures. It has been reassuring, however, to find that commercial mixtures closely resemble in their phase behavior the pure analog having the same number of oxyethylene groups as the average in the commercial sample [64].

It is well to keep in mind that the volume of the hydrophilic group in POE surfactants often exceeds that of the lipophilic group. In $C_{12}E_4$ the hydrophilic group and the lipophilic group occupy similar volumes, in $C_{10}E_4$ the hydrophilic group is larger than the lipophilic group, and in $C_{14}E_4$ it is smaller. In POE surfactants having large numbers of oxyethylene groups, the lipophilic "tail wags the dog" insofar as relative volume is concerned. This appendage also exerts a profound influence insofar as the solution and surface physical chemistry of the compound is concerned; the surfactant is profoundly different in many ways from the polyoxyethylene glycol. It has been suggested that the steric mechanism of colloid stabilization applies in colloids stabilized by the adsorption of POE surfactants [38,66]. These functional groups are perhaps unique (among the common surfactants) in this regard.

Role of the hydroxy group. The hydrophilicity of POE surfactants is varied by changing the number of ether groups, but the hydroxy group plays an extremely

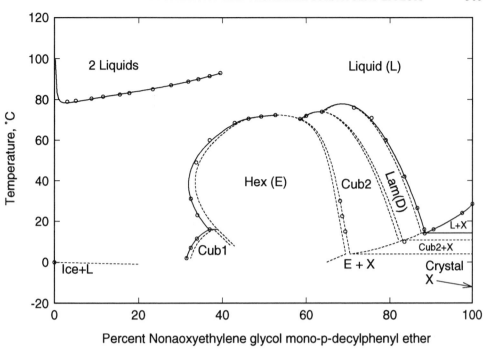

Fig. 10.23. The nonaoxyethylene glycol decylphenyl ether–water system [62].

important role in the hydration of these compounds. This is evident from comparisons between POE surfactants and related compounds in which the hydroxy group has been replaced by another functional group. The hydroxy group may be methylated to form an ether link, for example, which results in the "methyl-capped" POE surfactants. Since this structural change reduces the thermal stability of the liquid crystal regions and enlarges the miscibility gap (compare Figs 10.24 and 4.5), it is evident that methyl capping reduces the hydrophilicity of the parent surfactant.

   Another kind of structural modification is replacement of the hydroxy group with a nonoperative functional group. This modification has a profound impact on phase behavior, and if the number of ether groups is not sufficiently large may destroy hydrophilicity altogether. It is particularly interesting and significant to compare $C_{10}E_4$ with its isoelectronic fluorine analog:

$$CH_3CH_2CH_2CH_2CH_2CH_2CH_2CH_2CH_2CH_2OCH_2CH_2OCH_2CH_2OCH_2CH_2OCH_2CH_2OH$$

vs.

$$CH_3CH_2CH_2CH_2CH_2CH_2CH_2CH_2CH_2CH_2OCH_2CH_2OCH_2CH_2OCH_2CH_2OCH_2CH_2F$$

Replacement of the hydroxy group in $C_{10}E_4$ by fluorine destroys altogether the surfactant phase behavior displayed by the $C_{10}E_4$–water system. While $C_{10}E_4$ is a surfactant of moderate hydrophilicity (Fig. 4.5), the fluorine-capped analog is a water-immiscible oil. It follows, from this observation alone, that the hydroxy group plays a critically important role in the hydration of this surfactant. The influence of capping diminishes as the number of ether linkages increases, and with a sufficiently large number of ether

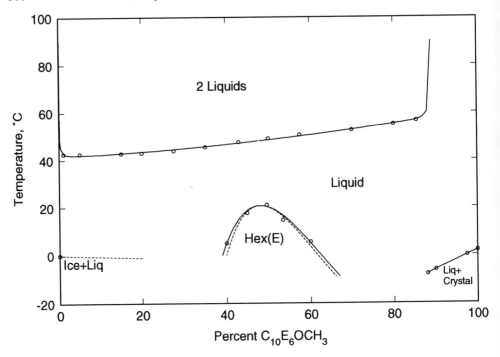

Fig. 10.24. The hexaoxyethylene glycol decyl methyl ether ($C_{10}E_6OMe$)–water system [65].

groups surfactant phase behavior would likely be restored in the fluorine compound just as occurs in methyl-capped POE surfactants. Nevertheless, one could legitimately describe POE surfactants as "hydroxy surfactants" whose hydrophilicity is enhanced by the ether groups also present. The enhancement depends on the number of ether groups.

If the OH group in POE surfactants is replaced by any of the functional groups in Table 9.1, a qualitatively similar alteration of phase behavior suggestive of reduced hydrophilicity (Section 9.5.1) occurs. If, however, a POE surfactant is capped with an operative hydrophilic group (Table 9.2), the hydrophilicity of the parent hydroxy surfactant is enhanced. This has been found to occur with capping groups such as 2,3-dihydroxypropyl, dimethylphosphinyl, and dimethylamine oxide (Fig. 10.25). All these compounds display phase behavior suggestive of stronger hydrophilicity than exists in the parent POE surfactant.

Role of the oxyethylene groups. The lone hydroxy group is especially important in determining the phase behavior of POE surfactants, but the oxyethylene groups do play a role. The impact of changing the number of oxyethylene groups is particularly strong if one varies "HLB" relative to that present in the pivotal structures (Section 5.10.7). Within the pivotal surfactants, the larger the lipophilic group the longer the POE group must be in order to maintain pivotal behavior. If the delicate HLB within these surfactants is altered by reducing the number of oxyethylene groups sufficiently, then monoglyceride-like phase behavior may be expected to result. If the HLB is altered by increasing the number of oxyethylene groups (even by one), then soluble surfactant behavior results. Further increasing the number of oxyethylene groups quantitatively changes the phase diagram and also influences the liquid crystal phase behavior of the system (see below), but the higher ethoxylogs are very soluble compounds. Commercial

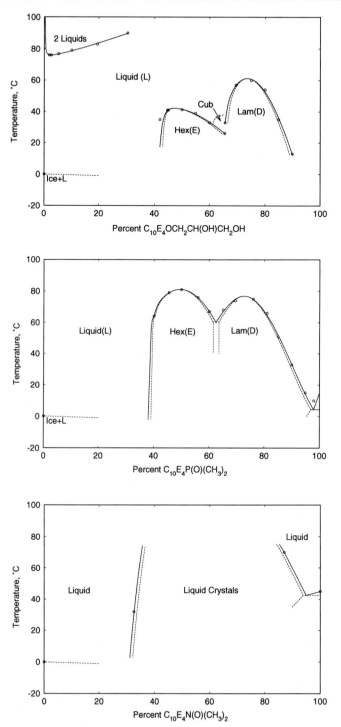

Fig. 10.25. The aqueous phase diagrams of $C_{10}E_4$ surfactants in which the $-OH$ group has been replace by 2,3-dihydroxypropyl- (upper), dimethylphosphinyl- (middle), and dimethyl amine oxide (lower) functional groups.

surfactants having very large numbers of OE groups (40 or 50) have been prepared, have very high cloud point boundaries, and are very soluble [67]. Since it is relatively easy to prepare AE_x compounds having small numbers of oxyethylene groups [68] and is very laborious to prepare standard compounds having very long POE chains [69], little information as to their physical behavior exists except on commercial mixtures.

The number of oxyethylene groups present significantly influences the relative thermal stability of the various liquid crystal phases that may be present. The observed variations in stability are qualitatively consistent with the geometric models relating phase structure to molecular structure that have been suggested (Section 8.8.1). Within the family of C_{10} homologs the thermal stabilities of the lamellar and the hexagonal liquid crystal phases are similar in the $C_{10}E_5$ ethoxylog. The hexagonal phase region is smaller (and the phase is less stable thermally) than is the lamellar phase in the $C_{10}E_4$ ethoxylog, and the hexagonal phase is the dominant liquid crystal in $C_{10}E_6$ and higher ethoxylogs [62]. In C_{12} chain homologs $C_{12}E_5$ displays pivotal behavior, the hexagonal phase is less stable thermally than is the lamellar phase in both $C_{12}E_5$ and $C_{12}E_6$, but (as in the C_{10} series) the hexagonal phase becomes the dominant liquid crystal in longer ethyoxylogs. Cubic phases may be found to either side of the hexagonal phases among AE_x surfactants having a lipophilic chain greater than C_{10}, but these phases have not so far been found in $C_{10}E_x$ surfactant systems. A more detailed review of the present status of our knowledge of the relationship between the phase behavior of AE_x non-ionic surfactants and their molecular structures has been published [70].

10.6   The concept of hydrophilicity

From the above it can be seen that considerable information exists regarding the influence of the molecular structure of the polar functional group on hydrophilicity. Since considerable knowledge of the electronic structure of these groups also exists, it is worthwhile to identify those structural features of hydrophilic groups which determine the existence and the magnitude of hydrophilicity.

The principal experimental observations which any model for hydrophilicity must explain include the following:

(1) All ionic functional groups are operative, regardless of whether they exist within the same molecule (zwitterionic surfactants) or in different molecules (ionic surfactants).

(2) Only a fraction of the dipolar functional groups that exist are operative.

(3) Strongly dipolar and hydrogen bonding functional groups may be either opera-tive or inoperative as hydrophilic groups. Some of the inoperative groups are more strongly dipolar and more strongly hydrogen bonding than are some of the operative groups.

(4) Virtually all dipolar functional groups that possess 2p multiple bonds are inoperative. The notable exceptions are the N,N-dimethylamides (which are borderline) and the ammonioamidates (which are moderately strongly hydro-philic).

(5) In most operative dipolar functional groups, either no back-bonding exists or the back-bonding occurs to 3d or 4d orbitals on the positive atom. Dipolar groups which display very strong back-bonding to 3d orbitals are not operative.

(6) Methyl substitution on *N*-terminal surfactants impairs hydrophilicity, and in some cases may eliminate it altogether.

(7) The aliphatic amine oxide group (which is strongly hydrophilic) accepts three strong hydrogen bonds from water, while the phosphine oxide and sulfoxide groups (which are less strongly hydrophilic) accept only two. No group that accepts only one hydrogen bond has been found to be operative – regardless of the thermodynamic strength of that bond.

10.6.1 Some important features of water

Hydrophilicity is intimately linked to the physical science of liquid water, and especially to its structure. A recapitulation of the molecular and phase structure and the miscibility properties of water is therefore appropriate [71].

The water molecule is bent; the H−O−H angle is 104.5°. The dipole moment of an isolated water molecule has been estimated to be about 1.84 D. At moderate temperatures extensive hydrogen bonding exists between water molecules in all its phase states, but these bonds are particularly important in ice. In an ice crystal at low temperatures each water molecule donates two hydrogen bonds and accepts two, and these four bonds are arranged tetrahedrally about the oxygen. Each hydrogen oscillates rapidly between two equilibrium positions, depending on to which nearby oxygen it is covalently bound and to which it is physically bound (via a hydrogen bond) [72]. Significant van der Waals' forces exist within water, but the phase structure of ice is dictated by the preferred hydrogen bonding geometry of water molecules. The unusually large heats of fusion and vaporization of water may be attributed to the fact that many hydrogen bonds are broken during these phase reactions. As noted elsewhere (Section 3.7), the changes that occur in the entropic free energy term with temperature dominate the dependence of the free energy of water on temperature, notwithstanding the substantial heat capacity and the large increases in enthalpy that occur as the temperature is increased.

At the melting point of ice (at one bar), this idealized structure must be partially disrupted (as occurs in all crystals [73] Section 8.3). In the coexisting liquid at the melting point the disruption is far more severe, but fragments of ice-like structure likely remain intact. As the temperature is raised these fragments are broken down to a progressively greater degree, and the miscibility of water with other compounds becomes more "normal" (in comparison to other materials) [74]. At its critical point ($T_c = 374°C$, $p^c = 218$ bars, $d_c = 0.32 \, \text{g cc}^{-1}$) the solvent properties of water are extremely different from those of liquid water at room temperature [75]. It is possibly significant that the temperature at which water becomes miscible in all proportions with both strongly and weakly hydrophilic surfactants is near its critical temperature (Section 3.12).

10.6.2 A model for hydrophilicity

In suggesting a model for hydrophilicity, it is first necessary to explain why it has not so far been possible to eliminate hydrophilicity in molecules containing formally charged functional groups ((1), Section 10.6). The reason for this may lie in the massive strength of the solvation of monopoles by water. The solvation of these groups may be so strong

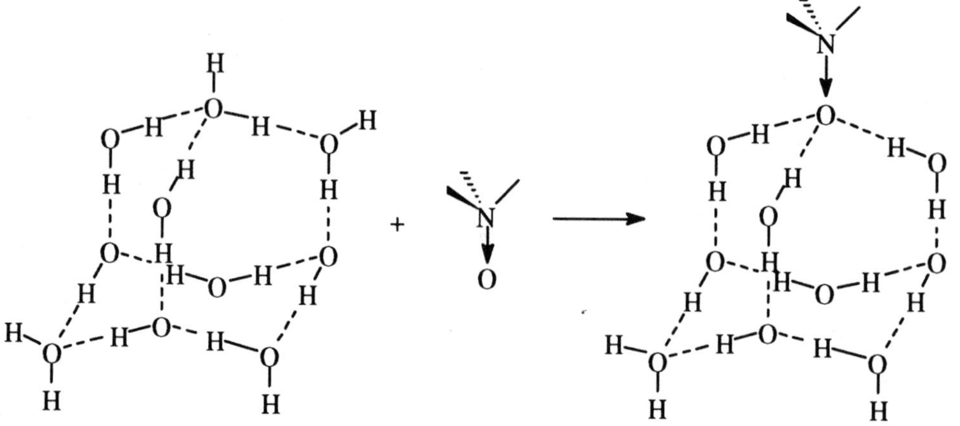

Fig. 10.26. The hydrophilicity equation, in which the oxygen atom of an amine oxide displaces a water oxygen in liquid water.

that it is impossible to eliminate hydrophilicity – although it is clearly possible (by structural variation) to modify its magnitude significantly.

The second major observation that must be rationalized is the result that many strongly dipolar groups are not hydrophilic at all ((2), Section 10.6). The strength of solvation of dipolar groups should be weaker than that of monopolar groups. That being the case, the details of solvation appears to be more important with respect to the existence of hydrophilicity in dipolar groups.

When the results are viewed as a whole, it is apparent that more than one criterion must be met in order for hydrophilicity to exist within dipolar groups. It is suggested that these are the following:

(1) A functional group must compete successfully with the water molecule itself, in a thermodynamic sense, for hydration within liquid water.

(2) Insertion of the functional group into liquid water must not substantially alter the phase structure of water.

If these two basic criteria are met, the further variation of intrinsic hydrophilicity results from quantitative variations in the free energy of hydration. Both aspects of hydrophilicity are expressed graphically by the reaction equation in Fig. 10.26.

10.6.3 The dimensions of hydrophilicity

The above model for hydrophilicity suggests that it has four dimensions:

(1) A thermodynamic dimension. The free energy of hydration must be sufficiently strong for the polar group to displace water molecules from within liquid water. That is, the reaction in Fig. 10.26 must proceed spontaneously to the right. This is a mandatory requirement. Regardless of whether or not other criteria are met, a functional group cannot be operative if the thermodynamic driving force for hydration is inadequate.

This hydration process must be favorable with the lipophilic group in place. All small dipolar molecules are miscible to a significant degree with water – irrespective of whether or not they are also hydrophilic groups. (An example is acetonitrile.) Those that possess sufficient hydrophilicity to carry a long chain into liquid water belong to an elite subset of functional groups.

(2) A multiple hydrogen bonding dimension. Another requirement is that multiple hydrogen bonding must occur to the functional group, or (more precisely) to an atom within the functional group. If only one hydrogen bond is formed, then that group is not a hydrophilic group regardless of how strong is the hydrogen bond.

(3) A geometry of hydrogen bonding dimension. The hydrogen bonds formed must display a three-dimensional, roughly tetrahedral, geometry about the hydrated atom. If this hydration geometry exists, then insertion of the hydrophilic group does not disrupt the phase structure of liquid water – indeed may strengthen it. If the strength of hydration in this geometry is weak, then hydrophilicity does not exist.

(4) A steric dimension. Hydrophilicity is impaired by the steric effects of substituents – especially by lipophilic substituents on the solvated atom (such as exist in *N*-terminal hydrophilic groups. (The term "steric" as used here refers to the space-occupying requirements of substituent groups, and not to the "steric" mechanism of colloid stabilization that exists when a hydrophilic polymer is adsorbed at an interface.) Hydrophilicity is also impaired (but to a lesser extent) by proximate substituents near to (but not on) the solvated atom.

In the hydrophilicity equation (Fig. 10.26) the amine oxide serves as a prototypical hydrophilic group, and Ice I as an idealized model for the structure of water. During the insertion process, the amine oxide oxygen atom replaces a water molecule within the phase structure without disrupting that structure. Obviously, additional water must also be displaced to provide space for the lipophilic group. In the case of the amine oxide group this reaction has a favorable free energy, three hydrogen bonds are formed to the $N \rightarrow O$ oxygen, these bonds are arranged in a tetrahedral geometry, and no substituents exist on the oxygen atom which sterically inhibit solvation. All the above criteria of hydrophilicity are very well met, and this explains why the aliphatic amine oxide group is the most strongly hydrophilic dipolar group known. Structural modifications of this group which interfere with any aspect of the insertion process first attenuate the hydrophilicity and, if carried sufficiently far, destroy it.

The perturbation of water by the lipophilic group during this process, is one of the factors that strongly influences the property of lipophilicity (Chapter 11).

10.6.4 Rationalizing the structure–hydrophilicity correlations using the hydrophilicity model

This model for hydrophilicity serves to rationalize, in a reasonable manner, all of the known correlations between structure and hydrophilicity.

Hydroxy groups. An important result of this survey was the demonstration that a single hydroxy group is inoperative as a hydrophilic group, while two or more hydroxy groups in the same molecule are operative. Viewed from the perspective of the above model, this suggests that the hydroxy group meets all the hydrophilicity criteria except the thermodynamic requirement. A lone hydroxy group may be properly hydrated numerically and geometrically, but it is simply too weakly hydrated to be (by itself) operative.

From the close structural relationship between fatty alcohols and water, and from numerous hydrogen bonding studies, it is evident that two or more hydrogen bonds can form to a hydroxy group, and that their geometric orientation is correct. However, no structural feature exists in a fatty alcohol which enhances its hydration thermodynamics relative to that of water, and the presence of the lipophilic group presents a barrier to this process. As a result, a lone hydroxy group is unable to compete effectively with water itself. Since the hydrophilic group of a surfactant molecule must drag the attached lipophilic group into water, it is possible that its hydration must be somewhat stronger than that of the water molecule.

In a dihydroxy (diol) compound, the additional hydration supplied by the second hydroxy group appears to overcome this marginal defect so that diols are operative. The introduction of additional hydroxy groups further strengthens in a progressive manner the hydrophilicity (Section 10.5.5.1).

Amino groups. The atomic dipole of the non-bonding pair in the primary amino group must be greater than those of the non-bonding pairs in the hydroxy group. This is to be expected because of the smaller nuclear charge in nitrogen. Both the basicity of the group, and the strength of hydrogen bonding to it, are strongly enhanced in the amino group in comparison to the hydroxy. The two N–H bonds are very weak hydrogen bond donors, but geometrically the hydration of the amino group must be similar to that of water itself. In spite of the fact that its hydration is essentially monofunctional, the net result is that the amino group is more hydrophilic than is the hydroxy. By analogy with diols 1,2-diaminoalkanes should be more strongly hydrophilic than are 1-aminoalkanes, but no information exists regarding their phase behavior.*

Steric effects on hydrophilicity. The sharp difference between the methylamino (–NHCH₃) and dimethylamino (–N(CH₃)₂) groups, on the one hand, and the primary amino (–NH₂) group on the other, suggests that steric inhibition of hydrophilicity exists. This is not due to lack of hydration energy, since the enthalpy of the hydrogen bond between a tertiary amine and phenol (9.5 kcal, 46 kJ) is almost twice that of the water/water hydrogen bond (*c.* 5 kcal, 24 kJ). This conclusion is also consistent with the unusual phase behavior of a dimethylamino-capped POE surfactant (Section 5.10.4).

Comparisons of methoxy and hydroxy compounds are also consistent with the idea that steric inhibition of hydrophilicity exists. It is likely that methylation of the –OH group reduces hydrophilicity just as does methylation of the –NH₂ group, but since neither the –OH nor the –OCH₃ groups are by themselves hydrophilic this effect is not apparent in monofunctional compounds. However, it *is* clearly evident in polyoxyethylene surfactants, since methylation of the hydroxy group in these compounds perceptibly reduces their hydrophilicity (Section 10.5.5.2).

Finally, comparing the phase behavior of dodecyl methyl sulfoximine (Fig. 10.15) and dodecyl *N,S*-dimethyl sulfoximine (Fig. 10.16)

Dodecyl Methyl Sulfoximine

Dodecyl Methyl N-Methyl Sulfoximine

---

*One may inquire as to the role that hydrolysis of the –NH₂ group to form the –NH₃⁺,OH⁻ salt has on its hydrophilicity. This question has been investigated using Raman spectroscopy of concentrated octyldimethylamine solutions. No evidence for hydrolysis could be detected using this method (W. Yellin, unreported work), and a similar conclusion was recently reported [51]. While hydrolysis must occur to some extent, it is not principally responsible for the hydrophilicity of the amino group.

also reveals the existence of steric barriers to hydrophilicity. This is evident from the effect of the N-methyl group on liquid crystal stability. Both of these surfactants display monoglyceride-like phase behavior, but the liquid crystal of the N–H compound extends to >100°C while that of the N-methyl compound is only stable to 56°C (a difference of >44°C).

The nitrile group. The strikingly weak hydrophilicity of the nitrile group is easily rationalized using the hydrophilicity model, since neither the thermodynamic, the multiplicity, nor the geometric requirements are met. Only one hydrogen bond is formed to the nitrile group, and its enthalpy is small. The phenol/nitrile hydrogen bond enthalpy is 3.2 kcal (15 kJ), which is substantially less than the minimal value of 5 kcal (24 kJ) required of operative hydrophilic groups. The strong polarity of the nitrile group (Fig. 10.1) belies the strength of its hydration – possibly because the hybridization of the non-bonding pair is sp. The electronegativity of carbon decreases in the order sp > sp$_2$ > sp$_3$, as indicated by the fact that the thermodynamic acidity of C–H bonds decreases in the order acetylene > ethylene > ethane. That being the case, the same should be true of nitrogen.

Carbonyl groups. In contrast to the nitrile group, the absence of hydrophilicity in most carbonyl groups *cannot* be rationally dismissed simply on the basis of inadequate hydration thermodynamics, since many carbonyl groups are strongly hydrogen bonding. For example, the hydrogen bond to phenol is 4.5 kcal (22 kJ) in ethyl esters, 5.3 kcal (25 kJ) in methylalkylketones, 6.4 kcal (31 kJ) in N,N-dimethylamides, and 6.5 (31 kJ) kcal in tetramethylureas. The first two are non-hydrophilic, while the amide is borderline. (Long-chain derivatives of tetramethylurea have not been examined.) The multiplicity of hydrogen bonding to carbonyl groups is not generally known, so this factor cannot be evaluated.

Failure to meet the geometric requirement is probably an important barrier to hydrophilicity among carbonyl groups. The axes of the nonbonding orbitals in these groups lie in the nuclear plane, which is also the nodal plane of the π orbitals (Fig. 10.27). Because of this, the angle between the nonbonding orbitals on oxygen is nominally 120°.

It is readily apparent, from inspection of molecular models, that this particular geometric arrangement of nonbonding orbitals makes it impossible for the carbonyl oxygen to participate in the structure of water. To do so would severely distort the

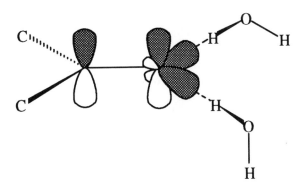

Fig. 10.27. The geometry of π-orbitals, of nonbonding orbitals, and of the hydrogen bonds formed to the nonbonding orbitals of carbonyl groups.

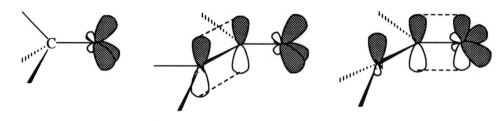

Fig. 10.28. Probable nonbonding orbital geometry in an alkoxide, the conjugated form of an amide, and the unconjugated form of an amide.

preferred collinear relationship between the axis of the non-bonding pair and water O−H bonds, and therefore severely weaken the hydrogen bonds that result. Alternatively, if two strong hydrogen bonds *were* to be formed to these orbitals by water, other hydrogen bonds within the water would be weakened. No satisfactory hydration geometry exists.

If the π bond order of the C=O bond of a carbonyl group is sufficiently weakened, for example by competitive conjugation with p-orbitals on a substituent atom, the hydrophilicity of the carbonyl group is measurably increased. Rudimentary hydrophilicity first appears in *N,N*-dimethylamides, and is quite strong in ammonioamidates.

Conjugation of orbitals on adjacent atoms with the carbonyl group probably alters the geometry of the nonbonding orbitals on the carbonyl oxygen. In esters and amides, it is well known that a nonbonding orbital on an oxygen or nitrogen attached to the C=O group is strongly conjugated with this group. In amides this flattens the nitrogen pyramid almost completely [76], significantly reduces the stretching frequency of the C=O bond, introduces a significant barrier to rotation about the C−N bond [77], and has other effects. However, the most important effect of conjugation with respect to hydrophilicity would be to shift the geometry of the nonbonding oxygen orbitals from planar trigonal towards a tetrahedral orientation. In one of the limiting canonical forms a negative charge exists on the oxygen, and the geometry of these orbitals should (in this limit) resemble that of the alkoxide group (Fig. 10.28).

This effect is presumably especially strong in the ammonioamidate group, which should be regarded as a formal derivative of an "amidate" anion rather than of an amide. "Amidates" are anions formed by abstraction of the N−H proton from amides. They are entirely analogous to "carboxylates", which are formed by abstraction of the O−H proton from carboxylic acids. This structural relationship is evident from the fact that the coordination number of the nitrogen bonded to the carbonyl carbon is two (as in the amide anion, NH_2^-), rather than three (as in amides and amines). (The inorganic and the organic nomenclature of these groups is not consistent with their structures.) Ammonioamidates may be viewed as derivatives of amidate anions by the substitution of a positive nitrogen group on the amidate nitrogen, and one of the canonical forms of this group is the dipolar structure shown in Fig. 10.29.

The fact that strong conjugation exists between the two-coordinate nitrogen and the carbonyl group is clearly evident from the exceptionally low stretching frequency in the carbonyl region (Table 10.5). This infrared band is similar in frequency to the antisymmetric mode of the carboxylate anion, but is higher than the mean frequency of the symmetric and antisymmetric modes. In the carboxylate anion too, the C=O π

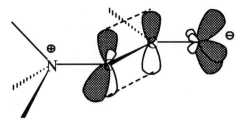

Fig. 10.29.   The most highly polarized canonical form of the ammonioamidate functional group, suggesting a tetrahedral arrangement of the nonbonding orbitals on the oxygen.

bond order is severely reduced by strong conjugation with a negatively charged oxide atom.

The limiting canonical structure (shown in Fig. 10.29) suggests that a high electron density exists on the oxygen in an ammonioamidate group. It is possible that the solvation of these groups occurs either exclusively on oxygen, or on both oxygen and nitrogen, but this matter has not been investigated. These compounds react with acids to form salts, and a similar uncertainty exists regarding the site of the proton in the salts. Two tautomeric structures (O—H and N—H) are possible, but which is preferred has not been determined. The preferred structure may depend on state.

Dipolar groups having back-bonding to oxygen. Dipolar groups in which back-bonding exists (such as the $P \rightarrow O$ and $S \rightarrow O$ groups) are particularly interesting with respect to structure–hydrophilicity correlations, because both hydrophilic and nonhydrophilic compounds exist. The amine oxide group is unique in that no back-bonding exists at all (Section 10.4).

The consequences as regards nonbonding orbital geometry of back-bonding between 2p orbitals on oxygen and 3d orbitals on the positive atom is not well defined. The geometry of non-bonding pairs cannot be directly ascertained by methods used to define the geometry of bonding orbitals because a nucleus exists only at one end of a nonbonding orbital.

What is known is that the nodal surfaces of d orbitals are geometrically different from the nodal plane of π orbitals [78]. More than one nodal surface exists and it may have various geometries (plane, sphere, cone). It seems likely, therefore, that conjugation of the 3d or 4d orbitals with oxygen 2p orbitals will not force the nonbonding orbitals of oxygen to lie in a planar trigonal pattern, as occurs in carbonyl groups. In fact, it has been suggested that the electronic structure of the $P \rightarrow O$ bond more closely resembles a triple than a double bond [79].

It is possible, then, that the hydration of dipolar bonds within which p–d back-bonding exists is three-dimensional, and that the orbital geometry problem (which is so important in reducing the hydrophilicity of carbonyl groups), is not a factor. The differences in hydrophilicity that exist may, as a result, stem primarily from differences in the thermodynamic strength of hydration. This, in turn, may influence the multiplicity of hydrogen bonding.

In passing from an amine oxide to a phosphine oxide the atomic moments of the nonbonding orbitals are reduced, but their spatial arrangement may not be fundamentally altered. The result is that the phosphine oxide remains a hydrophilic group, but is less basic than is the amine oxide (by 6 pKa units), is less strongly hydrated, and is therefore less strongly hydrophilic.

Alkoxy proximate substituent effects. If the alkyl groups on a phosphine oxide $(RP(\rightarrow O)(CH_3)_2)$ are replaced by alkoxy (as in alkyl dimethyl phosphate esters, $ROP(\rightarrow O)(OCH_3)_2)$, a further reduction in the atomic moments of the $P \rightarrow O$ oxygen non-bonding orbitals may be expected. This is because the electron withdrawing properties of oxygen are stronger than are those of carbon. Because of this, the phosphate ester group should be substantially less hydrophilic than the phosphine oxide – as is observed. It is amazing that hydrophilicity exists at all in phosphate esters, considering that their hydrogen bonding strength (phenol hydrogen bond 5.7 kcal, or 28 kJ) is comparable to or less than is that of carbonyl groups which are not hydrophilic. These comparisons illustrate with particular clarity the importance of orbital geometry to hydrophilicity.

A parallel effect is seen in comparing an alkane sulfonate $(-SO_3^-)$ salt with an alkyl sulfate $(-OSO_3^-)$ salt. These two anionic groups differ structurally in the same way as do phosphine oxides and phosphate esters. In both comparisons, the alkoxy (sulfate) derivative is less hydrophilic than is the alkyl (sulfonate) derivative (Sections 10.5.1, 10.5.3.1).

Effect of the formal charge on the heteroatom. In molecules where the electron-withdrawing properties of the positive atom are enhanced by the presence of a multiple formal charge (as in sulfones and dialkyl sulfates, where this charge is +2, Section 10.5.4.1), hydrophilicity is simply not observed. The reason for this may be that the extent of back-bonding is so strongly enhanced that the hydration thermo-dynamics are reduced below the critical level (see above). If that were to happen, hydrophilicity should vanish. If the formal charge on sulfur in the sulfone (+2) is reduced to +1 (as in the sulfoxide) the extent of back-bonding is reduced, the basicity is increased by 8–10 pKa units, the hydration is strengthened, and hydrophilicity is restored.

Dipolar groups having back-bonding from sulfur. In phosphine sulfides $(R_3P \rightarrow S)$ backbonding exists between filled 3p orbitals on sulfur and empty 3d orbitals on phosphorus, and in arsine sulfides $(R_3As \rightarrow S)$ 4d orbitals on arsenic are involved. The electronic structures of these groups are highly uncertain, but neither is likely to possess a structure resembling a carbonyl group. Both have large reduced dipole moments, but neither the thermodynamics nor the multiplicity of hydrogen bonding to these groups is likely to be favorable for hydrophilicity. A size problem may exist as well, since the nonbonding orbitals on sulfur are in the third electron shell and this atom is significantly larger than is oxygen.

10.6.5 Summary

The hydrophilicity model proposed in Section 10.6.2 allows an enormous amount of phase data (and other kinds of information related to hydrophilicity) to be rationalized in a credible manner. The success of this analysis lends support for the model, but does not firmly establish its validity. A rigorous test will require two additional kinds of information:

(1) quantitative determinations of hydrophilicity for representative structures, and
(2) incisive analysis of the electronic structure of both operative and non-operative functional groups, especially as regards the properties of non-bonding electron pairs.

Data of the first kind can (in principle) be obtained by Born cycle analysis of phase behavior in a system whose phase diagram is fully known, but such an analysis has not been attempted. The more refined analysis of electronic structure is presently feasible considering the advanced state of theoretical chemistry that presently exists, but this too has not so far been attempted. The ideas presented above may be viewed as a qualitative survey of the subject, which may possibly serve as a point of departure for more refined quantitative analyses of hydrophilicity.

Reassuring thermodynamic data have been obtained from studies of the solution thermodynamics of amine, phosphine, and arsine oxides [80]. The thermodynamics of micellization within dilute solutions of these compounds is consistent with the ranking of relative hydrophilicity proposed above.

10.7 The structural evolution of hydrophilicity

The correlations between molecular structure, physical data, and hydrophilicity presented above leave little doubt that both the thermodynamics of hydration and the structure of water are relevant to hydrophilicity. When hydration thermodynamics are overpowering the compatibility of a particular group with water structure is not terribly important, but when they are not the degree to which the geometry of hydration fits the structure of water is critically important.

The simplest surfactant imaginable, in terms of compatibility with water structure, is an amphiphilic derivative of water itself – a fatty alcohol. Fatty alcohols do not display surfactant behavior because the hydroxy group is too weakly hydrated (Sections 10.6.2, 10.6.4). (It is interesting, however, that the phase behavior of alcohols in ternary systems provides evidence of a substantial degree of hydrophilicity in combination with other surfactants (Sections 12.4, 12.6.1).) To increase their hydrophilicity, structural modifications of the hydroxy group must be introduced which strengthen the thermodynamics of hydration – without destroying its compatibility with water structure. The structural evolution of hydrophilic groups may readily be perceived by considering the different ways in which this task can be accomplished.

Transformation into the amino group. One way of strengthening the hydration of the hydroxy group is to transform it into the isoelectronic primary amino group. This particular isoelectronic group is indeed more hydrophilic than is the hydroxy group, while the other isoelectronic analogs (fluoro, methyl) are less hydrophilic. These two groups – the OH and the NH_2 – may therefore be regarded as the stem hydrophilic groups from which all others are derived. However, the hydration thermodynamics of both groups must be further strengthened by suitable structural modification before strong hydrophilicity results.

Introducing a formal charge. Another way in which hydration may be strengthened is to introduced a positive or a negative charge. This may be accomplished by reacting the hydroxy and amino groups with strong acids or bases to form salts. Reacting an alcohol with an acid leads to an oxonium salt, and reacting an amine with an acid leads to an ammonium salt. Reacting an alcohol with a strong base produces an alkoxide salt, and reacting an amine with a strong base leads to an *N*-alkylamide salt:

Stem molecule	Amphiphilic derivative	Amphiphilic derivative plus proton	Amphiphilic derivative minus proton
OH_2	Alkanol, ROH	Oxonium salt, ROH_2^+,X^-	Alkoxide salt, RO^-,M^+
NH_3	Alkylamine, RNH_2	Ammonium salt, RNH_3^+,X^-	N-Alkyl amide, RNH^-,M^+

Of these four compounds, only one (the ammonium salt) is stable in water. The oxonium, the alkoxide, and the alkamide salts are all so strongly acidic or basic that they are chemically unstable in water. Thus, the ammonium salt and its derivatives may be used directly, but the structures of the other salts must be modified still further so as to attenuate their chemical reactivity towards water, if useful hydrophilic groups are to result.

10.7.1   The consequences of proton gain

The formation of an oxonium salt by reacting an alcohol with an acid appears to be a dead end, insofar as the evolution of hydrophilic groups is concerned. The pKa values of protonated alcohols are about −6 [81], which signifies that an alcohol is half-protonated in c. 72% sulfuric acid. Because of their low basicity, they are not protonated to a stoichiometrically significant degree in aqueous media at measurable pHs. Moreover, oxonium (and sulfonium) salts having nucleophilic anions (such as chloride) are not highly stable chemically [82,83].

Salts of primary amines (RNH_3^+,X^-), in striking contrast, are stable, isolable compounds which are weakly acidic in water. Alkylammonium salts may be regarded as the parent structure of the entire cationic subclass of surfactant hydrophilic groups. All nitrogen-based cationic surfactants (including quaternary ammonium compounds, which no longer possess the acidity of the parent compound) may be viewed as substitution derivatives of this parent structure. Also, substitution of alkylidine groups (divalent R = radicals) onto nitrogen yields cationic surfactants having three-coordinate nitrogen, such as pyridinium and imidazolinium compounds.

Phosphonium compounds may also be regarded as analogs of ammonium compounds as well as derivatives of primary phosphines (RPH_2). Alkyl substitution on phosphine (which is extremely weakly basic) has a dramatic effect on the basicity of the phosphorus. Phosphine itself and primary phosphines are both extremely weak bases and virtually non-basic in water [84], but dialkylphosphines are more basic and tertiary phosphines are almost as basic as are tertiary amines. Quaternary phosphonium salts and phosphonio zwitterionics are chemically stable to both water and air, however, and may be strongly hydrophilic (depending on the structure of their anionic partner). They are somewhat more lipophilic than are ammonium ions (Section 10.5.3).

10.7.2   The consequences of proton loss

The formation of an alkoxide salt from an alcohol by reaction with a strong base is highly significant with respect to the evolution of hydrophilic group structure. The pKas of the lower straight chain alcohols are about 15, which means that they are only slightly less acidic than is water itself [85]. In strongly basic aqueous solutions of the lower water-soluble alcohols, a perceptible fraction of the alkoxide species exists. If a long-chain alkoxide salt is mixed with water, however, the hydrolysis reaction

$$RO^-,M^+ + H_2O \rightarrow ROH + HO^-,M^+$$

goes essentially to completion. More surprisingly, the same is true of long-chain phenoxide salts (Fig. 10.30). The pKa values of soluble alkylphenols fall within the range of 9–10, and they are extensively ionized in strongly basic solutions. Nevertheless, salts of long-chain alkylphenols undergo severe hydrolysis even in strongly basic aqueous solutions.

The reactivity towards hydrolysis of long-chain alkoxides and phenoxides is probably exaggerated by the insolubility of the hydrolysis product. The equilibrium constant for the hydrolysis reaction is the product of the hydrolysis constant (in solution) and the reciprocal of the solubility product, when precipitation of the product occurs:

$$
\begin{array}{ll}
RO^-,M^+ + H_2O \rightleftarrows ROH(\text{soln}) + HO^-,M^+ & K_h \\
ROH(\text{soln}) \qquad\quad \rightleftarrows ROH(\text{ppt}) & 1/K_{sp} \\
\hline
RO^-,M^+ + H_2O \rightleftarrows ROH(\text{ppt}) + HO^-,M^+ & K_{eq} \\
\quad K_{eq} = K_h/K_{sp}
\end{array}
$$

These two factors, working together, drive these hydrolysis reactions to a higher degree of completion than occurs with soluble homologs. A similar result will be found for salts of alkylamides (RNH^-,M^+), which are far more basic than are either alkoxides or phenoxides. (The pKa of ammonia as an acid is about 33, which signifies that it is perhaps 18 pKa units less acidic than is water.)

10.7.2.1 Influence of the basicity of anionic surfactants on phase behavior

The acidities of β-diketones may be varied over a wide range by altering the electron-withdrawing groups within these molecules. In 1964, a number of such molecules having a wide range of acidities were prepared and their pKa values determined (if not already known) [86]. In addition, long-chain homologs of several of these compounds (and of conventional acids) were prepared (together with their sodium salts), and the aqueous phase behavior of the salts was scanned. Three different kinds of behavior were observed:

(1) Salts of acids with pKa > 8 did not display surfactant-like phase behavior at all. Such severe hydrolysis occurred that the micellar solutions and liquid crystal phases characteristic of surfactant behavior were not observed.
(2) Salts of acids with pKa < 4 displayed normal surfactant phase behavior, without evidence of complications due to hydrolysis. Liquid crystal and micellar solution phases were shown to exist, no evidence for hydrolysis reactions was found, and violations of the Phase Rule for binary systems were not observed.
(3) Salts of acids whose pKas fell between 4 and 8 displayed complex, soap-like, phase behavior. Liquid crystal phases and concentrated micellar solutions

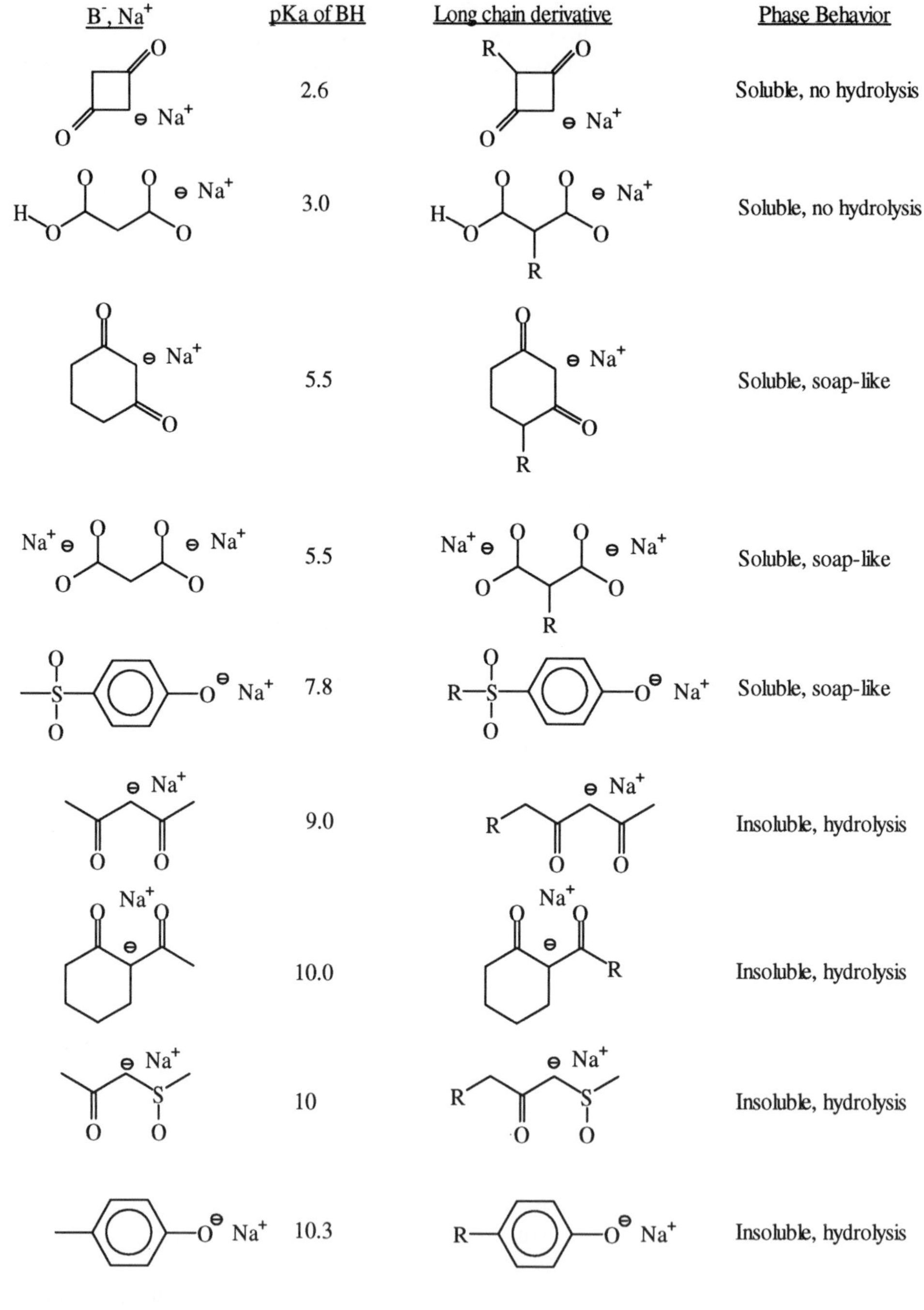

B⁻, Na⁺	pKa of BH	Long chain derivative	Phase Behavior
	2.6		Soluble, no hydrolysis
	3.0		Soluble, no hydrolysis
	5.5		Soluble, soap-like
	5.5		Soluble, soap-like
	7.8		Soluble, soap-like
	9.0		Insoluble, hydrolysis
	10.0		Insoluble, hydrolysis
	10		Insoluble, hydrolysis
	10.3		Insoluble, hydrolysis

Fig. 10.30. The structures of sodium salts of various amphiphilic compounds, the pKas of their methyl homologs, and the phase behavior of amphiphilic homologs.

were indeed found, but hydrolysis phenomena were also observed. The existence of significant hydrolysis was suggested by the form of the Krafft boundary which, in dilute mixtures within the plateau composition range, displayed a negative slope. A negative slope for this boundary is a violation of the Phase Rule, because the tie-lines that normally would link the coexisting liquid and crystal phases must pass through a second liquid region when this is true (Section 5.4.2).

The soaps behave in a qualitatively similar manner. It should be recognized that the published binary phase diagrams of soap–water systems (especially in dilute mixtures) are found only if sufficient base is present to suppress hydrolysis (*c*. 0.01 M). This was first noted in conversations with O. T. Quimby, and was later verified experimentally. Mixtures of pure soap and pure water should be analyzed as a ternary rather than a binary system, strictly speaking, but in studies limited to highly concentrated mixtures this complication is not strongly evident.

10.7.2.2 Attenuating the basicity of anionic groups

From the influence of the basicity of surfactants on their hydrolysis chemistry, it is apparent that the basicity of a hydrophilic group must be controlled else chemical hydrolysis intrudes and profoundly alters the physical science of the system. Basicity may be reduced by the formal insertion of an electronegative moiety between the carbon and the oxygen of the alkoxide group, and most hydrophilic groups can be viewed as having been derived in this manner (Fig. 10.31).

In the first four lines of Fig. 10.31, neutral electronegative groupings are inserted. These insertions produce many of the familiar anionic hydrophilic groups. The remaining groups (not shown) may be viewed as derivatives of these four groups which result from altering the structure of the attached *R* substituents. (For example, exchanging the alkyl group attached to a sulfonate anion with an alkoxy group yields an alkyl sulfate anion. Similarly, exchanging an alkoxy group for the alkyl group on the phosphonate anion yields the various known phosphorus anions.)

In the next five lines (Fig. 10.31), the moiety that is inserted is positive. If the charge on this moiety is +1, the resulting structure is electrically neutral. The semipolar hydrophilic groups may be regarded as having evolved in this manner. The same formal processes may be applied to the alkylamide anion so as to produce the *N*-terminal sulfoximine and sulfone diimine groups. If the charge on the group that is inserted is +2, the resulting functional group is cationic. The sulfoxonium ion is derived in this way.

Summary. The above model for hydrophilicity would suggest that the ultimate hydrophilic group is an alkoxide salt – except for the fact that this group is far too basic to exist in mixtures with water. Nevertheless, most actual hydrophilic groups may be regarded as structural derivatives of alkoxides whose basicity has been attenuated, in varying degree, by suitable structural modifications. A few hydrolytically stable groups are derived similarly from the *N*-alkylamide anion.

Hydroxy and ether groups may be viewed as derivatives of water itself, and lie at the low end of the hydrophilicity scale. In order for these compounds to display hydrophilicity, more than one such group must exist within the molecule.

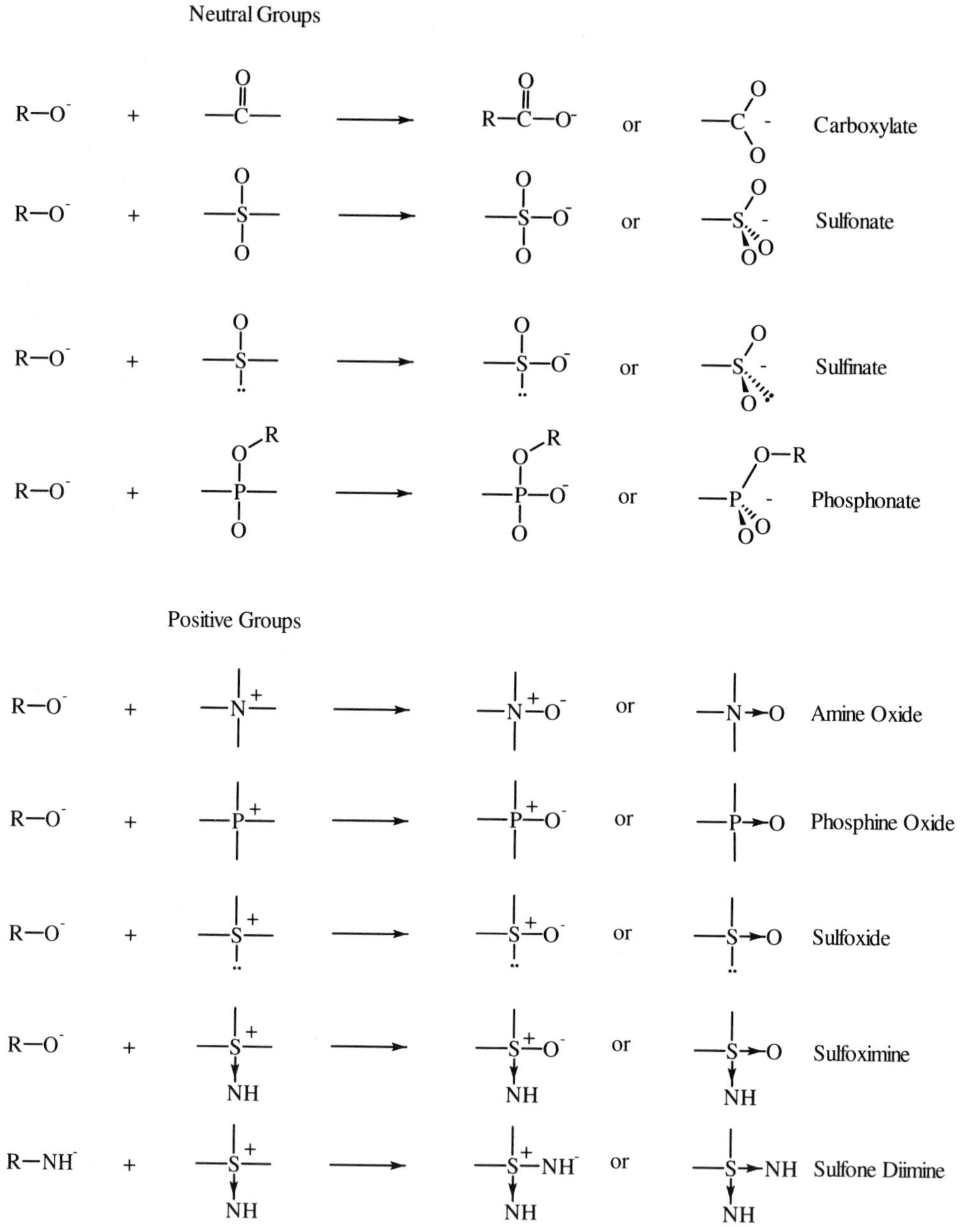

Neutral Groups

R—O⁻ + (C=O) ⟶ R—C(=O)—O⁻ or —C(=O)(O⁻) Carboxylate

R—O⁻ + (S with two O) ⟶ —S(=O)—O⁻ or —S(O⁻) Sulfonate

R—O⁻ + (S with O) ⟶ —S(=O)—O⁻ or —S(O⁻) Sulfinate

R—O⁻ + (P with O—R) ⟶ —P—O⁻ or —P(O—R)(O⁻) Phosphonate

Positive Groups

R—O⁻ + —N⁺— ⟶ —N⁺—O⁻ or —N→O Amine Oxide

R—O⁻ + —P⁺— ⟶ —P⁺—O⁻ or —P→O Phosphine Oxide

R—O⁻ + —S⁺— ⟶ —S⁺—O⁻ or —S→O Sulfoxide

R—O⁻ + —S⁺— (NH) ⟶ —S⁺—O⁻ (NH) or —S→O (NH) Sulfoximine

R—NH⁻ + —S⁺— (NH) ⟶ —S⁺—NH⁻ (NH) or —S→NH (NH) Sulfone Diimine

Fig. 10.31. Attenuation of the basicity of an alkoxide group by the insertion of neutral or positive electronegative groups into the C–O⁻ bond.

10.7.3 The limits to hydrophilicity

It is generally recognized, in considering the strength of acids and bases, that solvents exert a "leveling effect" on the range of acidity and basicity that can be observed [87]. A solvent like water may react as a base with an acid (HA), to form a hydronium salt H_3O^+, A^-. Up to a certain acidity of HA this reaction does not proceed to completion, and the equilibrium constant which governs the extent to which it does occur can be determined. When the acidity of HA exceeds a limiting value, however, the reaction to form the hydronium salt is complete and the strength of the acid can no longer be measured. An interesting example of this behavior is found in the crystalline hydrate of toluenesulfonic acid, which has been shown to be (in fact) the ionic salt, hydronium tosylate [88]. The same reaction likely occurs between dodecanesulfonic acid and water (Section 10.5.1).

Similarly, water reacts as an acid with bases (B) to form, in varying degree, hydroxide salts BH^+,OH^-. Up to a certain basicity of B this reaction also is incomplete and the extent to which it occurs can be determined. Beyond this limit the proton transfer process is complete and the strength of the base can no longer be discerned.

The upper limit of hydrophilicity. Water also exerts a leveling effect on hydrophilicity at its upper limit. The effect is qualitatively similar to its leveling effect on basicity, in that it too results from the intrusion of hydrolysis chemistry. If one starts with a functional group of moderate hydrophilicity (and basicity) and alters its structure so as to increase its hydrophilicity, this may proceed only up to a point. Past this point, hydrolysis chemistry intervenes and the structure no longer exists in water.

This phenomenon may be illustrated by the behavior in water of amine oxides ($R_3N \rightarrow O$) and hydrazinium salts ($R_3NNH_2^+,X^-$). The amine oxide (N $\rightarrow$ O) group is strongly hydrophilic, and one might expect that its isoelectronic nitrogen analog ($R_3N \rightarrow NH$) should be significantly more strongly hydrophilic. The amine oxide is only weakly basic in water, however, while its nitrogen analog is quantitatively hydrolyzed to the hydrazinium hydroxide salt:

$$R_3N \rightarrow NH + H_2O \rightleftarrows R_3N-NH_2^+ + OH^-$$

This is evident from the fact that hydrazinium salts are not measurably acidic in water [89]. In striking contrast, the hydrolysis of the amine oxide

$$R_3N \rightarrow O + H_2O \rightleftarrows R_3N-OH^+ + OH^-$$

proceeds to a moderate extent, the pKa of the R_3N-OH^+ species may be measured, and stable salts of amine oxides may be reversibly formed by reaction of amine oxides with acids [90]. The nitrogen analog is so basic that it reacts quantitatively with water to from the hydrolysis product, a hydrazinium salt. (The hydrazinium salt is, of course, a cationic hydrophilic group.)

The lower limit of hydrophilicity. The lower limit of hydrophilicity is dictated by different factors from those which dictate the upper limit. If one starts with a moderately hydrophilic functional group and reduces the thermodynamic strength of its hydration, hydrophilicity vanishes beyond a certain point. An example of this is found in comparing sulfoxides with sulfones. Sulfones are not hydrophilic, probably because they are simply too weakly hydrated. Hydrolysis chemistry does not play a role in defining the lower limit of hydrophilicity. Instead, the lower limit results from the inability of the group to interact sufficiently strongly with water to be operative.

Summary. Both lower *and* upper limits of hydrophilicity exist. The lower limit is determined by the requirement that a minimal thermodynamic strength of hydration must exist. The upper limit results from the intervention of chemical hydrolysis, and is related to the leveling effect of water on the strength of bases.

10.8   Phase behavior as a sensitive indicator of hydrophilicity

Emphasis to this point has been placed on the utilization of phase information to infer the existence of hydrophilicity, and the relative hydrophilicity of different functional groups. Having determined these relationships, it is worth considering the possibility that they provide information regarding a fundamental property of polar functional groups, and that this information should apply wherever these groups exist in aqueous media. For example, if it is true that a carboxylate anion is far more hydrophilic than is a sulfate ester anion with respect to the phase behavior of soaps and alkyl sulfate salts, then this difference should also exist with respect to their behavior within protein molecules, when these groups are components of bacterial cell walls (which contain chondroitin sulfate and carboxylate groups), and elsewhere. The situation as a whole must be considered (just as above), but the concept of hydrophilicity is surely not restricted to consideration of the phase behavior of surfactants.

It seems likely that phase information constitutes an extremely delicate "sensor" for the subtleties of hydrophilicity interactions. This sensor includes all the components of the thermodynamics of interaction – entropic as well as enthalpic. When real differences in hydrophilicity exist, there may be no better means to recognize and experimentally document these differences. Very real differences must exist in the hydration of a sulfate and a sulfonate ion exchange resin, for example, and these differences may be important with respect to selected properties of the resin (such as its swelling in water, which influences the kinetics of ion exchange). The above hydrophilicity data should be relevant wherever hydration is an important aspect of the property of interest.

The reason why phase behavior may serve as a particularly sensitive means of detecting subtle differences in hydrophilicity lies in the extremely small numerical values of the energies that accompany most phase transformations. The largest differences in thermodynamics of state that are encountered exist between crystal and fluid phases, but even these are small in comparison to the energetics of most chemical processes. The differences that exist between other states (coexisting liquid phases, for example) can be extremely small. It has been experimentally documented that some isoplethal phase discontinuities are, under some circumstances, calorimetrically invisible [91]. When two phases differ in composition by a fraction of a per cent, the free energy differences between them cannot possibly be very large. Nevertheless, such phase phenomena are very real, they are highly reproducible, and they correlate nicely with molecular structure via the hydrophilicity concept.

References

1. Laughlin, R. G. (1978). *Advances in Liquid Crystals*, Vol. 3 (G. H. Brown ed.), pp. 99–148, Academic Press, New York.
2. R. J. Blickenstaff, unreported work.
3. D. E. O'Connor, unreported work.
4. I. J. Goldstein (University of Michigan), unreported work.
5. Laughlin, R. G. (1990). *Cationic Surfactants Physical Chemistry*, 2nd ed., Vol. 37, pp. 1–40, Marcel Dekker, New York.
6. Dewar, M. J. S. (1969). *The Molecular Orbital Theory of Organic Chemistry*, pp. 463–468, McGraw-Hill, New York.

7. Joesten, M. D. and Schaad, L. J. (1974). *Hydrogen Bonding*, pp. 195–254, Marcel Dekker, New York.
8. Rochester, C. H. (1970). *Acidity Functions*, pp. 22–71, Academic Press, London.
9. Janata, J. and Jansen, G. (1972). *J. Chem. Soc., Faraday Trans.* **1**, 1656–1665.
10. Haake, P., Cook, R. D. and Hurst, G. H. (1968). *J. Am. Chem. Soc.* **89**, 2650–2654.
11. Kolp, D. G., Laughlin, R. G., Krauss, F. P. and Zimmerer, R. E. (1963). *J. Phys. Chem.* **67**, 51–55.
12. Ekwall, P. (1975). *Advances in Liquid Crystals*, Vol. 1 (G. H. Brown ed.), pp. 1–142, Academic Press, New York.
13. Fontell, K. (1990). *Colloid Polym. Sci.* **268**, 264–285.
14. Vold, M. J. (1941). *J. Am. Chem. Soc.* **63**, 1427–1432.
15. Kahn, A., Fontell, K. and Lindman, B. (1984). *J. Colloid Interface Sci.* **101**, 193–200.
16. Kahn, A., Fontell, K. and Lindman, B. (1985). *Prog. Colloid Polym. Sci.* **70**, 30–33.
17. Burkitt, S. J., Ingram, B. T. and Ottewill, R. H. (1988). *Prog. Colloid Polym. Sci.* **76**, 247–250.
18. Weil, J. K., Bistline, R. G., Jr and Stirton, A. J. (1957). *J. Am. Oil Chemists' Soc.* **34**, 100–103.
19. Jansson, M., Jonsson, A., Li, P. and Stilbs, P. (1991). *Colloid and Surf.* **59**, 387–397.
20. Reference 1, pp. 108–112.
21. Breslow, R. (1966). *Organic Reaction Mechanisms*, pp. 16–18, W.A. Benjamin, New York.
22. Kabachnik, M. I., Mastrukova, T. A., Shipov, A. E. and Melentyeva, T. A. (1960). *Tetrahedron* **9**, 10–28.
23. Wudl, F., Lightner, D. A. and Cram, D. J. (1967). *J. Am. Chem. Soc.* **89**, 4099–4101.
24. Ralston, A. W., Hoffman, E. J., Hoerr, C. W. and Selby, W. M. (1941). *J. Am. Chem. Soc.* **63**, 1598–1601.
25. Emge, T. J., Strickland, L. C., Oliver, J. D. and Laughlin, R. G. (1989). Presented at American Crystallographic Association Meeting, Seattle, WA, 23–29 July.
26. Okuyama, K., Soboi, Y., Hirabayashi, K., Harada, A., Kumano, A., Kaziyama, T., Takayanagi, M. and Kunitake, T. (1984). *Chem. Lett.*, 2117–2120.
27. Friedman, H. L. and Krishnan, C. V. (1973). *Water: A Comprehensive Treatise. Aqueous Solutions of Simple Electrolytes*, Vol. 3 (F. Franks ed.), pp. 26–30, Plenum, New York.
28. (1982). *Stability Constants of Metal-Ion Complexes. Part A: Inorganic Ligands. IUPAC Chemical Data Series.*, Vol. 21 (E. Hoegfeldt ed.), pp 164–165, Pergamon, Oxford.
29. Scott, W. D. (1980). *Sulfur Aust. Pap. Workshop*, Meeting Date 1978. (J. R. Freney and A. J. Nicholson eds), pp. 45–52, Australian Academy of Science, Canberra, Australia.
30. Ninham, B. W., Evans, D. F. and Wei, G. J. (1983). *J. Phys. Chem.* **87**, 5020–5025.
31. Evans, D. F. and Ninham, B. W. (1986). *J. Phys. Chem.* **90**, 226–234.
32. Laughlin, R. G., Munyon, R. L., Fu, Y.-C. and Fehl, A. J. (1990). *J. Phys. Chem.* **94**, 2546–2552.
33. Laughlin, R. G. and Munyon, R. L. (1990). Presented at the International Colloid and Surface Science Symposium, Compiegne, France, July 7–13, 1990.
34. Laughlin, R. G. (1991). *Langmuir* **7**, 842–847; Tausk, R. J. M., Oudshoorn, C. and Overbeek, J. Th. G. (1974). *Biophys. Chem.* **2**, 53–63.
35. Nilsson, P-G., Lindman, B. and Laughlin, R. G. (1984). *J. Phys. Chem.* **88**, 6357–6362.
36. McGrady, J. and Laughlin, R. G. (1984). *Synthesis*, 5426–5428.
37. T. W. Gibson, unreported work. The method for preparing the ultralong chain intermediate is reported in Gibson, T. and Tulich, L. (1981). *J. Org. Chem.* **46**, 1821–1823 and Gibson, T. W. (1981). Eur. Pat. Appl. EP 21496, 7 Jan 1981.
38. Jonsson, A., Bokstrom, J., Malmvik, A-C. and Wärnheim, T. (1990). *J. Am. Oil Chemists Soc.* **67**, 733–738.
39. Chevalier, Y., Germanaud, L. and Le Perchec, P. (1988). *Colloid Polym. Sci.* **266**, 441–448.
40. Faulkner, P. G., Ward, A. J. I. and Osborne, D. W. (1989). *Langmuir* **5**, 924–926.
41. (1976). *Handbook of Biochemistry and Molecular Biology*, 3rd ed., Vol. I, pp. 157–269, CRC Press, Cleveland, Ohio.
42. Weers, J. G., Rathman, J. F., Axe, F. U., Crichlow, C. A., Foland, L. D., Scheuing, D. R., Wiersema, R. J. and Zielski, A. G. (1991). *Langmuir* **7**, 854–867.
43. Laughlin, R. G. and Stewart, R. L. U. S. Patent 3925262, December 9, 1975; Heuring, V. P. and Laughlin, R. G. U. S. Patent 3929678, December 30, 1975.

44. Cope, A. C. and Trumbull, E. R. (1960). *Organic Reactions*, Vol. 11, pp. 317–493.
45. Ali, A. A. and Mulley, B. A. (1978). *J. Pharm. Pharmacol.* **30**, 205–213.
46. Naitoh, Kouichi, Ishii, Yasuo and Tsujii, Kaoru (1991). *J. Phys. Chem.* **95**, 7915–7918.
47. Prince, L. M. (1977). *Microemulsions*, pp. 6–11, Academic Press, New York.
48. Strey, R., Schomaeker, R., Roux, D., Nallet, F. and Olsson, U. (1990). *J. Chem. Soc. Faraday Trans.* **86**, 2253–2261.
49. Laughlin, R. G. and Yellin, W. (1967). *J. Am. Chem. Soc.* **89**, 2435–2443.
50. Ralston, A. W., Hoerr, C. W. and Hoffman, E. J. (1942). *J. Am. Chem. Soc.* **64**, 1516–1523.
51. Wärnheim, T. (1986). *Coll. Polym. Sci.* **264**, 1051–1059.
52. R. G. Laughlin, unreported work.
53. Lutton, E. S. (1965). *J. Am. Oil Chemists Soc.* **42**, 1068–1070.
54. R. G. Laughlin, unpublished work.
55. Benson, F. R. (1967). *Nonionic Surfactants*, Vol. 1 (M. J. Schick ed.), pp. 247–299, Marcel Dekker, New York; Satkowski, W. B., Huang, S. K. and Liss, R. L. (1967). *Nonionic Surfactants*, Vol. 1 (M. J. Schick ed.), pp. 142–174, Marcel Dekker, New York.
56. Fu, Y. C., Glardon, A. S. and Laughlin, R. G. (1993). Presented at the American Oil Chemists' Society Meeting, Anaheim, California, April 25–29.
57. Lutton, E. S., Stewart, C. B. and Fehl, A. J. (1970). *J. Am. Oil Chem. Soc.* **47**, 94–99; Seiden, P., Lutton, E. S., Sanders, R. A. and Laughlin, R. G. (1990). Paper presented at *8th International Symposium on Surfactants in Solution*, June 10–15, Gainesville, FL.
58. Loos, M., Baaeyens-Volant, D., David, C., Sigaud, G. and Achard, M.F. (1990). *J. Colloid Interface Sci.* **138**, 128–133.
59. Seiden, P., Lutton, E. S., Sanders, R. A. and Laughlin, R. G. (1990). *8th International Symposium on Surfactants in Solution*, 10–15 June, Gainesville, FA.
60. Sjoblom, J., Stenius, P. and Danielsson, I. (1987). *Nonionic Surfactants Physical Chemistry*, Vol. 23 (Martin J. Schick ed.), pp. 387–388, Marcel Dekker, New York.
61. Garti, N., Kaufmann, V. R. and Aserin, A. (1987). *Nonionic Surfactants. Chemical Analysis*, Vol. 19 (J. Cross ed.), pp. 225–284, Marcel Dekker, New York and Basel.
62. Laughlin, R. G. (1976). *J. Colloid Interface Sci.* **55**, 239–241.
63. Reference 60, pp. 369–434.
64. Bouwstra, J. A., Jousma, H., Van der Meulen, M. M., Vijerberg, C. C., Gooris, G. S., Spies, F. and Junginger, H. E. (1989). *Colloid Polym. Sci.* **267**, 521–538.
65. T. W. Gibson, unreported work.
66. Napper, D. H. (1983). *Polymeric Stabilization of Colloidal Dispersions*, Academic Press, London; Virdun, J. W. and Berg, J. C. (1992). *J. Colloid Interface Sci.* **153**, 411–419.
67. Reference 60, pp. 372–373.
68. Gibson, T. (1980). *J. Org. Chem.* **45**, 1095–1098; Gibson, T. W. (1981). *Eur. Pat. Appl.* EP 21497, 7 Jan 1981.
69. Rempp, P. (1957). *Bull. soc. chim. France*, 844–847.
70. Mitchell, D. J., Tiddy, G. J. T., Waring, L., Bostock, T. A. and McDonald, M. P. (1983). *J. Chem. Soc., Faraday Trans. 1* **79**, 975–1000.
71. (1972). *Water: A Comprehensive Treatise. The Physics and Chemistry of Water*, Vol. 1 (F. Franks ed.), Plenum, New York.
72. Hobbs, P. V. (1974). *Ice Physics*, pp. 18–43, Clarendon, Oxford.
73. Small, D. M. (1986). *The Physical Chemistry of Lipids. From Alkanes to Phospholipids*, Vol. 4, pp. 182–223, Plenum, New York.
74. Ramadan, M. S., Evans, D. F. and Lumry, R. (1983). *J. Phys. Chem.* **87**, 4538–4543.
75. Shaw, R. W., Brill, T. B., Clifford, A. A., Eckert, C. A. and Franck, E. U. (1991). *Chem. & Eng. News*, December, 26–39.
76. Costain, C. C. and Dowling, J. M. (1960). *J. Chem. Phys.* **32**, 158–165.
77. Roberts, J. D. (1959). *Nuclear Magnetic Resonance*, pp. 69–71, McGraw–Hill, New York.
78. Cotton, F. A. and Wilkinson, G. (1988). *Advanced Inorganic Chemistry*, 5th ed., pp. 58–76, John Wiley, New York.
79. Hays, H. R. and Peterson, D. J. (1972). *Organic Phosphorus Compounds*, Vol. 3 (G. M. Kosolapoff and L. Maier eds), pp. 385–388, John Wiley, New York.
80. Perron, G., Yamashita, F., Martin, P. and Desnoyers, J. (1991). *J. Colloid Interface Sci.* **144**, 222–235.

81. Arnett, E. M. (1963). *Progress in Physical Organic Chemistry*, Vol. 1 (S. G. Cohen, A. Streitweiser, Jr and R. W. Taft eds), pp. 223–403, Interscience, John Wiley, New York.
82. Perst, H. (1976). *Carbonium Ions*, Vol. 5 (G. A. Olah and P. v. R. Schleyer eds), pp. 1961–2047, John Wiley, New York.
83. Knipe, A. C. (1981). *Chemistry of the Sulfonium Group*, (C. J. M. Stirling, ed., S. Patai, Series ed.), pp. 313–386, John Wiley, Chichester.
84. Kosolapoff, G. (1950). *Organophosphorus Compounds*, p. 24, John Wiley, New York.
85. Olmstead, W. N., Margolin, Z. and Bordwell, F. G. (1980). *J. Org. Chem.* **45**, 3295–3299.
86. W. W. Bannister, unreported work.
87. Bell, R. P. (1973). *The Proton in Chemistry*, 2nd ed., p. 49, Cornell University Press, Ithaca, NY.
88. Dexter, D. D. (1971). *Z. Kristallogr.* **134**, 350–359.
89. R. G. Laughlin, unreported work.
90. Sheeran S. R. (1966). *U. S. Patent* 3267147, Aug. 16, 1966; this patent is not abstracted by Chemical Abstracts. See also Statlioraitis, J. S. and Redl, R. A. (1969). *Ger. Offen.* DE 1923582, 27 Nov. 1969.
91. G. G. Chernik, personal communication.

Chapter 11

Lipophilicity, Remote Substituent Effects, and HLB

11.1 Introduction

Being "lipophilic" implies that a molecule (or a structural fragment of a molecule) is miscible with oils and immiscible with water. The lipophilic groups of surfactants do indeed mix with oils, but perhaps it is their reluctance to mix with water that is more important with respect to their surfactant behavior [1,2]. Structural modifications that enhance the miscibility of a lipophilic group with water reduce (and may even destroy) surfactant behavior.

In many nonsurfactant systems, incomplete miscibility stems from the fact that the energies of interaction between like molecules of one compound differ greatly from the energies of interactions between like molecules of another [3,4]. The partial miscibility of carbon tetrachloride (*a*) and tin tetrachloride (*b*), for example, has been interpreted on this basis [3]. Because *a*–*a* interactions are much weaker than are *b*–*b*, *a* and *b* tend to remain segregated and the entropy of mixing is insufficiently large for mixing to occur. This is especially true at low temperatures, but the magnitude of this factor diminishes as the temperature is increased. When a disparity in intermolecular interaction energies between like molecules is the underlying cause of partial miscibility, miscibility increases as the temperature is increased. The phase diagrams of such systems display liquid/liquid miscibility gaps having upper consolute boundaries, as a result.

The underlying cause of the poor miscibility of alkanes and water is more complex. In addition to the dispersion forces that exist between these molecules, one must also allow for the energetics of the hydrogen bonding interactions that occur between water molecules. Additional factors come into play because of the preferred directionality of hydrogen bonds [5,6]. Regardless of the underlying reasons, not only alkane molecules, but also the nonpolar alkyl structural fragment of surfactant molecules, display a strong tendency to avoid mixing with liquid water. This tendency, coupled with the natural preference of the hydrophilic group within the molecule to mix with water, underlies many aspects of the physical science of surfactants.

11.2 The extensive variation of lipophilicity

It has been noted that hydrophilicity can be varied in either an intensive or an extensive manner (Section 10.6.2), and the same is true of lipophilicity. Lipophilicity is most commonly varied by changing the number of methylene groups, which represents the extensive mode of variation. However, the intrinsic nature of lipophilic groups may also be varied (as in perfluorocarbon or silicone surfactants, Section 11.5) and this represents the intensive mode. Most commercially and biologically important surfactants contain alkyl or alkylaryl lipophilic groups, so that the extensive variation of lipophilicity within these particular groups (by inserting or removing methylenes) is especially important.

The "lipophilicity" of lipophilic groups may be quantified in thermodynamic units, and is generally assumed to be proportional to their volume [1]. (Awkward assumptions are necessary in estimating their volume (Section 8.8.3), but these estimates are nevertheless quite useful.) It has been firmly established that the incremental change in free energy which occurs when a methylene group is transferred from a lipophilic to an aqueous environment is nearly constant, provided one stays within a particular homologous series of straight chain molecules [7]. This relationship holds over the entire range of chain-lengths commonly encountered in surfactants (C_8–C_{18}), and is especially useful in correlating the critical micelle concentrations (cmcs) of surfactants with their chain length. Since free energies are proportional to the logarithm of activities (Section 3.10) and activities in dilute ideal solutions are related to concentrations, the logarithms of cmcs vary inversely (and linearly) with the number of methylene groups in the lipophilic chain, n

$$\log_{10}(\mathrm{cmc}) = -k_{\mathrm{cmc}} \times n + C \qquad (11.1)$$

Numerous correlations between cmcs and chain-length have been determined [7]. The value of the constant C in eq. 11.1 depends on both the hydrophilic group structure and the way the number of carbons in the chain is defined. Therefore, while C is strongly influenced by the hydrophilic group, it remains in some degree an empirical constant. k_{cmc} is related to the above-mentioned incremental free energy change, but its value too is *not* dependent solely on the lipophilic group volume; it also depends on which hydrophilic group is present. The numerical value of k_{cmc} varies between about 0.35 and 0.5, so that for each two-carbon increment cmcs change by a multiplicative factor that ranges between 5 and 10.

The numerical values for the standard free energies of micellization for the C_6 through C_9 homologs of methyl alkyl sulfoxides and dimethylalkylphosphine oxides have been directly and accurately measured. This parameter (which is related to k_{cmc}) is perceptibly different even for hydrophilic groups as similar as are these; it is $-3.28\,\mathrm{kJ}$ per CH_2 for the sulfoxides and $-3.04\,\mathrm{kJ}$ per CH_2 for the phosphine oxides [8]. During these studies the mixtures were rigorously analyzed using the multiple equilibrium model for association in solution, rather than by using the cmc approximation.

The solubilities in water of alkanes [9] and alkanols [10,11] may be correlated with chain length using a similar algorithm, but the values of the slope (k) for these correlations fall outside the range of values found for k_{cmc}. In the case of liquid straight chain alkanes

$$\log_{10}(\mathrm{Soly}_{RH}) = -k_{RH} \times n + C_{RH} \qquad (11.2)$$
$$k_{RH} = 0.6473 \qquad C_{RH} = -0.0433$$

and for liquid straight chain alcohols

$$\log_{10}(\text{Soly}_{ROH}) = -k_{ROH} \times n + C_{ROH} \tag{11.3}$$
$$k_{ROH} = 0.604 \qquad C_{ROH} = 2.402$$

When the phase that coexists with the saturated solution is crystalline rather than liquid these correlations can no longer be expected to remain quantitatively valid. The magnitude of the deviation introduced by this change of state has not been determined.

The k which describes the dependence of the solubilities of alkanes on chain-length (k_{RH}) is the highest that is known, that which describes the solubility of alkanols (k_{ROH}) is somewhat smaller, and all of the ks which describe surfactant cmcs (k_{cmc}) are smaller still. These data indicate that the free energy gain per methylene group is larger for removing a hydrocarbon from water (to form an alkane phase) than it is for removing an alkanol from water (to form an alkanol phase), which in turn is larger than is that for the "removal" of a surfactant lipophilic group from water (to form micelles within the aqueous liquid phase). The conclusion that the lipophilicity of a compound is measurably influenced by its hydrophilicity is inescapable.

11.2.1   Lipophilicity, surface activity, and phase equilibria

Surfactants get their name from contraction of the phrase "surface active agent". The determination of surface activity may be based on surface tension data at the air–water interface; a reduction in surface tension that is caused by the presence of a solute signifies that the solute is surface active. For this reason, a particularly useful measure of surface activity is "surface pressure". Surface pressure is defined by the equation

$$\pi = \gamma_W - \gamma_{soln} \tag{11.4}$$

where π is the surface pressure, γ_W is the surface tension of water, and γ_{soln} is the surface tension of the solution. The numerical value of the surface pressure is a function of solute concentration (*if* composition is a degree of freedom), and often displays a plateau above a certain concentration. A useful measure of the efficiency with which a particular compound lowers surface tension is the concentration at which this plateau is attained, which we will term C_{min}.

Comparisons of long chain and nonlong chain compounds. It has long been evident, using the criterion of reduced surface tension or surface pressure, that surfactants are highly surface active in comparison to compounds that retain the hydrophilic group but lack the lipophilic group. Examples of such comparisons include sodium soaps vs. sodium formate, sodium alkyl sulfates vs. sodium methyl sulfate, alkyldimethylamine oxides vs. trimethylamine oxide, etc. In all cases the surface pressure of the surfactant solution at its cmc concentration is very large, in comparison to the surface pressure of solutions of the corresponding nonlong chain compound at the same molar concentration.

Comparisons of surfactants with other long-chain compounds. It is also of interest to compare surfactants with other long-chain compounds which either lack a polar functional group altogether, or which possess functional groups that are not operative hydrophilic groups (Section 9.5). Such comparisons can be based both on the numerical value of the maximum surface pressure attainable, and also on C_{min}. Table 11.1 provides π and C_{min} data for a variety of amphiphilic octyl derivatives.

For the alkane and the alcohol the plateau in tension is presumed to occur at the solubility boundary, while for the surfactant compounds the plateau concentration is taken to be the cmc (but see later). From consideration of the maximum attainable surface pressure, it can be seen that all the surfactants are indeed highly surface active. The magnitude of their surface activity varies significantly; the larger surface pressures are produced by the less hydrophilic surfactants.

C_{min} increases smoothly with the hydrophilicity of the polar functional group, without regard for whether or not surfactant phase behavior exists. While the alkane (which lacks a hydrophilic group altogether) is the least surface active of the above compounds in terms of the maximum surface pressure (π) criterion, its C_{min} occurs at extremely low concentrations.

Octanol (which is also not a surfactant) displays a higher level of surface activity than does octane in terms of surface pressure (comparable to that of many surfactants), but its value of C_{min} is much higher than is that of octane. C_{min} for the alkanol is lower, however, than are those of any of the surfactant compounds. It is reassuring to find that the relative cmcs of the various amphiphilic compounds fall in an order that is fully consistent with their relative hydrophilicities as inferred from other data (Section 10.5).

Phase considerations. Discontinuities of state should impose discontinuities on the surface properties of mixtures as well, and the existence of these discontinuities should be recognized. If, for example, a mixture of octane and water were to be equilibrated at 25°C in a closed container, an octane-saturated liquid water-rich phase, a water-saturated liquid octane-rich phase, and the vapor phase may coexist:

$$\text{Liq Water(octane)} \leftrightarrows \text{Liq Octane(water)} \leftrightarrows \text{Vapor(water and octane)}$$

The condition of equilibrium of state requires that the chemical potential and activity of octane must be the same in all three phases (Section 4.17.1). The activity of octane is related to its partial pressure, which for this situation is virtually identical to that of

TABLE 11.1
Surface activity of octyl compounds

Compound	C_{min} (minimum concentration (M) required to achieve maximum surface pressure)	Maximum surface pressure, π (mN m^{-2})
$C_8H_{17}-H^*$	6.0×10^{-6}	1 [12] – 20 [13]
$C_8H_{17}-OH$	3.7×10^{-3}	37 [14]
$C_8H_{17}-E_6OH$	9.9×10^{-3}	— [15]
$C_8H_{17}-S(\rightarrow O)Me$	2.3×10^{-2}	46 [16]
$C_8H_{17}-P(\rightarrow O)Me_2$	4.1×10^{-2}	— [17]
$C_8H_{17}-N(\rightarrow O)Me_2$	1.5×10^{-1}	— [18]
$C_8H_{17}-OSO_3^-,Na^+$	1.3×10^{-1}	38 [19]
$C_8H_{17}-CO_2^-,Na^+$	1.6×10^{-1}	— [20]
$C_8H_{17}-N(Me)_3^-,Br^-$	2.8×10^{-1}	37 [19]

*In a careful study of the tensions at the octane–water interface [13] it was found that the maximum surface pressure that could be achieved was about 1 mN m^{-2}, but the interfacial tension between liquid octane and water phases is 20 mN m^{-2} lower than is that of water [14]. Regardless of the true value, the maximum surface pressure that can be achieved by octane is less than is that which is attainable using any amphilic molecule (whether surfactant or nonsurfactant).

pure octane ($a_{oct} \approx 1$). This must be true because one of the phases present is a very dilute solution of water in octane ($x_W < 0.001$ [21]), and for this phase the octane (as the solvent component) is expected (from experience) to obey Raoult's Law [22]. The water in this octane-rich phase, on the other hand, may be expected to follow Henry's Law. The reverse situation holds in the water-rich phase, where the water is expected to obey Raoult's Law and the octane to obey Henry's Law. The Henry's Law constant, K_H, is defined by the equation

$$K_H = \frac{a_i}{x_i} \simeq \frac{p_i/p_i^0}{x_i}$$

where a_i, x_i, and p_i equal the activity, mole fraction, and partial pressure, respectively, of component i.

The mole fraction of octane in the octane-saturated water phase is 1×10^{-7} [9]. If this aqueous phase were ideal Raoult's Law would apply to both components, and the partial pressure of octane should be smaller than that of pure octane by a factor of 1×10^7. Since the partial pressure of octane is, in fact, virtually the same as that of pure octane, it follows that the Henry's Law constant, K_H, for octane in a saturated solution of octane in water equals 1×10^7. This means that in a one-phase mixture containing one molecule of octane for each ten million molecules of water, the partial pressure of octane is essentially as large as if the water were absent! The mixture behaves as though the surface of the water phase were pure octane, although no liquid octane phase exists.

The values for the Henry's Law constants of alkanes in aqueous solutions are equal, to a good approximation, to the reciprocal of their solubilities (in mole fraction units). The solubilities of liquid n-alkanes in water have been experimentally determined up to C_8 [9], and from the linear regression analysis of these data estimates of the Henry's Law constants for the higher alkanes may be calculated. These are plotted in Fig. 11.1; all of them are extremely large.

Equilibrium of state with respect to bulk phases also signifies equilibrium with respect to the interfaces between phases. If just enough octane were added to liquid water to saturate this liquid phase, the interface between the liquid and the gas phases must also be saturated with octane. Such a mixture is bivariant in general, or monovariant if either temperature or pressure is defined.

Viewing the situation from the perspective of its phase behavior thus suggests that a coincidence is to be expected in the octane compositions at which saturation of the bulk phase and saturation of the gas/liquid interface occurs. A plateau in the surface pressure may also be exected to occur at this composition, since if temperature and pressure are specified the system is invariant. This coincidence is presumably not restricted to alkane–water systems, but is a universal characteristic of mixtures.

Strictly speaking, this same correlation must hold true for surfactant solutions. Consideration of phase behavior would suggest that the surface of aqueous solutions of surfactants is fully saturated only when the solution phase is itself saturated, and that this will occur at the solubility boundary concentration. Since for insoluble surfactants micellar solutions do not exist (Sections 5.4.1, 5.6.8), one would expect the concentration at which the surface pressure plateaus to equal the solubility limit of these compounds – just as is found in nonsurfactant compounds (above).

In the case of soluble surfactants, which display micellar structure within the solution phase, the situation is more complex. For soluble surfactants a large surface pressure

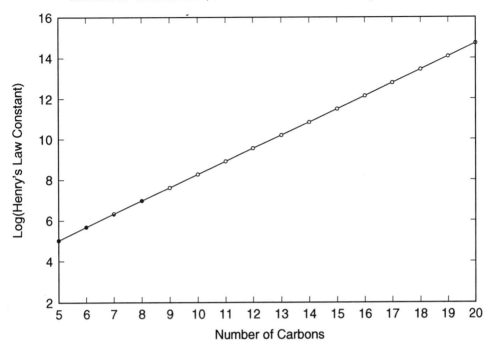

Fig. 11.1. The $\log_{10}$ of the dimensionless Henry's Law constant (defined as the ratio of activity in the vapor phase to mole fraction in the liquid phase) for the liquid *n*-alkanes at 25°C. The filled triangles are the data points of McAuliffe [9] from which the calculated values (open circles) were derived.

exists and the surface is apparently saturated (or nearly so) at the cmc, which is far below the solubility limit of these surfactants. It is to be expected (on the basis of the principles of phase science) that surface pressures should rise as the concentration in solution is increased, until the solubility limit of the surfactant in liquid water is reached. The air/liquid tension should then plateau within the composition span of the liquid/liquid crystal miscibility gap. (Other interfaces will exist whose presence must be taken into account.)

No record of experimental investigations of the tensions of surfactant solutions of soluble surfactants having been carried to the solubility limit could be found [23]. Instead, such investigations have typically focused on determining the concentration at which the abrupt plateau in surface pressure at the cmc occurs, on those concentrations which lie just below the cmc (from which the area/molecule can be obtained), and on concentrations within the plateau a sufficient distance above the cmc to define the cmc.

11.3 Structural variations among hydrocarbon lipophilic groups

11.3.1 The influence of chain-length

Extensive information regarding the dependence of phase behavior on chain-length exists for straight-chain aliphatic lipophilic groups, and the influence of this parameter on the various kinds of surfactant phase boundaries will now be considered.

The Krafft boundary. It has been firmly established that increasing the chain-length raises the temperature of the Krafft boundary, regardless of the structure of the hydrophilic group [24]. Some data which document this correlation are shown in Fig. 11.2. Since the form of this boundary is determined mainly by hydrophilic group structure (Section 5.4.2), the boundaries of different homologs within a series lie nearly parallel to one another. The plateau region of the Krafft boundary of ionic surfactants is typically more steeply sloped than is that of nonionic surfactants, and in many zwitterionic surfactant–water diagrams is nearly horizontal (Section 5.4.2).

The dependence of the temperature of the Krafft discontinuity on chain-length is not linear (Fig. 11.2). The data on sodium soaps indicate that odd–even alternation occurs – just as occurs in the melting points of many fatty crystals. (Alternation is to be expected when the chains in the crystal phase are tilted with respect to the bilayers planes, which is usually found.) Both strongly and weakly hydrophilic surfactants display Krafft boundaries that fall within the temperature span of liquid water.

The temperature of the Krafft boundary, by itself, has no fundamental significance with respect to hydrophilicity; the difference in temperature between the melting point of the dry crystal and the Krafft eutectic temperature (the ΔT parameter, Section 9.5.1) is far more significant (Section 5.11). The fact that both strongly and weakly hydrophilic surfactants display roughly similar Krafft discontinuities suggests that "compensation" exists. That is, altering the structure of the hydrophilic group so as to increase the melting point (such as enhancing the polarity or introducing formally

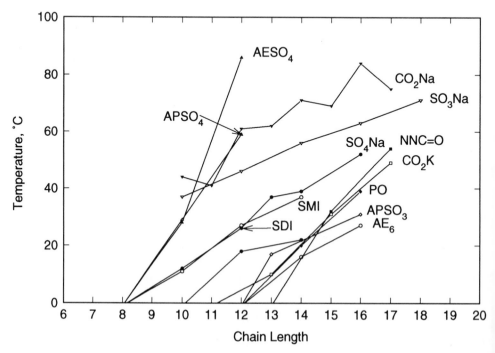

Fig. 11.2. The dependence of the Krafft boundary on lipophilic group chain-length in several families of surfactants. The acronyms for the hydrophilic groups are defined in Appendix 1.

charged substituents) also tends to increase the hydrophilicity, and therefore the magnitude of ΔT. Both the melting point and ΔT are increased, and this compensating effect of one parameter on the other attenuates the influence of hydrophilicity on the Krafft boundary.

Liquid crystal solubility at the Krafft discontinuity. As the chain-length is increased, the concentration of the dilute liquid at the Krafft discontinuity (which is also the solubility of the coexisting liquid crystal at this temperature, Section 5.4.3) slowly decreases. The solubility of a sodium soap having a C_{10} chain (sodium undecanoate) is at this point 34%, while that of a soap having a C_{17} chain (sodium stearate) is 19%. As above, the available data suggest that the dependence on chain-length is nonlinear.

It has been noted (Section 5.6.5) that the composition of the liquid crystal phase at the Krafft discontinuity provides a measure of the water free energy ($n_w\mu_w$) required to form that phase (Section 5.1). The same should be true of the liquid phase. Since the surfactant composition decreases as the molecular weight increases, the mole fraction of water increases (and probably also its chemical potential). A decrease in the solubility of the liquid crystal as the chain is lengthened thus implies that progressively greater water free energy is required for the saturated liquid phase to exist for the longer chain surfactants, which is reasonable.

Thermal stability of liquid crystal phases. Increasing the hydrophilicity clearly increases the thermal stability of liquid crystal phases (that is, the upper temperature limit of their existence), but no simple correlation exists between this parameter and lipophilicity. Again using the soap data, it can be seen from Fig. 11.3 that as the chain-length is increased (starting at C_{11}), the thermal stability of the hexagonal phase first

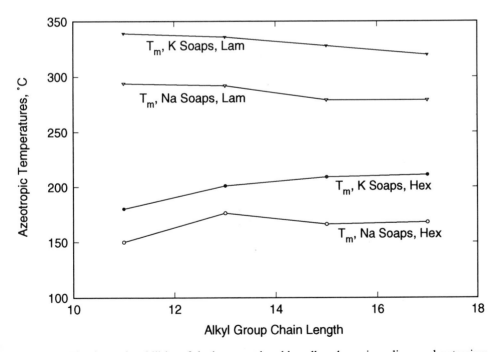

Fig. 11.3. The thermal stabilities of the hexagonal and lamellar phases in sodium and potassium soaps.

increases slightly, then slowly decreases. Increasing the chain-length slowly and steadily *decreases* the stability of the lamellar phase. It is clearly evident, from these results, that the most important parameter with respect to the thermal stability of liquid crystal phases is hydrophilicity, rather than lipophilicity.

The magnitude of ΔT. The magnitude of ΔT (Section 9.5.1) in soaps is reduced by lengthening the lipophilic chain. As seen in Table 10.1, this parameter decreases by 54°C (from 269 to 215°C) in sodium soaps with a change in the lipophilic group from 11 to 17 carbons, and by more than 79°C (from >375 to 296°C) in the potassium soaps within the same span of chain-lengths. The melting points of these compounds vary in a complex manner, perhaps (in part) because the phase reaction which occurs at the "melting point" is not a crystal $\rightleftarrows$ liquid reaction. Partial "melting" occurs at a temperature below that at which the liquid exists, which leads to dry liquid crystal phase states.

The lower consolute boundary. Increasing the chain length of a surfactant which has a lower consolute boundary enlarges the area of this miscibility gap, and thereby lowers its critical temperature (Section 10.2.1). Data which support this correlation are shown in Fig. 11.4. Data are shown only for specific chain-lengths; at lower chain-lengths than those shown this miscibility gap does not usually exist, while at higher chain-lengths the gap is obscured by interference from another phase boundary. Very often this interference is due to insolubility of those liquid crystal phases which lie at higher compositions (monoglyceride-like behavior), but some-times it is due to pivotal or monoglyceride-like phase behavior (Section 5.10.7). To illustrate, in alkyl methyl sulfone diimines this miscibility gap does not exist in the C_{10} homolog below 100°C, it does

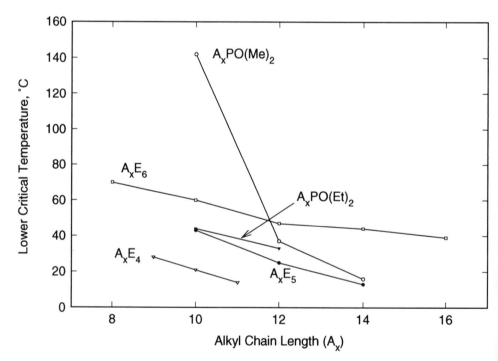

Fig. 11.4. The lower critical temperatures of homologous series of several surfactant–water systems.

exist in the C_{12} homolog (lower critical temperature 27°C), and it is obscured in the C_{14} homolog by the insolubility of the liquid crystal.

In alkyl methyl sulfoximines the miscibility gap is found in the C_8 homolog (lower critical temperature 37°C) but is obscured in higher homologs by the insolubility of the liquid crystal. The observed differences between these surfactants are consistent with their presumed relative hydrophilicities (Section 10.5.4.3).

In the alkyldimethylphosphine oxides the lower critical temperature for the C_{10} homolog is 142°C, while for the C_{12} it is 37°C. In the C_{14} homolog the critical composition region is obscured by interference from the Krafft discontinuity. Pivotal phase behavior is displayed by the C_{14} system (Section 5.10.7).

The upper consolute boundary. Increasing the chain length of a surfactant which has an upper consolute boundary enlarges the area of this miscibility gap (just as in the case of the lower consolute boundary), and therefore *raises* the upper critical temperature [25,26]. Because the location of this feature is very sensitive to hydrophilic group structure, and because other phase boundaries often interfere (Section 10.5.3.1), it is usually impossible to compare the critical temperatures of the same homolog for surfactants having different hydrophilic groups.

The data presented in Fig. 11.5 illustrate the progressive increase in the upper critical temperature that occurs as the chain length is increased. These data also illustrate the large difference in the area of this gap between ammonio sulfonate and ammonio sulfate surfactants (Section 10.5.3.1). The absence of the liquid/liquid miscibility gap phenomenon in ammoniocarboxylates (even at exceptionally long chain lengths) has been noted (Section 10.5.3.1).

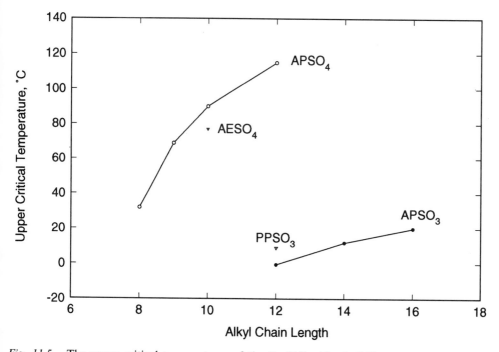

Fig. 11.5. The upper critical temperatures of the liquid/liquid miscibility gap in zwitterionic surfactants.

11.3.1.1 Hydrotropes

It was known to McBain that the chain-length of soap molecules must exceed a minimal number of carbons (C_5) in order for the characteristic features of soap–water solution chemistry and phase behavior to exist [27]. Recent experience suggests that weak surfactant behavior exists for a wide range of hydrophilic groups at C_8, but is either nonexistent or exceedingly weak for C_6 lipophilic groups. It is therefore worthwhile to consider the consequences of shortening the lipophilic chain of surfactants below this limit.

Sodium formate has the same hydrophilic group as do sodium soaps, the phase diagram of the sodium formate–water system (Fig. 11.6) qualitatively resembles that of the sodium chloride–water system (Fig. 1.1). The melting point is high, weakly stable crystal hydrates exist, and the compound is highly soluble [28]. Presumably, no micellar structure exists in sodium formate solutions and, by analogy with alkali metal carbonate salts, this compound is expected to be a "salting-out" electrolyte [29]. Sodium methyl sulfate and methanesulfonate presumably display similar aqueous phase behavior.

If one or more methylene groups are inserted into the carbon–hydrogen bond of formate to form the acetate anion or higher homologs, the aqueous phase behavior is not at first qualitatively altered. However, the properties of solutions of these low homologs do *not* remain constant as the chain-length is varied. At a chain-length intermediate between that of sodium formate and a soap, these compounds are expected to become "salting-in" electrolytes. Such amphiphilic compounds are widely used in industry for various purposes, and are commonly termed "hydrotropes" [30].

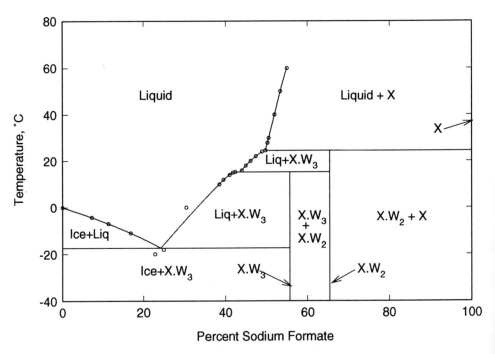

Fig. 11.6. The sodium formate–water system [31].

Hydrotropes are highly soluble in water, they do not form aggregates in solution of a size comparable to micelles, and their aqueous mixtures do not exist as liquid crystalline states. However, they *do* enhance the miscibility of weakly polar and otherwise water-insoluble molecules with aqueous solutions [30]. They also influence the ionic strength of their solutions. Hydrotropes must be present in substantial concentrations (1% or more) in order to display their "hydrotropic" properties. Surfactants display surface activity at much lower concentrations; the exact concentration depends strongly upon their chain-length (Section 11.2.1).

Hydrotropes constitute a distinctive class of amphiphilic compounds which may be regarded either as surfactants that have too short a lipophilic group to display surfactant phase behavior, or as hydrophilic compounds that have too large a lipophilic group to be regarded as being purely hydrophilic. Almost any surfactant hydrophilic group could probably serve as the functional group of a hydrotrope, but the most commonly encountered hydrotropes are the aryl sulfonate salts (sodium or potassium benzene-, toluene-, xylene-, or cumenesulfonates) [30].

Given the relationship between liquid crystal stability and thermal energy (Section 5.6), it is likely that those "hydrotropes" having the larger lipophilic groups would display surfactant phase behavior if one could determine their behavior far below the freezing point of water. The distinction between the two classes is arbitrary, from this perspective, in that it is dictated in part by the phase behavior of water itself.

11.3.2   The structural variation of hydrocarbon lipophilic groups

11.3.2.1   Insertion of aromatic rings

For many decades, sodium "*Linear Alkylate Sulfonate*" (LAS) surfactant mixtures have been important commercial surfactants [32]. Before the mid-1960s the more highly branched sodium tetrapropylenebenzenesulfonate salts (sodium ABS, or *Alkyl Benzene Sulfonates*) were important [32], and these materials are still used to some extent. The replacement of ABS by LAS was spurred by the fact that LAS undergoes biodegradation in the environment faster than does ABS [33]. Surprisingly little phase information exists on these compounds, considering their immense industrial importance.

The Krafft eutectic of sodium dodecanesulfonate in water is 49°C, and solubility studies of sodium *p*-dodecylbenzenesulfonate vs. temperature [34] suggest that its Krafft discontinuity lies above 66°C . The same relationship holds for the decyl homologs (38°C for sodium decanesulfonate, and >50°C for sodium decylbenzenesulfonate). The balance of the phase behavior of these compounds is unknown, but these data suggest that inserting a benzene ring between the chain and the hydrophilic group increases the temperature of the Krafft discontinuity. The influence of this structural variation on liquid crystal regions has not been reported.

11.3.2.2   Branching via hydrophilic group position

Isomers of terminally substituted surfactant structures in which the hydrophilic group is positioned at different sites along the chain ("branched" compounds) are also important surfactants. "Branching" may be viewed as a structural variant of lipophilic groups in which volume is held approximately constant while shape (conformational structure) is altered. Volume is *not* truly constant among branched and straight-chain isomers, because the incremental contributions of methyl (CH_3-), methylene ($-CH_2-$), and methinyl ($=CH-$) groups to molecular volumes are not the same [35]. The contributions of methyl and methylene groups to the molar volumes of the normal alkanes can be derived from a plot of molar volume vs. number of methylene groups; the slope may be regarded as the contribution per methylene, and the intercept as the contribution of the two methyl groups. From data on the normal alkanes (both equilibrium and metastable liquid phases), values at 25°C of 16.413 cm^3 per CH_2 and 32.788 cm^3 per CH_3 may be derived. The published data at 25°C [36] fit a linear model (with an r^2 of 0.999995) over an extraordinary range of chain-lengths, from C_5 through C_{40}.

Melting point data reveal that any deviation from the straight chain structure in acyclic molecules within any series of aliphatic compounds lowers the melting point. This trend is severely violated in the case of rigid polycyclic molecules in which the number of conformational states does not change on melting (Section 4.1). Branching presumably diminishes the enthalpic stability of the crystal phase by impairing the close packing of molecules and reducing the density of the phase [36]. Since branching in the lipophilic group is not expected to greatly alter the hydration of the hydrophilic group, the ΔT hydrophilicity parameter (Section 9.5.1) is not expected to be strongly affected. One might expect on this basis that the Krafft discontinuity will be lower in branched than in straight chain surfactants, and this is invariably observed.

Systematic and thorough phase studies of a series of branched surfactants have not been performed, so the full impact of this structural variation on phase behavior remains unknown. Some information of uncertain quality exists on sodium *x*-dodecyl-benzenesulfonates which differ in that the phenylsulfonate moiety is attached at various positions along a straight dodecyl chain [37]. The available information is summarized in Table 11.2.

The Krafft boundary is progressively lowered as the phenylsulfonate moiety is moved towards the center of the dodecyl chain, and at the 4-position this boundary becomes metastable. The coexisting phase at the solubility limit above the Krafft discontinuity was reported to be the hexagonal phase in all the sodium *x*-dodecylbenzenesulfonates, but recent studies [38] suggest that it may be, instead, the lamellar phase. The latter result is consistent with independent evidence that the saturating phase of sodium 8-hexadecylbenzenesulfonate (the mid-chain isomer of the C_{16} homolog of the above compounds) is also the lamellar phase [39].

Enlarging the C_{12}Ph lipophilic group by four methylene groups (while retaining the cental substitution along the chain) dramatically reduces the solubility in liquid water.

TABLE 11.2
Phase information on sodium *x*-dodecylbenzenesulfonates

Compound (Na$^+$ salt)	Krafft discontinuity (°C)	Coexisting phase at 25°C
1-Dodecylbenzenesulfonate	> 66	Crystal
2-Dodecylbenzenesulfonate	46	Crystal
3-Dodecylbenzenesulfonate	28	Crystal
4-Dodecylbenzenesulfonate	< 0	Hexagonal liquid crystal (?)
5-Dodecylbenzenesulfonate	< 0	Hexagonal liquid crystal (?)
6-Dodecylbenzenesulfonate	< 0	Hexagonal liquid crystal (?)

The solubility of the C_{16} compound in water at 25°C is 0.06%, while in 0.51 M sodium chloride it is 0.0002% [39]. The effect of salt on the composition of the coexisting liquid crystal phase is unknown. Owing to its poor solubility, colloidal dispersions of the C_{16} compound in salt solutions are highly stable.

It has recently been suggested that an upper consolute boundary and critical phenomena exists within the lamellar phase in the sodium 5-dodecylbenzene-sulfonate–water system (see Section 11.3.2.3).

Branched surfactants also exist in which the hydrophilic group is attached directly to the chain, rather than being separated from the chain by a phenylene group. The sodium x-alkyl sulfates (x-R-OS_3^-,Na^+ where x is greater than one) exemplify such structures. Renewed interest has developed in these compounds recently. They may be expected to undergo solvolysis to olefins and sodium bisulfate faster than do sodium 1-alkyl sulfates [40].

Such data as are available suggest that the same trend in phase behavior exists for salts of sodium secondary alkyl sulfates as is found among the isomers of sodium alkylbenzenesulfonate salts. For example, the phase behavior of sodium 9-heptadecyl sulfate [41] qualitatively resembles that of sodium 8-hexadecylbenzenesulfonate (above). The aliphatic sulfate salt is poorly water soluble, its Krafft discontinuity lies < 0°C, and the coexisting phase at the solubility boundary is (from its texture) the lamellar liquid crystal. Sodium 1-heptadecyl sulfate, in striking contrast, may be expected to have a high Krafft boundary, to be highly soluble above the Krafft boundary, and the saturating liquid crystal phase will almost certainly be the hexagonal phase. (This prediction is based on interpolation of data on the C_{16} and C_{18} homologs.)

11.3.2.3   Branching via multiple lipophilic groups

Amine oxides. A different kind of "branching" exists in dihexylmethylamine oxide, $(C_6H_{13})_2N(\rightarrow O)CH_3$. The nitrogen lies within the longest chain in this compound, so that only the dipolar oxygen atom is pendant to the chain. This compound differs by only one methylene group from dimethyldodecyl- or dimethyldecylamine oxide, but its phase behavior is profoundly different from that of both of these surfactants [42]. The dihexyl compound is simply a highly water-soluble compound whose phase diagram displays no evidence whatsoever of surfactant behavior. The same is true of tributylamine oxide. Such comparisons demonstrate that redistributing the methylene groups of surfactant molecules symmetrically among two (or three) lipophilic groups dramatically influences their phase behavior.

Phosphine oxides. A similar "redistribution" of methylene groups has been investigated in phosphine oxides. Phase data on these compounds provide information as to the influence of this structural variation on the lower consolute boundary, as well as on liquid crystal regions. The structures, and a general description of the phase behavior of the compounds studied are shown in Table 11.3. The diagrams are shown in Fig. 11.7. These results show that the surfactant structure displays the lower consolute boundary having the smallest area, and that symmetrically redistributing the methylene groups among the substituents enlarges this miscibility gap. Doing so also eliminates liquid crystal phases. The data on both amine and phosphine oxides suggest that an asymmetric distribution of methylene groups, terminal substitution, and minimal branching favor surfactant phase behavior. Similar data have recently been reported in the literature [43].

TABLE 11.3

Effects of redistributing methylene groups in posphine oxides

Compound	Number of carbons	Phase behavior
$C_{12}H_{25}P(\rightarrow O)(CH_3)_2$	14	Highly miscible, surfactant behavior
$C_{10}H_{21}P(\rightarrow O)(C_2H_5)_2$	14	Highly miscible, no liquid crystals
$C_8H_{17}P(\rightarrow O)(C_3H_7)_2$	14	Poorly miscible, visible critical point
$(C_6H_{13})_2P(\rightarrow O)CH_3$	13	Less miscible, large miscibility gap
$C_5H_{11}P(\rightarrow O)(C_5H_{11})_2$	15	Still less miscible, large miscibility gap

Ionic surfactants. Sodium dialkyl phosphates are branched anionic surfactants which (at the C_{16} chain length) are poorly soluble in water. They have been widely studied as analogs of the phosphatidic acid salts (Section 11.3.2.7), as anionic vesicle-forming surfactants, and as compounds which serve to introduce negative charges in the vesicles formed by other surfactants [44]. However, except for the temperature of their "chain-melting transition", their aqueous phase behavior is unknown.

Dihexyldimethylammonium halides would be expected to behave much like dihexylmethylamine oxide (above) in water, but as the chain is lengthened within this class of compounds surfactant behavior is introduced. Didodecyldimethylammonium chloride and bromide have attracted considerable interest because they have been reported to form two coexisting lamellar liquid crystal phases of different compositions (Fig. 11.9) [45,46]. This result suggests the possibility that a miscibility gap (together with the associated critical point) exists within the lamellar liquid crystal phase region. If true this miscibility gap persists to high temperatures, however, and critical behavior has not so far been established to exist in this system. An upper consolute boundary miscibility gap and critical point *has* recently been reported to exist within the lamellar phase of the sodium 5-dodecylbenzenesulfonate–water system (Section 11.3.2.2), and it had been suggested earlier that a miscibility gap and critical phenomena could exist within the hexagonal-monoclinic ($H_\alpha–M_\alpha$) phase region of the SDS–water system [47].

Didodecyldimethylammonium bromide probably has low (but finite) water solubility, and dispersions of the coexisting liquid and liquid crystal phases of this compound display complex colloidal behavior [48]. The C_{18} homolog must display vanishingly low solubility, however, and colloidal dispersions of this surfactant (if properly stored) are extremely long-lived [48]. The hydroxide salts of soluble single-chain quaternary ammonium surfactants display highly unusual solution chemistry (Section 10.5.2), and it has been reported that didodecyldimethylammonium hydroxide tends to form hollow spherical micellar aggregates [49]. Further investigations of both the solution chemistry and the phase behavior of these interesting compounds appear to be warranted, as it is impossible to exceed the hydrophilicity of the hydroxide counterion in ionic surfactant salts (Sections 10.5.2, 10.7.3).

11.3.2.4   Sodium sulfosuccinate diesters (Aerosols)

"Aerosol OT" (AOT) and its analogs represent another important class of branched commercial surfactants. These compounds are salts of diesters of succinic acid which contain a sulfonate substituent group within the succinate moiety. The anionic sulfonate substituent group is pendant to the longest chain in these compounds, and approximately centered in it. The phase behavior of the sodium salt of the 2-ethylhexyl ester homolog (Fig. 11.8) has been determined [50], and found to deviate significantly from that of soap-like compounds (including the sodium n-alkanesulfonates themselves). The solubility of AOT in liquid water is real but modest, but the separating phase at the solubility boundary is (in striking contrast to surfactants having soap-like molecular structure) lamellar rather than hexagonal. A cubic and a hexagonal phase are found in more concentrated mixtures; the hexagonal phase has an inverted structure. The magnesium and calcium salts qualitatively resemble the sodium salt insofar as phase behavior is concerned, but differ strikingly in the span of composition of the lamellar phase. A fraction of water of about 25% is required for the lamellar phase to exist in all three salts, but this phase is saturated with water at 85, 50, and 35% water in the sodium, magnesium, and calcium salts, respectively. The phase behavior of the barium salt (which displays a large liquid/liquid miscibility gap) is altogether different from all of these (Fig. 10.7).

Aerosol surfactants have played an important role in the development of geometric theories of surfactant phase behavior (Section 8.8.1), and have been widely studied as the surfactant component of microemulsions [51].

A systematic screening of the phase behavior of sulfosuccinate esters having n-alkyl ester groups has been performed, and surfactant phase behavior has been shown to exist at ester chain-lengths as short as di-C_4 [52]. The Krafft boundary appears as an equilibrium boundary at di-C_6, but it is pushed below 0°C by insertion of a 2-ethyl substituent into each hexyl group. The phases and phase sequences are qualitatively similar among the various homologs, which serves to establish the generality of the earlier results on sodium 2-ethylhexyl AOT itself. These results also demonstrate that the observed phase sequence does not hinge on the presence of the unusual 2-ethyl substituent in the lipophilic groups. The role of the ester groups is uncertain.

11.3.2.5 Unsaturation

One of the most important variations of lipophilic group structure is the introduction of unsaturation. The most important form of unsaturation is the cis-ethylenic (double) bond, which is widely found in natural fats [53]. The $trans$-ethylenic bond and the acetylenic (triple) bond are also encountered, however, and the cyclo- propane ring may be regarded as a form of unsaturation [54]. Most of the ethylenic structures encountered are 1,2-disubstituted ($RCH=CHR'$). As many as five ethylenic bonds may exist within the same chain in fatty compounds of biological origin, but the monoethylenic (oleate) derivatives are the most important commercially.

It is worth noting that while introducing one double bond does not dramatically alter the stability in air of a hydrocarbon surfactant, introducing two (or more) introduces considerable reactivity towards autoxidation [55]. This is especially true when the bonds are located in 1,4 positions relative to each other ($-CH=CH-CH_2-CH=CH-$) so that they are separated by a single methylene group. If autoxidation of such compounds occurs, the polar autoxidation products that are formed profoundly alter the physical behavior [56].

Introducing an ethylenic bond has very important consequences with respect to the conformational structures allowed of a chain. The reason is that four of the atoms in the chain (the two ethylenic carbons plus the two attached carbons) are constrained within a single plane. In spite of this constraint, the fluidity of unsaturated chains in monoene structures such as oleates (which have one cis-$\Delta_{9,10}$ double bond in a C_{17}

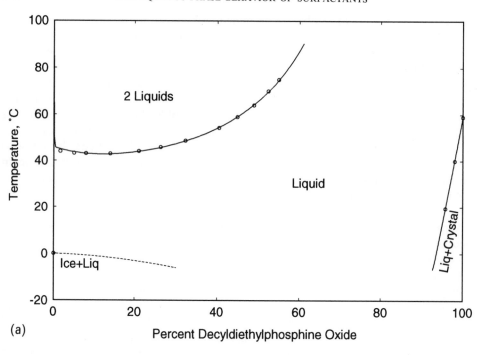

(a)

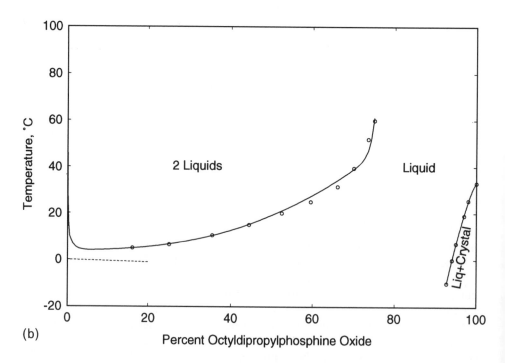

(b)

Fig. 11.7. The decyldiethyl-, octyldipropyl-, dihexylmethyl-, and tripentylphosphine oxide–water systems.

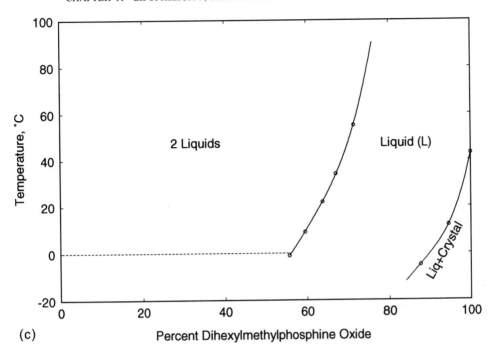

(c)

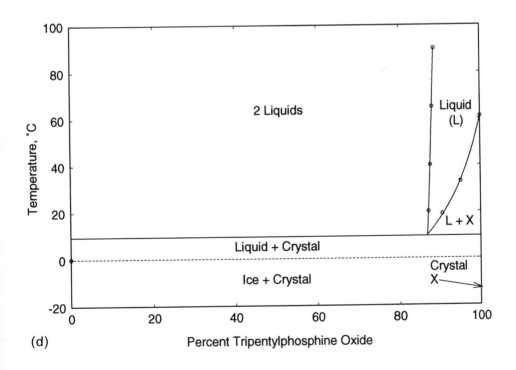

(d)

Fig. 11.7 Continued.

chain) is apparently not grossly different from that of saturated chains at temperatures where both are in fluid states [57]. The most important consequence of unsaturation, by far, is to lower the lowest temperature at which fluid states may exist (the Krafft discontinuity).

The melting points of oleic (cis-$\Delta_{9,10}$), elaidic ($trans$-$\Delta_{9,10}$), and stearic (saturated) acids are 16.3, 46.5, and 69.6°C, respectively [58]. It is relevant that the order in which these melting points increase is the same as the order in which the temperatures of the Krafft discontinuities of the sodium soaps lie (30, 50, and 77°C). Phase data on salts of stearolic acid (the $\Delta_{9,10}$ alkyne) or sterculic acid (the cis-$\Delta_{9,10}$ cyclopropane) do not exist, but both the acetylenic and the cyclopropane groups may be expected to influence phase behavior in a manner that is qualitatively similar to that of the ethylenic bond.

An ethylenic bond in a molecule that exists within a crystal phase must be in register with the ethylenic bonds of adjacent molecules [59]. Also, it would be expected that the translation of a molecule within the crystal along the direction of the chain axis should be very disruptive in such crystals. Because of these factors, solid solutions within the crystal phase are perhaps less likely in unsaturated crystalline chain compounds than in saturated analogs. (It was found during a study of triacontanol that triacontanol itself was inseparable from its near homologs by any process involving crystal ⇆ liquid equilibria (including zone refining). In striking contrast, homologous impurities could easily be removed from 15-triacontene (the intermediate in the synthesis) by ordinary recrystallization. These results suggest that the saturated compound is more prone to form solid solutions than is the olefin.)

It was suggested by McBain that the cis-double bond in oleate has a profound effect on the Krafft boundary, but otherwise exerts little influence on soap–water phase behavior [60]. While differences do exist, it is well-documented that both saturated and unsaturated chains are effective lipophilic groups. This is true both for mid-chain unsaturated compounds (such as oleate soaps) and for ω-unsaturation at the far end of the chain from the hydrophilic group, as in sodium 10-undecenyl-1-sulfate. The phase behavior of this alkyl sulfate may be expected to be similar to that of the saturated analog, as a similar result has been found in nonionic surfactants containing biphenyl moieties within the lipophilic group [61].

The introduction of unsaturation does not appear to have a substantial influence on liquid/liquid miscibility gaps, but good information regarding this correlation does not exist.

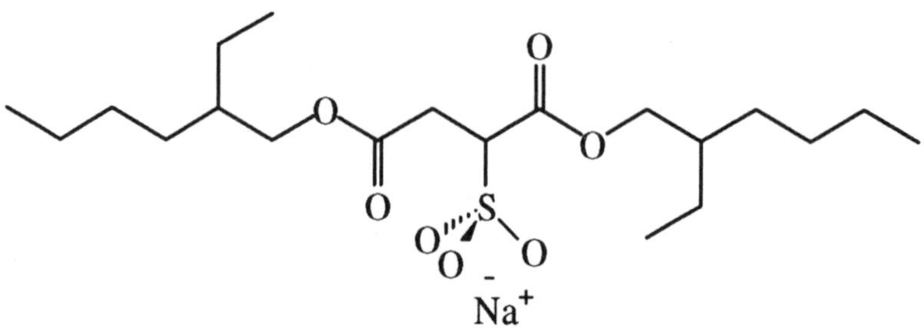

Fig. 11.8. The molecular structure of sodium bis(2-ethylhexyl) sulfosuccinate (Aerosol OT, or AOT). A family of related surfactants exists that differ in the structure of the ester group.

11.3.2.6   Alkyl substituents

Methyl substitution in the lipophilic group of sodium octadecyl sulfate has been observed to decrease the temperature of the Krafft boundary at 15% composition [62]. The magnitude of the decrease becomes larger as the methyl group is moved from the proximate 2-position (near the hydrophilic group) to more remote positions (Fig. 11.9). A minimum necessarily exists just as in the case of hydroxyl substitution (Section 10.2.2), since the Krafft boundary temperatures of the midchain substituted methyl hexadecyl compounds are far below that expected of sodium heptadecyl sulfate.

Such structural modifications are not expected to influence the hydration of a surfactant, but they should disrupt the packing of the chain within a crystal lattice [36]. It therefore seems likely that they exert their effect on phase behavior principally by raising crystal energies and lowering melting points, rather than by altering the value of ΔT.

The substitution of two ethyl groups α to the carboxylate groups of soaps is known to alter their phase behavior profoundly, but the nature of the alterations has not been determined (Section 10.2.2) [63]. The substitution of two n-butyl groups in this position also has a profound influence on phase behavior; the hexagonal phase is eliminated, and the phase behavior more closely resembles that of sodium AOT than it does a typical soap [64]. *Both* crystal stability and the hydration of the carboxylate anion are almost certainly directly affected by the large proximate lipophilic substituents in these compounds.

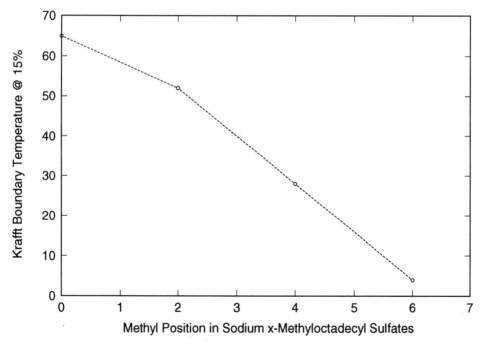

Fig. 11.9. The Krafft boundary temperatures at 15% of some sodium *x*-methyloctadecyl-sulfates.

11.3.2.7　The biological polar lipid surfactants

Molecular structures. The biological polar lipids represent a special class of branched surfactants which are important components of cell membranes [65], and are of commercial value as well [66]. A large fraction of the polar lipids in living cells are derivatives of "phosphatidic acids". Phosphatidic acids are 1-monophosphoric acid esters of 2,3-diglycerides, and the polar lipids are derivatives of phosphatidic acids in which a second phosphate ester link is formed to one of several small alcohols (Fig. 11.10):

Alcohol	*Hydrophilic group class*	*Common name*
None	Anionic	Phosphatidic acid salts
Inositol	Anionic	Phosphatidylinositol
Ethanolamine	Zwitterionic or anionic	Phosphatidylethanolamine
Serine	Zwitterionic or anionic	Phosphatidylserine
Choline	Zwitterionic	Phosphatidylcholine, lecithin

Lecithins are nonionic (zwitterionic) surfactants that are neither acidic nor basic, but the others are acidic and/or basic so that their molecular structure is pH dependent. In all these compounds the longest extended chain is the diglyceride moiety; the hydrophilic group is actually pendant to this longest chain in its extended conformation, and is separated from it by the 1-carbon of the glycerol moiety. In all the known phase states, however, the glycerol moiety is actually folded (as shown in Fig. 11.10) so that the two long chains lie parallel to one another [67]. In this conformation the lipophilic groups lie at one end of the molecule and the hydrophilic group at the other, which is why these compounds are usually regarded as "double chain" or "branched" surfactants (Section 8.3.1.1).

The glycosides of diglycerides constitute an important subgroup of 1,2-diglyceride-based polar lipids [68]. In these compounds the diglyceride moiety is linked to a mono- or polysaccharide group by a glycosidic (acetal) bond, so that the hydrophilic group is a cyclic ether polyol. Examples include the diglyceride monogalactosides and digalactosides found in bread flour lipids [68].

A second major group of polar lipids are the derivatives of ceramides. Ceramides are fatty acid amides of sphingosine [65], which is an unsaturated C_{18} dihydroxy amine

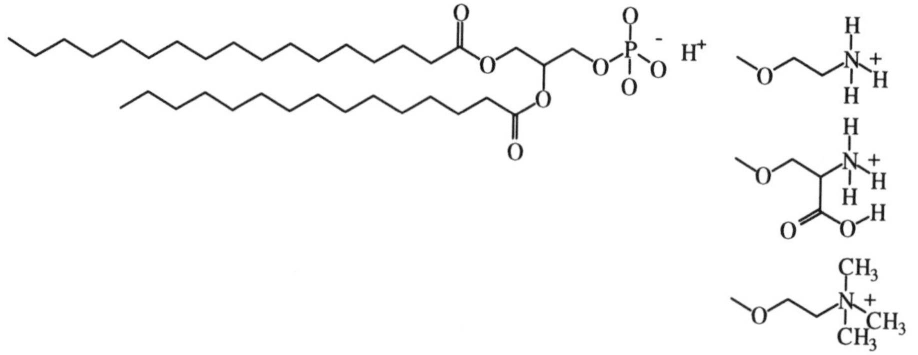

Fig. 11.10. Molecular formulae of a salt of phosphatidic acid, of phosphatidylethanolamine, phosphatidylserine, and phosphatidylcholine (lecithin).

(2-amino-4-*trans*-octadecene-1,3-diol) whose structure is shown in Fig. 11.11. The same family of hydrophilic groups that are found in the diglyceride polar lipids (above) may be coupled to ceramides (via the 1-hydroxy group). The phosphocholine ester of a ceramide has a hydrophilic group structure that is analogous to that found in lecithins, and is termed a "sphingomyelin". The galactosyl acetal of a ceramide is termed a galactoylcerebroside.

The presence of the free 3-hydroxy substituent, and of the amide group, in ceramide derivatives will significantly enhance their hydrophilicity and (probably also their crystallinity), relative to their 1,2-diglyceride analogs.

A third structurally distinctive group of polar lipids (the plasmalogens) exists in which the lipophilic groups are linked to the glycerol moiety via a vinyl ether (rather than via carboxylate ester) bonds [69]. The terminal carbon is in the aldehyde oxidation state in these compounds, which are less hydrolytically reactive than are their ester analogs in neutral or weakly basic media.

Finally, surfactants that have a wide range of still more complex hydrophilic groups are found in small amounts in living cells [65]. It is likely that the surfactant nature of these compounds is responsible for anchoring them to the cell membrane, as this would be a natural place for them to reside from a purely physical point of view. These molecules are likely involved in cell–cell communication and other external interactions of the cell.

Lysopolar lipids. Two isomeric lysopolar lipid analogs exist for each of the above polar lipids [65]. The lysopolar lipids are simply the monolong chain derivatives that result from the hydrolytic cleavage of one long chain from the parent polar lipid. These compounds are intermediates in the biosynthesis of the polar lipids, and are present in small regulated amounts in cell membranes. They resemble in their physical behavior other monolong chain surfactants [70].

It can be seen from this brief summary that considerable complexity exists in the molecular structures of polar lipid surfactants. All of them may be regarded as being polyfunctional surfactants; the roles that the individual functional groups play in determining their physical behavior are for the most part unclear.

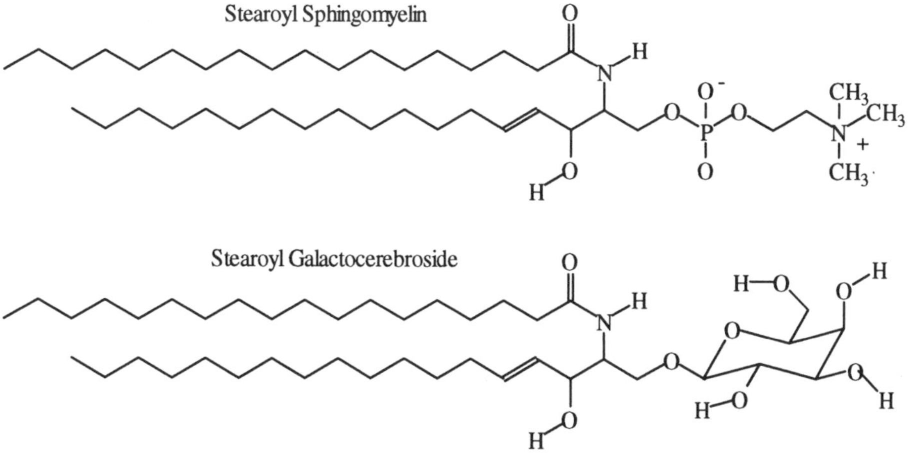

Fig. 11.11.   The molecular structure of a sphingomyelin and a cerebroside polar lipid.

Phase behavior of the polar lipids. An enormous amount of effort has been invested in the study of these materials during the past 40 years, and most of this information has recently been compiled into a computer database [71].

The phase behavior of polar lipids may be described in general as being "monoglyceride-like" (Section 5.10.7); their solubility in liquid water (at the usual chainlengths found in cells) is invariably low [72]. Both Ekwall and Small describe the polar lipids as insoluble but "swelling" surfactants, which is an apt description of their physical behavior. While insoluble in liquid water, they are highly miscible with water as liquid crystal phase states. If an interface is created between water and a dry polar lipid (at a temperature above the Krafft discontinuity), they do indeed swell by the uptake of water (as do all surfactants). The swelling of polar lipid surfactants often leads directly to colloidally structured liquid crystal dispersions (myelinic textures) [73], and is therefore considerably more complex than is the swelling of most surfactants [74].

Polar lipids form a wide variety of liquid crystal phase structures. The lamellar phase (usually termed the L_α phase in this field) is very prominent, and cubic phases of differing structures (excepting the discontinuous structure, Section 8.4.10) are widely encountered.

A variety of phase states exists within which the chains are considerably more highly ordered than is found in the lamellar phase. Examples include the L_β, L_β', and P_β' phases (Section 8.4.7) [75]. Luzzati has proposed a systematic nomenclature for these liquid crystal phases that is based on their phase structure (see Table 8.1) [75].

Liquid/liquid miscibility gaps are not typically encountered in naturally occurring polar lipid–water systems, but partial phase studies of very short-chain (C_6, C_7, and C_8) lecithins have revealed the existence of an upper consolute boundary (Section 5.10.2) [76]. This boundary qualitatively resembles the probably related boundary found in ammoniosulfate and ammoniosulfonate–water systems (Section 10.5.3.1) [26].

The coexisting phase at the solubility boundary is the lamellar liquid phase at temperatures above the Krafft discontinuity temperature for most polar lipids. However, it is worth noting that this coexisting phase in monogalactosyl diglycerides is, instead, an inverted hexagonal phase [77]. (The corresponding phase for digalactosyl diglycerides is the lamellar phase.) This qualitative difference in phase behavior is possibly important with respect to their stabilization of the foam structure of bread [68].

Status of the aqueous phase science of polar lipids. In spite of the massive effort that has gone into physical studies of the polar lipids, the present knowledge of the phase science of these materials is (as with many surfactants) incomplete. The most extensively studied polar lipid is dipalmitoylphosphatidylcholine (DPPC). Considerable information has been obtained regarding its physical science [78] and a partial phase diagram has been determined (Fig. 8.11) [79,80]. Major gaps in our knowledge of the DPPC–water phase diagram remain, however.

As to the phases that exist, several crystal hydrates have been characterized and the structures of some have been determined using single crystal X-ray methods [67]. The coexistence relationships of these phases is unclear. The existence of the lamellar (L_α) phase has been established beyond doubt, and this phase (more precisely, dispersions of it within the coexisting liquid phase) has been exhaustively investigated. Yet, the boundaries of the L_α phase region have not been firmly established.

At its lower temperature limit, it has been shown that the L_α phase disproportionates at a eutectic to the P_β' phase and dilute liquid. This phase reaction is highly unusual relative to surfactant behavior in general, because the concentrated product phase is not a crystal. Instead, it is another fluid (or liquid crystal) phase (Section 8.4.7). The manner in which the L_α phase decomposes at the upper limit of its temperature range is also unknown; this may be inaccessible due to hydrolytic cleavage of ester groups at high temperatures.

In addition to the L_α liquid crystal phase, the L_β (or L'_β) and the P'_β phases have been prepared and described structurally in considerable detail [81]. The coexistence relationships of the P'_β phase with the surrounding phases have been determined, originally by use of deuterium NMR (Fig. 8.11) [79] and later using calorimetric methods [80]. This part of this diagram is probably well established. It has been suggested that cubic bicontinuous phases exist in many polar lipid–water systems, but this phase does not appear in this DPPC phase diagram.

A vast amount of physical information of various kinds has been obtained, especially on colloidal (vesicular or liposomal) dispersions of DPPC. Calorimetric studies of dilute mixtures of DPPC (in "excess water" [71,78]) have revealed that at least five isothermal discontinuities exist in this system (including the freezing of water). This information has been incorporated in Fig. 8.11.

The dilute liquid phase at these discontinuities is essentially pure water for DPPC and other naturally occurring polar lipids. Evidence exists that the solubility in liquid water of polar lipids having two C_{18} chains at temperatures above the "chain-melting" temperature is $< 10^{-15}$ M [72]. (It is plausible that the solubility of such compounds may be precisely zero; the various forms of the free energy of mixing function which would lead to this result are described in Section 3.20.) The water solubility of homologous surfactants increases exponentially as the chain is shortened, however (Section 11.2), so that the solubility may be appreciable and significant in, for example, di-C_{12} homologs.

The gaps that exist in the information on the DPPC–water system preclude the drafting of a comprehensive and rigorous phase diagram. A particularly important uncertainty is which of the phases that have been discovered are equilibrium states, and which are metastable. Phase equilibrium is quickly attained when the lamellar liquid crystal phase is the equilibrium state, but when the lamellar phase is cooled phase states result that have been described as being "gel" or "coagel" states [82]. This may be an apt description of their properties [83], but leaves entirely open the answers to basic questions such as whether or not they are one- or two-phase mixtures, whether the phases present are equilibrium or are metastable states, etc.

In the physically related DODMAC–water system, it has been established that the state described earlier as a "gel" state [84] is not a single phase at all, but a colloidally structured dispersion of a stoichiometric crystal dihydrate $(X \cdot W_2)$ in the dilute liquid phase [85]. Similarly, the "coagel" state in this system [84] was shown to be a colloidal dispersion of the crystal monohydrate $(X \cdot W)$ in the dilute liquid. Mainly on the basis of these results, it would appear that the equilibrium phase behavior of those mixtures in polar lipid–water systems that are described as "gel" or "coagel" remains highly uncertain. The information as to their structure remains valid; the uncertainties that exist relate to the phase behavior of the systems.

A second major uncertainty is the qualitative classification and the quantitative dimensions of the isothermal discontinuities that have been discovered (Section 4.14). With some exceptions, it has not even been firmly established which of these are eutectic, which are peritectic, and which (if any) are polytectic phase transformations. This being the case, it follows that the stoichiometric phase reactions that occur at each of these discontinuities remain undetermined.

As regards the quantitative description of these discontinuities, it will be recalled that eight parameters must be determined in order to fully define 2–3–2 phase discontinuities (Section 4.14). These include (at a defined pressure):

(1) the classification (as eutectic, peritectic, or polytectic),
(2) the temperature,
(3) three phase compositions, and
(4) three phase structures.

Of these parameters the temperature, and the composition and structure of one of the phases involved (the dilute liquid) are always known. The other five are, to varying degrees, uncertain. The heats of the phase reactions that occur at these discontinuities are fundamental thermodynamic features of these systems (Section 3.21), but since these phase reactions remain undetermined the interpretation of the calorimetric data in terms of the relevant phase reaction equation is clearly impossible. Such data are typically expressed per unit quantity of polar lipid, which has no fundamental significance with respect to phase chemistry.

The best-studied phase discontinuity in polar lipid–water systems, by far, is the "chain-melting" transformation [79]. (The term "chain-melting" for this transition is unfortunate, as it implies that the latent heat of the phase reaction serves only to "melt" the chains and does not alter the balance of the phase. It is more than mere semantics to recognize that whole "phases" melt, not molecular fragments of the molecules within phases.) This important discontinuity corresponds to the eutectic that exists at the lower temperature limit of the lamellar (L_α) phase (see above and Section 5.6.2). Carefully executed calorimetric studies strongly suggest that the phase transition that occurs at this discontinuity is first order [86]. At this eutectic a dilute liquid (of well-defined structure and composition), a lamellar liquid crystal (of well-known structure but highly uncertain composition), and the P'_β phase (of known composition), coexist. Even for this discontinuity, then, one of the eight parameters required (the composition of the L_α phase) remains uncertain. At other discontinuities the situation is far worse.

The "chain-melting" eutectic may be regarded as being equivalent to the Krafft discontinuity of surfactant systems in general, but differs from the familiar Krafft eutectic in that the concentrated phase is (presumably) not a crystal. This inference is compromised by uncertainty as to the nature of the adjacent discontinuities. However, it has been firmly established that as the chain-length is increased the temperature of the "chain-melting" transition is also increased, and that introducing unsaturation lowers this transition [71]. Both of these correlations are also found in the case of the Krafft discontinuity of soluble surfactants, and support this classification of this discontinuity.

11.3.2.8   Remote polar substituents

Polar substituents in an aliphatic straight chain have the effect of shortening the effective chain-length, so that the resulting surfactant behaves (in the first approximation) as if the methylene groups between the hydrophilic group and the substituent were absent. The ethoxylated anionic surfactants can be viewed in this manner, for example. From the perspective of molecular structure, the sodium sulfate ester of $C_{12}E_1S$ ($C_{12}H_{25}OCH_2CH_2OSO_3^-$,$Na^+$) may be viewed as either a C_{12} surfactant, or as a C_{15} surfactant having a 3-oxa "substituent" group.* The oxa group enhances the hydrophilicity of a surfactant, however, so that the phase behavior of the oxa compound more closely resembles that of a C_{12} surfactant than it does a C_{15} surfactant. To illustrate, the temperature of the Krafft discontinuity of sodium dodecylsulfate is 26°C while that of sodium 3-oxapentadecyl sulfate ($C_{12}E_1SO_4Na$ above) is 8°C. Since that of sodium tetradecylsulfate is 38°C and that of sodium hexadecylsulfate is 52°C, the Krafft discontinuity of sodium pentadecylsulfate (which has not been determined) will be in the vicinity of these two and far above that of the oxa compound.

In an ethoxylated surfactant the methylene groups of the oxyethylene chain may contribute a real lipophilic effect, but this effect is attenuated by the fact that they lie between the two polar groups. The same phenomenon occurs within the tether of zwitterionics (Section 10.5.3.2) [87]. The reality of their lipophilic effect is suggested by the fact that the cmcs of sodium alkyl sulfates are progressively *lowered* by the insertion of oxyethylene groups [88]. They are not raised, as one might have expected on the basis that only the hydrophilicity is enhanced by the presence of the additional ether group. Other explanations for this effect are possible, however, and the truth of the matter remains uncertain.

During an exploratory study of ammoniohexanoate surfactants, the effects of inserting various sulfur functionalities into the lipophilic chain of pentadecyldimethyl-ammoniohexanoate ($C_{15}AH$) on its phase behavior were determined [89]. The qualitative results of this study are tabulated below.

Surfactant behavior retained		*Surfactant behavior destroyed*	
Functional group	*Position*	*Functional group*	*Position*
$-S-$	6	$-SO_2-$	12
$-S-$ (2)	6,11	$-SO_2-$ (2)	6,11
$-SO_2-$	6		

These results show that one may have a sulfide atom in the chain as far away as the 11-position (leaving a pendant butyl group), and still retain surfactant phase behavior. Clearly, the sulfide group is more lipophilic than hydrophilic. If the sulfide is oxidized to the sulfone group (or especially to the much more strongly hydrophilic sulfoxide group), however, the surfactant phase behavior of the compound is destroyed. If no polar functional group exists past the 6-position (leaving a terminal nonyl chain which, by itself, is an operative lipophilic group) then surfactant behavior is retained in all these derivatives.

Limited phase and solution chemical data on prostaglandin $F2_\alpha$ demonstrate that the ring and the polar substituents in this compound profoundly alter the physical science of this "sodium soap" [90]. These substituents likely increase the solubility in water of the acid as well.

---

*In naming compounds such as these it is very convenient to select as the parent compound the hydrocarbon analog and name the compound by designating the position at which a methylene group is replaced by an oxygen atom. This "oxa" group may be regarded from perspective of structural correlations as a "substituent" group, even though it is not pendant to the parent chain but a part of it.

11.4   HLB: hydrophilic–lipophilic balance

In 1949 the formal concept of HLB (hydrophilic–lipophilic balance) was introduced by Griffin as a parameter that enabled the effectiveness of polyol and polyoxyethylene surfactants as emulsifiers to be correlated with their molecular structures [91]. This important concept has been useful not only for this purpose, but it also caught the fancy of many surfactant chemists and has been applied to many different situations [92]. Numerous attempts have been made, since HLB was originally devised, to place this concept on a less empirical basis, for example by consideration of Hildebrand or other solubility parameters [93]. It is therefore worthwhile to consider the relevance of HLB to the phase behavior of surfactants.

If one is focused entirely on the behavior of surfactant molecules at an interface, then HLB is an extremely useful concept. Viewed from a qualitative perspective, it suggests that a change in the lipophilicity of a surfactant must be balanced by a corresponding change in the hydrophilicity if one is to preserve the interfacial behavior of the original surfactant. This qualitative concept is valid with respect to selected surface properties, but maintaining such a balance is not always the most important aspect of surfactant physical science. Other important aspects of physical behavior may vary wildly among compounds which have the same HLB, and in these cases its value as a device for correlating information with structure in a useful manner is lost.

To illustrate, the cmcs of surfactants will vary considerably within a family of homologs all of which have the same HLB. As the cmc varies so does the concentration in solution at which surfaces are saturated, the kinetics of diffusion within bulk phases and to an interface [94], the dynamics of reorganization of micellar species in solution [95], and many other properties. If any of these properties are relevant to the utility of the surfactant, then utility cannot reasonably be expected to correlate with HLB.

11.4.1   Scaling HLB

One of the most serious difficulties with HLB is the problem of scaling the hydrophilicity and the lipophilicity properly [96]. As noted in Section 11.2, lipophilicity can be handled in a comparatively straightforward manner due to the fact that it is often varied in an extensive manner and is linear with chain-length. For surfactants in which hydrophilicity is also varied in an extensive manner (the polyethers and polyols), the relative hydrophilicities of different surfactants can also be scaled on an empirical basis, simply by counting the number of functional groups present (Section 10.5.5.2).

Ideally one would like to scale both hydrophilicity and lipophilicity on a thermodynamic basis, but while this may be possible for lipophilicity it has not so far been accomplished for hydrophilicity. In Chapter 10 an attempt was made to rank different functional groups in a reasonable order with respect to relative hydrophilicity, yet these results represent but a first crude attempt to accomplish this task. It is therefore clear that, at the present time, it is impossible to scale hydrophilicity quantitatively on a thermodynamic basis over a wide range of hydrophilic structures. While possible in principle, doing so will require full and complete knowledge of selected phase diagrams, plus carefully designed and executed scanning and isothermal calorimetric studies. Such work has not so far been attempted.

11.4.2   HLB and pivotal phase behavior

In the area of surfactant phase behavior, the HLB concept is most clearly relevant to the concept of pivotal phase behavior. As noted in Section 5.10.7, several families of surfactants exist whose phase behavior is delicately balanced, insofar as the spatial relationship in the diagram between the miscibility gap and the liquid crystal regions is concerned. When the hydrophilicity and the lipophilicity are just right with respect to each other, these phase regions interfere with one another in a characteristic manner. Both the upper and the lower parts of the miscibility gap, and liquid micellar phases of intermediate composition, are often visible. A corridor (of variable width) cuts through the miscibility gap, and within this corridor phases having unusual structures and properties may exist which are not found in the diagrams of surfactants which do not display pivotal behavior. The "anomalous" (L_3) phase of $C_{10}E_4$ is an example of such a phase.

If one starts with a surfactant displaying pivotal phase behavior and alters either its hydrophilicity or its lipophilicity by a small amount, dramatic qualitative changes in phase behavior result. An increase in hydrophilicity causes the miscibility gap and the liquid crystal regions to disengage, and soluble surfactant behavior is observed (as in $C_{10}E_5$ and higher ethoxylogs). A decrease in hydrophilicity (as in $C_{10}E_3$) causes the solubility in liquid water to vanish, and eventually the surfactant is expected to display monoglyceride-like phase behavior. When this is true, miscibility of surfactant and water exist only within the liquid crystal phases, as the lower part of the miscibility gap (and the associated liquid region) no longer exist and the characteristic features of pivotal phase behavior are absent.

In considering hydrophilicity one must consider all the polar functional groups that exist – both operative and non-operative. Similarly, in considering lipophilicity one should also consider the molecule as a whole. The methylene groups within poly-oxyethylene chains, and proximate substituents, must be included along with those that exist within the principal lipophilic group. As has been noted, the lipophilic contributions of these different structural elements are real but vary significantly.

11.5   The intensive variation of lipophilicity

We turn now to the variation of lipophilicity by changing the chemical structure of the lipophilic structural fragments, a process that may be viewed as an "intensive" variation of lipophilicity. Surfactants which possess nonhydrocarbon lipophilic groups will be considered in this section.

11.5.1   Fluorocarbon lipophilic groups

Fluorocarbon chains represent an important class of non-hydrocarbon lipophilic groups [97]. The fluorine atom has a nuclear charge of $+9$ and an atomic mass of 19, and is thus far heavier than a hydrogen atom ($+1$ and 1). A fluorine substituent also has three non-bonding electron pairs in the second electron shell, while a hydrogen sub-stituent has no second electron shell at all. Perhaps because of the larger size of the fluorine atom, the conformational chemistry of fluorocarbons is considerably different from that of hydrocarbons [97]. Also, perfluorocarbon polymers do not crystallize in the all-*trans* conformation that is characteristic of crystalline hydrocarbons; there tends to be a twist along the chain of fluorocarbons in the crystal state, so that a characteristic periodicity exists in the long direction of the chain [97].

Fluorine is the most electronegative of the elements; it is far more electronegative than hydrogen. Because of this, the fluorination of a hydrocarbon group significantly alters the chemical reactivity of polar functional groups. Fluorination may also reduce the hydrophilicity of a hydrophilic group via inductive effects, unless the fluorines are widely separated from it. (Trifluoromethanesulfonic acid, CF_3SO_3H, is a significantly stronger acid than is methanesulfonic itself, for example.) It is therefore important, in comparing hydrocarbon and fluorocarbon surfactants, to consider the influence of fluorination on *both* the chemical and the physical aspects of surfactant physical science (Section 2.2).

Both fluorocarbons and hydrocarbons are strongly hydrophobic, but hydrocarbons are lipophilic while fluorocarbons are, in some degree, lipophobic. Alkanes and fluorocarbons are often only partially miscible (as illustrated in Fig. 11.12) [98]. Both classes of compounds were utilized by Hildebrand in preparing a mixture having 10 coexisting liquid phases [3]. It has been suggested that fluorocarbon and hydrocarbon surfactants in the same solution do not mix completely within micellar aggregates [99]. It is even possible that two discrete micellar species coexist within such solutions – one rich in the hydrocarbon and one rich in the fluorocarbon surfactant. It is important to recognize, however, that the two kinds of molecules are *not* completely immiscible. Also, the fact that hydrocarbon surfactants readily adsorb onto fluorocarbon surfaces, and vice versa, clearly indicates that hydrocarbons and fluorocarbons are not entirely incompatible.

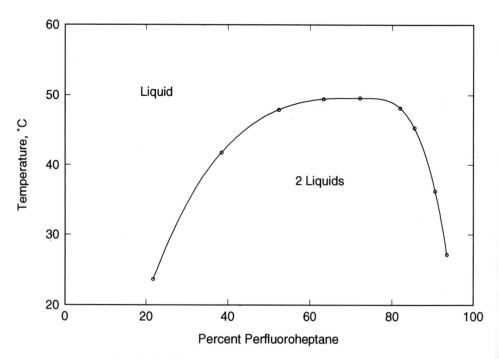

Fig. 11.12. The heptane–perfluoroheptane system [98].

Cohesive energy densities provide a useful measure of the energetics of interactions between molecules [3], and one component of these interactions is the polarizability. Polarizability may be inferred from refractive index vs. frequency data, and the group polarizabilities of hydrocarbons and fluorocarbons have been estimated [100]. These data pose an interesting puzzle, because the group polarizabilities of fluorocarbons are measurably *greater* than are those of hydrocarbons – not less.

Structure	Molar refraction ($m^3 \times 10^{30}$)
$-CH_2-$	1.800
$-CF_2-$	1.917
$-CH_3$	2.200
$-CF_3$	2.395

Phase behavior. From the limited data available, fluorocarbon surfactants appear to display more or less conventional aqueous phase behavior. Perfluorocarboxylic acids are strong acids in water [97]. This indicates not only that the carboxylate anion of a perfluoroacid is very weakly basic, but suggests as well that it may also be weakly hydrophilic in comparison to the carboxylate of a soap salt (Section 10.4). Because of their strong acidity, perfluoroalkanoic acids are anionic surfactants in their own right and (like alkanesulfonic acids, Section 10.5.1) may be expected to exist in water as hydronium perfluoroalkanoate salts. The phase behavior of perfluorononanoic acid has been investigated and its diagram is shown in Fig. 11.13. It is highly unusual as regards the span of the lamellar liquid crystal phase. Several ammonium salts of this compound have also been prepared, and some of these display similar behavior (see Appendix 2 for compounds and references).

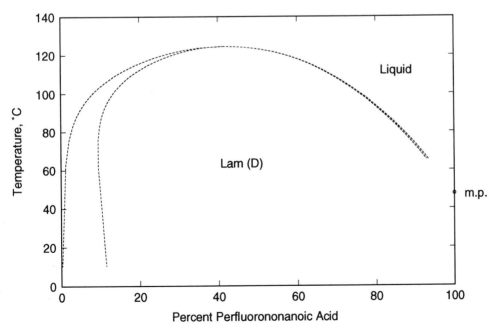

Fig. 11.13. The perfluorononanoic acid–water system [101].

Perfluoroalkanoic acids may be neutralized to form metal perfluoroalkanoate salts, and a partial phase study of one such compound (potassium perfluorooctanoate, $C_7F_{15}CO_2^-,Na^+$) has been performed [102]. The dilute end of the Krafft discontinuity lies at 59% and 52°C, and the coexisting liquid crystal phase above the Krafft discontinuity is the expected hexagonal phase. The concentration at this discontinuity is somewhat higher than might be expected, but the temperature is considerably higher. Since the melting point is not known ΔT is also unknown, but it is tempting to speculate (on the basis of the relative basicities, above) that the high value of the Krafft discontinuity is due, at least in part, to the weak hydrophilicity of the carboxylate group in this fluorocarbon salt.

The phase behavior of sodium 1,1-dihydroperfluorooctylsulfate ($C_7F_{15}CH_2OSO_3^-$, Na^+) in water has also been scanned [103]. (A synthetic route to the perfluoro compound cannot be envisioned, because the alkanol required as an intermediate is unstable due to facile elimination to form hydrogen fluoride and an acid fluoride ($RC(=O)F$). The 1,1-dihydroperfluoroalkanol, on the other hand, is both chemically stable and readily synthesized via lithium aluminum hydride reduction of the perfluoroalkanoic acids.) As found in potassium perfluorooctanoate, the Krafft discontinuity of this sulfate also lies at a very high temperature, considering its chain-length (50°C). The solubility at this temperature (50%) is somewhat lower than is found in the perfluoroalkanoate salt.

1,1-Dihydroperfluorooctyltrimethylammonium iodide ($C_7F_{15}CH_2N(CH_3)_3^+,I^-$) has also been prepared and its phase behavior investigated [103]. Since the hydrophilicity of cationic surfactants resides primarily in the anionic counterion (rather than in the amphiphilic ion), one would not expect perfluorination to strongly affect the phase behavior of this compound. However, the Krafft boundary of this ammonium iodide is 59°C, and its solubility at the (hexagonal) liquid crystal phase boundary at this temperature is 59%. The solubility value is not unexpected for a C_8 surfactant, but (again) the Krafft boundary is surprisingly high.

1,1-Dihydroperfluorooctyldimethylamine oxide has also been prepared and its phase behavior investigated [103]. Its hydrophilicity too may be expected to have been significantly reduced by fluorination, and its phase behavior is, in fact, somewhat unusual. The Krafft boundary in this instance is metastable (which is no surprise), but the solubility at 25°C is only about 5% (which is highly unusual). The liquid crystal coexisting with the saturated liquid is reported to be the lamellar phase, which is also surprising. The phase behavior of this compound is sufficiently unexpected to deserve reexamination before it can be accepted.

Partial fluorination. Two cationic surfactants which are fluorinated in the last three carbons of the hydrocarbon chain (to form a heptafluoro derivative) have been prepared, and their phase behavior scanned [103]. In these compounds the fluorine substituents are sufficiently far removed from the hydrophilic group that they are not likely to influence the hydrophilicity of the molecules, especially since they are cationic surfactant salts (above). Any perturbation of their phase behavior is therefore presumably due to perturbation of the hydrophobicity of the compounds.

The solubility of the heptafluoro C_8 trimethylammonium bromide is remarkably high (85%), and it was suggested that the lamellar phase existed (as in the amine oxide, above). Both of these results represent unusual phase behavior compared to that expected of their hydrocarbon analogs. The solubility of the heptafluoro C_{10} compound is more usual (55%), and the more familiar hexagonal and lamellar phases were observed. Both compounds displayed metastable Krafft discontinuities, as expected on the basis of their normal hydrophilicities and short chain-lengths.

11.5.2 Silicone lipophilic groups

Silicones represent another important non-hydrocarbon lipophilic moiety. Possibly the lowest air–water interfacial tension recorded to date ($c.\,16\,\text{mN m}^{-1}$) is produced by the action of a fluorocarbon surfactant [104], but the next lowest value ($c.\,20\,\text{mN m}^{-1}$) results from the action of a silicone surfactant [105]. Both fluorocarbon and silicone surfactants are capable of lowering surface tensions considerably more than are hydrocarbon surfactants. Solutions of nonionic surfactants do not typically attain surface tensions below $28\,\text{mN m}^{-1}$, and the surface tensions of solutions of ionic surfactants plateau at still higher values (35–$40\,\text{mN m}^{-1}$). Whether or not the fluorocarbon and silicone surfactants will also be more surfactive at oil–water interfaces is uncertain. Ultralow interfacial tensions ($c.\,10^{-4}\,\text{mN m}^{-1}$) are known for hydrocarbon surfactant–oil–salt–water systems (Section 12.3.3) [106].

The structures of some of the commercially important silicone surfactants, at the present time, are shown in Fig. 11.14.

Because Si–H bonds are susceptible to autoxidation, silicon analogs of hydrocarbon chains in which silicon simply replaces carbon (and Si–H bonds are retained) are not stable in air [108]. The silicon molecules that are encountered as lipophilic groups are, instead, "silicones". Silicones may be viewed as condensation polymerization products of silane diols ($R_2\text{Si(OH)}_2$. Silicone groups contain alternating silicon and oxygen atoms that are for the most part connected in a straight chain, but the chain may be branched if a trifunctional silicon atom is present. Silicone chains are typically capped with a silicon atom that bears a trimethylene group. The trimethylene group serves to link the silicone chain to the hydrophilic group. The substituents on the silicon are

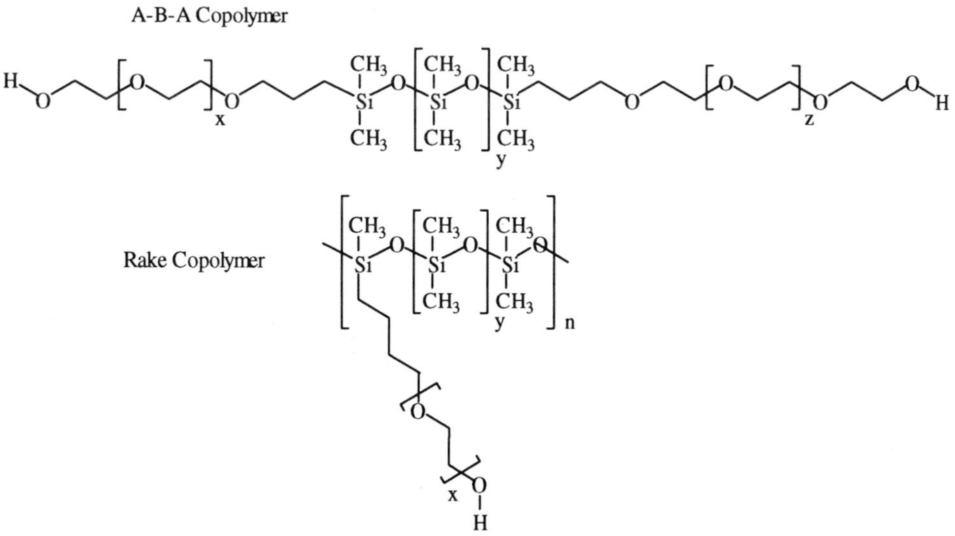

Fig. 11.14. The molecular structures of A–B–A and "rake" copolymer silicone surfactants [107].

almost invariably methyl groups, but on occasion aryl groups may be encountered. The presence of large numbers of aryl substituents in place of methyl would have an enormous effect on their physical behavior.

A silicone chain presents a "methyl" face to the world. Methyl groups have no non-bonding orbitals and occupy a significantly larger volume than do methylene groups (Section 11.3.2.2) [35]. The molecular structure of a silicone is therefore significantly different from that of an aliphatic lipophilic group, and also differs from that of a fluorocarbon group (Section 11.5.1).

The pattern of substitution in methyl silicones corresponds to an aliphatic chain that has gem-dimethyl substituents on every third carbon atom, neglecting the fact that the Si–O bond is slightly longer than the C–C bond. It is possible that such hydrocarbon chains would closely resemble silicones in their lipophilic and hydrophobic properties, but this hypothesis has not been tested. Such chains would not be expected to possess the flexibility that characterizes n-alkyl lipophilic chains.

Because silicon is in the third row in the period table (next to phosphorus), strong back-bonding occurs between the non-bonding orbitals on oxygen and the 3d orbitals of both silicon atoms to which it is attached (Section 9.4) [110]. This spreads the bond angle at oxygen from $110°$ (in a dialkyl ether) to $150 \pm 10°$ in acyclic silicones. The back-bonding also greatly reduces the electron density on oxygen [110]. These structural changes reduce the basicity of silicone oxygens, and also would be expected to reduce drastically the hydrophilicity of silicone oxygens. These electronic effects (in combination with the methyl shielding) easily rationalize the fact that the compounds are on the whole lipophilic rather than hydrophilic.

Chemical stability. Because each silicon atom (except the terminal atom) is bonded to two oxygen atoms, the chemistry of silicones is in some degree analogous to that of polyoxymethylene compounds (formaldehyde polymers). The latter compounds

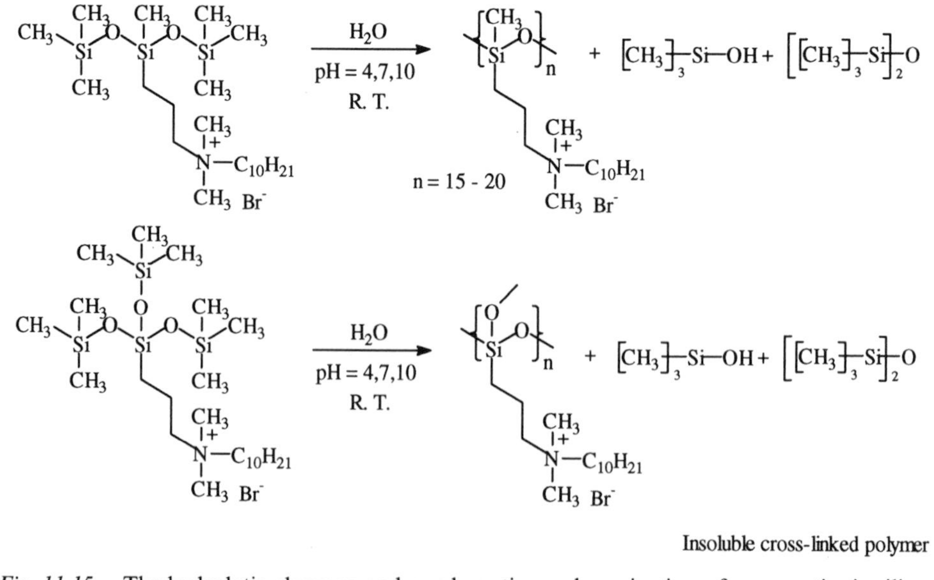

Insoluble cross-linked polymer

Fig. 11.15. The hydrolytic cleavage and condensation polymerization of some cationic silicone surfactants [109].

are stable as high molecular weight crystalline polymers, but as small oligomers are water-soluble (as in formalin) and chemically labile in aqueous solution [111]. The unsaturated electronic structure (d orbitals) of the silicon atom facilitate the hydrolytic cleavage reactions of silicones, so that both silicones and polyoxymethylene compounds are potentially unstable chemically if they are actually soluble in aqueous solution (Section 5.2) [109,111]. In striking contrast, isolated ether groups (as in polyoxyethylene compounds) are inert to hydrolytic cleavage. Perhaps due in part to the chemical stability issue, silicone surfactants are presently utilized predominantly in nonaqueous applications.

Several water-soluble cationic hybrid silicone/hydrocarbon surfactants have been prepared in which either two or three trimethylsiloxy groups are coupled to a central silicon atom [109]. The central silicon is (as above) connected by a trimethylene chain to an alkyldimethylammonio substituent group. The compounds were prepared as the bromide salts. All of these compounds were shown to be both soluble in water, and unstable as well. Hexamethyldisiloxane and trimethylsilanol result from their hydrolytic cleavage, which is complete at pH 4, 7, and 10 within a fraction of a minute (Fig. 11.15). The hydrolytic cleavage reaction is reversible, however, so that the silane di- or triols initially formed undergo condensation to form oligomeric polycationic surfactants having 10–15 monomer units. The phase behavior of these oligomeric surfactants was examined, and found to qualitatively resemble that of conventional quaternary ammonium salt surfactants of similar lipophilic chain length. Strictly speaking these compounds do not have silicone lipophilic groups. Rather, they contain a silicone backbone that links together several hydrocarbon surfactant units.

11.5.3 Polyether lipophilic groups

Triblock copolymers of propylene and ethylene oxide represent a class of surfactants that are both structurally unusual and possess distinctive properties [112]. In these compounds (Fig. 11.16) the terminal polyoxyethylene (POE) blocks serve as the hydrophilic group, while the polyoxypropylene (POP) block (usually the central block in a triblock polymer) serves as the lipophilic moiety!

The fact that these compounds are surfactants at all indicates that the substitution of a single methyl group in the ethylene moiety of a POE chain transforms this hydrophilic group into a lipophilic group. The group has to be rather large to function in this capacity. The effect is somewhat analogous to the methylation of a primary amino ($-NH_2$) group to form a methylamino ($-NHCH_3$) group, which also transforms an operative hydrophilic group into a nonoperative polar functional group (Section 10.5.5).

Phase studies of POE–POP surfactants have not been reported, but their surface activity is distinctive in at least one respect. That is the fact that they are far more compatible with living cells than are conventional surfactants having a soap-like structure [113]. Because of this property, they are used to emulsify the liquid fluorocarbons that serve to carry oxygen in blood extenders. In recent years it has been shown that perfluoroalkyl surfactants also display minimal ability to cause cell lysis, even at concentrations at which they stabilize aqueous dispersions of fluorocarbons [114].

Oligomers of tetrahydrofuran have been shown not to be hydrophilic groups, from phase data on poly(tetrahydrofuran) monoalkyl ethers [115]. The utility of these groups

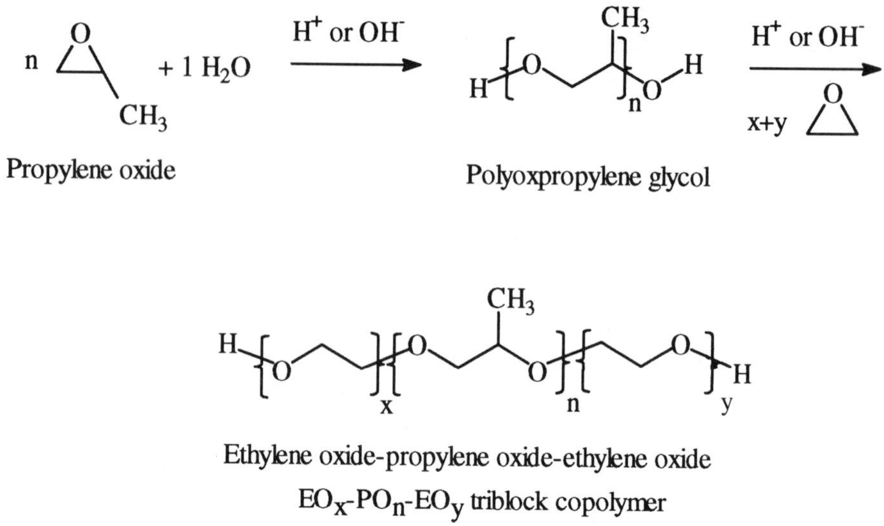

Propylene oxide   Polyoxpropylene glycol

Ethylene oxide-propylene oxide-ethylene oxide

EO_x-PO_n-EO_y triblock copolymer

Fig. 11.16 The structures of POE–POP–POE triblock surfactants.

as lipophilic groups has not been investigated, but from the investigations of other polar substituents in aliphatic chains they are probably also very poor or inactive as lipophilic groups (Section 11.3.2.8). A substituent adjacent to the ether oxygens which offers steric hindrance to the hydration of these oxygens may be required for such aliphatic polyethers to behave as lipophilic groups.

11.5.4 Polyester lipophilic groups

In recent years an interesting family of block copolymers have been reported in the patent literature which possess ethylene terephthalate and polyoxyethylene moieties, or polyoxypropylene terephthalate and sodium sulfonate moieties [116]. These materials are highly surface active, and may possibly be regarded as a distinctive class of surfactants that possess unusual lipophilic groups. Their physical science has not been firmly established by appropriate physical studies of standard materials, but if the above assumption is true the ethylene or polyoxypropylene terephthalate moiety should be regarded as an operative lipophilic group.

It is relevant in this connection that the accumulation of ester or nitrile groups in a polymer (as in polyacrylonitrile and polyacrylate esters) renders such molecules progressively more hydrophobic and less miscible with water [117]. The accumulation of ether or hydroxy substituents, in striking contrast, renders molecules more hydrophilic [118]. (This hydrophilicity may be frustrated by the existence of strong crystallinity, as in pure polyvinyl alcohol. Weakening this crystallinity, for example by partial acetylation, allows the hydrophilicity of this compound to be revealed by its high water solubility.) Also, the surfaces of polyester and polynitrile plastics tend to be hydrophobic, while those of polyhydroxy compounds (cellulose) tend to be hydrophilic. These qualitative observations regarding the consequence of polyfunctionality on physical behavior further support the classification of ester and nitrile groups as non-hydrophilic dipolar functional groups.

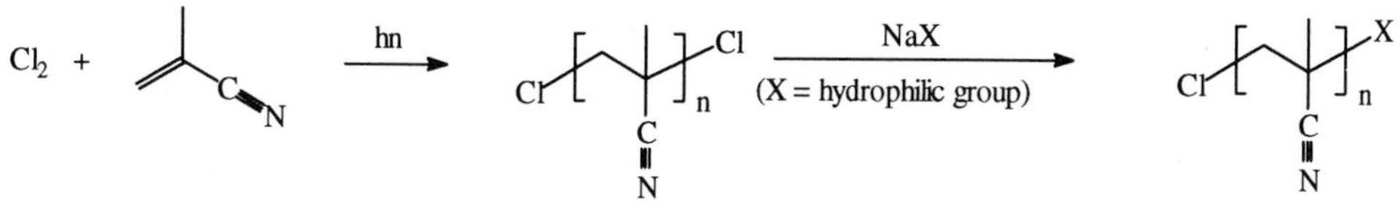

Fig. 11.17. A possible synthesis of poly(methacrylonitrile) hydrophobic groups. At $n = c.4$ or greater, the dichloro compounds produced are immiscible with liquid water [102].

11.5.5 Polynitrile lipophobic/hydrophobic groups

The possibility that a polynitrile chain might serve as the "hydrophobic" group of a surfactant has been briefly explored. It was anticipated (from the correlations between hydrophilicity and structure) that polynitriles should be both hydrophobic and lipophobic, and that surfactants constructed from these groups might display distinctive surface properties. The accumulation of many nitrile groups in close proximity to one another dramatically enhances the chemical reactivity of the resulting molecule, so that very few compounds can be expected to be sufficiently inert chemically to be of interest [119]. Exploratory studies of telomers of chlorine and methacrylonitrile nevertheless produced polynitrile compounds that lacked hydrogen substituents α to the nitrile groups (Fig. 11.17), and whose physical properties proved to be interesting [102].

It was found that these telomers did indeed become progressively less soluble in water as the number of nitrile groups increased (up to about six). A polynitrile surfactant having an operative hydrophilic was never actually synthesized, but this remains an intriguing area for future investigation.

References

1. Tanford, C. (1973). *The Hydrophobic Effect: Formation of Micelles and Biological Membranes*, pp. 74–79, John Wiley, New York.
2. Ben-Naim, A. (1980). *Hydrophobic Interactions*, Plenum, New York.
3. Hildebrand, J. H. and Scott, R. L. (1950). *The Solubility of Nonelectrolytes*, 3rd ed., Reinhold, New York.
4. Scott, R. L. (1987). *Accts. Chem. Res.* **20**, 97–107.
5. Kjellender, R. (1982). *J. Chem. Soc., Faraday Trans. 2* **78**, 2025–2042; Andersen, G. R. and Wheeler, J. C. (1978). *J. Chem. Phys.* **69**, 2082–2088, 3403–3413.
6. Sjoblom, J., Stenius, P. and Danielsson, I. (1987). *Nonionic Surfactants. Physical Chemistry*, Vol. 23 (Martin J. Schick ed.), pp. 380–381, Marcel Dekker, New York; Saeki, S., Kuwahara, N., Makata, M. and Kaneko, M. (1976). *Polymer* **17**, 685–689.
7. Kresheck, G. (1975). *Water: A Comprehensive Treatise*, Vol. 4 (F. Franks ed.), pp. 95–167, Plenum, New York.
8. Clint, J. H. and Walker, T. (1975). *J. Chem. Soc., Faraday Trans. 1* **71**, 946–954.
9. McAuliffe, C. (1966). *J. Phys. Chem.* **70**, 1267–1275.
10. Kinoshita, K., Ishikawa, H. and Shinoda, K. (1958). *Bull. Chem. Soc. Japan* **31**, 1081–1082.
11. Laughlin, R. G., Munyon, R. L., Ries, S. K. and Wert, V. F. (1983). *Science* **219**, 1219–1221.
12. Jones, D. C. and Ottewill, R. H. (1955). *J. Chem. Soc.*, 4076–4088; Hauxwell, F. and Ottewill, R. H. (1970). *J. Coll. Interface Sci.* **34**, 473–479.
13. Kloubek, J. (1989). *Collect. Czech. Chem. Commun.* **54**, 3171–3186.
14. Platford, R. F. (1980). *Can. J. Chem. Eng.* **58**, 393–395.
15. Corkill, J. M., Goodman, J. F. and Harrold, S. P. (1963). *Trans. Faraday Soc.*, 202–207.

364 THE AQUEOUS PHASE BEHAVIOR OF SURFACTANTS

16. Corkill, J. M., Goodman, J. F., Robson, P. and Tate, J. R. (1966). *Trans. Faraday Soc.*, 987–993.
17. Herrmann, K. W., Brushmiller, J. G. and Courchene, W. L. (1966). *J. Phys. Chem.* **70**, 2909–2918.
18. Corkill, J. M. and Herrmann, K. W. (1963). *J. Phys. Chem.* **67**, 934–937.
19. Haydon, D. A. and Taylor, F. H. (1962). *J. Chem. Soc. Faraday Trans.* **58**, 1233–1250.
20. Markina, Z. N., Tsikurina, N. N., Kostova, N. Z. and Rehbinder, P. A. (1964). *Koll. Zh.* **26**, 76–82.
21. Black, C., Joris, G. G. and Taylor, H. S. (1948). *J. Chem. Phys.* **16**, 537–543.
22. Lewis, G. N. and Randall, M. (revised by K. S. Pitzer and L. Brewer) (1961). *Thermodynamics*, 2nd ed., pp. 224–241, McGraw-Hill, New York.
23. J. Lyklema, personal communication.
24. Madelmont, C. and Perron, R. (1976). *Colloid and Polymer Sci.* **254**, 6581–6595.
25. Laughlin, R. G. (1978). *Advances in Liquid Crystals*, Vol. 3 (G. H. Brown ed.), pp. 99–148, Academic Press, New York.
26. Nilsson, P-G., Lindman, B. and Laughlin, R. G. (1984). *J. Phys. Chem.* **88**, 6357–6362.
27. McBain, J. W., Bunbury, H. M. and Martin, H. E. (1914). *J. Chem. Soc.* **105**, 419–433.
28. (1965). *Solubilities*, 4th ed., Vol. II, pp. 851–852, American Chemical Society, Washington, DC.
29. Ray, A. and Nemethy, G. (1971). *J. Am. Chem. Soc.* **93**, 6787–6793.
30. Schwartz, A. M. and Perry, J. W. (1978). *Surface Active Agents: Their Chemistry and Technology*, pp. 307–313, R. E. Krieger, Huntington, New York; Saleh, A. M., Badwan, A. A. and El-Khordagui, L. K. (1983). *Int. J. Pharm.* **17**, 115–119; Friberg, S. and Chiu, M. (1989). *J. Dispersion Sci. Tech.* **9**, 443–457.
31. (1965). *Solubilities* (W. F. Linke ed.), 4th ed., Vol. II, pp. 851–852, American Chemical Society, Washington, DC.
32. Goldstein, R. F. and Waddams, A. L. (1967). *The Petroleum Chemicals Industry*, 3rd ed., pp. 285, 458–459, E. & F. N. Spon, London; Feighner, G. C. (1976). *Anionic Surfactants*, Vol. 7 (W. M. Linfield ed.), pp. 288–314, Marcel Dekker, New York.
33. Swisher, R. D. (1987). *Surfactant Biodegradation*, 2nd ed., Vol. 18, p. 25, Marcel Dekker, New York.
34. O. T. Quimby, unreported work.
35. Dreisbach, R. R. (1959). *Advances in Chemistry Series, Physical Properties of Chemical Compounds – II*, Vol. 22, American Chemical Society, Washington, DC.
36. Small, D. M. (1986). *The Physical Chemistry of Lipids. From Alkanes to Phospholipids*, Vol. 4, pp. 23–25, 191–196, Plenum, New York.
37. D. F. Searle, California Research Corporation, unreported work.
38. Ockelford, J., Tminini, B. A., Narayan, K. S. and Tiddy, G. J. T. (1993). *J. Phys. Chem.* **97**, 6767–6769.
39. Franses, E. I., Puig, J. E., Talmon, Y., Miller, W. G., Scriven, L. E. and Davis, H. T. (1980). *J. Phys. Chem.* **84**, 1547–1556.
40. March, J. (1992). *Advanced Organic Chemistry. Reactions, Mechanisms, and Structure*, 4th ed., pp. 339–340, 352–357, John Wiley, New York.
41. J. S. Clunie, unreported work.
42. R. E. Zimmerer, unreported work.
43. Doerfler, H.-D. (1992). *Tenside Surf. Det.* **29**, 351–358.
44. Streefland, L., Wagenaar, A., Hoekstra, D. and Engberts, J. B. F. N. (1993). *Langmuir* **9**, 219–222; Fonteijn, T. A., Hoekstra, D. and Engberts, J. B. F. N. (1992). *Langmuir* **8**, 2437–2447; Okahata, Y., Ando, R. and Kunitake, T. (1981). *Ber. Bunsen-Ges. Phys. Chem.* **85**, 789–798.
45. Warr, G. G., Sen, R., Evans, D. F. and Trend, J. E. (1988). *J. Phys. Chem.* **92**, 774–783.
46. Fontell, K., Ceglie, A., Lindman, B. and Ninham, B. (1986). *Acta Chem. Scand. A* **40**, 247–256.
47. Kekicheff, P., Grabielle-Madelmont, C. and Ollivon, M. (1989). *J. Coll. Int. Sci.* **131**, 112–132.
48. Laughlin, R. G., Munyon, R. L., Burns, J. L., Coffindaffer, T. W. and Talmon, Y. (1992). *J. Phys. Chem.* **96**, 374–383.

49. Evans, D. F. and Ninham, B. W. (1986). *J. Phys. Chem.* **90**, 226–234.
50. Rogers, J. and Winsor, P. A. (1969). *J. Coll. Interface Sci.* **30**, 247–257.
51. Middleton, M. A., Schechter, R. S. and Johnston, K. P. (1990). *Langmuir* **6**, 920–928; Robertus, C., Joosten, J. G. H. and Levine, Y. K. (1988). *Prog. Colloid Polym. Sci.* **77**, 115–119; Dutkewicz, E. and Robinson, B. H. (1988). *J. Electroanal. Chem. Interfacial Electrochem.* **251**, 11–20; Easthoe, J., Fragneto, G., Robinson, B. H., Towey, T. F., Heenan, R. K. and Leng, F. J (1992). *J. Chem. Soc., Faraday Trans.* **88**, 461–471; Kunieda, H. and Shinoda, K. (1980). *J. Colloid Interface Sci.* **75**, 601–606; Barelli, A. and Eicke, H. F. (1986). *Langmuir* **2**, 780–786; Magid, L. J. and Martin, C. A. (1984). *Reverse Micelles [Proc. Eur. Sci. Found. Workshop], 4th*, (P. L. Luisi and B. E. Straub eds), pp. 181–193, Plenum, New York.
52. J. M. Corkill and J. F. Goodman, unreported work.
53. Swern, D. (1964). *Bailey's Industrial Oil and Fat Products*, 3rd ed., pp. 3–55, Interscience, John Wiley, New York.
54. Reference 40, pp. 151–152, 755–758.
55. Frankel, E. N. (1985). *Flavor Chemistry of Fats and Oils. AOCS Monograph No. 15.*, (D. B. Min and T. H. Smouse eds), pp. 1–37, American Oil Chemists' Society, Champaign IL.
56. Hamburger, R., Azaz, E. and Donbrow, M. (1975). *Pharm. Acta Helv.* **50**, 10–17; Donbrow, M., Hamburger, R. and Azaz, E. (1975). *J. Pharm. Pharmacol.* **27**, 160–166; Lin, Z. (1986). *Yaoxue Tongbai* **21**, 626–627.
57. Emsley, J. W. (1985). *Nuclear Magnetic Resonance of Liquid Crystals (NATO ASI Series)*, D. Reidel, Dordrecht.
58. Small, D. M. (1986). *The Physical Chemistry of Lipids. From Alkanes to Phospholipids*, Vol. 4, pp. 585–602, *Handbook of Lipid Research*, Plenum, New York.
59. H. L. Strauss, personal communication.
60. McBain, J. W. and Elford, W. J. (1926). *J. Chem. Soc.*, 421–438; Laing, M. E. and McBain, J. W. (1920). *Trans. Chem. Soc.* **117**, 1506–1528.
61. Luehmann, B. and Finkelmann, H. (1966). *Colloid Polym. Sci.* **264**, 189–192.
62. A. O. Snoddy and R. J. Blickenstaff, unreported work.
63. Hauser, C. R. and Chambers, W. J. (1956). *J. Am. Chem. Soc.* **78**, 3837–3841; Hill, D. G., Burkus, J., Luck, S. M. and Hauser, C. R. (1959). *J. Am. Chem. Soc.* **81**, 2787–2788.
64. Kahn, A., Das, K.P., Eberson, L. and Lindman, B. (1988). *J. Coll. Interface Sci.* **125**, 129–138.
65. Lenhinger, A. L. (1975). *Biochemistry*, 2nd ed., pp. 279–308, Worth, New York.
66. Prosise, W. E. (1985). *AOCS Monogr. 12(Lecithins)*, Vol. 12, pp. 163–182, American Oil Chemists' Society.
67. Pascher, I., Lundmark, M., Nyholm, P.-G. and Sundell, S. (1992). *Biochim. Biophys. Acta*, **1113**, 339–373.
68. Larsson, K. (1986). *Chemistry and Physics of Baking, Special Publication No. 56*, (J. M. V. Blanshard, P. J. Frazier, and T. Galliard eds), pp. 62–74, The Royal Society of London, Burlington House, London W1V 0BN.
69. Klenk, E. and Debuch, H. (1963). *Progr. Chem. Fats Lipids*, Vol. 6 (R. T. Holman, W. O. Lundberg, and T. Malkin eds), pp. 1–29, Pergamon, London.
70. Arvidson, G., Brentel, I., Kahn, A., Lindblom, G. and Fontell, K. (1985). *Eur. J. Biochem.* **152**, 753–759.
71. NIST Standard Reference Database 34, *Lipid Thermotropic Phase Transition Database*, available from the National Institute of Standards and Technology, 221/A320, Gaithersburg, MD 20896.
72. Evans, E. and Needham, D. (1987). *J. Phys. Chem.* **91**, 4219–4228.
73. Hyde, A. J., Langbridge, D. M. and Lawrence, A. S. C. (1954). *Discuss. Faraday Soc.* **18**, 239–257.
74. Laughlin, R. G. (1992). *Adv. Coll. Interface Sci.* **41**, 57–79.
75. Luzzati, V. (1968). *Biological Membranes, Physical Fact and Function*, (D. Chapman and D. F. H. Wallach eds), pp. 71–124, Academic Press, New York.
76. Tausk, R. J. M., Oudshoorn, C. and Overbeek, J. Th. G. (1974). *Biophys. Chem.* **2**, 53–63.
77. Shipley, G. G., Green, J. P. and Nichols, B. W. (1973). *Biochim. Biophys. Acta* **311**, 531–544.

78. Albon, N. (1983). *J. Chem. Phys.* **78**, 4676–4686.
79. Ulmius, J., Wennerstrom, H., Lindblom, G. and Arvidson, G. (1977). *Biochemistry* **16**, 5742–5745.
80. Grabielle-Madelmont, C. and Perron, R. (1983). *J. Coll. Interface Sci.* **95**, 471–482.
81. Seddon, J. M., Cevc, G. and Marsh, D. (1983). *Biochemistry* **22**, 1280–1289.
82. Vincent, J. M. and Skoulios, A. (1966). *Acta Cryst.* **20**, 432–440, 441–447, 447–451.
83. Barnes, H. A., Hutton, J. F. and Walters, K. (1989). *An Introduction to Rheology*, Elsevier, Amsterdam; Ferry, J. D. (1980). *Viscoelastic Properties of Polymers*, 3rd ed., p. 529, John Wiley, New York.
84. Kawai, T., Umemura, J., Takenaka, T., Kodama, M., Ogawa, Y. and Seki, S. (1986). *Langmuir* **2**, 739–743.
85. Laughlin, R. G., Munyon, R. L., Fu, Y.-C. and Fehl, A. J. (1990). *J. Phys. Chem.* **94**, 2546–2552.
86. Albon, N. and Sturtevant, J. M. (1978). *Proc. Natl. Acad. Sci. USA* **75**, 2258–2260.
87. Chevalier, Y., Germanaud, L. and Le Perchec, P. (1988). *Coll. Polym. Sci.* **266**, 441–448.
88. Weil, J. K., Stirton, A. J. and Wrigley, A. N. (1968). *C. R. Congr. Int. Deterg., 5th*, 9 Sept.–13 Sept., 1968 , Vol. 1, pp. 45–50, Ediciones Unidas, S.A., Barcelona.
89. E. P. Gosselink, unreported work.
90. Roseman, T. J. and Yalkowsky, S. H. (1973). *J. Pharm. Sci.* **62**, 1680–1685.
91. Griffin, W. C. (1947). *J. Soc. Cos. Chemists* **1**, 311–326.
92. Becher, P. (1965). *Emulsions: Theory and Practice*, 2nd ed., Reinhold, New York.
93. Becher, P. and Griffin, W. C. (1974). *Detergents and Emulsifiers/North American Edition*, McCutcheon, Glen Rock, New Jersey.
94. Knight, P., Wyn-Jones, E. and Tiddy, G. J. T. (1985). *J. Phys. Chem.* **89**, 3447–3449.
95. Kahlweit, M. (1982). *J. Colloid Interface Sci.* **90**, 92–99.
96. Laughlin, R. G. (1981). *J. Soc. Cosmet. Chem.* **32**, 371–392.
97. Patrick, C. R. (1962). *Advances in Fluorine Chemistry*, Vol. 2 (M. Stacey, J. C. Tatlow, and A. G. Sharpe eds), p. 1, Butterworths, London; Banks, R. E. (1964). *Fluorocarbons and Their Derivatives*, pp. 12–15, Oldbourne, London.
98. Hildebrand, J. H., Fisher, B. B. and Benisi, H. A. (1950). *J. Am. Chem. Soc.* **72**, 4348–4351.
99. Tamori, K., Ishikawa, A., Kihara, K., Ishii, Y. and Esumi, K. (1992). *Colloids Surf.* **67**, 1–8; Pletnev, M. Yu., Remizov, Yu. V. and Soya, A. V. (1989). *Kolloid Zh.* **51**, 1025–1029.
100. Clint, J. H. and Walker, T. (1974). *J. Colloid Interface Sci.* **47**, 172–185.
101. Fontell, K. and Lindman, B. (1983). *J. Phys. Chem.* **87**, 3289–3297.
102. R. G. Laughlin, unreported work.
103. J. S. Clunie *et al.*, unreported work.
104. Burkitt, S. J., Ingram, B. T. and Ottewill, R. H. (1988). *Prog. Colloid Polym. Sci.* **76**, 247–250.
105. Ananthapadmanabhan, K. P., Goddard, E. D. and Chandar, P. (1990). *Colloids Surf.* **44**, 281–297.
106. Langevin, D. (1987). *Microemulsions: Structure and Dynamics*, (S. E. Friberg and P. Bothorel eds), pp. 173–196, CRC Press, Boca Raton, FL.
107. Gruening, B. and Koerner, G. (1989). *Tenside, Surfactants, Deterg.* **26**, 312–317; Hill, R. M., He, M., Davis, H. T. and Scriven, L. E. (1993). *Langmuir* (in press); Gradzielski, M., Hoffmann, H., Robisch, P. and Ulbricht, W. (1990). *Tenside Surf. Det.* **27**, 366–379; Lin, Z., He, M., Davis, L. E. and Snow, S. A. (1993). *J. Phys. Chem.* **97**, 3571–3578.
108. Rochow, E. G. (1951). *Chemistry of the Silicones*, 2nd ed., pp. 5–6, John Wiley, New York.
109. J. J. Yetter, unreported work.
110. Cotton, F. A. and Wilkinson, G. (1988). *Advanced Inorganic Chemistry*, 5th ed., pp. 58–76, John Wiley, New York; West, R., Whatley, L. S. and Lake, K. J. (1961). *J. Am. Chem. Soc.* **83**, 761–764.
111. Walker, J. F. (1944). *Formaldehyde (American Chemical Society Monograph)*, Vol. 98, Reinhold, New York.
112. Karpinska-Smulikowska, J. (1984). *Tenside, Detergency* **21**, 243–246; Szymanowski, J. (1980). *Fette, Seifen, Anstrichm.* **82**, 59–63; Nuernberg, E. and Friess, S. (1990). *Pharm. Acta Helv.* **65**, 105–112.
113. Clark, L. C., Jr., Becattini, F. and Kaplan, S. (1972). *Ala. J. Med. Sci.* **9**, 16–29.

114. Mathis, G., Leempoel, P., Ravey, J. C., Selve, C. and Delpuech, J. J. (1984). *J. Am. Chem. Soc.* **106**, 6162–6271; Nivet, J. B., Le Blanc, M. and Riess, J. G. (1992). *J. Dispersion Sci. Technol.* **13**, 627–646.
115. J. M. Janusz, unreported work.
116. Gosselink, E. P. (1987). *Eur. Pat. Appl.* EP 241984, A2, 21 Oct. 1987; Scheibel, J. J. and Gosselink, E. P. (1990). *Eur. Pat. Appl.* EP 357280, A2, 7 Mar. 1990.
117. Hsieh, Y. L. and Yu, B. (1992). *Text. Res. J.* **62**, 677–685; Yamaishi, K., Ayame, T. and Sanuki, H. (1978). *Sen'i Gakkaishi* **34**, T49–T52.
118. Modi, T. W. (1980). *Polyvinyl Alcohol. Handbook of Water-Soluble Gums and Resins, 20/1–20/32*, (R. L. Davidson ed.), McGraw-Hill, New York; Lockhead, R. Y. (1992). *Cosmet. Toiletries* **107**, pp. 131 ff.
119. Ciganek, E., Linn, W. J. and Webster, O. W. (1970). *The Chemistry of the Cyano Group*, (Z. Rappoport ed. (S. Patai, Series editor)), pp. 423–638, Interscience, John Wiley, London.

Chapter 12

The Influence of Third Components on Aqueous Surfactant Phase Behavior

12.1 Fundamentals of ternary phase science

The focus to this point has been almost entirely on binary systems of a surfactant and water and on relevant unary systems. Understanding these is absolutely essential if one is to comprehend surfactant phase science, but the influence of third components on aqueous surfactant phase behavior is also very important. One reason is that no industrial system contains only two components (a pure surfactant and water). Another is that many aspects of the physical science of surfactants are explored by considering the influence of a third component (a salt or an oil, for example) on selected properties of the system.

Comprehensive investigations of phase behavior which span the entire composition range and also a range of temperatures have been conducted only on a tiny number of ternary surfactant systems; a large fraction of these are considered later in this chapter. Also, no comprehensive ternary diagram has been determined at a temperature that is below the Krafft eutectic of the binary aqueous surfactant system. To do so would require confronting the crystal chemistry of surfactants, which has not been an object of serious research in this area.

The influence of pressure on the solution chemistry of surfactants has been investigated [1], and limited investigations of the influence of pressure on the Krafft boundary have been performed (although not described as such) [2]. It is therefore impossible to consider seriously the influence of this variable.

Many aspects of phase science are independent of the number of components present. Among these are the process of defining systems (Section 3.6), the free energy of mixing concepts (Section 3.10), the conditions which define equilibrium of state (Section 3.15), the meaning of "phase behavior" (Section 4.1), and the concepts of phase and interface (Section 4.18). The premise that partial miscibility arises from anomalies in the free energy of mixing of homogeneous states was developed for binary systems in Chapter 3, and there is every reason to believe that this premise applies equally well to systems of more than two components.

To depict free energies of mixing graphically for a ternary system requires a three-dimensional figure. A particularly useful geometric form would be a triangular prism, in which composition is defined by the coordinates within a triangular cross-section and free energy (at a particular temperature and pressure) is scaled along the prism axis. The "double-tangent" and "triple-tangent" lines of binary systems (Sections 3.14, 3.18) become "double-tangent" and "triple-tangent" planes within such a prism, but the slopes of these planes would be defined in a manner very similar to that in binary systems. Miscibility gaps should result from anomalies in the free energy surface that are entirely analogous to those which occur in the free energy curve in binary systems.

12.1.1   Ternary diagrams

12.1.1.1   Reading compositions

In binary diagrams composition can be depicted using only one dimension (a line), while in ternary diagrams two dimensions (a plane) are required. (Three (a solid figure) are required in the diagrams of quaternary systems.) In order to depict composition in undistorted form within a ternary isotherm, an equilateral triangular plot (sometimes referred to as a "Gibbs triangle" [3]) is used (Fig. 12.1). The phase behavior at a particular temperature (or pressure) may be expressed as a function of composition within such a triangle, and from a stack of triangular diagrams generated at different temperatures a triangular prism diagram may be constructed. Temperature is scaled along the prism axis in such a figure. Just as pressure is presumed to be constant (or its variation insignificant) in binary diagrams, so it is in ternary prisms. Phase regions which span an area within a binary diagram span the volume within a solid figure in a ternary diagram, and the boundaries of one-phase regions are surfaces instead of lines. One could equally well imagine prisms that describe phase behavior at constant temperature, in which case pressure would vary along the prism axis.

The lines at the vertices of ternary prisms have the dimension of temperature and constitute the unary diagrams of the three components (Section 4.4). The face planes have the dimensions of composition and temperature, and display the three possible binary diagrams (from components $1 + 2$, $2 + 3$, and $1 + 3$). The core of the prism displays the phase behavior of mixtures containing all three components.

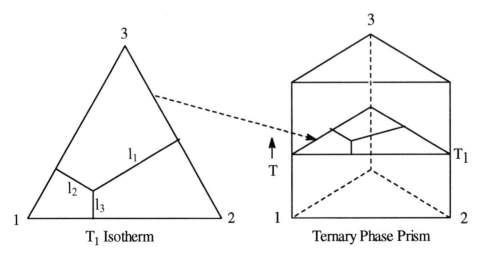

Fig. 12.1. Depiction of the triangular prism diagram of a ternary system (right). One of the triangular ternary isotherms (at T_1) is extracted from the prism, and displayed within the plane of the paper to the left. The lengths of the three perpendiculars l_1, l_2, and l_3 are proportional to the fraction of each component (w_1, w_2, and w_3) in the designated mixture.

A mixture of defined composition corresponds to a particular point within any triangular cross-section. The lengths of the perpendiculars dropped from the co-ordinates of a mixture to each of the three sides (l) are proportional to the fractions of the components present (Fig. 12.1). This is also true in ordinary Cartesian graphs, but reading the triangular plot can be confusing because the angle between its axes is 60° rather than 90°. In Fig. 12.1 the fraction of component 1 (w_1) is proportional to the length of the line l_1, which is the perpendicular dropped from the coordinates of the mixture to the side of the triangle opposite vertex 1 (side 2–3). The fractions of components 2 and 3 are depicted similarly.

One often sees in the literature ternary diagrams in which the vertices are not labeled, the fraction of each component is designated along one of the sides, and the tick marks are angled so as to be parallel to a particular side. For example, the fraction of component 2 might be designated by a label below the base of the triangle (side 1–2) in Fig. 12.1, and the tick marks oriented parallel to side 1–3. The fraction of component 3 might be designated along the right side and the tick marks oriented parallel to side 1–2, and the fraction of component 1 indicated along the left side with the tick-marks oriented parallel to side 2–3. This method of designating composition is equivalent to that described above; the choice is a matter of personal preference.

Just as the sum of the three perpendiculars equals the altitude of an equilateral triangle (from a purely geometric perspective), the sum of the weight fractions of the three components equals one (from consideration of mass balance). If the "mixture" lies at a vertex, the perpendicular to the opposite side *is* an altitude of the triangle, the numerical values of the other two perpendiculars equal zero, and such a mixture is a pure component. If the coordinates of a mixture fall on a side of the triangle, the value of one perpendicular equals zero but the other two are finite and these (binary) mixtures include two of the three components. If the coordinates of a mixture lie within the triangle, all three perpendiculars have a finite value and such (ternary) mixtures contain all three components [3].

It is desirable, on occasion, to distort the equilateral triangular plot in order to expand selected regions [4]. This is usually accomplished by plotting the data within a right triangle; an example is the phase diagram of the sodium palmitate–salt–water system (Figs 3.2 and 12.5). In such diagrams the angle between the axes at the water vertex is 90°, and the scale of composition along one axis is expanded relative to the scale along the other axis. Expanding the salt axis allows one to perceive much more easily the influence on phase behavior of small fractions of salt, but one must recognize the scaling of compositions is distorted in such diagrams. A locus of mixtures of constant salt composition is a vertical line in Fig. 3.2 while a locus of mixtures of constant soap composition is a horizontal line. (The reverse is true in Fig. 12.5.) The fraction of each of the three components may be inferred from the length of the perpendiculars to each side just as in Fig. 12.1, but the scaling factor (f_i) relating composition to length is different for each perpendicular. For a non-equilateral triangle

$$w_i = \frac{f_i \times l_i}{(f_1 \times l_1 + f_2 \times l_2 + f_3 \times l_3)} \tag{12.1}$$

It is worth noting that some plots of phase data in the literature do not conform to either of the above patterns. An interesting example is McBain's depiction of the liquid phase region of the sodium palmitate–sodium chloride–water system (Fig. 12.6). Here the fractional molar quantities of sodium palmitate or sodium chloride present are scaled linearly with respect to the perpendiculars to the sodium chloride–water or sodium palmitate–water sides, respectively, but the quantity of water present is constant (1 kg). Therefore, water composition is *not* scaled in a linear manner in this plot. While such plots may serve specific purposes, they are difficult to read and are to be avoided in general.

12.1.1.2   Isotherms and isopleths

Isotherms. An isotherm in any phase diagram is the locus of points that have the same temperature but vary in composition. In the usual binary diagram an isotherm is a horizontal line that has the dimension of composition. In a ternary prism an isotherm is a triangular plane right cross-section of the prism; this plane figure provides the two dimensions required to depict composition graphically at a particular temperature.

In a binary system isothermal mixing processes follow a path that is defined within the phase diagram merely by stating temperature (and pressure), but in a ternary system defining the path followed during isothermal mixing requires more information. The compositions of both of the two mixtures that are being mixed must be stated in order to define isothermal mixing paths in ternary systems (Section 12.1.1.3).

Isopleths. An isopleth in any phase diagram is the locus of points that have the same gross composition but vary in temperature. In a binary diagram an isopleth is a vertical line corresponding to a particular composition. In a ternary prism, an isopleth is also a vertical line (parallel to the vertices of the prism) which passes through the prism at the coordinates of the stated composition. Isopleths in all systems have only one dimension, and that dimension is temperature. (If composition and temperature were to be held constant and pressure varied, such a path would be termed an "isobar".)

An isoplethal process path follows an isoplethal line regardless of the number of components present. The isoplethal process (heating and cooling mixtures without changing composition) is surely the most commonly performed process in all of experimental science, and it is puzzling that the word itself is relatively unfamiliar.

12.1.1.3   Loci of compositions

In addition to specifying the composition of a specific mixture, it is often useful to define the locus of a range of compositions that are constrained in some manner. For example, all mixtures that have a fixed composition of component 2 (w_2), but variable fractions of components 1 and 3, will fall on line $a–b$ in Fig. 12.2A. This line is parallel to side 1–3, and is separated from this side by that distance l_2 which corresponds to the weight fraction w_2. A line describing mixtures that are constant in component 3 (w_3), but variable in 1 and 2, would fall on the line $c–d$, which is parallel to side 1–2 at a distance l_3. The point where these two loci cross (e) defines the specific mixture whose composition is w_2, w_3. For this mixture the value of w_1 is defined by consideration of mass balance, is not independently variable, and is not required in order to define composition fully. Since any two loci may serve to uniquely define composition, any of the three vertices may serve as the origin in triangular graphs. If compositions are expressed as the fraction of the surfactant and a third component, it is convenient to

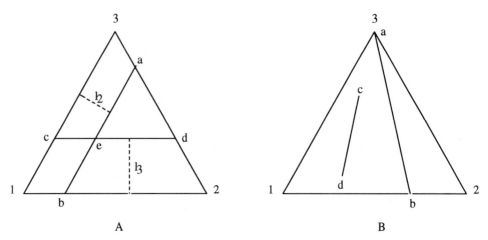

Fig. 12.2. A. A figure showing the locus of lines *a–b* (of constant fraction of component 2) and *c–d* (of constant fraction of component 3), which define the composition of mixture *e*. B. The line *a–b* is the locus of mixtures of *a* with *b* and the line *c–d* is the locus of mixtures of *c* with *d*.

place water at the lower left vertex. However, this is an arbitrary choice and all possible orientations of ternary phase triangles are found in the literature.

Another important kind of locus is the one which is followed by mixing together two different mixtures in varying proportions. To illustrate, a mixture having a particular ratio of components 1 and 2 would fall at a specific point (e.g. *b*) on the 1–2 border (Fig. 12.2B). Adding to this mixture a second "mixture" consisting of the pure component 3 (point *a*) may only produce those mixtures which fall on a straight line between points *a* and *b*. Similarly, if the ternary mixture *c* is mixed with mixture *d* (Fig. 12.2B), only those mixtures which fall on the line *c–d* may be formed. These observations illustrate the general principle that *mixing any two different mixtures, a and* b, *may only produce those mixtures which fall on the straight line between the coordinates of* a *and* b *in the phase diagram.* This rule applies only to gross compositions – *not* to the compositions of coexisting phases. It is usual for the compositions of phases to lie off the mixing path (Section 12.2.1).

It is evident that the projection of any of these loci over a span of temperatures (or pressures) will result in a vertical plane surface within the composition/temperature (pressure) prism. To illustrate, the locus of mixtures which conform to the condition that w_2 is constant will fall within a plane surface parallel to the 1–3 face of the ternary prism at a distance, l_2, from this face corresponding to the numerical value of w_2 (line *a–b*, Fig. 12.2A).

While data within such sections can provide useful information [5,6], (to quote Masing) "in general it is impossible to reach any conclusions concerning composition by means of plane section diagrams" [7]. That is, knowledge of the compositions of coexisting phases is lacking in these sections. Diagrams loosely related to such plane sections are often plotted for ternary mixtures of two polar lipids in water, except that when the mixtures have been prepared in "excess water" the actual location of the data within the prism is not even defined. (It is also of interest that mixtures of two polar lipids in water are often described as "binary" systems, which is a serious (and puzzling) error in lieu of the fact that water is present in all the liquid crystal and in many of the crystal phases that these compounds form. This tendency to ignore the water composition in studies of polar lipid systems may be attributed to reference [8].)

12.1.1.4 Tie-lines in two-phase mixtures

In binary diagrams the limits of two-phase regions are designated by stating the compositions (at a particular temperature and pressure) of the coexisting phases, which define the ends of tie-lines (Section 4.8.2). The horizontal orientation of such tie-lines is dictated by the condition of equilibrium, since both phases must be at the same temperature. Sometimes the coexisting phases are part of the same one-phase region (as in the liquid/liquid miscibility gap), but often they are not because the two phases have different structures.

Comparable situations exist in isothermal ternary diagrams. Sometimes two coexisting phases have similar phase structures and lie within the same phase region. This situation is extremely common in the case of liquid phases, but may (in principle) also occur with other kinds of phases. The coexisting phases may also differ in structure, in which case each is found within discrete and unconnected regions.

When coexisting compositions do lie on the same phase boundary, the orientation of tie-lines within the miscibility gap is undefined except in the limit that the fraction of one component equals zero. Such boundaries extend from one end of a binary miscibility gap (along the edge of a triangle) into the center of the triangle, and then return to the binary composition of the coexisting phase at the other end of the same binary miscibility gap. (No rule requires that biphasic regions must touch the binary boundaries of ternary diagrams. Such behavior is not often encountered in surfactant systems, but see Section 13.3.2.) In Fig. 12.3 the tie-lines within the propanol–heptane–water system fan out from the heptane corner of the triangle, and at the position of point (a) shrink to zero length and vanish. The position on the miscibility gap boundary at which this occurs is termed a "plait point". Plait points are isothermal critical points, and phases near plait points display critical behavior (Section 5.10.1) [9].

12.1.1.5 Tie-triangles and three-phase mixtures

Three-phase mixtures in ternary systems are designated by "tie-triangles" (Fig. 12.4). Each of the three coexisting phases (a, b, and c) lie at a particular point, and the lines connecting these points define a (usually irregular) tie-triangle. By analogy with the process of defining composition (Section 12.1.1.1), mixtures at the vertices of tie-triangles consist of a single phase, those on a side consist of two phases, and those inside the triangle consist of three phases.

The alternation rule (Section 4.12) applies universally to phase behavior, irrespective of the number of components [11]. Therefore, two-phase regions necessarily exist adjacent to all three sides of such triangles, one-phase regions exist next to two-phase regions and touch the vertices of the triangle (Section 12.2.2), and the progression of phases may only occur by the insertion or deletion of phases in a logical manner (Section 5.6.10).

12.1.1.6 Phase ratios and the lever rule in ternary diagrams

While the lever rule is necessarily applicable to the calculation of phase fractions in ternary mixtures, its form is somewhat different than in binary systems (Section 4.9)

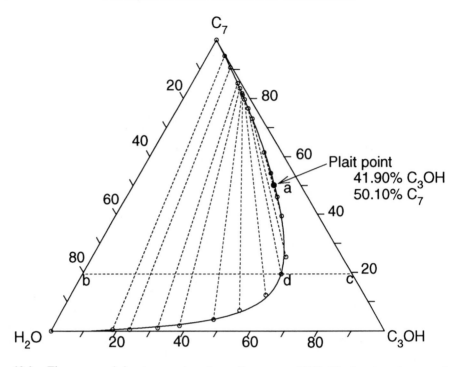

Fig. 12.3. The propanol–heptane–water phase diagram at 25°C [9], showing the two-phase liquid/liquid miscibility gap region. The plait point is at *a*, and the locus of points for which $w_{\text{heptane}} = 19.4\%$ is line *b–c*. The mixture on the boundary for which this is true is point *d*.

and it is more complex to apply. It is useful to consider first the calculation of phase fractions when three phases (each of defined composition) coexist.

Three-phase mixtures. To calculate the fraction of each phase present in mixture *m* within the tie-triangle *a–b–c* (Fig. 12.4), the area of the inner triangle opposite each vertex is divided by the total area of the triangle [12]. That is,

$$w^a = \frac{A(m-b-c)}{A(a-b-c)}, \quad w^b = \frac{A(m-c-a)}{A(a-b-c)}, \quad w^c = \frac{A(m-a-b)}{A(a-b-c)},$$

where "*A*" symbolizes area. This procedure is directly analogous to that utilized to calculate phase ratios in binary systems (Section 4.9) where the fraction of each phase equals the opposite lever arm divided by the length of the tie-line, except that areas are utilized instead of lengths and three "lever triangles" must be considered instead of two "lever-arms".

Performing this calculation is in principle straightforward, but is in practice laborious. Therefore, a computer program for calculating phase fractions in tie-triangles is provided in Appendix 3. The inputs are two composition parameters for each of the three coexisting phases (*a*, *b*, and *c*) plus the mixture (*m*). The outputs are the fractions of the three phases that are present. A test is also provided (before the calculation is executed) as to whether or not a particular mixture actually falls within the tie-triangle. Ascertaining whether or not this is true by inspection is difficult when a mixture is near the side of a tie-triangle, and is virtually impossible in the case of thin triangles (such as the "soap-boiler's triangle" (Fig. 2.4, Section 14.4)).

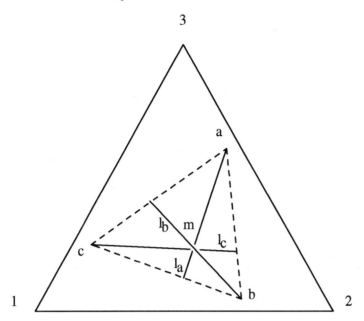

Fig. 12.4. A three-phase tie-triangle $a-b-c$ in a system of components 1, 2, and 3, showing the lever arms used to calculate the ratios of the phases (a, b, and c) that coexist within the triangle for the mixture m.

Two-phase mixtures. The approach to the calculation of phase fractions in two-phase mixtures in ternary systems is very similar to this calculation in binary systems in principle (Section 4.9), but is actually less straightforward than is the calculation of phase ratios in three-phase mixtures (above). The problem is that the calculation requires knowledge of the compositions of the coexisting phases, and ascertaining these compositions is difficult. This is true even in those systems for which the boundary is well documented [9], as in Fig. 12.3. Unless one is exactly on an experimentally defined tie-line (which also cannot be ascertained simply by inspection), the relevant tie-line cannot be easily extracted.

No facile means presently exists for quantitatively extracting the compositions of coexisting phases within miscibility gaps, simply by specifying the composition of a mixture within these regions. The matter has been addressed extensively in the context of liquid/liquid equilibria, and some approaches to the interpolation of tie-line positions are discussed by Francis [13].

The experimental investigator has no such problem during determination of the phase diagram, because the gross composition of each mixture is an independent variable that is known *a priori*. The experimental data on the compositions of the coexisting phases (which may be regarded as dependent variables) are used to define the boundary position.

12.1.2   The Phase Rule in ternary systems

One-phase mixtures. Every aspect of the Phase Rule that is considered in Chapter 5 applies to ternary systems, until one encounters variance. When $C = 3$ instead of 2, the quantity $P + F$ equals 5. If $P = 1$ then $F = 4$, so that if temperature and pressure are specified *two* additional variables must be stated in order to define the state of a mixture (rather than one, as in binary systems). The two variables are typically compositions, but other combinations (such as water activity plus one composition, Section 7.3) may be selected instead.

Two-phase mixtures. When two phases exist ($P = 2$) then $F = 3$. Now, only one composition variable is required (in addition to temperature and pressure) to define state. This statement might appear at first glance odd, since two compositions must always be known to define a ternary phase. It is nevertheless valid, and its validity can be illustrated by consideration of the propanol–heptane–water system (Fig. 12.3).

Suppose that one selects as the composition parameter with which to define state to be $w_{heptane} = 19.4\%$. In Fig. 12.3 the locus of mixtures having the composition $w_{heptane} = 19.4\%$ is the horizontal line b–c. This line passes through the boundary of the miscibility gap at point d, which is one end of an experimentally determined tie-line; the compositions of the two coexisting phases are $w_{PrOH} = 59.4\%$, $w_{heptane} = 19.4\%$ and $w_{PrOH} = 21.2\%$, $w_{heptane} = 76.6\%$. The intersection of this locus ($w_{heptane} = 19.4\%$) with the phase boundary does, in fact, fully define the compositions of both of the coexisting phases. The state of a two-phase mixture for which $w_{heptane} = 19.4\%$ in one of the coexisting phases is therefore fully defined simply by specifying one of the composition parameters in one of two coexisting phases: the temperature is 25°C, the pressure is 1 atm, there are two phases whose compositions are as indicated above, and they both have liquid phase structures.

Had the composition variable $w_{propanol} = 30\%$ (for example) been selected, there would have been two points where the locus defined by this constraint crosses the miscibility gap boundary. This case differs only in that one or the other of the two crossing points must be selected, and once that is done the analysis is the same as above.

Three-phase mixtures. When three phases coexist $F = 2$, so that specifying only temperature and pressure is sufficient to fully define state. (Specifying pressure and one composition variable would also define state – including temperature.) In contrast to three-phase mixtures in binary systems, *both* temperature and pressure may be independently varied in ternary systems while retaining three qualitatively similar phases. The variance of three-phase regions in ternary systems thus parallels that of two-phase regions in binary systems, in that both regions span a range of temperatures.

Three-phase mixtures in ternary diagrams define the vertices of tie-triangles within the diagram (Section 12.1.1.5). The alternation rule (Sections 4.12, 5.6.7 and 5.6.11) applies to ternary as well as binary diagrams, and requires that two-phase regions lie next to all three sides of tie-triangles. In addition, it was recognized by Schreinemakers [14,15,16] that constraints also exist as regards the manner in which the boundaries of the one-phase regions (that the triangle touches) approach the triangle. Both of these restrictions as to the form of phase diagrams stem from the Phase Rule, and are addressed in Section 12.2.2.

Four-phase mixtures. If three condensed phases and a gas phase were to coexist in ternary systems ($P = 4$), only one degree of freedom remains. Stating temperature or pressure automatically defines the other, and either temperature or pressure (but not both) may be varied while retaining four phases. This situation parallels that found at three-phase discontinuities in binary systems (Section 4.8.3).

Five-phase mixtures. Five phases may (in principle) coexist in a ternary system, in which case the mixture is completely invariant. Altering any system variable will then eliminate a phase (Section 4.8.5). While this situation is rarely encountered, it is entirely possible. Six or more phases may not coexist in equilibrium in a ternary system.

12.2   The influence of electrolytes

To analyze the influence of a third component on aqueous surfactant phase behavior, it is useful to first extract from the binary surfactant–water diagram the isotherm at the temperature of interest, and form the triangular composition figure by inserting the third component. The influence of sodium chloride on selected soap–water systems was thoroughly documented by McBain and coworkers over a wide range of temperatures above the Krafft discontinuity, and is worthy of analysis [17,18]. However, one must recognize that McBain was not aware of the existence of the intermediate phases (between the hexagonal and lamellar liquid crystal regions, Section 8.4.14), so the regions of existence of these phases is not indicated. Considerable uncertainty as to the crystal chemistry of soap/water systems also exists.

Most (if not all) ternary phase studies of aqueous surfactant–water systems have been performed at temperatures above that of the Krafft discontinuity. As a result, little information exists as to the influence of third components on crystal/liquid phase equilibria. It has been shown that the crystal solubility of SDS is sharply reduced by the addition of salt [19], but the phase behavior of the SDS–salt–water system has not been reported.

12.2.1   The sodium palmitate–sodium chloride–water system

In considering the influence of sodium chloride on the aqueous phase behavior of sodium palmitate soap (NaP) at 90°C the isotherm extracted from the NaP–water binary diagram at this temperature becomes one side of the phase triangle, while the other two sides are the isotherms at 90°C of the salt–water and salt–NaP diagrams. The salt–water isotherm may be extracted from the known salt–water phase diagram (Fig. 1.1), but the phase diagram of the NaP–salt system has not been determined. This situation is often encountered, so it is worth considering what may reasonably be predicted regarding phase behavior from knowledge of the unary phase behavior of the individual components under these circumstances.

Both NaP and salt are high melting crystals; the melting point of NaP is 290°C (Section 10.5.1) while that of salt is 801°C [20]. The compounds are not expected to be miscible in the solid state because they are so dissimilar in their molecular structures. They *are* likely to be miscible within the liquid phase, however, because they are both salts. When compounds are miscible within the liquid phase but not within their respective crystal phases, a eutectic inevitably results [21].

Because of the wide difference in their melting points, the eutectic liquid is expected to contain predominantly the low-melting component (which in this case is NaP). Further, the temperature of this eutectic will lie just below the melting point of NaP [21]. Since a temperature of 90°C is expected to lie far below the temperature of the salt/NaP eutectic, it is reasonable to assume that these compounds will coexist as immiscible crystal phases at 90°C. Insofar as is known, that is the case.

The phase diagram of the NaP–salt–water system at 90°C is shown both as an equilateral triangular plot and as a right triangular plot in Fig. 12.5. The broad picture

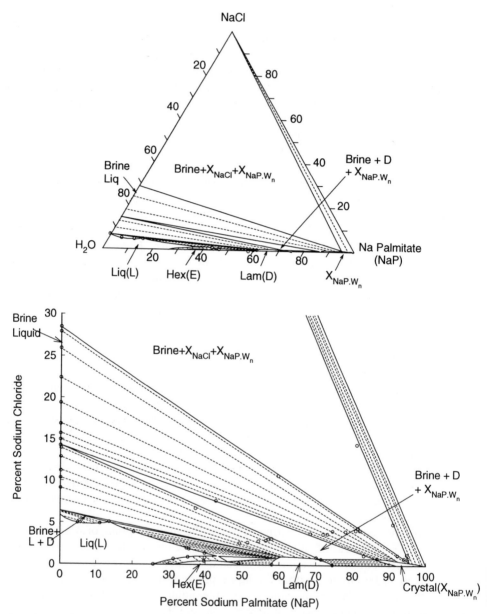

Fig. 12.5.   The phase diagram of the sodium palmitate–sodium chloride–water system at 90°C. The upper equilateral triangular diagram presents an undistorted view of the phase behavior, while the salt composition scale is expanded in the lower right triangular diagram to show the phase behavior at low salt compositions more clearly. The lower diagram is related to Fig. 3.2 by interchanging the sodium palmitate and salt axes.

may easily be visualized in the upper equilateral triangular plot in Fig. 12.5, but the details are more clearly revealed in the lower right triangular plot.

It is useful to approach this analysis by selecting the different phase regions that exist in the soap–water system, and considering the consequences of adding salt to these regions. The kind of mixing path that is followed, in so doing, corresponds to the line $a \rightarrow b$ in Fig. 12.2B.

Salt is ordinarily soluble to a finite extent in water-containing fluid phases, so the boundary of the one-phase region extends a finite distance into the center of the diagram. When the solubility of salt is just exceeded, the phase that separates is designated in exactly the same manner as in binary diagrams – except that the orientation of the tie-lines is not constrained to any particular direction (as is the case in binary diagrams) (Section 5.3). One aspect of phase behavior has been defined (there are two phases) and the two remaining questions are, as always:

(1) what are the compositions of the coexisting phases?, and
(2) what are their structures?

Dilute liquid phases. A 5% NaP water mixture (w_{NaP}) is a liquid phase having micellar structure. As salt is added to this mixture, a second liquid phase separates when the fraction of salt reaches about 5%. The compositions of the coexisting phases that result as more salt is added do *not* lie on the mixing path. Instead the tie-lines within the liquid/liquid miscibility gap *cross* the mixing path, almost at right angles. Their orientation signifies that salt causes disproportionation of composition (with respect to soap) to occur. One of the phases formed is rich in soap (compared to the gross composition of the mixture) while the other is poor. For example, at a salt fraction of 6% the fraction of NaP in one liquid phase is about 10%, while that in the other is approximately 0%.

In effect, the addition of salt to a 5% soap solution "salts out" the soap from the solution as a more highly concentrated liquid soap solution. Both of the two phases that result contain similar fractions of salt; their exact compositions vary with the total salt composition. (The name given this region in the days of soap-boiling is "lye-nigre bay". The lye phase is the salt solution and the nigre phase is the soap solution.)

Tie-line crossing. The phenomenon of "tie-line crossing" exists along the above mixing path, and is commonly encountered along isothermal mixing paths in ternary mixtures [22]. (Tie-line crossing does not exist during isothermal mixing in binary systems, because the reactant and the product phases necessarily lie on the mixing path.) When tie-line crossing occurs, the mixing process creates a thermodynamic driving force for changes of state that are often dramatic (Section 5.6.7). In the above instance extensive mass transport is required to maintain equilibrium, although in this particular case both the reactant and the product phases retain a liquid structure. In other cases (below) not only mass transport, but also structural changes, are required of one or both of the product phases.

If the change in entropy in either of the phase states that are formed when tie-line crossing occurs is unfavorable (negative), then it may be anticipated that a substantial length of time will be required for the mixture to reach equilibrium (Section 6.3). Also, a multistep phase reaction mechanism may be invoked during the process of equilibration (Section 6.5). The barriers to attaining equilibrium of state encountered along these isothermal paths are not unlike those encountered along isoplethal paths in binary systems (Section 5.6.11).

Concentrated liquid phases. Continuing, the addition of salt to a 20% NaP solution also causes disproportionation to occur, but at a salt fraction of about 4% the soap-rich phase that precipitates is the *lamellar* (neat) liquid crystal phase. It is interesting that the lamellar phase precipitates, in spite of the fact that the hexagonal phase exists at intermediate compositions. Since the precipitated phase contains a NaP fraction of about 60% and a salt fraction of about 1%, the NaP composition of the remaining liquid phase is necessarily lowered and the salt composition increased. The further addition of salt carries the gross composition into the narrow "soap-boiler's tri-angle" (Section 14.4), and with still more salt the curd phase is formed. Analysis of the diagram as a whole reveals that under no circumstances does addition of salt to the liquid phase precipitate the hexagonal phase.

The hexagonal phase. Salt is soluble in the hexagonal (middle) phase to the extent of about 0.5%, which is much less than its solubility in the liquid phase. Also, except at its most concentrated limit, the phase which separates when the solubility of salt is exceeded is a liquid phase of very similar soap composition. The further addition of salt to the liquid causes separation of the lamellar phase, just as occurred starting from the 20% liquid (above). The response of the hexagonal phase to added salt is thus rather different from the response of the liquid phase.

The addition of salt near the concentrated limit of the hexagonal phase leads initially to separation of the lamellar phase, and then quickly to a hexagonal/lamellar/liquid three-phase tie-triangle. At salt fractions beyond this triangle the hexagonal phase disappears and the phase behavior is as described above.

The lamellar phase. Salt appears to be slightly more soluble in the lamellar than in the hexagonal phase (1% vs. 0.5%), even though the water composition and the activity of water is greater in the hexagonal phase. (The intermediate phases likely exist in between the hexagonal and lamellar phases, but their existence was not recognized at the time that these studies were performed.) When the solubility of salt in the lamellar phase is exceeded, an entirely different phase reaction ensues than is found in more dilute phases. The phase which separates is a concentrated salt solution that contains virtually no soap; the addition of salt has, in effect, dehydrated the liquid crystal phase. (The micellar soap solution is *not* a product of this phase reaction; it is out of the picture. If one were to measure the activity of water in these salt solutions at this temperature, one would also know the activity of water in the coexisting lamellar phase.) With the further addition of salt a curd phase is formed and the three-phase lamellar(neat)/lye/curd tie-triangle is entered. When still more salt is added, the lamellar phase vanishes and a curd + lye region is entered.

At the most dilute limit of the lamellar phase, the addition of salt actually causes separation of the more *dilute* hexagonal phase. However, this is true only within an extremely narrow span of compositions.

The curd (crystal?) phases. At >90% NaP soap "curd" (crystal hydrate?) phases exist, which behave towards added salt in much the same manner as does the lamellar phase; a soap-free lye (or brine) phase separates. The salt concentration of the lye phase becomes progressively higher as more salt is added, until finally the lye phase itself is saturated with salt and the large "curd–lye–salt crystal" three-phase tie-triangle is entered.

This "curd–lye–salt crystal" triangle spans about two-thirds of the diagram, and the addition of sufficient quantities of salt to any NaP–water mixture leads eventually to this dominant tie-triangle. Since the further addition of salt to any mixture within this triangle is inconsequential insofar as the state of the system is concerned, the phases that exist at its corners may be viewed as the inevitable end result of adding salt to all NaP/water mixtures. Adding more salt to any mixture within this triangle simply increases the fraction of salt crystals present.

12.2.2   The alternation rule and Schreinemakers' rule

The alternation rule is followed rigorously in the sodium palmitate–salt–water diagram. Along any mixing path, odd–even alternation in the number of phases present occurs systematically, by insertion or deletion of a phase and in logical increments of one (Section 5.6.10). One-phase regions lie adjacent to two-phase regions, and two-phase regions abut three-phase tie-triangles. The middle/neat region is separated from the nigre/neat region by a nigre/middle/neat triangle, the nigre/neat region from the neat/lye region by a neat/nigre/lye triangle (the "soap-boiler's triangle", Section 14.4), the neat/lye region from the curd/lye region by a curd/neat/lye triangle, and the curd/lye region from the curd/salt crystal region by a curd/lye/salt crystal triangle. The areas of these triangles vary enormously, but they are there.

The form of this graph also adheres to Schreinemakers' Rule, which governs the manner in which the boundaries of the three one-phase regions at the vertices of tie-triangles approach the triangle. Each corner of a tie-triangle touches a one-phase region, the two boundaries of which intersect at the vertex of the tie-triangle to form a cusp. Schreinemakers' Rule requires that tangents to the two boundaries of each one-phase region must either both lie inside the tie-triangle or both lie outside (within the adjacent two-phase regions). One tangent may not lie inside the triangle and one lie outside it. A lucid treatment of this rule is presented by Francis [23].

12.2.3   The influence of temperature on sodium palmitate–salt–water phase behavior

The phase behavior of the sodium palmitate–salt–water system has been explored at 80, 84, 90, 150, 180, 210, 230, and 250°C [18], and McBain's plot of these data (at 90°C and above) is presented in Fig. 12.6. *This is not a usual ternary phase diagram plot.* The quantities of sodium palmitate and salt are scaled linearly in this diagram as the fraction of a mole present per kilogram of water, so that it represents a kind of ternary "molality" plot. Such a plot facilitates extracting data regarding the molar quantities of sodium palmitate and soap that are present, but water composition is not scaled linearly in such a plot.

The data presented in Fig. 12.6 show that the area of the liquid phase region increases as the temperature is increased (Fig. 12.6). The liquid/liquid miscibility gap ("lye-nigre bay"), found at low fractions of soap, persists throughout this wide temperature range.

The phase behavior at low temperatures differs qualitatively from that at higher temperatures with respect to the saturating phase when salt is added. At 80°C the addition of salt to the liquid phase precipitates soap crystals (curd) rather than the neat phase, while at 84°C and higher temperatures the addition of salt precipitates the neat phase. The reason for this is evident from the binary diagram; 80°C is just below the eutectic at the lower limit of the neat phase. Again, the reluctance of the hexagonal

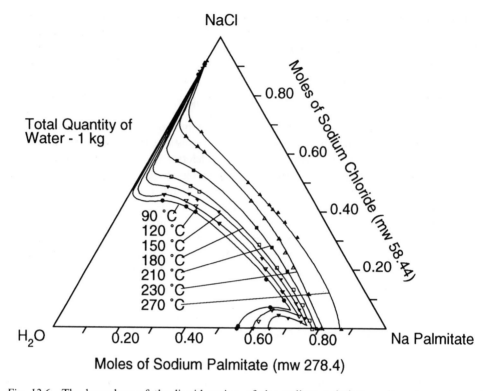

Fig. 12.6 The boundary of the liquid region of the sodium palmitate–salt–water system at various temperatures [18]. *This plot is not a usual ternary phase diagram.* The total quantity of sodium chloride, per kilo of water, is scaled along the perpendicular to the water–sodium palmitate border, and the total quantity of sodium palmitate, per kilo of water, is scaled along the perpendicular to the water–sodium chloride border. The composition of water is not scaled linearly at all in this diagram.

phase to precipitate upon the addition of salt (noted above in this section) is evident. These data further indicate that the temperature range of the lamellar region is not significantly lowered by the addition of salt.

Most of these studies of sodium palmitate have been duplicated in the sodium laurate–salt–water system [24]. The phase behavior observed closely resembles that found during the sodium palmitate study (above), if allowance is made for differences that exist in the phase behavior of the two binary soap–water systems.

12.2.4   The effects of other salts

The influence of other kinds of salts on aqueous surfactant phase behavior is important, but very little has been published on this subject. However, it has been shown that in the case of decyldimethylphosphine oxide ($C_{10}PO$) a salting-out salt (sodium sulfate) enlarges the miscibility gap defined by the lower consolute boundary so that it extends to lower temperatures in the presence of salt [25]. The effect of a hydrotrope (sodium cumenesulfonate) on the phase behavior of the dodecyl homolog ($C_{12}PO$) is exactly the opposite; the miscibility gap shrinks (and eventually vanishes) with increasing concentrations of hydrotrope [26].

Dodecyldimethylammoniopropane sulfonate ($C_{12}APS$) is influenced at 26.7°C by sodium chloride in much the same way as is sodium palmitate at 90°C; the form of the liquid region is remarkably similar, and the liquid/liquid miscibility gap is observed [26]. In a survey of various kinds of electrolytes, it was found that they could be classified into two distinct groups insofar as their influence on the phase behavior of $C_{12}APS$ is concerned [26]. One group displayed an influence on phase behavior that is qualitatively similar to that of sodium chloride on sodium palmitate. This group included tetrapotassium pyrophosphate, tripotassium hydrogen pyrophosphate, tetrapotassium ethylenediamine tetraacetate (K_4EDTA), trisodium phosphate, and sodium sulfate – but, interestingly, not sodium chloride. The other group did *not* induce liquid/liquid coexistence when added to dilute solutions of $C_{12}APS$; instead, crystals of the added salt separated from solution when its solubility in the surfactant solution was exceeded. The salts that behaved in this manner included sodium chloride, sodium iodide, potassium iodide, sodium perchlorate, and sodium acetate.

It has been reported in the patent literature that the presence of hydrotrope salts (plus urea) convert the lamellar phase of anionic surfactants (such as sodium linear alkylbenzenesulfonate (LAS)) into the hexagonal phase [27], which represents a dramatic influence on phase behavior.

12.3 The influence of nonpolar oils

In this section the influence of "nonpolar" oils on the phase behavior of ionic surfactants will be considered. The term "nonpolar" is used to signify molecules that are not amphiphilic (Section 9.1), but may contain weakly dipolar nonhydrophilic functional groups (such as chloro substituents).

12.3.1 Soap–nonpolar oil–water systems

If one allows that liquid crystal phases constitute regions of miscibility, an ionic surfactant is at sufficiently high temperatures miscible with water over a large fraction of the composition range. At and above the melting point, strongly hydrophilic surfactants are miscible with water in virtually all proportions (Section 3.12), but as the temperature is reduced below the melting point their miscibility is reduced due to separation of a crystal phase. The amount of water required to dissolve this phase progressively increases as the temperature is lowered (Section 5.6.2).

Crystalline ionic surfactants are typically immiscible with nonpolar oils over a wide range of temperatures, and oil and water are also immiscible. In considering the phase behavior of soap–nonpolar oil–water systems, therefore, one is dealing with a ternary system in which substantial miscibility exists only on one face of the ternary prism – the surfactant–water face. Virtually no miscibility exists on the other two faces.

Given these circumstances, it is hardly surprising to find that the influence of nonpolar oils on the aqueous phase behavior of these surfactants is minimal. Extensive data exist, thanks to studies by Ekwall and his associates, of ternary systems that contain sodium octanoate and water plus a wide range of nonpolar oils (Appendix 2). Data exist at 20°C where the third component is an alkane (octane), an alkylbenzene (xylene), an alkyl chloride (octyl chloride), and a chlorinated solvent (carbon tetrachloride). The same qualitative result is found in all these systems. The sodium octanoate–octane–water diagram (Fig. 12.7) serves to illustrate the phase behavior that is encountered.

At 20°C, the octane–water and the octane–sodium octanoate borders are (as indicated above) spanned by large miscibility gaps. Octane is weakly soluble in the hexagonal liquid crystal phase (the fraction of octane reaching only 3%), and even less so in the micellar liquid phase. When the solubility in either phase is exceeded nearly pure octane separates, so that the tie-lines within this large miscibility gap extend from the boundaries of the liquid crystal regions almost to the octane vertex. In this system the dilution of aqueous surfactant mixtures with octane does *not* result in tie-line crossing (Section 12.2.1). In striking contrast to salt, the saturation of these phases with octane does not alter the surfactant:water ratio in the phase, nor does it cause disproportionation reactions to occur.

The most unusual feature of the octane diagram is the formation of a tiny cubic phase region, which lies between the micellar liquid and the hexagonal phase. This cubic phase does not extend to the soap–water border, which signifies that the presence of octane is required for it to exist. Further, from the tie-line orientations it can be seen that the addition of octane produces this phase only within a narrowly defined range of soap:water ratios. It is also apparent that this phase is quickly saturated with octane so that, just as with the liquid and hexagonal phases, long tie-lines connect the cubic phase region with the octane vertex.

The existence as small islands of phases such as this cubic phase signifies that they are delicately balanced thermodynamically. The addition of even small amounts of any of

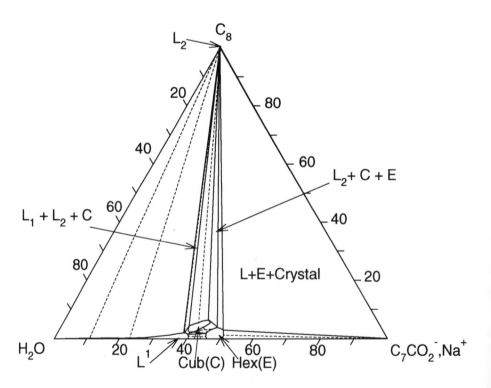

Fig. 12.7. The sodium octanoate–octane–water system at 20°C [28].

the three components to such a phase causes separation of another phase. They are not robust in terms of their ability to dissolve any of the three components.

The influence of octanenitrile ($C_7H_{15}CN$), methyl octanoate, and octanal on the phase behavior of the sodium octanoate–water system is similar to that of the above compounds, but differs in some respects. Somewhat more extensive miscibility exists within the hexagonal phase, which will dissolve oil up to an oil fraction of about 20%. Also, it is interesting that tiny regions of lamellar phase are created at slightly higher surfactant:water ratios than exist in the hexagonal phase (near 60:40). These data suggest that the nitrile, ester, and aldehyde groups are not entirely devoid of the property of hydrophilicity – in spite of the fact that they do not reveal this property in their binary aqueous systems (Section 9.6.1).

12.3.2 Cationic surfactant–nonpolar oil–water systems

The strong current interest in the physics and structure of microemulsion phases has spawned research on phase equilibria in dilong chain quaternary ammonium salt–hydrocarbon–water systems. Extensive studies of didodecyldimethylammonium bromide (DDAB)–water systems in combination with a wide range of hydrocarbon oils have been conducted [29], and the exploration of dilong chain surfactants having two different lipophilic groups has been reported [30].

As noted elsewhere (Section 11.3.2.3), the binary DDAB–water system is of interest because two lamellar liquid crystal phases of widely differing composition coexist. Ternary diagrams that include DDAB, water, and either hexane or dodecane are shown in Fig. 12.8. In both systems the alkane is only weakly miscible with the water-rich lamellar phase. Hexane is weakly soluble in the water-poor lamellar phase (w_{hex} = about 3%), but dodecane is considerably more soluble in this phase (w_{dodec} = about 20%).

The crystalline DDAB salt is insoluble in both hydrocarbons, but the addition of water to alkane–DDAB mixtures results in the formation of an isotropic liquid when the water:DDAB ratio is about 15:85. Curiously, the miscibility of the dodecane phase with water is greater than is the miscibility of the hexane phase; the upper limit of w_{water} in the C_{12} liquid is about 62%, while that in the C_6 phase is only 42%. Starting from the oil–water border, it can also be seen that the effectiveness with which DDAB causes water and alkane to mix is greater in the case of dodecane than it is in the case of hexane. At a 50:50 ratio of alkane to water, the fraction of DDAB required to enter the liquid phase region is only 15% for dodecane, while it is 24% for hexane.

An exploration for liquid phase regions in dialkyldimethylammonium salt–alkane–water systems in which the ammonium salt has two different lipophilic groups has revealed two distinctive features of their phase behavior. One is that *two* liquid regions are often (but not always) found in such systems. One liquid phase is tied to the oil corner and occupies the usual composition range. The other occurs as an island at surfactant:oil ratios centered about 50:50 for the C_8C_{16} homolog with hexane, and at about 60:40 ratios for the $C_{10}C_{12}$ homolog with hexane, octane, or decane. This liquid region does not extend to the oil vertex, requires extremely small amounts of water ($w_{water} \approx 1\%$) to exist, and is restricted to a narrow range of water compositions (<10%).

The second unusual feature of the phase behavior of these surfactants is that sometimes a three-component liquid region exists that is isolated as an island; it is not

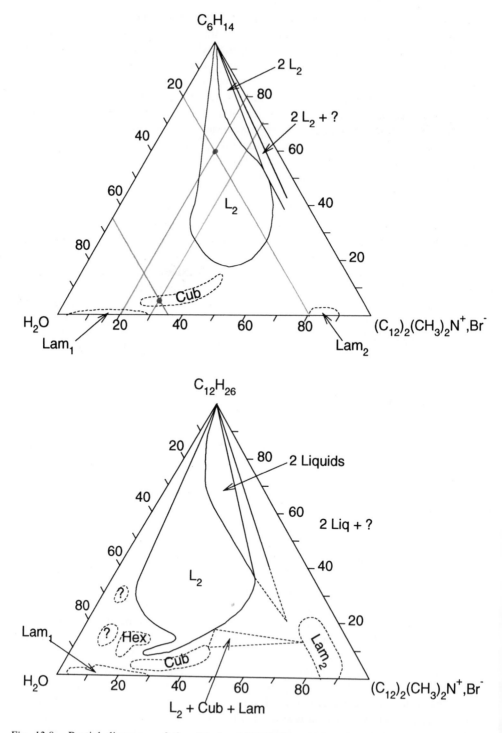

Fig. 12.8. Partial diagrams of the didodecyldimethylammonium bromide (DDAB)–hexane–water (upper) and DDAB–dodecane–water (lower) systems at 20°C [29]. The large microemulsion liquid region is evident, as is the miscibility gap that exists within this region.

connected to the oil vertex (in no case is the liquid connected to the water vertex in these systems, but it is often connected to the oil vertex). In the case of the $C_{10}C_{12}$dimethylammonium bromide, the liquid region is connected to the oil vertex in the cyclohexane and hexane ternary systems, but not in the octane, dodecane, or dodecane systems. The center of gravity of these isolated liquid regions does not shift dramatically with the molecular weight of the oil. Isolated liquid islands are also seen in the di-C_{14}–hexadecane–water and C_8C_{16}–decane–water systems; the former example indicates that having two different chains is not absolutely required for a liquid phase island to exist. Phase information on states other than the liquid were not obtained during these studies.

12.3.3   The $C_{10}E_4$–hexadecane–water system

Serious studies of the $C_{10}E_4$–hexadecane–water system [31] have provided significant information regarding the phase behavior of a system containing a low-melting non-ionic surfactant, a nonpolar oil, and water. Such systems differ fundamentally from the ionic surfactant–nonpolar oil–water systems (described above) in that the surfactant and the oil are miscible in all proportions over a wide temperature range, while the miscibility of the surfactant with water varies strongly with temperature. Only the behavior of the oil–water binary system remains the same as in the soap–oil–water systems considered above.

The $C_{10}E_4$–hexadecane–water system has been studied between 19 and 60°C. The temperatures of the studies were selected (using the binary diagram as a guide) to depict the various distinctive kinds of phase behavior that are encountered as the temperature is varied. The data [31,32] are presented in Figs 12.9, 12.10, and 12.11.

19°C. This temperature is below the closed loop of coexistence, and the phase sequence along the binary surfactant–water isotherm (starting from the water border) is $L/D/L$ (L = liquid phase, D = lamellar phase) and the surfactant is very soluble in the water-rich L phase. Except within the tiny miscibility gaps to either side of the D phase, the surfactant and water are miscible in all proportions (as either a liquid or a liquid crystal) at this temperature. If one adds hexadecane to dilute aqueous liquid solutions of the surfactant at 19°C, the oil is miscible with this liquid until the oil:surfactant ratio reaches about 35:65; additional oil separates as a nearly pure oil phase. This behavior is reflected by the large miscibility gap that lies against the oil–water border, within which the tie-lines fan out from the oil corner.

The oil is also soluble in the lamellar phase at this temperature; the fraction of oil that can exist in this phase is extremely sensitive to the surfactant:water ratio. At a ratio of 75:25 the value of w_{oil} in this phase reaches an astounding 80%, while at an 80:20 ratio w_{oil} is about 30% and at 65:35 it is about 25%. The lamellar phase region at this temperature has the form of a long, narrow finger that extends towards the oil corner.

A second small liquid/liquid miscibility gap exists to the oil:surfactant side of the lamellar finger, within which an oil phase coexists with ternary mixtures just as in the case of the larger miscibility gap. The overall appearance of the diagram suggests that a single large miscibility gap would have existed had the lamellar phase not been formed. The intrusion of the lamellar finger into this gap has split it into two discrete gaps (of unequal size), which lie to either side of the finger. This phenomenon, which has occurred within ternary composition space, is qualitatively analogous to intrusion

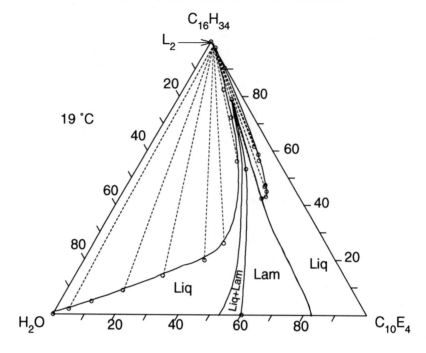

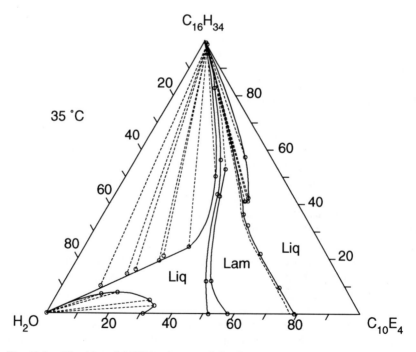

Fig. 12.9.   The 19 and 35°C isotherms of the $C_{10}E_4$–hexadecane–water system.

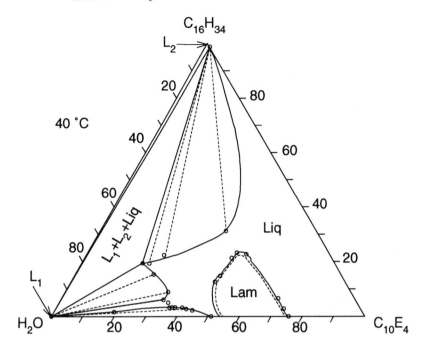

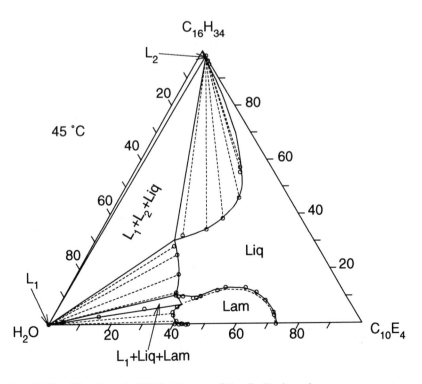

Fig. 12.10.   The 40 and 45°C isotherms of the $C_{10}E_4$–hexadecane–water system.

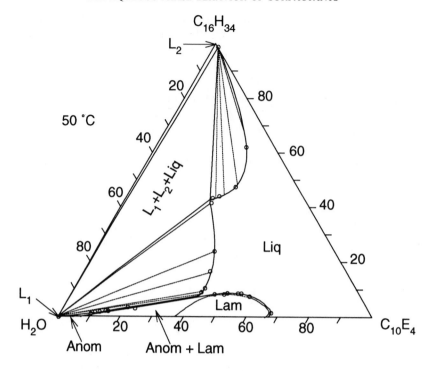

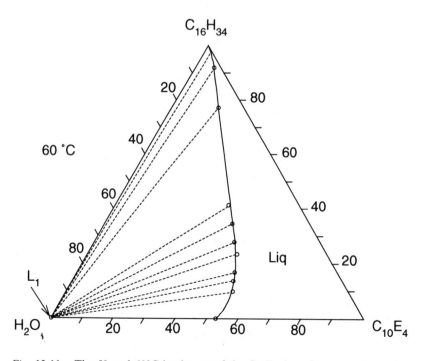

Fig. 12.11.   The 50 and 60°C isotherms of the $C_{10}E_4$–hexadecane–water system.

of the anomalous phase corridor into the closed loop of coexistence in the binary $C_{10}E_4$–water diagram [32]. However, this latter intrusion occurs within composition–temperature space rather than within composition–composition space (Section 5.10.6).

35°C. This temperature is above the lower critical temperature of the closed loop. A new liquid/liquid miscibility gap is introduced into the surfactant–water face, the binary phase sequence is $L/L/D/L$, and a new liquid/liquid miscibility gap intrudes into the ternary diagram from the surfactant–water border. The liquid region of ternary mixtures remains, but its area is reduced by this new feature.

40°C. At 40°C the surfactant–water behavior is qualitatively similar to that at 35°C, but a dramatic change has occurred in the liquid/liquid equilibria. The miscibility gap that existed along the oil–water border at 35°C is, at 40°C, split into three discrete miscibility gaps. This results from the intrusion of a large three-phase triangle. Within this triangle the coexisting phases are a nearly pure oil phase, a nearly pure water phase, and a third phase that contains significant fractions of all three components. It is this latter phase that is commonly termed the "microemulsion" phase. The composition of the microemulsion phase at this temperature is $w_{surf} = 19\%$, $w_{oil} = 20\%$, and $w_{water} = 61\%$. During the balance of this discussion any part of the major liquid region that contains significant quantities of all three components (lies within the central part of the diagram) may be termed a "microemulsion" phase. From inspection of the various diagrams, it is evident that this central liquid region may extend towards one or more of the water, the oil, or the surfactant vertices at any given temperature. From this fact alone, it is evident that the structure and properties of this central liquid region must vary widely and smoothly with composition.

Two of the three miscibility gaps adjacent to the triangle are remnants of the large miscibility gap that existed along the oil–water border at 35°C. One of these lies between this border and the triangle, and the coexisting phases within this gap are a nearly pure oil phase and a nearly pure water phase. These are *phases* – not components; each phase contains all three components.

The larger part of this original gap now exists to the side of the three-phase triangle that lies towards the surfactant–oil border (away from the water vertex). The coexisting phases are a nearly pure oil phase (as in the first-mentioned gap), plus a microemulsion phase. The tie-lines fan out from the oil corner, as before.

The composition of the microemulsion phase at different points along the boundary of these miscibility gaps is widely variable. At 35°C, the oil composition (w_{oil}) of this phase spans a range of 0–99%, and the water composition (w_{water}) spans a similar range of 0–100%. However, at 40°C the range of oil compositions is now restricted to $w_{oil} > 20\%$ and the range of water compositions to $w_{water} < 61\%$.

The third miscibility gap is entirely new. It lies below the triangle towards the surfactant–water border (away from the oil vertex), and represents an extension of the closed loop of miscibility in the binary surfactant–water diagram into the three-component prism. A nearly pure water phase coexists with three-component liquid mixtures of variable composition within this gap. The tie-lines fan out from the *water* vertex, and the fraction of water in all the coexisting phases is >61%.

These data signify that increasing the temperature from 35 to 40°C results in the loss of water (a kind of "dehydration") of a three-component phase. (A more exact description of the phenomena that actually occurred between 35 and 40°C in terms of "critical tie-line" and "critical end-point" concepts is provided in Section 13.5.3.) The water that is "lost" separates as a nearly pure water phase, so that the phase left behind is richer in both oil and surfactant.

The microemulsion region that was present at 35°C still exists at 40°C, but part of it is crowded between the new miscibility gaps attached to the three-phase triangle, on the one hand, and the enlarged miscibility gap attached to the surfactant–water border that results from the closed loop of coexistence. One result of this is that, at high water fractions, this part of the microemulsion phase also exists as an extremely long, narrow finger. Within this finger the surfactant : oil ratio is about 85 : 15. If one were to produce this water-rich phase by mixing an 85 : 15 surfactant : oil mixture with water, one would pass through the lamellar liquid crystal region in the interval $w_{water} = 25$–43%. The gaps in miscibility that exist along this path are very small, but the changes in phase structure that occur in passing from liquid to liquid crystal and back to liquid are dramatic.

Under conditions such that the microemulsion phase coexists with an oil-rich and a water-rich phase, and contains roughly equal quantities of surfactant and oil, the numerical values of the interfacial tensions between the coexisting phases are extremely low. As the conditions are changed so that the microemulsion phase becomes imbalanced with respect to either water or oil, the interfacial tensions rise. The attainment of low tensions may be accomplished by the addition of salt to the mixture as well as by adjusting the temperature (as in this ternary system).

It is apparent from consideration of phase behavior at lower temperatures that phases of comparable composition (and presumably structure) to the microemulsion phase at 40°C also exist at lower temperatures – well below the span of temperatures within which the three-phase triangles exist. For example, the 40°C microemulsion phase also exists at 19 and at 35°C, but at these temperatures the water-rich phase does not exist. As a result, at these lower temperatures the reduction of tensions between an oil and a water phase is a moot issue.

Another effect of increasing the temperature to 40°C is to reduce dramatically the miscibility of the lamellar phase with oil. Whereas at lower temperatures an oil-rich lamellar phase intruded deeply into the principal miscibility gap, at 40°C the capacity of this phase to dissolve oil is significantly reduced and it no longer intrudes into the liquid/liquid miscibility gap.

45°C. At 45°C the binary phase behavior remains qualitatively the same as at 40°C, since this temperature is still below the anomalous phase corridor (Fig. 4.5) or its projection into the three-component prism. The ternary phase behavior is also qualitatively similar to that at 40°C, but the water fraction in the microemulsion phase is reduced from 61 to 44% while the oil fraction (31%) and surfactant fraction (25%) are increased. The ratio of oil to surfactant in the phase has also increased. With the increase in temperature loss of water from the microemulsion phase has progressed to a larger degree, miscibility gaps are larger, and the area of the liquid region is smaller than at 40°C.

The trend towards reduced capacity of the lamellar phase region to dissolve oil with increasing temperature that was evident at 40°C continues at 45°C. This temperature is near the upper limit of the existence of this phase in the binary aqueous system.

50°C. At this temperature the binary surfactant–water phase behavior is qualitatively different from that which exists at 45°C in that the anomalous phase is present. The phase sequence on the surfactant–water face is now $L/An/D/L$ (An = "anomalous phase"). Introduction of the anomalous phase region along the surfactant–water border is an important change, since this phase is not only far more dilute than the liquid which it replaced but is also structurally different. The anomalous phase is unable to dissolve much oil; its extension into the core of the triangle is almost imperceptible.

The presence of the anomalous phase greatly complicates the phase behavior insofar as adherence to the Phase Rule is concerned, but its presence appears not to have a profound influence on the liquid/liquid equilibria. The form of the phase behavior in this region has been determined [31], and is consistent with the Phase Rule.

The solubility of oil in the lamellar phase, which was inordinately high at 35°C, steadily decreases as the temperature is increased to 50°C. The maximum fraction of oil present in the phase is 80% at 35°C, 24% at 40°C, 14% at 45°C, and 9% at 50°C. The form of this region at 40°C and higher is smoother, and more closely resembles that seen in other surfactant–oil–water diagrams. At 60°C, the lamellar phase does not exist in either the binary or the ternary system.

The water content of the lamellar phase increases as the temperature increases. The limits of existence of this phase span a range of from 20:80 to 55:45 water:surfactant ratio at 35°C, and from 33:67 to 63:37 at 50°C. It may be inferred from these data that more water (and therefore greater hydration energy, Section 5.6.2) is required for this phase to exist at 50°C than at 35°C; the increased level of thermal energy is not helping the situation. This shift in phase composition suggests that the presence of the nonpolar oil places a stress on the stability of the lamellar phase, because in the absence of an oil *less* hydration energy is typically required for a liquid crystal phase to exist at high temperatures than at low (Section 5.6.9).

60°C. At 60°C the liquid crystal phase no longer exists, the complexity introduced by the anomalous phase corridor has vanished, and the coexistence of three liquid phases has also vanished. An extremely simple phase situation exists as a result of these truly profound changes of state.

The phase behavior at 60°C is dominated by the fact that the surfactant and water are only partially miscible at this temperature, while the surfactant and the oil remain miscible in all proportions. The solubility of water in the surfactant phase remains substantial, but the solubility of surfactant in the water phase is vanishingly small. The dilute surfactant solutions have, in effect, been severely dehydrated; only concentrated surfactant solutions remain.

The tie-lines within the single large miscibility gap now fan out from the *water* vertex, whereas at 19°C they fanned out from the oil vertex. Their orientation at 60°C signifies that the addition to surfactant–oil mixtures of water beyond its solubility limit results in the separation of a nearly pure water phase. As the proportion of oil in liquid oil–surfactant mixtures is increased, the solubility of water in this liquid is smoothly decreased. The form of the boundary indicates that the ratio of surfactant to water in the phase which coexists with the water phase remains nearly constant over a wide range of oil composition.

At 70°C (graph not shown), the situation is qualitatively similar to that which exists at 60°C, but the solubility of water in the surfactant is further diminished at all oil:surfactant ratios. To illustrate, the fraction of water in the water-saturated surfactant phase in the absence of oil is 47% at 60°C and 35% at 70°C. In a 50:50

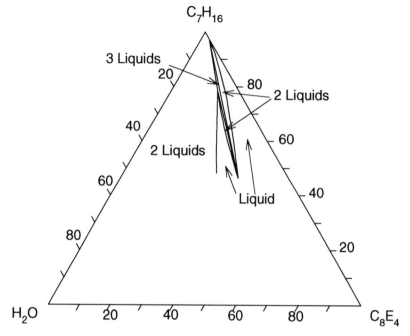

Fig. 12.12. A partial phase diagram of the C_8E_4–heptane–water system at 25°C [6].

oil: surfactant mixture the fraction of water in the water-saturated phase is 23% at 60°C and 15% at 70°C.

At 300°C $C_{10}E_4$ and water are miscible in all proportions (Fig. 5.17), and the phase behavior of the surfactant–oil–water system may be expected to differ dramatically from that which is described above. The behavior of hexadecane and water at temperatures near the critical point of water is not known. Recent literature [33] suggests that the miscibility of oils with water is substantially greater at temperatures near the critical point of water than near room temperature – a trend that is already apparent from studies in the vicinity of 150°C [34].

12.3.4   The C_8E_4–heptane–water system

The overall pattern described above for the $C_{10}E_4$–hexadecane–water system is likely to apply to the C_8E_4–heptane–water system over a wide temperature range. The phase behavior within the intermediate temperature region (where microemulsion phases exist) has been investigated over part of the composition range [6].

The phase behavior of this system appears to be more complex than is that of $C_{10}E_4$–hexadecane–water. Two "three-phase bodies" (the solid described by the locus of three-phase triangles within which an oil, a water, and a microemulsion phase coexist) are found instead of one [35]. This study was performed by carefully determining only one isotherm, and then scanning various isoplethal sections (either at constant surfactant fraction or at constant surfactant: oil ratio) through the phase prism. Isotherms at different temperatures do not exist, but the temperature limits of particular phenomena within the planes of the isoplethal scans are perhaps better defined using this approach.

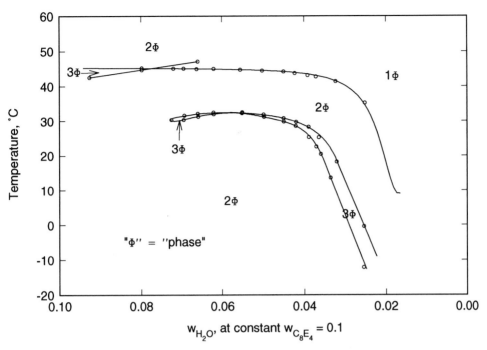

Fig. 12.13. An isoplethal scan of the C_8E_4–heptane–water system at a constant $w_{surf} = 0.1$, and variable w_{water}. While such a scan does not reveal coexistence relationships, it does suggest that the two different three-phase bodies merge as the temperature is increased [6].

A partial 25°C isotherm is shown in Fig. 12.12, and an isoplethal scan within the plane defined by the condition $w_{surf} = 0.1$ in Fig. 12.13. It is well to keep in mind that one may *not* draw tie-lines within figures such as Fig. 12.13 with impunity. While possible, it is unlikely that the tie-lines connecting coexisting phases will fall within the plane of this figure. Figure 12.12 shows that a tiny three-liquid-phase triangle intrudes into the liquid phase region. The miscibility gap to the left side of this triangle is the large (central) oil–water miscibility gap, which emanates from the oil–water border. The other two required (and observed) miscibility gaps intrude a tiny distance from the edges of the triangle into the liquid region. A narrow spike of the liquid phase lies between one of these miscibility gaps and the central miscibility gap, as a result of this behavior.

The temperature scans reveal the existence of *two* discrete three-phase triangles within this system at low temperatures. One evolves from the central miscibility gap, while the other evolves in an independent manner. The two coalesce as the temperature is increased, and both vanish at a sufficiently high temperature.

It is evident from the above two nonionic–oil–water studies that extremely complex phase equilibria may exist within such systems. This complexity has only recently been revealed by phase studies; probably, most of the phase science of these systems remains to be determined.

Important studies of similar mixtures have been reported that will not be discussed here [35]. Also, it is interesting to consider the perspective regarding the complex phase behavior of these systems that is provided by viewing the system in "field-space" (Section 3.5 [32]).

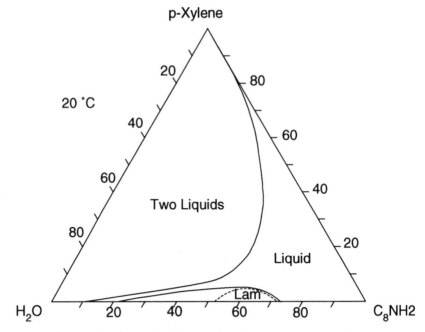

Fig. 12.14. The octylamine–*p*-xylene–water system at 20°C [36].

12.3.5 Miscellaneous nonionic–oil–water systems

Primary amines contain one of the most weakly hydrophilic groups that exist (Section 10.5.5), but these compounds can still serve to bring oil and water together with modest efficiency [36]. The diagram of the octylamine–*p*-xylene–water system is shown in Fig. 12.14. It displays a large xylene–water miscibility gap, and along the surfactant–water border the lamellar liquid crystal phase. Xylene is soluble in the lamellar phase to a small extent, but the miscibility is less than that in the liquid phase at the surfactant : water ratios within which the liquid crystal exists. As a result, a narrow liquid phase lies next to the large miscibility gap that exists along the oil–water border. This liquid phase strip connects the liquid region that lies to the dilute side of the liquid crystal in the aqueous system with the large water-poor liquid that lies along the surfactant–oil border.

Other homologs of the amine, ranging from C_6 through C_{10}, have been investigated with *p*-xylene as the oil component [37]. The area of the liquid region becomes progressively smaller as the chain-length of the surfactant is increased.

12.4 The influence of amphiphilic oils

We now turn to the influence of amphiphilic oils, such as fatty alcohols or acids, on aqueous surfactant phase behavior. Several such systems have been investigated in detail at selected temperatures, and the influence of temperature on their phase behavior has been explored using isoplethal studies along selected mixing paths.

12.4.1 Soap–medium-chain alcohol–water systems

The best studied surfactant–alcohol–water system for all is the sodium caprylate (octanoate)–decanol–water system at 20°C (Fig. 12.15, [38]). The phase behavior of this "holy" system provides a well-documented prototype for the behavior of strongly hydrophilic crystalline surfactants in combination with water and a medium chain-length amphiphile.

Analysis of the influence of decanol on soap–water phase behavior may be approached in a manner similar to that described above for sodium chloride. This soap has a short (C_7) lipophilic group compared to sodium palmitate (C_{15}), (C_{11}), so that the only liquid crystal phase that exists in its aqueous mixtures at the temperature of this study is the hexagonal phase. Dry sodium octanoate is crystalline and poorly miscible with decanol. It is likely that this soap forms crystal hydrates, but they have not been defined and their role in the phase behavior of this system is unknown.

Decanol is weakly soluble in the hexagonal phase of sodium octanoate. The phase that separates when the solubility of decanol in the hexagonal phase is exceeded is the lamellar liquid crystal. From the direction of the tie-lines, one may infer that the lamellar phase that separates is not only richer in decanol, but also has a noticeably higher surfactant : water ratio than does the hexagonal phase from which it was formed. The decanol not only induces the hexagonal phase to reorganize into a lamellar structure, but replaces some of the water in the process. Complex tie-line crossing phenomena (along a diagonal) occur during this mixing process.

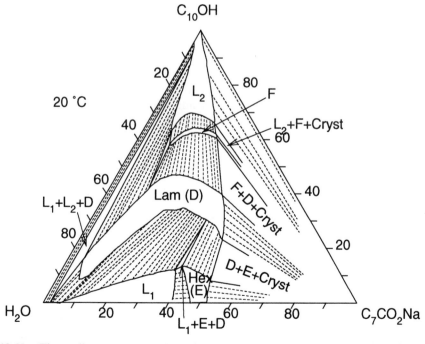

Fig. 12.15. The sodium octanoate (caprylate)–decanol–water system at 20°C [38]. This is without doubt the best-documented soap–oil–water system that exists, yet it remains incomplete as regards the crystal phase chemistry.

Over most of its composition range, adding decanol to the liquid phase also induces separation of the lamellar phase. The soap:water ratio in the lamellar phase that separates is substantially higher than in the liquid phase from which it was formed. The presence of decanol clearly favors formation of the lamellar phase structure.

Decanol has a small solubility in water, but the miscibility of decanol with solutions of sodium octanoate is not greatly altered by the presence of the soap at soap concentrations below its cmc (Section 12.4.1). Because of this, the liquid phase boundary does not depart significantly from the soap–water border until the fraction of soap reaches about 6% (the cmc of sodium octanoate at 20°C is 0.36 M (c.5.9%)). Once micelles exist, decanol is solubilized into sodium octanoate solutions at a ratio of about 30:70 decanol:sodium octanoate over a wide range of soap compositions. The further addition of decanol to lamellar (D) phase mixtures which have sodium octanoate:water ratios between 25:75 and 60:40 leads to separation of the inverted hexagonal (F) phase. However, in this case the coexisting F phase lies almost on the mixing path; tie-line crossing (almost) does not occur. The soap:water ratio in the F phase that is formed is approximately the same as in the lamellar phase from which it was produced.

The F phase spans an extremely narrow range of decanol compositions. The further addition of decanol (once the phase is saturated with decanol) results in separation of the liquid L_2 (microemulsion) phase. Again, the ratio of soap to water in the L_2 phase closely resembles that of the F phase from which it is formed. Within soap:water ratios between 25:75 and 60:40, not only the lower boundary of the L_2 phase, but both boundaries of the F phase and both boundaries of the lamellar phase, are convex upward. This suggests that somewhat less decanol is required for the $D \rightarrow F \rightarrow L_2$ succession of phases to occur at each end of this composition range than is required near the middle.

If the soap:water ratio is less than 25:75, the capacity of the D phase (formed by adding decanol to the aqueous soap solutions) to dissolve additional decanol is smaller than at higher soap compositions. Also, the addition of decanol past the miscibility limit does not, in these mixtures, lead to separation of the F phase; the L_2 phase is formed instead. Viewed from another perspective, this aspect of the phase behavior indicates that the F phase is destroyed by the addition of too much water. This phase disproportionates into a D phase (of progressively higher water content), and an L_2 phase (of progressively lower water content), when water is added. Ultimately, the capacity of the lamellar phase itself to accommodate water is exceeded and, after passing through a $L_1/L_2/D$ tie-triangle, the region within which L_1 and L_2 liquids coexist is entered.

The $L_1/L_2/D$ tie-triangle within these systems is of considerable importance with respect to the stabilization of emulsions of L_1 and L_2 phases. When the compositions of mixtures lie within this tie-triangle the D phase exists, and its presence strongly promotes emulsion stability [39]. If the composition lies outside this triangle (within the L_1/L_2 two-phase region) the D phase is absent, and emulsion stability is extremely poor. Similar considerations apply to the analysis of foam stability [40]. It is thus possible, merely from a consideration of phase behavior, to understand how a given fatty alcohol may under some conditions stabilize foams, and under others kill foams.

The boundaries of the L_2 region extend from the decanol corner towards the water–soap border in an almost linear manner, so as to form a roughly triangular liquid phase region. The form of this region signifies that the stability limits of the L_2 phase are defined largely by the ratio of soap to water. Within these limits the fraction of decanol in this phase may vary substantially, but it is noteworthy that the phase does not exist in the absence of water.

If one starts inside the lamellar phase at a 30:70 ratio of soap:decanol and dilutes this mixture with water, the water dissolves in the phase up to a water fraction of about 85%. At this optimal soap:decanol ratio, this phase will tolerate this exceptionally high fraction of water and still remain intact. At higher and lower ratios, however, it cannot dissolve nearly so much water. The lamellar phase containing a 50:50 soap:decanol ratio is saturated with water at a water fraction of about 35%, for example.

Potassium octanoate behaves with decanol and octanol in a manner similar to sodium octanoate [41], but potassium oleate represents a fundamentally different type of binary phase behavior in some respects (Fig. 12.16 [41]). One difference arises from the fact that the lamellar liquid crystal phase *does* exist along the 20°C isotherm in this system, whereas it is absent in the potassium octanoate–water system. As a result, the lamellar region in the oleate system is not truncated below a minimal level of decanol, but extends to the surfactant–water border. The form of the lamellar phase region in the oleate system is similar to that found in the octanoate system, although the weight fraction of water that can exist in the oleate lamellar phase is increased to about 95%. That is a remarkably large fraction of water for any liquid crystal phase to tolerate; in binary systems this phase is generally unstable at such high water fractions in all surfactant systems (Section 5.6.6). An important effect of the presence of decanol, therefore, is to stabilize the lamellar phase against excessive hydration and render it far more tolerant of large fractions of water than is the case in its absence.

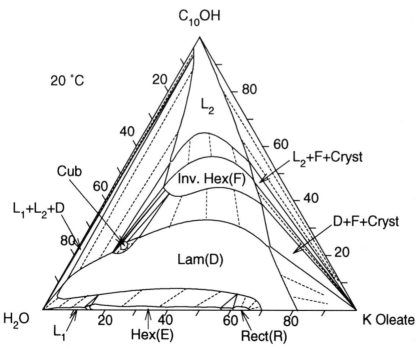

Fig. 12.16. The potassium oleate–decanol–water system at 20°C [41].

The inverted hexagonal (F) phase region is substantially enlarged in the oleate in comparison to the octanoate, and a small cubic phase region exists to the low-soap side of the lamellar region that is not found in octanoate systems.

12.4.2 Soap–short-chain alcohol–water systems

Since water and decanol are only slightly miscible with each other, a miscibility gap spans almost the entire length of the water–decanol border (Fig. 12.15). The lower straight-chain alcohol–water systems display an upper consolute boundary whose critical temperature rises as the chain length of the alcohol is increased [42]. At C_3 (n-propanol) the critical temperature is very low so that this compound is miscible with water in all proportions at 20°C, while at C_4 (n-butanol) the upper critical temperature is 125°C [42] and the miscibility gap at 20°C spans 70% of the composition range.

From data in the Ekwall compilation [43], it can be seen that the alcohol can be varied from decanol to hexanol without qualitatively altering the L_2 region of the phase diagram. The F phase vanishes in the octanol and shorter alcohol systems. The L_2 and the lamellar phase regions are slightly larger in the hexanol than in the decanol diagram, but the diagrams are otherwise similar. When the alcohol is pentanol substantial enlargement of the L_2 region occurs, in the direction of the water corner.

Butanol has a very different influence on the phase behavior of aqueous sodium octanoate systems than does decanol [38], as shown in Fig. 12.17. The principal difference is that the addition of sodium octanoate to butanol–water mixtures, at a sodium octanoate fraction of about 2%, causes the L_1 and L_2 phase regions to merge.

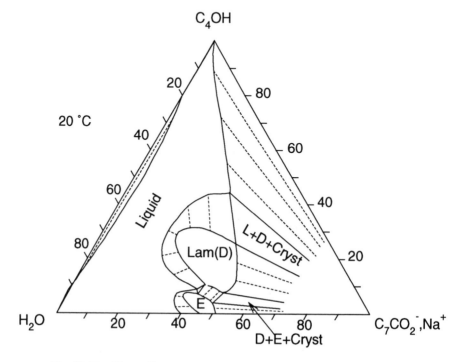

Fig. 12.17.   The sodium octanoate–butanol–water system at 20°C [41].

As a result, a liquid region which spans most of the decanol–water composition range exists at sodium octanoate fractions between about 2 and 20%. Increasing the sodium octanoate composition beyond this limit results in precipitation of a structured phase, whose structure and composition depends on the ratio of butanol to water in the mixture. Above a 65:35 butanol:water ratio, the phase that separates is a crystal, between about 15:85 and 65:35 it is a lamellar phase, and below a ratio of 15:85 it is a hexagonal phase. Cubic phases are not found in this system, but a tiny cubic region is found between the lamellar and hexagonal phase regions in the sodium octanoate–propanol–water and sodium octanoate–ethanol–water systems.

As the length of the alcohol is shortened still further, the area of the lamellar phase is steadily decreased and, at methanol, it vanishes altogether. The hexagonal phase will dissolve about as much methanol as decanol, but methanol and ethanol destroy liquid crystal phases altogether when present in sufficiently large fractions below C_4. (They also destroy micellar structure in surfactant solutions [45].)

12.4.3 Soap–long-chain alcohol–water systems

Having considered the influence of medium- and short-chain alcohols on the phase behavior of an aqueous soap system, it would be logical (considering their industrial importance) to analyze the influence of longer (C_{12} and greater) alcohols on longer chain surfactants. Unfortunately, no substantial data exist and such an analysis is presently impossible. Alcohols having C_{12} or longer chains are crystalline at 20°C, so consideration would have to be given to the crystal phase of the alcohol. Since longer chains tend to favor the formation of more highly structured fluids, it might be expected that liquid crystal regions (especially the lamellar phase) should persist. Opposing this trend is the fact that the optimal interaction of two different long-chain compounds often occurs when the lipophilic chains of the two compounds are of comparable length [44]. (An example of this is found in the solubilization efficiency of homologous ammoniohexanoate surfactants for DSPC [44]. This efficiency, which is the ratio of DSPC to surfactant found in the mixed micelle at the solubility boundary, is optimal for $C_{18}AH$ and decreases at both higher and lower chain-lengths.) One might expect, on this basis, to find that if the chain-length of the alcohol were much greater than that of the soap, the degree of miscibility in all the phases would be diminished. A perfect match in chain-length is not required within fluid phases, of course; the alkyl chain-length of decanol is, after all, three methylene groups longer than is that of sodium octanoate.

The longer chain alcohols will likely form relatively viscous liquid and liquid crystal phases which would make studies such as those described above very difficult. Short-chain alcohols yield highly fluid phases which separate and coalesce relatively quickly, and are much easier to study.

12.4.4 Other ionic surfactant–fatty alcohol–water systems

Cationic surfactants. The above pattern of phase behavior applies to cationic surfactant–alcohol–water systems, but in modified form. The reason is that these compounds are far more soluble in dry fatty alcohols than are the soaps. As a result, the L_2 region in the octylammonium chloride–decanol–water system [43] (Fig. 12.18) is pushed against the surfactant–decanol border and little or no water is required to form this phase.

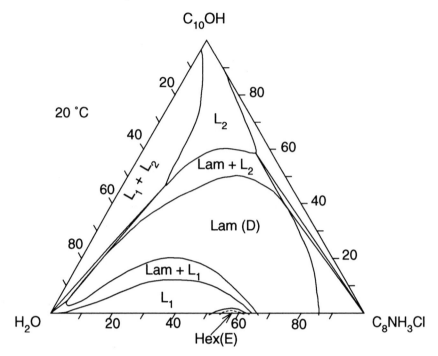

Fig. 12.18. The octylammonium chloride–decanol–water system at 20°C [48]. Compare Fig. 12.24.

Even with its short chain, octylammonium chloride displays both the hexagonal and the lamellar phase in the binary aqueous system [43]. The addition of decanol to the hexagonal phase produces the micellar liquid, rather than the lamellar phase. As a result, the liquid phase region extends completely around the hexagonal phase region. This is unusual (but not unique) phase behavior. Similar behavior is also found in the potassium laurate–lauric acid–water system [46] and in the octylammonium chloride–octylamine–water system [47] (Fig. 12.24, Section 12.6.2).

The addition of decanol to the lamellar phase significantly extends this phase region. When this phase contains a 60:40 ratio of decanol:octylammonium chloride, the maximum w_{water} that can exist in the phase is $c.\,94\%$! This represents an even higher capacity to dissolve water than that of the potassium oleate–decanol lamellar phase (Section 12.4.1).

Nonsoap anionic surfactants. Sodium alkanesulfonates and alkyl sulfates, in combination with fatty alcohols and water, behave in a manner that is qualitatively similar to that of the soaps. The diagram of the sodium octanesulfonate–decanol–water system at 20°C [48] (Fig. 12.19), for example, shows a remarkable resemblance in its overall form to that of the sodium octanoate–decanol–water system (Fig. 12.15). An interesting feature of this diagram is that no liquid crystals at all exist along the binary isotherm at this temperature. Strictly speaking, this compound is not a surfactant by the phase criteria, but a hydrotrope (Section 10.5.1). The presence of decanol induces the formation of both the hexagonal and lamellar phases, however.

A partial study of the SDS–pentanol–water system [49] (Fig. 12.20) reveals that liquid crystal regions exist that qualitatively resemble those found in the soap–butanol

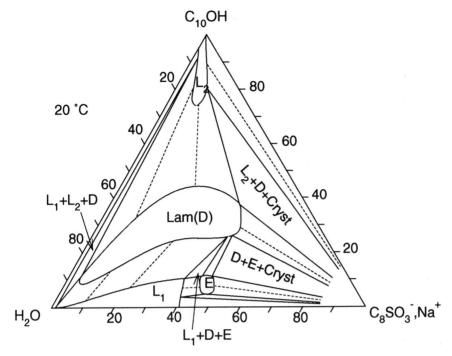

Fig. 12.19.   The sodium octanesulfonate–decanol–water system at 20°C [48].

system (Fig. 12.17), but that additional complexity exists with respect to liquid/liquid/ lamellar three-phase equilibria. These complications have not so far been found in soap systems (Section 12.4.1). No studies of this system have taken into account the complex liquid crystal phase equilibria that are now known to exist above the Krafft disconti- nuity; the temperature of this study is almost exactly at the discontinuity, so that only the hexagonal phase is shown to exist [50]. At higher temperatures considerably more complex behavior can be expected.

12.4.5   Nonionic surfactant–fatty alcohol–water systems

In soap–alcohol–water systems the fatty alcohol is poorly miscible with both the crystalline soap and water, while the soap and water are extensively miscible. Most nonionic surfactants are weakly crystalline and low-melting, and therefore miscible with fatty alcohols to a considerable degree. This fact may be expected to exert a profound influence on the ternary phase equilibria of aqueous nonionic surfactant systems containing fatty alcohols.

Only limited data exist on nonionic–oil–water systems, but one isotherm of the monooctanoin (monocaprylin)–decanol–water system at 20°C has been reported [51] (Fig. 12.21). In this system a dilute aqueous liquid phase that is predominantly water exists along the monoglyceride–water border, a lamellar liquid crystal, and the dry liquid are encountered ($L/D/L$ phase sequence). Decanol is poorly miscible with water, but is extensively soluble in the lamellar phase ($w_{decanol} = 25\%$). Along the oil surfactant border, it is noteworthy that decanol and monooctanoin are not miscible in all proportions. Monooctanoin is soluble in decanol up to $w_{monooctanoin} = 63\%$, but

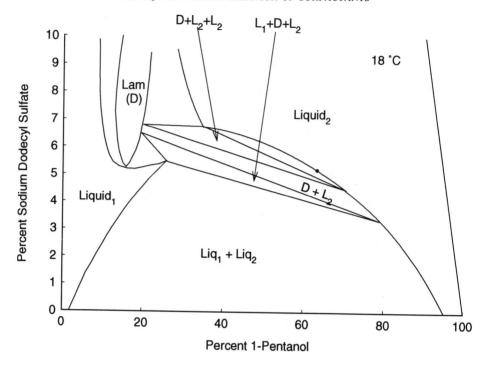

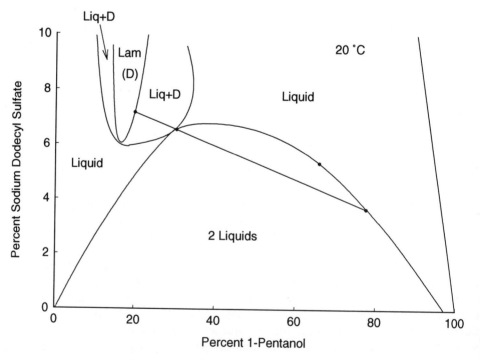

Fig. 12.20. The sodium dodecyl sulfate–pentanol–water system at 18°C (upper) and at 20°C (lower) [49].

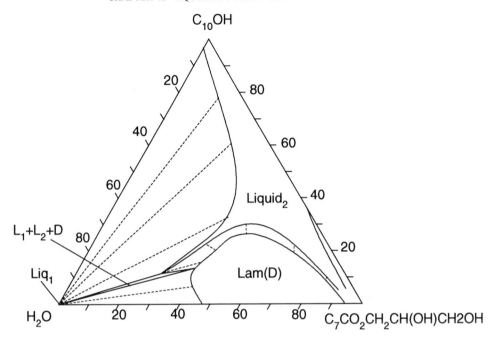

Fig. 12.21.   The monooctanoin (monocaprylin)–decanol–water system at 20°C [51].

decanol is poorly soluble in monooctanoin. Water dissolves in the decanol–mono-octanoin liquid to a modest extent over a wide range of decanol : monooctanoin ratios ($w_W \approx 10\%$), but at a ratio of about 40 : 60 decanol : monooctanoin the fraction of water in the liquid phase increases sharply (to about 60%). This produces another narrow finger in the liquid phase, which points (approximately) towards the water vertex.

12.4.6   Soap–fatty acid–water systems

Fatty acids represent a particularly important kind of amphiphilic oil, inasmuch as they are the conjugate acids of the soaps (Section 2.2.1). The sodium octanoate–octanoic acid–water system [51] (Fig. 12.22) has been well studied, and the resulting diagram resembles in many respects the corresponding soap–octanol–water diagram. The same phases are found in both the fatty alcohol and the fatty acid systems, but there are two important differences. One is that the liquid L_2 region displays a sharp spike which extends to extremely high water fractions (to about 98% water [51]) within a narrow range of octanoic acid/sodium octanoate ratios (about 75 : 25 octanoic acid : sodium octanoate ratio).

The second important difference is that crystalline phase compounds containing both fatty acids and soaps (acid–soaps) exist as equilibrium phases in this system. A dry acid soap having the formula Na_2HOc_3 (where "Oc" is the $C_7H_{15}CO_2^-$ anion) is seen in Fig. 12.22 along the NaOc–HOc border, and it is this phase (rather than a pure NaOc or hydrated soap crystal phase) that coexists with the L_2 region at low water fractions. The behavior suggested in the crystal coexistence regions of this diagram represents a reasonable hypothesis (insofar as the crystal chemistry is concerned), rather than a well-supported experimental result.

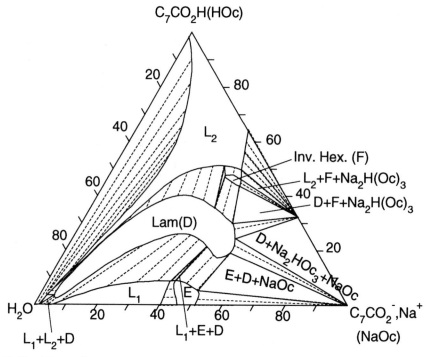

Fig. 12.22. The sodium octanoate–octanoic acid–water system at 20°C, using the published liquid and liquid crystal data and incorporating plausible hypotheses as to the nature of the crystal phase equilibria [43].

The form of the L_2 region seen in sodium octanoate–octanoic acid–water, with the sharp spike extending towards the water vertex, is also found in the analogous C_6, C_7, and C_{12} systems of a soap, its conjugate fatty acid, and water. The full diagrams of these other systems have not been determined, however.

12.4.7   Nonionic surfactant–fatty acid–water system

The diagram of the $C_{12}E_{10}$–caprylic acid (octanoic acid)–water system [43] has been determined, and provides information regarding a nonionic surfactant–fatty acid–water system. This surfactant exists as a low-melting crystal, so that the surfactant and oil are miscible over most of the composition range (above $w_{oil} = 7\%$). The addition of water to liquid mixtures of these two components has an effect that is remarkably similar to that observed in the soap–fatty acid–water systems. The solubility of water in this phase is modest at most oil:surfactant ratios, but displays a sharp increase within a narrow span of compositions. In this system, this spike is centered at about a 40:60 oil:surfactant ratio. Neglecting this spike, the fraction of water in the water-saturated oil:surfactant mixture is roughly proportional to the fraction of surfactant present, and attains a value of 25% in the absence of oil.

The phase sequence along the surfactant–water border is $L/C/E/L/X$. Oil is miscible in all proportions with the concentrated liquid phase, but adding oil to the hexagonal (E) or the dilute liquid (L) phase (at a surfactant:water ratio >20:80) induces formation of the lamellar (D) phase. The narrow spike of the liquid phase (noted above) lies just above the *D* phase, and extends to about the same water:surfactant ratio as does the *D* phase itself.

Overall, the form of the liquid and liquid crystal regions (and the coexistence relationships among these phases) are remarkably similar for the soap–fatty acid–water and the nonionic surfactant–fatty acid–water systems, except for the region where surfactant crystals may exist.

12.5   The general influence of nonsurfactant third components

It would be desirable to develop correlations between the molecular structure of added third components and their influence on aqueous surfactant phase behavior that are of predictive value. To do so is presently difficult, because comprehensive ternary phase information is available for only a few types of systems. Also, much of this is incomplete or has been defined only at a single temperature. Another problem is that uncertainties exist with respect to our knowledge of the component binary phase diagrams.

12.5.1   Miscibility within the limiting binary systems

Surfactant–water miscibility as a limiting factor. It is tempting to approach this matter by considering first the phase behavior, in a broad sense, of the three binary systems. Whether or not the surfactant itself is miscible with water at the temperature of interest is very important, for example. For this purpose it is useful to ignore, for the moment, the structure of the phases within which the two components exist and focus instead on the span of compositions occupied by miscibility gaps. In taking this approach a low-melting nonionic surfactant (like $C_{10}E_4$) is regarded as being miscible with water in almost all proportions below 20°C, while its miscibility at 60°C and higher temperatures is far more limited (Fig. 4.5).

A strongly hydrophilic surfactant, on the other hand, is miscible with water in all proportions at and above its melting point, but not below this temperature. As the temperature falls below the melting point the miscibility gap between the crystal phase and hydrated fluid phases becomes progressively larger. When the temperature falls below the Krafft discontinuity the miscibility of surfactant and water drops precipitously and, below the knee of this boundary, surfactants and water are essentially immiscible. Because temperature strongly influences the miscibility of surfactants with water, and this aspect of miscibility is extremely important with respect to the influence of third components, temperature is clearly an extremely important factor to consider during analysis of the influence of third components on aqueous surfactant phase behavior.

Third component–water miscibility. A second consideration is whether or not the third component is miscible with water. It may be extensively soluble in water (as in the case of sodium chloride or ethanol), or it may be insoluble (as in the case of alkanes, or fatty alcohols or acids). This fact, by itself, dictates the phase behavior in the limit that the fraction of surfactant approaches zero. To the extent that the miscibility of these two components is also sensitive to temperature, temperature also influences the limiting behavior of these two components.

Third component–surfactant miscibility. Finally, there is the miscibility of the third component with the surfactant to consider. If the two are miscible, then one situation applies in the limit that the fraction of water approaches zero. If they are immiscible, then another will hold. Since surfactants are amphiphilic molecules, most of the compounds with which they are miscible are either amphiphilic or lipophilic.

12.5.2　An overview

Third component immiscible with both water and surfactant. If the third component is immiscible with both water and the surfactant, then it should not influence the phase behavior of the aqueous surfactant system. Examples would be very high melting crystalline hydrocarbons, fatty compounds such as triacontane, behenic (C_{22}) acid or behenyl alcohol, inorganic salts such as calcium sulfate or carbonate, alumina, titania, metals, glass, etc. Regardless of the degree of miscibility of the surfactant and the water, the behavior of this system would not be greatly influenced by the presence of such third components.

Third component miscible with water, immiscible with surfactant, surfactant and water miscible. An example of this situation is found in the soap–salt–water system at high temperatures. Since salt and water are miscible to a significant degree, one effect of adding salt is to reduce the activity of water in aqueous mixtures. Salt does not mix well with aqueous surfactant phases unless they are rich in water, as in the dilute liquid phase. This phase extends a significant distance into the phase prism, as a result, while the more concentrated (liquid crystal) phases exist as thin regions plastered against the surfactant–water face of the prism.

All components are fully or partially miscible with each other. An example of this situation would be ethanol in combination with most liquid nonionic surfactants and water. While phase data on such a system have not been produced, its form may be anticipated with considerable confidence and is supported by casual observations. The addition of ethanol to the phases that exist at various surfactant:water ratios will transform them, at low fractions of ethanol, into a liquid solution. A featureless diagram, spanned over most of its area by a liquid phase, would result.

A system in which the miscibility of the third component with the surfactant and water is extensive, but not complete, is the sodium octanoate–butanol–water system. Partial miscibility exists at the butanol–water border, but small amounts of soap produce a liquid band that spans the entire region between the surfactant–oil and oil–water borders. As more soap is added to these liquid mixtures a soap phase that contains variable fractions of the other two components separates. Which phase separates depends on the soap:water ratio (Section 12.4.2).

Third component immiscible with water, miscible with surfactant, and surfactant and water are miscible. This situation is encountered in the $C_{10}E_4$–hexadecane–water system at low temperatures (Section 12.3.3), and leads to miscibility as a liquid phase along both the surfactant–oil border and the surfactant–water border. Along the latter, the phase structure is transformed from liquid to lamellar liquid crystal, and then back to liquid, as the water:surfactant ratio increases, but the miscibility gaps separating these different phase structures are small (a few per cent). Along the oil–water border, however, a large liquid/liquid miscibility gap exists within which a nearly pure oil phase coexists with surfactant–oil–water mixtures of widely varying composition.

This miscibility gap is split into two separate regions by intrusion of a finger of the lamellar liquid crystal phase.

Third component immiscible with water, miscible with surfactant, surfactant and water immiscible. This situation is encountered in the $C_{10}E_4$–hexadecane–water system at high temperatures (Section 12.3.3). It leads to a miscibility gap found not only along the oil–water border, but also along part of the surfactant–water border as well. Miscibility exists only within mixtures that are rich in both surfactant and oil, and these mixtures coexist with a nearly pure water phase.

12.5.3 The transition from conditions of surfactant–water miscibility to surfactant–water immiscibility in surfactant–oil–water systems

It has been recognized only during the past decade or so [33,52] that the evolution of liquid/liquid miscibility, from the condition of surfactant–water miscibility to the condition of surfactant–water immiscibility, follows a clearly recognizable pattern (Fig. 12.23). If one starts with ternary mixtures that lie within the central miscibility gap at a temperature where a state of surfactant–water miscibility and aqueous surfactant–oil immiscibility exists (e.g. at low temperatures in the $C_{10}E_4$–hexadecane–water system) and increases the temperature, it has been found that a point is reached at

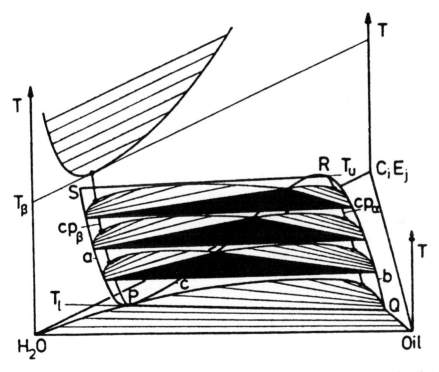

Fig. 12.23. The evolution of the three-phase body as a function of temperature. Starting from low temperatures, the three-liquid-phase tie-triangles (in black) first appear at a lower critical end-point, grow in size with increasing temperature while at the same time changing in shape, and then shrink in size and vanish at an upper critical end-point. Reproduced with permission from reference [35c].

which a tie-line near the oil–water border splits in two. A three-phase tie-triangle is formed as a result. By analogy with critical phenomena, in which single phases split into two phases at a "critical point", the phenomenon in which two phases split into three is termed a "critical end-point". Since a three-phase state exists *above* the temperature at which this occurs in this particular instance, this critical end-point is, by analogy with "lower critical points", termed a "lower critical end-point".

The new third phase which is created at the lower critical end-point is water-rich, but is also comparatively surfactant-rich. With the creation of this third phase a new miscibility gap must also be created; this gap is found along the new side of the tie-triangle which connects the two water-rich phases. The tie-lines within this new miscibility gap fan out from the most water-rich vertex of the tie-triangle, signifying that the phase behavior within this new miscibility gap differs profoundly from that which is found at lower temperatures.

As the temperature is increased above the lower critical end-point, the newly created surfactant–oil–water phase becomes progressively richer in both oil and surfactant, and consequently poorer in water. It swings in a shallow arc from a position near the water vertex, to a position at which oil and water are present in comparable proportions, and finally to a position near the oil vertex. At an intermediate temperature the fraction of oil and water are similar. The fraction of surfactant that is present is close to its maximum value, and extremely low tensions between coexisting phases exist when this is true. (The temperature at which this situation obtains was earlier termed the "phase inversion temperature" [53].)

At the upper temperature limit of existence of the tie-triangle, the critical end-point phenomena described above again occur but in the opposite sense with respect to temperature. Two of the coexisting phases are now very oil-rich, and these two phases (while coexisting with a third water-rich phase) become identical at an "upper critical end-point". The three-phase triangle vanishes, and the miscibility gap within which a nearly pure water phase coexists with surfactant–oil–water phases is now the dominant miscibility gap along the oil–water border.

As the temperature is further increased, the extent of miscibility of water with the surfactant–oil mixtures can be expected to decrease over a finite span of temperatures. At a sufficiently high temperature the miscibility of surfactant and water will begin to increase, however, and the situation may again change. The temperature at which this reversal of surfactant–water miscibility occurs in those systems that have been studied is very high, and no information exists at these temperatures.

The above picture provides a rational phenomenological description of the changes in phase behavior that occur with temperature in surfactant–oil–water systems. In some cases the behavior may be still more complex insofar as liquid/liquid equilibria are concerned, though still related to the above relatively simple picture (Section 12.3.4). The existence of liquid crystalline phase structures in the surfactant–water mixtures will complicate the picture. However, it is worth noting that the above pattern of the evolution of liquid/liquid phase behavior as conditions are changed appears to be superimposed upon the behavior of the liquid crystal regions. The liquid/liquid coexistence, and the liquid crystal aspects of phase behavior, develop independently of one another (Section 5.10.6).

It is very important to note that the evolution of liquid/liquid behavior described above is encountered in systems containing amphiphilic molecules having a wide range of molecular structures. The existence of surfactant behavior, as reflected by the phase behavior of amphiphile–water systems (Chapter 9), is *not* a prerequisite for this pattern to exist.

Finally, it is also noteworthy that the above phenomena may be caused to occur isothermally if the salt composition in quaternary amphiphile-nonpolar oil–salt–water systems is varied [54]. Doing this represents another means by which the miscibility of the amphiphile with water can be influenced, and it is possibly this feature of the binary phase behavior that is important. The analysis in the salt-containing systems may be related to that found in the ternary systems as temperature is varied, but the phase behavior itself is complicated by the presence of a fourth component and has not been determined.

12.5.4 The influence of amphiphilic structure in the oil

When the oil component is amphiphilic (as in the case of a fatty alcohol) it interacts with water and surfactants far more strongly than when it is not, even though it may be poorly miscible with water (Section 9.6.1). When the surfactant is highly crystalline (as in a soap), the influence of an alkanol on the behavior of the aqueous system is *profoundly* different from that of nonpolar oils (compare Sections 12.3.1 and 12.4.1). Fatty alcohols can participate in the formation of liquid crystal structure far more easily, without disrupting this structure, than can nonpolar oils. This is understandable on the basis that the hydroxy group possesses finite (if weak) hydrophilicity (Section 9.6.1). An alcohol may not be able to stand alone insofar as the formation of liquid crystal states is concerned, but in mixtures with a more strongly hydrophilic surfactant it does not impair the capacity to form these states. In fact, when the surfactant by itself is unable to form the lamellar or the inverted hexagonal phase structure, the addition of a fatty alcohol often induces their formation.

The presence of a fatty alcohol within a lamellar phase invariably enhances – sometimes dramatically – the miscibility of this phase with water.

12.6 The influence of another surfactant

12.6.1 Homologs

It has been emphasized throughout, especially in Section 12.5, that the phase behavior of binary systems strongly influences that of ternary systems. Since the phase behavior of surfactant–water systems differs qualitatively from that of nonsurfactant–water systems, the phase behavior of a mixture of two surfactants and water would not be expected to resemble that of a surfactant–oil–water system. While probably true, the amount of actual phase information upon which this premise is based is extremely small.

Long ago, McBain noted that the presence of homologs in commercial soaps did not seem to alter qualitatively the phase behavior of the commercial system relative to that of binary soap–water systems [55]. This conclusion was based on the observation that no new phase phenomena were found in complex mixtures that did not exist in simpler systems. As noted in Section 12.1.2, phase transitions that are isothermal in binary mixtures are expected to be non-isothermal in ternary (and higher order) systems. If allowance is made for this fact, mixtures of homologous soaps appear to behave in a manner which parallels that of binary soap–water mixtures. This constitutes an example of "parallel" behavior of a simple and a complex system. An example of "non-parallel" behavior would be the sodium octanoate–decanol–water system, in which the presence of the decanol induces the formation of phase states that do not exist in the binary soap–water system.

Very few studies of ternary mixtures of two surfactants and water have been reported, but it has been suggested from isoplethal studies of aqueous mixtures that the Krafft boundary of ternary systems of two related (but different) ionic surfactants displays a eutectic-like form [56]. This was found with mixtures of sodium dodecyl sulfate and calcium dodecyl sulfate ($NaC_{12}SO_4$ and $Ca(C_{12}SO_4)_2$) at 97% water composition, with sodium tetradecyl sulfate and calcium tetradecyl sulfate ($NaC_{14}SO_4$ and $Ca(C_{14}SO_4)_2$) at 99% water composition, and with calcium dodecyl sulfate and calcium $C_{12}E_1$ sulfate ($Ca(C_{14}SO_4)_2$ and $Ca(C_{14}E_1SO_4)_2$) at 99% water composition. These results are to be expected when the surfactants are miscible within the liquid and immiscible within the crystal state.

12.6.2   The octylammonium chloride–octylamine–water system

The phase behavior of this system of a cationic surfactant salt plus a nonionic surfactant plus water (Fig. 12.24) is of particular interest because the amine is not just another surfactant; it is also the conjugate base of the cationic surfactant. A lamellar phase exists in both binary surfactant–water systems, and spans a roughly similar range of compositions in each. It is hardly surprising, therefore, to find that a band of lamellar phase spans the entire range of compositions between these two lamellar phase regions. The

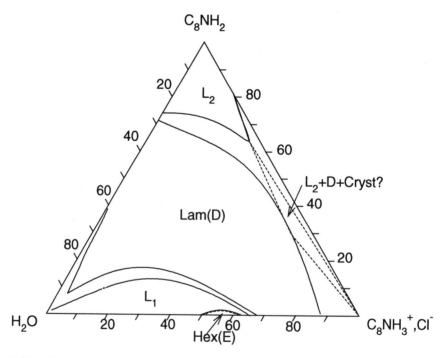

Fig. 12.24. The octylammonium chloride–octylamine–water system at 20°C [47]. Compare Fig. 12.18.

upper limit of the range of water composition does not vary smoothly within this band, however. The presence of small amounts of ammonium salt in the amine significantly increases the fraction of water that may exist in the amine–water lamellar phase.

A hexagonal liquid crystal phase exists within the octylammonium chloride–water diagram, but octylamine is barely miscible with this phase. The addition of amine past its solubility limit transforms the hexagonal into the liquid phase, then into the lamellar phase, and finally into the L_2 liquid region appended to the amine vertex. If the surfactant is replaced by octyldimethylammonium chloride, qualitatively similar phase behavior is observed except that the liquid phase no longer envelops the hexagonal region. In this case the diagram more closely resembles that of the sodium octanoate–decanol–water system (Fig. 12.15) [38].

Comparison of the octylammonium chloride–decanol–water diagram (Fig. 12.18) with that of the octylammonium chloride–octylamine–water system is interesting because the amino and the hydroxy functional groups are isoelectronic. These two systems are, as a result, extremely similar except for the fact that the chain-lengths of the alcohol differ from that of the alkylamine (Section 11.3.1). Even with this difference, it is apparent that striking similarities exist in their phase behavior. Decanol may replace octylamine virtually without altering the phase states that exist, except very close to the decanol–water border. The same would be expected of octanol.

12.7 Summary

Since the binary aqueous phase behavior of two different soluble surfactant–water systems is qualitatively similar, the ternary behavior of two surfactants in water can be expected to be relatively straightforward. The octylammonium chloride–octyl-amine–water diagram provides an example. The data in this area are extremely limited, however, and the consequences of mixing two surfactants whose aqueous phase behavior differs substantially remain uncertain.

If the two surfactants are both insoluble within the dilute liquid phase (as in the case of two polar lipids in water), then the facts of the matter are rarely known. However, there is convincing evidence that if both form a lamellar phase, the two surfactants are miscible with one another in the ternary lamellar phase. Binary polar lipid–water phase behavior has yet to be defined in a rigorous and comprehensive manner (Section 11.3.2.7), so it is evident that our knowledge of ternary systems is still more primitive.

If one surfactant is soluble and the other insoluble, then the soluble surfactant may solubilize the insoluble surfactant in dilute mixtures up to a finite limit. At extremely low compositions, the ratio of the two surfactants in the saturated micellar solution remains constant within a finite range of compositions [44]. It has also been established that the soluble surfactant may extensively partition into coexisting liquid crystal phases [44], but little information as to the actual compositions of such coexisting phases exists. It has been suggested that the dissolution of a cationic surfactant (CTAB) in the lamellar phase of a lecithin dramatically extends the miscibility of the lamellar phase with water [57]. From this observation (and in Section 12.6), it is apparent that anomalies may be expected to exist as regards the composition ranges that liquid crystal phases span in ternary two-surfactant-plus-water systems.

Finally, if a chemical reaction occurs between two surfactants then complexity in the phase behavior of the "ternary" two surfactant-plus-water system is inevitable. An example of this is the ion exchange reaction that may occur between a cationic surfactant salt and an anionic surfactant salt. If either surfactant is present in large excess then it will solubilize such "catanionic" surfactants [58], but if the two are present in approximately stoichiometric ratio then the miscibility of the catanionic surfactant with water is poor. (Advantage is taken of this phenomenon, in combination with ion exchange chemistry with dyes, to analyze anionic or cationic surfactants [58].) It is possible to synthesize the "catanionic surfactant salt" in which both ions are amphiphilic [59], and this salt in combination with water constitutes a binary system. If the two chain-lengths combined exceed about 20 carbon atoms then these materials display monoglyceride-like phase behavior, and the lamellar phase is the coexisting phase at the solubility boundary [59]. The catanionic surfactant salt may be viewed as a phase compound that exists in the ternary system, but may alternatively be viewed as a separate component.

References

1. Nishikido, N., Kobayashi, H. and Tanaka, M. (1982). *J. Phys. Chem.* **86**, 3170–3172; Tanaka, M., Kaneshina, S., Sugihara, G., Nishikido, N. and Murata, Y. (1982). *Solution Behavior of Surfactants: Theoretical and Applied Aspects*, Vol. 1 (K. Mittal and E. J. Fendler eds.), pp. 41–71, Plenum, New York.
2. Hubner, W., Wong, P. T. and Mantsch, H. H. (1990). *Biochim. Biophys. Acta*, 229–237; Fotland, P. (1987). *J. Phys. Chem.* **91**, 6396–6400; Wong, P. T. T., Chagwereda, T. E. and Mantsch, H. H. (1987). *J. Chem. Phys.* **87**, 4487–4497.
3. Findlay, A. (and Campbell, A. N.) (1938). *The Phase Rule and its Applications*, 8th ed., p. 213, Dover, New York.
4. Rosevear, F. B. (1968). *J. Soc. Cosmetic Chemists* **19**, 581–594.
5. Kahlweit, M., Strey, R. and Firman, P. (1986). *J. Phys. Chem.* **90**, 671–677.
6. Findenegg, G. H., Hirtz, A., Rasch, R. and Sowa, F. (1989). *J. Phys. Chem.* **93**, 4580–4587.
7. Masing, G. (1944). *Ternary Systems*, Translated by B. A. Rogers, p. 29, Reinhold, New York.
8. Lee, A. G. (1977). *Biochim. Biophys. Acta* **472**, 285–344.
9. Vorob'eva, A. I. and Karapet'yants, M. Kh. (1967). *Russ. J. Phys. Chem.* **41**, 602–605.
10. Glasstone, S. (1946). *Textbook of Physical Chemistry*, 2nd ed., pp. 792–800, Van Nostrand, New York.
11. Reference 7, pp. 34–36.
12. Palatnik, L. S. and Landau, A. I. (1964). *Phase Equilibria in Multicomponent Systems*, Hold, Rinehart, and Winston, New York.
13. Francis, A. W. (1961). *Critical Solution Temperatures*, Vol. 31 (R. F. Gould ed.), pp. 32–38, American Chemical Society, Washington, DC.
14. Schreinemakers, F. A. H. (1911). *Die Heterogene Gleichgewichte vom Standpunkte der Phasenlehre*, Vol. 3 (H. W. B. Roozeboom ed.), pp. 6–17, Friedr. Vieweg und Sohn, Braunschweig.
15. Prince, A. (1966). *Alloy Phase Equilibria*, pp. 143–145, Elsevier, Amsterdam.
16. Marsh, J. S. (1937). *Principles of Phase Diagrams*, pp. 122–170, McGraw-Hill, New York.
17. McBain, J. W. and Burnett, A. J. (1922). *J. Chem. Soc.* **121**, 1320–1333.
18. McBain, J. W., Lazarus, L. H. and Pitter, A. V. (1926). *Z. physik. Chem.*, **A147**, 87–117.
19. Nakayama, Haruo, and Shinoda, Kozo (1967). *Bull. Chem. Soc. Jpn.* **40**, 1797–1799; Shinoda, K., Yamaguchi, N. and Carlsson, A. (1989). *J. Phys. Chem.* **93**, 7216–7218.
20. (1991). *Handbook of Chemistry and Physics* (D. R. Lide ed.-in-chief), 72th ed., p. 4-98, CRC Press, Boca Raton, FL.
21. Oonk, H. A. J. (1981). *Phase Theory: The Thermodynamics of Heterogeneous Equilibrium*, Elsevier Scientific, New York.
22. Laughlin, R. G. (1992). *Adv. Coll. Interface Sci.* **41**, 57–79.

23. Reference 13, pp. 72–76.
24. McBain, J. W., Brock, G. C., Vold, R. D. and Vold, M. J. (1938). *J. Am. Chem. Soc.* **60**, 1870–1876.
25. Shinoda, K. (1967). *Solvent Properties of Surfactant Solutions*, Vol. 2 (K. Shinoda ed.), pp. 57–58, Marcel Dekker, New York; Shinoda, K. and Sagitani, H. (1978). *J. Coll. Interface Sci.* **64**, 168–171.
26. K. W. Herrmann, unreported work.
27. Leng, F. J., Machin, D., Reed, D. A. and Erkey, O. (1985). *Eur. Pat. Appl.* EP 153857 A2 4 Sept. 1985.
28. Ekwall, P., Mandell, L. and Fontell, K. (1969). *Mol. Cryst. Liq. Cryst.* **8**, 157–213.
29. Fontell, K., Khan, A., Lindstrom, B., Maciejewska, D. and Puang-Ngern, S. (1991). *Coll. Polym. Sci.* **269**, 727–742; Fontell, K. and Jansson, M. (1988). *Prog. Coll. Polym. Sci.* **76**, 169–175; Fontell, K., Ceglie, A., Lindman, B. and Ninham, B. (1986). *Acta Chem. Scand. A* **40**, 247–256.
30. Warr, G. G., Sen, R., Evans, D. F. and Trend, J. E. (1988). *J. Phys. Chem.* **92**, 774–783.
31. Lang, J. C. (1984). *Proc. Int. Sch. Phys. "Enrico Fermi" (Phys. Amphiphiles)*, Vol. 90, pp. 336–375.
32. Lang, J. C. and Morgan, R. D. (1980). *J. Chem. Phys.* **73**, 5849–5861.
33. Shaw, R. W., Brill, T. B., Clifford, A. A., Eckert, C. A. and Franck, E. U. (1991). *Chem. & Eng. News*, December, 26–39.
34. Ramadan, M. S., Evans, D. F. and Lumry, R. (1983). *J. Phys. Chem.* **87**, 4538–4543.
35. Kahlweit, M., Strey, R. and Busse, G. (1990). *J. Phys. Chem.* **94**, 3881–3894; Kahlweit, M. and Strey, R. (1987). *J. Phys. Chem.* **91**, 1553–1557; Kahlweit, M., Lessner, E. and Strey, R. (1983). *J. Phys. Chem.* **87**, 5032–5040.
36. Wärnheim, T., Bergenstahl, B., Henriksson, U., Malmvik, A. C. and Nilsson, P. (1987). *J. Coll. Inter. Sci.* **118**, 233–242; Wärnheim, T. (1986). *Coll. Polym. Sci.* **264**, 1051–1059.
37. Wärnheim, T. (1986). *Coll. Polym. Sci.* **264**, 1051–1059.
38. Friman, R., Danielsson, I. and Stenius, P. (1982). *J. Coll. Interface Sci.* **86**, 501–514; Ekwall, P., Danielsson, I. and Mandell, L. (1960). *Kolloid Z.* **169**, 113–124; Ekwall, P. (1963). *Finska Kemistsamfundets Medd.* **72**, 59–89; Fontell, K., Mandell, L., Lehtinen, H. and Ekwall, P. (1968). *Acta Polytech. Scand., Chem. Incl. Met. Ser.* **74**, 1–116.
39. Friberg, S. (1979). *J. Soc. Cos. Chemists* **30**, 309–319; Friberg, S. E. and El-Nokaly, M. A. (1985). *Surfactants in Cosmetics*, (Martin J. Schick ed.), pp. 55–86, Marcel Dekker, New York.
40. Friberg, S. (1978). *Advances in Liquid Crystals*, Vol. 3 (G. H. Brown ed.), pp. 149–165, Academic Press, New York.
41. Ekwall, P., Mandell, L. and Fontell, K. (1969). *J. Colloid Interface Sci.* **31**, 508–529.
42. Lawrence, A. S. C. and McDonald, M. P. (1967). *Liquid Crystals*, (G. H. Brown, G. J. Dienes and M. M. Labes eds), pp. 1–19, Gordon & Breach, New York.
43. Ekwall, P. (1975). *Advances in Liquid Crystals*, Vol. 1 (G. H. Brown ed.), pp. 1–142, Academic Press, New York.
44. Laughlin, R. G. and Munyon, R. L. (1984). *Chem. Physics Lipids* **35**, 133–142.
45. Herrmann, K. W. and Benjamin, L. (1967). *J. Colloid Interface Sci.* **23**, 478–486.
46. McBain, J. W. and Field, M. C. (1935). *J. Am. Chem. Soc.* **55**, 4776–4793; Ekwall, P. (1975). *Advances in Liquid Crystals*, Vol. 1 (G. H. Brown ed.), p. 123, Academic Press, New York.
47. Wärnheim, T., Bergenstahl, B., Henriksson, U., Malmvik, A. C. and Nilsson, P. (1987). *J. Coll. Inter. Sci.* **118**, 233–242.
48. Ekwall, P., Mandell, L. and Fontell, K. (1969). *Mol. Cryst. Liq. Cryst.* **8**, 157–213.
49. Guerin, G. and Bellocq, A. M. (1988). *J. Phys. Chem.* **92**, 2550–2557.
50. Kekicheff, P., Grabielle-Madelmont, C. and Ollivon, M. (1989). *J. Coll. Int. Sci.* **131**, 112–132.
51. Ekwall, P. and Mandell, L. (1969). *Kolloid Z.* **233**, 938–944.
52. Davis, H. T., Bodet, J. F., Scriven, L. E. and Miller, W. G. (1989). *Physica A (Amsterdam)* **157**, 470–481; Kilpatrick, P. M., Davis, H. T., Scriven, L. E. and Miller, W. G. (1987). *J. Colloid Interface Sci.* **118**, 270–285.
53. Reference 25, pp 60–61.

54. Bourrel, M. and Schechter, R. S. (1988). *Microemulsions and Related Systems*, pp. 127–206, Marcel Dekker, New York; Valint, P. L., Jr., Bock, J., Robbins, M. L. and Zushma, S. (1987). *Coll. Surf.* **26**, 191–203.
55. McBain, J. W. (1926). *Colloid Chemistry*, Vol. I (J. Alexander ed.), pp. 137–164, The Chemical Catalog Co., New York.
56. Hato, M. and Shinoda, K. (1973). *J. Phys. Chem.* **77**, 376–381.
57. Rydhag, L. (1979). *Fette, Seifen, Anstrichm.* **81**, 4168–4173.
58. Heinerth, E. (1977). *Anionic Surfactants – Chemical Analysis*, Vol. 8 (J. Cross ed.), pp. 221–233, Marcel Dekker, New York; Cross, J. T. (1970). *Surfactant Science Series, Cationic Surfactants*, Vol. 4 (E. Jungermann ed.), pp. 419–488, Marcel Dekker, New York; Reid, V. W., Longman, G. F. and Heinerth, E. (1967). *Tenside* **9**, 292–304.
59. Jokela, P., Jonsson, B. and Kahn, A. (1987). *J. Phys. Chem.* **91**, 3291–3298; Jokela, P. and Jonsson, B. (1988). *J. Phys. Chem.* **91**, 1923–1927; Friberg, S. E., Sun, W. M., Yang, Y. and Ward, A. J. I. (1990). *J. Coll. Interface Sci.* **139**, 160–168.

Part V

MISCELLANEOUS

The Relationship of the Physical Science of Surfactants to Their Utility

13.1 "Nondurable goods", "products", "technologies", and other terms

In Chapter 2 a broad picture of the various elements of physical science was presented in order to place in perspective the role of phase science within physical science. In this chapter the field of view will be enlarged still further, so that the relationship of physical science as a whole to technology may be perceived. The focus will be on technologies underlying the nondurable goods category of consumer products [1]. As always, it is useful to define the words and concepts that will be used.

Nondurable goods. "Nondurable goods" (earlier, "soft goods") encompasses household cleaning products, personal care products and toiletries, paper products, food products, cosmetics, etc. Surfactants play either a major or a minor role in practically all of these products. It is probably impossible to devise a hard and fast definition of nondurable goods that cleanly distinguishes them from "durable goods" (formerly "hard goods") such as automobiles, but many nondurable goods possess the following characteristics:

(1) they are purchased directly by consumers for their personal use (as contrasted with materials purchased by manufacturers that are used to manufacture consumer products, for example),

(2) they are identified by their brand name, and the selection of a particular brand by the consumer is based on subjective judgment and/or experience rather than on technical data,

(3) they are purchased, consumed, and then repurchased at relatively frequent intervals (which vary widely), and

(4) their unit price is sufficiently small that their purchase is not a major investment – but they are often viewed by consumers as "necessities". For this reason, the purchase of nondurable goods is not usually postponed for significant lengths of time during conditions of economic stress (except under dire circumstances).

Durable goods also possess some of the above characteristics, but are distinguished from nondurable goods by others.

Products. A "product" will be regarded as the sum of the physical material that the consumer purchases (a bar of soap, a bottle of shampoo, etc.), plus a set of subjective attributes (an image) associated with the physical product [2].

$$\text{Product} = \text{Physical object} + \text{Subjective attributes (image)}$$

Products are identified by their brand names. They are marketed (advertised, distributed, and sold) in such a way as to take on their image, which becomes recognizable both to the consumer and to marketing personnel. It is mainly for this reason that a "product" is more than a physical mixture of the raw materials from which it is made. The chemical composition of products typically changes significantly over time as a result of evolutionary improvement and refinement. The brand name and image tend to persist much longer. Companies that own brand names (which are legally the property of the company and can be kept, if desired, for indefinitely long periods) may of course consciously choose to alter the image associated with the name.

To illustrate, Tide and other laundry brands introduced just after World War II contained sodium tetrapropylenebenzene sulfonate (ABS) as a major surfactant raw material, but this material was replaced around 1965 by straight chain linear alkyl benzene sulfonates (LAS) due to environmental pressures on ABS [3]. The attributes that Tide had developed during the preceding years were jealously guarded during its transition from one formula to the other.

From the perspective of the analytical chemist, products consist of *raw materials*. A raw material is a characteristic mixture of chemical compounds produced commercially by a *supplier*, who manufactures and sells it to those companies who manufacture consumer products. The manufacturer may incorporate the raw material directly into a product, or may transform it chemically into another raw material before that happens.

Two examples of raw materials are coconut oil, and the "coconut fatty acids" produced from the triglycerides found in coconut oil. Chemically, "coconut fatty acids" are a mixture of homologous (mostly saturated) fatty acids that contains principally dodecanoic (lauric) acid, smaller amounts of tetradecanoic (myristic) acids, and minor amounts of lower and higher homologs [4]. Tallow and "tallow fatty acids" are raw materials obtained from the tallow fat of cattle or other animals. Tallow fatty acids contain homologs that are longer in chain-length, and are (in part) unsaturated. The statistically weighted chain-length, including the carboxylate carbon, is 12.5 carbons in coconut and 17.3 carbons in tallow fatty acids. The chemical composition of raw materials varies from batch to batch, but usually within narrow limits.

A product is fabricated from raw materials by use of a *manufacturing process*. This process may involve simply mixing the raw materials, but is usually more complex. *Process development* is concerned with the definition of manufacturing processes, and is largely the province of the engineering disciplines. Process development is strongly influenced by the physical science of the raw materials. Engineers are perhaps the major "consumers" (or users) of physical science data, and nowadays must often produce it as well.

Technologies. A "technology" may be regarded as the objectively measurable dimensions of a product or a process. The two principal dimensions of product technology are the raw materials in the product and their chemical composition, but the process by means of which raw materials are fabricated into the physical product is another aspect of a technology. A technology is not a product, as it does not yet have a brand name or image.

13.2   The roles of industrial technologists

It is obvious that the introduction of a new product by a company requires the development of technology. It is less obvious, but still true, that the refinement of technologies also requires the development of technology. Similar functions are performed during both kinds of technology development activities; the difference lies in the degree of novelty or "newness" that exists.

Three basic functions must be performed during every technology development activity, but the relative emphasis on each varies widely from one project to another. One function is selection of the raw materials to be used in the prospective product, and determination of the amounts of each to be used. This may be termed the "*raw material selection*" function. Because of the extremely limited period of time normally allocated for this function, the raw materials selected are ordinarily chosen from among those that are commercially available at the time.

A second basic function is definition of the process that will be used to combine the selected raw materials into a product – the "*raw material integration*" or "*process development*" function. This is an altogether nontrivial function that entails taking the raw materials in, fabricating the product, and putting out a packaged product that is ready for sale on a commercial scale at an acceptable cost.

On rare occasions, raw materials that are not commercially available are identified and selected for use, and the "*raw material acquisition*" function must also be pursued. This requires the development of a practical synthetic method for commercial production of the raw material, and the engineering of a production process. Synthetic method development is best performed by those chemists who specialize in the quantitative treatment of reaction equilibria and rates and the execution of large scale syntheses, working closely with chemical engineers.

Obviously, the acquisition of raw materials must precede the integration of the selected raw materials into a product. Fortunately, the evaluation of performance on which preliminary selection decisions can be based (Section 13.3.3) may often be accomplished using laboratory size samples. The physical science of surfactant raw materials is important to those engaged in raw material acquisition, especially during isolation of the refined reaction product from the crude reaction mixture. This subject will not be treated here, however.

After technology development is complete the product must be manufactured, and this typically involves a wide range of engineering disciplines. Since the role of physical science during the manufacturing process is similar to its role during development of the process, the manufacturing process itself does not require separate consideration, except in so far as the conditions in the plant process differ from the conditions during laboratory or pilot scale experiments.

13.3 Raw material selection: correlating molecular structure with sales

Everyone in business is interested in sales, for their livelihood depends on it. Sales are governed by which way "purchase/no purchase" decisions go while consumers are shopping, and these decisions are influenced (in part) by the success with which raw materials have been selected and integrated into a product.

In its simplest form, the development of technology involves the selection of raw materials and the design of a manufacturing process, which is then taken to the manufacturing plant without intervening steps. In this primitive model the success of the development is evaluated on the basis of the sales that result. The approach to the raw material selection process that applies in this instance may be depicted schematically as follows:

Sales ($)

⇕

Raw materials selected

Sales data may be regarded as information. The decisions made during the raw material selection function may be viewed as a different kind of information, which defines the molecular composition of the raw materials selected. The double-headed arrow symbolizes the correlation, or relationship, that exists between these two extremely different kinds of information.

The success of the selection process hinges to a significant degree on the quality of this correlation. If the assumed correlation is good, then acceptable sales *may* result. If the correlation is poor, then poor sales will likely result. (It is assumed that all other aspects of the manufacturing and marketing of the product have been well handled. It is evident that this may or may not be true, but we are not here concerned with these factors.) The essential point, for the present purpose, is that the selection of raw materials influences, in some degree, the sales results.

Selecting product raw materials entirely on the basis of sales data constitutes an empirical approach to the selection process, since the correlation is not supported by any underlying principle that (cor)relates sales to the nature of the raw materials selected. Such a process may actually be utilized during the introduction of entirely new or very small technology, and is still widely used (when coupled with reliance on historic experience) as the basis for the componment selection decision. It is no longer entirely empirical when this experience is factored into the decision, of course, but neither is it likely to be a "rational" decision when made in this manner.

When the scale of a technology is large, or for technologies that have reached an advanced stage of development, the above approach is a highly inefficient way to select raw materials. More refined methods must be utilized, and the basic question that arises is "how may the raw material selection process be rendered less empirical, more rational, and more effective?". This issue is extremely important, since the degree to which product development (of which technology development is a part) is successful is a matter of life or death (figuratively speaking) to companies, and also to the people in companies.

13.3.1 The concept of "utility"

The purchase/no purchase decision made by consumers (including technologists, when they function as consumers) is purely subjective. In one frequently exercised scenario the consumer scans the store shelf, reacts to the display that is presented, and (within a very short space of time) makes a decision on every product that is on display. He or she decides to purchase one of the products, and in so doing also decides not to purchase all the others. The factors that influence this decision have been the subject of intense scrutiny, but we will focus only upon the decision itself.

"Utility" and "value" are among the important factors that influence this decision and are particularly relevant to the product component selection process. "Utility" may be regarded as *the capacity of a product to perform a useful function*, and "value" as *the ratio of utility to price.*

$$\text{Value} = \frac{\text{Utility}}{\text{Price}}$$

Price is dictated by many economic factors (with which we are not presently concerned), and strongly influences sales.

One of the parameters in the value equation (price) is easily quantifiable, while the other (utility) is not. This does not invalidate the concept, however, so long as both utility and price are "scalar" quantities. That is, for two products having the same utility the one with the lower price will represent the better value. For utility to be used in this manner, one need only assume that its magnitude varies from one product to another. It is possible in principle, but very difficult (in fact), to scale utility numerically.

Introducing the concept of "utility" considerably simplifies analysis of the raw material selection process. Directly correlating sales with the nature of the raw materials selected amounts to directly correlating sales with molecular structure, which is a mind-boggling task. Relating molecular structure to utility is considerably more feasible than is relating molecular structure to sales, because sales are influenced by so many factors other than the molecular structures of the compounds in the product.

"Sales" may be regarded as information whose level of "complexity" (Section 2.4) is extremely high. (In Chapter 2 the concept of "relative complexity" was defined and utilized to place in perspective various kinds of structural information. In this chapter the same basic concept will be applied to a wider range of information.) Since utility is only one of the factors that influence sales, it is clear that the relative complexity level of sales information is considerably greater than is that of utility information. "Molecular structure" information lies at the opposite end of our complexity scale; it is, by comparison, extremely simple information. As a result, the gap in the level of complexity between sales and molecular structure is extremely large. In fact, it is unmanageably large.

13.3.2 Rational, empirical, direct, and indirect correlations

The value of introducing the concept of "utility" lies in the fact that it breaks the enormous complexity gap that exists between sales information and molecular structure information into two smaller gaps, each of which is far more manageable than is the sales-structure gap. This step is mandatory if one is to develop rational correlations between structure and sales. If the correlation of sales with utility, and also the correlation of utility with molecular structure, both make sense, then the overall correlation will be improved. It will be useful at this point to review the intrinsic characteristics of "rational correlations" and of "empirical correlations" as these words are used here, and the distinctions between them. Rational correlations may also be regarded as being either direct or indirect, and the implied distinction between these two variants is also worth considering.

A correlation may be regarded as a defined relationship (ideally, a simple mathematical equality) that exists between two kinds of information that differ in their level of complexity. Correlations may be sought and found between any two kinds of information, but they differ enormously in their significance. It is convenient to divide them into two broad categories – empirical and rational. The distinction between these categories resides principally in the basis upon which the correlation rests. It resides to a far lesser degree on the statistical level of predictability that exists.

Empirical correlations. The direct correlation of sales with molecular structure is here viewed as being inherently empirical. The basis for this classification is that many factors strongly influence sales that have nothing at all to do with molecular structure. These include the market situation, the effectiveness with which the product is marketed, economic policies within the company, external economic factors, political factors, regulatory agency policies, environmental issues, etc.

Whether a correlation is empirical or rational is *not* perceptible simply on the basis of the quality of fit to a model. The field of statistical analysis is littered with correlations for which the value of r^2 is high, but which are nonsensical (r^2 is the mean value of the square of the residuals found during the calculation of standard deviations). If no rationale for a correlation exists, then that correlation is empirical no matter how close to one r^2 is. Providing a rationale may (or may not) transform an empirical into a rational correlation.

Rational correlations. A particularly useful kind of correlation is the "rational correlation". A rational correlation may be thought of as *a defined relationship between two or more kinds of information that is supported by substantial relevant experimental data, is understandable by reference to established fundamental principles, and has a significant degree of predictability*. If any of these three criteria are not met, then the correlation is (in some degree) empirical.

Viewing a correlation as rational or empirical is not necessarily a value judgment. Virtually all the major discoveries of science were, at the time of their discovery, necessarily empirical. However, with the passage of time (and usually with the investment of considerable work) a rationale for the early empirical correlation was developed and it came to be viewed, instead, as a rational correlation.

Innumerable rational correlations exist within the physical sciences. For example, chemical potential is correlated with thermodynamic activity, variance is correlated with the number of phases present, the acidities of acids and bases and reaction rate constants are correlated with the Hammett σ (substituent) parameters, etc.

In each of these examples the level of complexity of the two kinds of information that are correlated differs. One kind of information (chemical potential) is more complex than the other (activity); the more complex information is usually the dependent variable. The second is less complex, is one (but not the only) factor which directly influences the first, and is usually the independent variable. The more complex dependent variable is influenced by factors other than the simpler independent variable.

To illustrate using the above examples, chemical potentials are influenced not only by thermodynamic activity, but also by temperature and pressure. Variance is influenced not only by the number of phases, but also by the number of components. Reaction rates and equilibria are influenced not only by the σ parameter, but also by ρ (the reaction) parameter.

A rational correlation is not only a statistically good correlation, but one that is supported by extensive data of high quality and is consistent with independently established fundamental principles. The correlation of chemical potential with activity, by way of eq. 7.3, is based on the fundamental laws of thermodynamics and is supported by massive amounts of experimental data. The same is true of the Phase Rule, and of the various correlations that may be extracted from it (such as the alternation rule). The Hammett correlations are perhaps of comparatively lesser quality than these two, but they too are supported by a substantial body of data and by theoretical analysis based on chemical physics.

It is important to recognize that the distinction between empirical and rational correlations is not black and white. Shades of gray exist, and the degree to which a correlation is empirical or rational varies. Correlations also typically evolve with time and experience. Experience suggests that they simply do not emerge full-blown and in a final form as initially conceived. Instead, a recently developed correlation is to be regarded at first as a hypothesis; stating this hypothesis serves to expedite its further testing and refinement.

This situation is reminiscent of the process by means of which phase diagrams are determined (Appendix 4.1). The first diagram produced during a preliminary study is almost invariably wrong in some way. Only after having been severely tested may it be regarded as "final", and even "final" diagrams have had to be altered as a result of new information or the development of new methods (Section 8.4.14).

Exactly the same situation exists with correlations. They may be stated at some point in time on the basis of the information available, but are typically refined as new information is obtained. They are a living, evolving idea, not a concept that is cast in stone at the moment of its inception.

Direct and indirect correlations. Within the category of rational correlations, it is also useful to recognize that correlations may be either "direct" or "indirect". When variation of the independent variable has a direct or immediate consequence as to the numerical value of the dependent variable, the correlation is not only rational but direct. Beer's Law may be used as an example; the variation of concentration (or path length, or molar absorbance) has a direct influence upon the absorbance of solutions that follow Beer's Law. The correlations between absorbance and concentration (or path length) are therefore direct. Varying the structure of the molecule, on the other hand, has an indirect influence upon the absorbance that is exercised by way of the molar absorbance. It is not true that no correlation exists between molecular structure and the absorbance properties of solutions, but this correlation is indirect. It is necessary to insert an intermediate correlation (which invokes the variables of concentration and path length) in order to perceive this relationship. The overall correlation process may thus be depicted as

Observed absorbance

⇕

Molar absorbance

⇕

Molecular structure

The direct correlation of molecular structure and absorbance is not by itself rational, but neither is this an empirical relationship (in the sense that the correlation between sales and molecular structure is empirical). Rather, the correlation between molecular structure and absorbance is indirect, and a sequence of two rational correlations are required in order to describe fully the overall relationship.

13.3.3 The concepts of "performance" and "performance data"

In order to relate the nature of raw materials to utility, it is usual to measure within the controlled environment of the laboratory the utility of prospective formulations, and to correlate these laboratory data with the nature and amount of the raw materials selected. A convenient term for this information is *"performance data"*, since it is intended that the prospective formulation will "perform" during the laboratory experiment in a manner that will reflect the utility of the ultimate product as seen by the consumer at some future time. The process of correlating utility with the nature and amount of the raw materials selected has, by this process, been further subdivided as follows:

Utility

⇕

Performance data

⇕

Raw material selection

Inserting this step has proven to be an advantage, but it is also dangerous. If the technologist is able to anticipate utility accurately from tests performed in the laboratory, then the work is efficient and the likelihood that the product will be successful is enhanced. If the technologist is not successful and the performance data do *not* reflect utility, then the project is likely doomed from the outset and the effort will have been wasted. However, the risk of success or failure during this process is far less than that incurred in taking an inherently empirical approach, because raw material selection in the laboratory is considerably cheaper than is manufacturing and selling an untested product. Also, one never goes directly from the laboratory data to the manufacturing process; intermediate stages of development, at progressively increasing scales of use, are required before a mature product is manufactured and sold.

In evaluating utility, one simulates, insofar as possible, the manner in which the product will actually be used by the consumer. Laboratory tests are designed that utilize small amounts of experimental raw materials. Some literature exists regarding these tests for various technologies. In the case of laundry detergents, for example, a "miniwasher" may be purchased that is designed to simulate the conditions that exist within home washing machines [5]. Test swatches (small pieces of fabric) may also be purchased that are soiled in a reproducible manner with a standardized soil [6].

After the swatches are washed in the miniwasher in different test formulas, the differences in performance are assessed in some manner. Statistical evaluation of the resulting data is a critically important aspect of the assessment, and consideration of this factor should be included in the design of the test [7]. Performance in a detergency test has historically been assessed either by measuring the fraction of the soil on the fabric that is removed, or by determining the appearance of the swatch (for example by using a reflectance meter) [8].

Performance tests and their selection constitute valuable "property". They are jealously guarded by the firms that develop them, and will not be explicitly discussed. The important aspect of these tests to recognize, for our purpose, is that they produce information intended to measure how well a test product performs the task for which it was designed. This information constitutes "performance data", and it is intended that these performance data should correlate with utility as will be perceived by the consumer.

13.3.4   The concept of "performance-determining properties"

The next step is the interpretation of performance data, and the extraction of the correlation between these data and the identity and amount of raw materials selected for use in the product. It will be assumed that performance is determined largely by a few select properties of the experimental product. These key properties may be termed the *"performance-determining properties"*. Introducing the concept of performance-determining properties further refines the process of correlating the nature and amount of raw materials used with utility, as follows:

Utility

⇕

Performance data

⇕

Performance determining properties

⇕

Raw material selection

It is finally clear, at this point, just where the physical science of raw materials fits into the correlation, for *these performance-determining properties are dictated by the physical science of the mixture and the conditions of the performance test*. It remains only to identify (for each product) which properties, among many possible choices, actually constitute the "performance-determining properties", to measure these properties, and to perceive the relationship that exists among these various kinds of information. This process is the essence of basic industrial research. It may be illustrated using familiar examples.

One aspect of the performance of laundry detergents is the removal from fabric (the "detergency") of oily soils. Considerable literature exists regarding the composition of these soils, which have been found to contain a variety of fatty compounds such as triglycerides, hydrocarbons, fatty acids, sterols, etc. [9]. It has long been recognized that the tendency of soil to wet the fibers within a fabric, and of the detergent solution to cause dewetting, plays an important role in the ease with which this soil can be removed [8]. Performance can therefore be measured by determining the fraction of oily soil that is removed during a washing process. Gravimetric data can be obtained either on actual fabrics washed under realistic conditions, or on synthetically soiled swatches (small squares of fabric) washed in a miniwasher (a small washing machine).

Alternatively, detergency performance can be measured by assessing the appearance of the fabric using a reflectance meter [8]. It is also possible to observe directly the physical process by means of which oily soil is removed by observing a single soiled fiber that is suspended within a flowing detergent solution [8]. This method provides the most penetrating insight into the details of the process.

Each of these tests measures a different dimension of performance. Each provides a different level of insight as to the actual process, uses different quantities of materials, and requires different amounts of time (and therefore money).

Do such tests really measure utility? It is generally assumed that they do in some degree, but the consumer will of course *not* analyze the amount of oily soil left on a garment after it is washed, or place it in a reflectance meter. Instead, he/she will wash the clothes in a washing machine, dry them (perhaps in a dryer), pick the clothes out of the dryer, fold and store them, and (after a period of time has lapsed) select and wear a garment. The garment will (or will not) have an attractive appearance. It will (or will not) rustle nicely, and drape softly on the body during wear. It will not (or it will) develop static charge and cling to the body during wear. It will not smell bad (and may even smell good), and so on. Perhaps these are really the performance-determining properties, rather than the oily soil tests. More likely, the oily soil detergency tests described above measure one or the other aspect of these performance-determining properties, but not usually the composite result.

The performance of other familiar consumer products may be similarly analyzed. In a shampoo, the performance-determining properties might be the appearance and sheen of the hair, its "body", the effort required to comb it, its tendency to crackle, pop, and show sparks during combing (due to development of a static charge), etc.

In a solid food shortening, a variety of performance-determining attributes can be imagined to exist. The uniformity of the product when the consumer opens the container, and the stability of its macroscopic (colloidal) structure (and rheology) during storage may be important. So may its smell; it may have a pleasant smell, and it must *not* have the smell of rancid fat. Of particular interest is the texture of a pastry or pie-crust made using the product, the sound that the crust makes when the pie is eaten, etc

In a peanut butter, the stability of the product against oil separation during storage, the appearance and smell encountered upon opening the jar, the manner in which the peanut butter spreads on bread, and the way that it feels and sounds while chewing a peanut butter sandwich are possibly important.

13.3.5 The characteristics of performance-determining properties

If one considers a wide array of consumer products from the above perspective, it becomes apparent that the properties of products which influence their performance are the ones that our five senses respond to:

(1) The eyes perceive optical properties: the appearance of fabrics, the color and sheen of hair, the color and texture of peanut butter, etc.
(2) The hands (or tactile sensors at other points on the body) detect rheological properties: the handle of fabrics, the softness of hair, the ease with which peanut butter spreads, etc.
(3) The ears detect acoustic properties: the sounds made by fabrics during handling and wear, the snap when a spark jumps, the sound of food being eaten, etc.
(4) The nose detects odors: the smell of fabrics, of hair, of peanut butter, etc.
(5) The mouth detects flavors: the taste of peanut butter, of a toothpaste, etc. The mouth may also detect aerosolized salt particles (Section 1.7) or fine particles of a food sweetener dispersed into air during its use.

Inspection of this list further suggests that there are two broad classes of performance-determining properties which differ in the kind of structural information (Section 2.4) which correlates rationally with performance. Those dimensions of performance that are detected by the sense of smell and taste are probably determined principally at the molecular level of structure. Those detected by sight, feel, and hearing, on the other hand, are probably more directly influenced by the phase and colloidal levels of structure. While smell and taste are correlated in a very direct way with molecular structure, these other performance-determining properties are only indirectly correlated with molecular structure.

13.3.5.1 Performance related to smell and taste

The molecular level of structure is surely a dominant factor insofar as smell is concerned. Molecules must be volatile in order to get to the sensors in the nose. The physical science of odor-producing materials (such as their partial pressures) are obviously important and should not be ignored. Still, the response of the nose depends critically upon whether or not the molecule has a particular molecular structure (or a narrow range of molecular structures). Molecules having exactly the same partial pressure (but different molecular structures) are *not* expected to necessarily smell the same. Subtle structural differences (such as stereoisomerism) may strongly influence smell, while hardly affecting other aspects of physical science.

Underlying these correlations is the hypothesis that the olfactory response results directly from physical binding of the stimulus molecule to a receptor site in the nose. Only molecules that bind to the receptor induce the olfactory response, and the capacity to bind is strongly correlated with molecular structure.

To the extent that taste is dictated by binding to receptor sites on the tongue, this property too is expected to be strongly influenced by molecular structure. Data from the development of synthetic sweeteners clearly illustrate this principle, since the sweetness of compounds is strongly influenced by features of molecular structure that do not strongly influence their physical science [10].

During the process of eating food a particular molecule may activate both the smell and the taste response, but the two responses may still be considered separately. To the extent that taste is influenced by texture (which is a macroscopic property of foods, Section 13.3.5.2) the classification of taste as a property that is determined by molecular structure information is also compromised. Perhaps both a tactile and a molecular taste response occur in the mouth, but the distinction between taste and texture remains intact.

The underlying assumption in the whole of drug design is that molecular structure correlates both rationally and directly with "performance", which in this instance constitutes a particular biological response [11]. Given an effective drug molecule one must still control the dose size, particle sizes in a tablet, the rate of dissolution of a crystal to form a solution, and many other aspects of the physical science of a drug to have a good product. Control over these parameters is meaningless unless the molecule used is intrinsically active, however, so that (just as with smell) it is very clear that molecular structure is the uppermost factor with respect to drug performance. The applicability of computer-assisted methods in the design of drugs hinges on the validity of this premise.

13.3.5.2 Performance related to seeing, feeling, and hearing

Responses to the use of a product from the other three senses (seeing, feeling, and hearing) are *not* determined solely by molecular structure. The truth of this statement is most clearly revealed by the fact that changes of state within the product (during which molecular structure remains constant) may profoundly alter the response of these particular senses to the product.

Appearance and optics. The visual appearance of an object depends upon the nature of the interactions of photons in the visible region of the spectrum with its surface, as described using optical physics. These interactions may be analyzed as a combination of absorption and scattering phenomena. The dependence of absorption on the wave-length of the light dictates the hue. Changes of state that alter the absorption spectrum also alter the hue. This phenomenon is particularly noticeable in compounds that display charge transfer bands in the solid state, but not in solution [12].

Scattering depends on the wave-length of light, the physical dimensions of optically homogeneous regions, and the numerical value of the refractive index differences that exist at the boundaries of these regions. Scattering intensities are strongly influenced by the relationship between the wave-length of the radiation and the structural dimensions of the scattering sites. Since the wave-length of visible light is 0.4–0.7 μm, scattering by visible light occurs only from macroscopic features that span a distance of hundreds of molecular diameters. Materials having entirely different molecular structures will display similar scattering properties if these structural dimensions are similar. For example, the optical textures of liquid crystal phases are *not* influenced by molecular structure.

It can be seen that both absorption and scattering are rationally and directly correlated with the physics of light and the optical properties of matter, but only indirectly correlated with the molecular structures of the compounds present.

Feel and rheology. Tactile responses to solid surfaces, such as roughness and smoothness, likewise depend on both the intrinsic mechanical properties of matter and its macroscopic structure, rather than directly on molecular structure [13]. Similar roughness or smoothness may result from materials having totally different molecular structures – provided the macroscopic structures and the mechanical properties of the material are similar. Changes of state dramatically alter mechanical properties.

The "feel" of fluids may be related in a similar manner to their rheology. So long as two different peanut butters (which are mixtures of solids in a fluid) have identical rheological profiles, the mechanical response and the force required to spread them with a knife will be identical. The flow properties of fluids is thus dictated directly by their rheological behavior, and only indirectly by molecular structure – again as illustrated by the fact that a change of state profoundly alters rheological properties.

Sound and acoustics. Sound is an oscillating longitudinal pressure wave that is produced by a variety of mechanical processes. As with rheology, the acoustic properties of materials are directly related to the mechanical properties of the phase in which the compound exists, and are dramatically altered by a change of state.

Sound is a dimension of performance in perhaps unsuspected places. Considerable research has been invested in the sound of food being chewed, which (for some foods) provides a measure of the freshness and quality of the food [14]. The sound produced when fabrics are bent, the sharp report that occurs when a static spark jumps, the sound from brushing ones teeth, and many other areas in which sound is produced illustrate the relevance of acoustic phenomena to performance.

It has recently been discovered that ultrahigh frequency sound is produced by a great many physical processes such as the hydration of minerals, the frying of potato chips, ion exchange reactions, etc. [15]. It has further been suggested that such acoustic effects are produced by all processes that occur in solution during which a volume change occurs – which means virtually all processes. Whether or not such phenomena are in any way related to performance is unclear because so little is known of them.

13.3.6   The role of phase science in the raw material selection process

The role of phase science in raw material selection can now be approached from the above analysis, in terms of its influence on the performance-determining properties. We may therefore consider those aspects of the physical science underlying these properties to which phase science is relevant. It is important to recognize that this is a large subject that is constantly undergoing evolutionary change, so none of the following discussion should be regarded as the "truth" of the matter. In fact, since the analysis of the information regarding performance is inherently complex, the certainty of the analysis is inherently poor. Nevertheless, some areas where phase behavior probably has a strong influence on performance can be seen and are listed.

13.3.6.1 Rheology

The phase state of a mixture is important with respect to its rheology, but the extent of its influence depends on whether or not the mixture is a single phase. If it is, then a direct and rational correlation exists between phase behavior and rheology. If a mixture consists of more than one phase it also likely displays colloidal structure. In multiphase colloidally structured mixtures (which are extremely common), both the intrinsic rheology of the phases present and the colloidal structure of the mixture influence its rheology [16].

It is evident that phase behavior dictates the selection of the components of products that exist in various states. The components of a liquid product must be miscible in a liquid phase at ambient temperatures, which rules out a large number of possible raw materials. Some might separate as a crystalline phase under these conditions, while others might separate as a liquid crystal. The phase state of the product dictates what can be used, and binary aqueous phase data constitute a boundary condition that governs the behavior of each component. It is also desirable for liquids to survive freeze–thaw tests, as freezing may occur during shipping and warehouse storage. This is evidently purely a phase issue which entails consideration of whether or not the liquid product is an equilibrium state.

The intrinsic rheology of surfactant liquid phases, which is relevant to the rheologies of liquid dishwashing and laundry detergent products, is considered in Section 8.4.2. The variation of rheology with composition and phase structure is important to this area. Normal micellar solutions are only slightly more viscous than is water itself, but solutions containing rod-shaped or entangled flexible micelles are considerably more viscous and are not Newtonian.

If a mixture exists as a liquid crystal phase the viscosity of the mixture will typically be much higher than that of a liquid, and the phase is not likely to display Newtonian rheological behavior (Section 8.4.2). It is very important to keep in mind the grossly nonmonotonic dependence of the rheology of liquid crystal phases on their composition. The lamellar phase is in most aqueous surfactant systems both the most concentrated phase, and also the liquid crystal phase of lowest viscosity; the more dilute hexagonal and cubic phases are substantially more viscous. Some cleaning products that are viscous pastes exist in the hexagonal phase [17].

This dependence of the viscosities of phases on composition comes into play when a particle containing a surfactant that is miscible with water at the temperature of the experiment is placed in water. The swelling of the particle by intrusion of water produces, at an intermediate time, layers of the various phases that exist along that isotherm. The most dilute phase is to the outside, and progressively more concentrated phases exist towards the core of the particle. Such phenomena influence the rate of dissolution of the particle [18].

If a mixture is a solid rheologically (which typically means a crystal structurally), then a high yield stress exists (Section 8.4.2). This will obviously have a profound influence upon all those aspects of performance that are influenced by rheology.

13.3.6.2 Optics

The phase state of surfactants has only an indirect influence on the optical properties of a product. Most liquid phases are clear to the naked eye, although not to a sensitive light scattering instrument if they display micellar structure. Even liquids that are highly viscous may not scatter light strongly.

The natural turbidity of most liquid crystals, and the unusual clarity of cubic phases, have been noted (Section 8.4.3). Since most surfactant compounds do not absorb light in the visible region of the spectrum, their crystal phases individually are optically clear and transparent, while masses of their crystals are white (due to total reflection). The refractive indices of surfactants are determined principally by the refractive index of the lipophilic group, and their absolute values typically fall within the range of 1.46 to 1.5. As a result, surfactant powders lack the opacity and brilliance of inorganic whitening powders, such as titania, and more closely resemble colorless aliphatic compounds (such as waxes) in this regard.

The appearance of a product, or of an object such as a fabric that has been treated with a surfactant-containing product, is not grossly influenced by the surfactant. The appearance of fabrics that have been washed is more strongly influence by the soils that exist on the fabric such as carbon, colored clays and minerals, hemoglobins and chlorophylls, etc. [8]. Properties such as greyness are often invisible on magnification, which suggests that they originate from scattering phenomena dictated by ill-defined macroscopic structural features of the fabric. Oils such as linseed oil undergo yellowing during the process of autoxidation (to form hydroperoxides), decomposition (to form carbonyl compounds), and polymerization. The same may be true of the olefinic oils on fabric that are deposited from sebum, etc. It is therefore likely that surfactants do not, *per se*, affect the color of fabrics, so much as indirectly affect it by contributing to the removal of pigmented soils.

The optical clarity of a product that is a mixture of phases is governed, to a large degree, by the difference in refractive index that exists between water and lipophilic groups. Technology for making transparent soap bars has existed for a long time, and hinges on increasing the index of aqueous phases by adding a water-soluble organic component.

13.3.6.3   Acoustics

As in the case of optical phenomena, the influence of surfactants on sound and acoustics is also probably indirect. The study of the influence of surface physics on the fracture of materials was pioneered by Rehbinder, who demonstrated that the adsorption of surfactants at a fracture could alter phenomena such as stress-cracking. If this is true surfactants might also alter the acoustic phenomena that accompany fracture, but this too is an indirect consequence of their presence.

13.3.6.4 Colloidal phenomena

The capacity of surfactants to modify the mechanical properties of surfaces, by way of their adsorption at interfaces, is the principal reason why they are important industrially. For the most part these effects result from the adsorption of surfactant molecules from solution. The phase from which they adsorb is the liquid, and the solution and surface chemistry of the liquid phase is the relevant discipline with respect to these phenomena.

There are specific areas of the surface property modification by surfactants to which their phase behavior is directly relevant. The aqueous phase behavior of surfactants may be more relevant to areas such as the chemistry of black films than is generally recognized.

The thinning of a film formed from a very dilute surfactant solution proceeds in discrete stages [19]. The first stage occurs by flow of the liquid phase through the film, which is governed by its viscosity (among other factors). If the film exists within a foam, the rate of flow through the Plateau junctions at the perimeters of the individual facets is the determining factor. If the film has been drawn using a wire frame this factor is absent, but thinning still occurs via flow processes.

When a film reaches a particular thickness (10–30 nm), a stationary state is reached that is called a "common black film" at which the forces that exist within the film are in balance. It is black because it is too thin to display interference phenomena using visible light. Its thickness varies depending on the presence and concentration of added electrolyte. Its composition is not terribly different from that of the bulk liquid phase from which it was drawn.

With time, the common black film may undergo an abrupt change to a "Newton black film". The Newton black film has a bilayer structure that is only a few nm thick, and the molecules within this film are densely packed – as in a crystal. In the SDS–water system the phase that coexists with the saturated liquid is reported to be the dihydrate crystal $X \cdot W_2$ [20]. One wonders what relationship, if any, exists between the structures and compositions of the Newton black film, and those of the equilibrium crystal.

It has been shown that films suspended in wire frames that have been drawn from concentrated aqueous solutions of SDS thin in stages, and that as many as four or five steps may exist in the thickness of the film [21,22]. The mechanical properties of films formed from concentrated SDS solutions thus differ significantly from those of films formed from dilute solutions.

The Friberg correlations. In 1978, Friberg noted that if a small amount of a lamellar liquid crystal phase was present in a mixture, then the stability of foam produced from the mixture was dramatically enhanced [23]. These ideas were later extended to consideration of emulsion stability; if a lamellar liquid crystal phase was present, then emulsions of the two liquids in one another were remarkably stable [24]. These results suggest that the lamellar liquid crystal phase is a surface active *phase*, and evidence that this phase tends to exist at interfaces has been produced [25]. It is presumed that its effect on foam stability, for example, is to retard the rate of drainage through the foam due to its high viscosity (relative to the liquid phase).

During the same period this concept was found to apply to the baking of bread [26] Larsson has investigated both the phase behavior and the bread-baking performance properties of the monogalactosylmonoglycerides and the digalactosyldiglycerides which are polar lipids that occur in wheat flour. The coexisting phase of the former surfactant in aqueous dispersions is the inverted hexagonal, and in the latter the lamellar phase [27]. The digalactosyl diglyceride provides superior foam structure during the baking of bread. It must be recognized that other components (such as the gluten) also play an important role in stabilizing foam structure in bread.

The emulsion-stabilizing effect of the lamellar phase also plays a role in the behavior of fatty alcohols (such as dodecanol) as both a foam-killer and a foam-stabilizer in the same ternary system [23]. If one adds droplets of dodecanol to foams formed from SDS solutions the foam is killed, perhaps because the L_2 phase is formed at the locally high dodecanol and low surfactant fractions. If, on the other hand, one were to dissolve dodecanol in a dilute surfactant solution and create a foam from this solution, foam stabilization should result. A sufficiently high ratio of surfactant to dodecanol should exist that the lamellar phase forms.

13.3.6.5   Solubilization

If an oil were to be solubilized by any surfactant phase, significant changes in the physical behavior of the phase would certainly result. One occasionally encounters in the literature the idea that solubilization plays a role in detergency [28], but its role is uncertain. For one thing, it is well known that detergency typically plateaus near the critical micellar concentration (cmc) of the surfactant, whereas solubilization has not yet commenced at this concentration [29]. Second, large molecules (such as triglycerides) are difficult to solubilize. The capacity of surfactants to solubilize oils falls of rapidly as the molecular weight of the oil increases, and there are presently no known surfactants that will effectively solubilize a triglyceride in water. It presently appears that solubilization is not a major factor in the detergency of oily soils, but its precise role remains murky.

A perfume is just the kind of oil one might expect to be solubilized in fluid surfactant phases, because perfume molecules are small (mono- or sesquiterpene) compounds and amphiphilic as well. Definitive studies of the phase behavior of a surfactant–perfume–water system are lacking, however.

13.3.6.6   The role of structural information

An extremely large fraction of the work on surfactants today is concerned with defining various aspects of the structures of surfactant states. Molecular structure is not usually a major uncertainty, since so many methods exist for quickly and easily determining molecular structure. Molecular structure does become an important concern, however, when chemical reactivity exists (Section 2.2).

Specific conformational structures cannot be identified in fluid states because of the vast number of discrete conformational structures that exist, but concern with better defining this level of structure is increasing as methods for investigating it become available [20].

An enormous amount of interest exists in the phase level of structure at this point in time. The history of this area extends back to the earliest days of surfactant research, and with the advent of modern scattering methods (based not only upon the use of electromagnetic radiation but upon neutrons as well) this area has flourished.

Just how important to utility is structural information? A possible answer is apparent from consideration of the above dissection of the relationship between molecular structure and performance into the relationship between molecular structure and performance-determining properties. If this analysis is correct, then structural information is, in fact, not *directly* relevant to performance and utility at all. That is not to say that it is unimportant; only that it is important only insofar as it influences those properties that do directly influence performance, and thus utility. Considerably greater emphasis on this relationship would be of considerable value.

13.4 Raw material integration: correlating process variables with phase behavior

The conditions selected for a manufacturing process must be tailored to all aspects of the natural science of the raw materials that have been selected. The system variables exert a major influence on the physical aspect of this science, and for this reason on the whole of process development. The chemical aspect must be taken into consideration if only because products are likely to be exposed to oxygen, water, and light and may be chemically altered by the action of these factors. Some functional groups (such as the ester group) may be hydrolytically cleaved by water, others (such as olefinic, amino, and ether groups) may react with oxygen, and oxidation may be accelerated by the action of light [30]. Both direct photochemical autoxidation and sensitized autoxidation, during which light is initially absorbed by sensitizer impurities, must be considered. The biological aspect of natural science must be considered even for products whose performance-determining properties are non-biological, since biological degradation due to the growth of microbes must be arrested [31]. This may occur even in products containing antimicrobial surfactants.

13.4.1 The process path

The path followed with respect to composition, temperature, and time can be directly plotted for manufacturing processes. Such a plot might, for example, display composition and temperature along the x- and y-axes and time along the z-axis. From such a plot a profile in composition–temperature space that is characteristic of the process can be produced, and this profile could be overlaid onto the phase diagram of the system if it were available (Section 4.2). From such a composite figure the equilibrium state could be perceived at any point along the process, and by consideration of the time dimension consideration could be given to the kinetics of equilibration.

It is not presently possible to do this for any real manufacturing process, because real systems have so many components that the phase diagram is not known and cannot be determined. This fact is often used as an excuse to forget the entire matter and proceed on an entirely empirical basis, but there are other courses of action. Reasonable simplifying assumptions can often be made, and these permit the application of existing data to provide insight that would not otherwise exist. A superb model for this process exists, since most of these assumptions can be perceived by reference to McBain's study of the basic science underlying the soap-boiling process (Section 14.4). This study provides several important lessons in how to apply phase science to manufacturing processes, and is all the more remarkable when one considers the state of knowledge that existed at the outset of McBain's study. McBain's work occurred before the advent of chromatographic and spectrographic analysis, and he was literally on his own at every turn. In striking contrast, we have 70 years of experience and powerful analytical tools upon which to draw at the present time.

13.4.2 Simplification of the analysis

First, one may consider the formula of the product and focus on those components of the raw materials present that are miscible with water (within any phase). Those components that are *immiscible* with water under a particular set of conditions may be safely ignored insofar as their influence on phase behavior is concerned (Section 3.20). In analyzing soap-boiling, one must include all the soap molecules present at the high temperatures that exist at the outset of the process. As the temperature is lowered the miscibility of each of the various soaps changes in a different manner, sometimes dramatically. Those components that must be included for consideration depend on the conditions. Also, it is very important to remember that *none* can safely be ignored with respect to nonequilibrium aspects of the physical science (kinetic and colloidal phenomena).

Those water-miscible components present in the largest amounts will dominate the phase behavior, and must be given special attention. McBain may have chosen sodium palmitate for serious study in part because its physical science is typical of the major components of soap products. (The focus on dipalmitoylphosphatidyl choline in the polar lipid area probably exists for exactly the same kind of reason. This compound is both an important component of membranes, and its physical behavior is rather similar to that of naturally occurring mixtures.) Those compounds present in small amounts (fractions of a per cent) may influence colloidal structure or kinetic and surface phenomena, but will not dramatically alter the equilibrium phase state.

This illustrates a particular advantage that the phase science approach to the analysis of industrial problems has over the purely surface chemistry approach. It is difficult to simulate in the laboratory the composition and mechanical properties of the surfaces that are encountered in real life, but it is comparatively easy to extend laboratory phase data to realistic situations. The preferred approach is to incorporate both kinds of information in the analysis, of course.

Third, it is often useful to group together compounds that are closely related physically, and consider them at the outset as a single entity. Water-soluble electrolytes, homologous members of the various subclasses of surfactants (anionic, zwitterionic, single bond, etc.), nonionics that vary in the number of oxyethylene groups present, and the two classes of oils (nonpolar and amphiphilic) can legitimately be considered, in the first approximation, to be a single component. Further, a specific molecule can often be identified which resembles these groups of molecules [32]. This aspect of simplification is particularly important, since the number of truly different kinds of molecules present is usually quite small.

In the soap area, it is interesting that McBain chose sodium palmitate as the first model system to be studied and later potassium laurate as another [33]. Having worked out the sodium palmitate–salt–water system, the effort required to produce comparable phase information on the potassium laurate–salt–water system must have been considerably smaller. In soap products sodium palmitate typifies in its physical science those components that are insoluble at room temperature, while potassium laurate (or oleate) typifies those that are soluble.

All the above steps amount to selecting a model system that has a reasonable chance of behaving like the commercial product. Having done this, one may extract from the literature the phase information that exists and overlay the plot of the process (above) on this diagram. This may be done in the above example for both sodium palmitate and potassium laurate, and useful information often results. During the 1920s McBain likely took exactly this approach using data on the sodium palmitate–salt–water diagram and later using data on the potassium laurate–salt–water system, but he first had to determine these diagrams.

13.4.3 The reintroduction of complexity

The last step in the analysis is to consider the consequences of the presence of the other components on the behavior of the product mixture. This requires judgment based upon a sound knowledge of the principles of physical chemistry, of phase science, and usually of colloid science. This extrapolation to a still more realistic analysis is still not possible for real soap products, unfortunately, because the necessary basic studies have not been performed. To illustrate, a stoichiometric excess of fatty acid over the total base present exists in most soap products, but no studies have been performed that provide information regarding such ternary systems for soaps of commercial importance (Section 12.4.6). No extrapolation has a significance comparable to good data, of course.

13.4.4 The various roles of phase behavior in manufacturing processes

We may now consider the various ways in which the phase behavior of a formula influences manufacturing processes. Actually, it is quite difficult to identify aspects of manufacturing processes that are *not* directly influenced by phase behavior.

Isothermal mixing. Many processes commence with mechanical mixing of the raw materials, and all the phenomena described in Section 5.6.7 come into play. These include the alternation of stoichiometric phase reactions and dilution processes, changes of state, the fact that mixing only occurs at the surface of particles, etc.

Experience suggests that the time required for mixing depends on the magnitude of the departure from equilibrium at the outset of the process. It takes much longer to create an equilibrium mixture from a dry surfactant and water, for example, than it does to dilute an aqueous mixture with water or to enrich it with surfactant. This is true in spite of the fact that the magnitude of the driving force for physical change is greater in mixtures far from equilibrium than in mixtures that are close to equilibrium at the outset.

Isoplethal paths. Isoplethal are paths that are extremely common during manufacturing processes. Phase behavior exerts a profound, and often largely unrecognized, influence on isoplethal processes (Section 5.6.11). The changes in heat capacity required to change the temperature are widely recognized, and it is also usually apparent that changes of state (melting of crystals, formation of liquids from liquid crystals, etc.) typically ensue as a result of changing the temperature.

The factor that is least often appreciated is that *changes in temperature usually create a driving force for mass transport* within mixtures. The extent of mass transport required to maintain equilibrium may be quite extensive, and represent the most important kinetic barrier to attaining equilibrium. Diffusion processes simply cannot transport mass over long distances in a short time, and if convective mixing is inadequate then equilibrium of state cannot be achieved.

A further complication is the fact that the mechanisms of phase reactions for which the entropy change is negative and large (principally crystal formation) are usually complex. The formation of crystals usually requires nucleation, if it is to proceed at a reasonable pace, and if nucleation sites do not exist the process can be extremely slow. It is also clear that cooling a liquid crystal phase into a liquid crystal-plus-crystal region invariably results in supercooling.

13.4.5   The formation of colloidal structure by phase reactions

The interaction of nonequilibrium phases with water may produce colloidally structured mixtures, and (if the surfactant is poorly soluble) this structure can be quite stable (Section 2.7.1). The principle that applies in the analysis of these processes is clear: if the reacting phase is also the equilibrium phase that exists when mixing is complete, then only the ratio of the two phases changes and the reaction typically occurs by quiet convection and diffusion. An example is the dissolution of sodium chloride by water; nothing spectacular happens during this process. If the reacting phase is *not* the equilibrium phase, then profound changes of state may occur, the kinetics may be anomalously fast, and complex colloidal structured mixtures may result. The most dramatic example to date is the explosive disintegration of the DODMAC monohydrate crystal by the action of water to produce a vesicle dispersion [34,35]. This transforms a fluid slurry of crystals into a mixture that behaves rheologically as a "gel" in seconds, and one should in general suspect that phenomena of this sort occur any time that a crystal phase reacts with water to form a "gel"-like product. The simple dissolution of crystals does not alter their crystal habit and does not result in the development of colloidal structure.

The severe stress that may be induced in a crystal hydrate, when it undergoes a peritectic reaction, is also worthy of note [34].

13.5   Some roles of phase science in medicine

There are areas of medicine that are strongly influenced by phase behavior and involve surfactant materials, but these will not be seriously treated. The precipitation of cholesterol from bile, which may be treated as an aqueous mixture of phospholipid, bile salts, and sterol, represents one example [36]. Both the equilibrium and the kinetic aspects of phase science come into play in this area, since the degree of supersaturation (as well as other factors) not only influence whether or not the sterol will actually precipitate from its solutions, but the course that the precipitation takes. Precipitation of crystal phases is involved in the formation of calcium oxalate kidney stones [37].

Several different kinds of stones form that differ as to which of the various crystal hydrates of calcium oxalate have precipitated [38].

13.6 Summary

One aspect of the development of technology to which the phase behavior of surfactants is relevant is the raw material selection function. Selecting raw materials requires that one correlates molecular structure with sales, but this correlation is inherently empirical. More rational correlations exist if the gap in complexity between sales and molecular structure is broken into smaller gaps. Introducing the concept of "utility", and recognizing that utility is governed by key "performance-determining properties", is particularly helpful. Further, if one assumes that "performance-determining properties" are governed by the physical science of the product formula, then the possible relationship between utility and the various aspects of physical science may be visualized. Since physical science is rationally correlated with molecular structure, the insertion of these additional steps transforms an inherently empirical correlation between sales and molecular structure into a series of rational correlations, and renders the overall correlation possible.

The physical science of a formulation has a direct and readily discernible effect on manufacturing processes. In principle knowledge of the phase diagram of the product would allow one to describe quantitatively the changes of state that occur during the process, but in fact real systems are too complex to be analyzed in this manner. Reasonable simplifying assumptions can be made, however, which may permit the available phase information to be utilized during the analysis of manufacturing processes.

References

1. Moffat, D. W. (1983). *Economics Dictionary*, Elsevier, New York.
2. (1979). *Dictionary of Advertising Terms*, (L. Urdang ed.), p. 140, Tatham-Laird and Kudner, Chicago, IL.
3. Goldstein, R. F. and Waddams, A. L. (1967). *The Petroleum Chemicals Industry*, 3rd ed., pp. 285, 458–459, E. & F. N. Spon, London; Swisher, R. D. (1987). *Surfactant Biodegradation*, 2nd ed., p. 25, Marcel Dekker, New York; Feighner, G. C. (1976). *Anionic Surfactants*, Vol. 7 (W. M. Linfield ed.), pp. 288–314, Marcel Dekker, New York.
4. Swern, D. (1964). *Bailey's Industrial Oil and Fat Products*, 3rd ed., pp. 3–55, Interscience, John Wiley, New York.
5. Spangler, W. G. (1972). *Detergency. Theory and Test Methods*, Vol. 5 (M. J. Schick ed.), pp. 413–449, Marcel Dekker, New York.
6. Neiditch, O. (1972). *Detergency. Theory and Test Methods*, Vol. 5 (M. J. Schick ed.), pp. 5–30, Marcel Dekker, New York.
7. Scheffé, H. (1952). *J. Am. Statistical Assoc.* **47**, 381–400; Törnquist, L., Vartia, P. and Vartia, Y. O. (1985). *The American Statistician* **39**, 43–46.
8. (1981). *Detergency. Theory and Test Methods*, Vol. 5 (W. G. Cutler and R. C. Davis eds, M. J. Schick series ed.), pp. 31–64, Marcel Dekker, New York; (1975). *Detergency. Theory and Test Methods*, Vol. 5 (W. G. Cutler and R. C. Davis eds, M. J. Schick series ed.), pp. 31–64, Marcel Dekker, New York; (1972). *Detergency. Theory and Test Methods*, Vol. 8 (W. G. Cutler and R. C. Davis eds, M. J. Schick series ed.), pp. 31–64, Marcel Dekker, New York.

9. Powe, W. C. (1972). *Detergency. Theory and Test Methods*, Vol. 5 (W. G. Cutler and R. C. Davis eds, M. J. Schick series ed.), pp. 31–64, Marcel Dekker, New York; Linfield, W. M. (1976). *Anionic Surfactants*, Vol. 7 (W. M. Linfield ed.), pp. 1–10, Marcel Dekker, New York.

10. (1991). *ACS Symposium Series. Sweeteners: Discovery, Design and Chemireception*, Vol. 450 (D. E. Walters, G. E. Dubois and F. T. Orthoefer eds), American Chemical Society, Washington, DC.

11. Richards, D. G. (1983). *Quantum Pharmacology*, 2nd ed., Butterworths, London.

12. Rao, C. N. R. (1967). *Ultra-Violet and Visible Spectroscopy. Chemical Applications*, 2nd ed., pp. 147–163, Butterworths, London.

13. Hoffman, R. M. and Beste, L. F. (1951). *Text. Res. J.* **21**, 66–77; Ellis, B. C. and Garnsworthy, R. K. (1980). *Text. Res. J.* **50**, 231–238; Laughlin, R. G. (1990). *Cationic Surfactants Physical Chemistry*, 2nd ed., Vol. 37, pp. 449–468, Marcel Dekker, New York.

14. Seymour, S. K. and Hammann, D. D. (1988). *Journal of Texture Studies* **19**, 79–95.

15. Betteridge, D., Joslin, M. T. and Lilley, T. (1981). *Anal. Chem.* **53**, 1064–1073.

16. Goodwin, J. W. (1984). *Surfactants*, (Th. F. Tadros ed.), pp. 133–152, Academic Press, New York.

17. Leng, F. J., Machin, D., Reed, D. A. and Erkey, O. (1985). *Eur. Pat. Appl.* EP 153857 A2 4 Sept. 1985.

18. Laughlin, R. G. (1992). *Adv. Coll. Interface Sci.* **41**, 57–79.

19. Mueller, H. J., Balinov, B. B. and Exerowa, D. R. (1988). *Colloid Polym. Sci.* **266**, 921–925.

20. Kekicheff, P., Grabielle-Madelmont, C. and Ollivon, M. (1989). *J. Coll. Int. Sci.* **131**, 112–132.

21. Bruil, H. G. and Lyklema, J. (1971). *Nature* **233**, 19–20.

22. Wasan, D. T., Nikolov, A. D., Lobo, L. A., Koczo, K. and Edwards, D. A. (1992). *Prog. Surf. Sci.* **39**, 119–154; Nikolov, A. D. and Wasan, D. T. (1992). *Langmuir* **8**, 2985–2994.

23. Friberg, S. (1978). *Advances in Liquid Crystals*, Vol. 3 (G. H. Brown ed.), pp. 149–165, Academic Press, New York.

24. Friberg, S. (1979). *J. Soc. Cos. Chemists* **30**, 309–319.

25. Friberg, S., Jansson, P. O. and Cederberg, E. (1976). *J. Colloid Interface Sci.* **55**, 614–623.

26. Larsson, K. (1986). *Chemistry and Physics of Baking, Special Publication No. 56*, (J. M. V. Blanshard, P. J. Frazier and T. Galliard eds), pp. 62–74, The Royal Society of London, Burlington House, London.

27. Shipley, G. G., Green, J. P. and Nichols, B. W. (1973). *Biochim. Biophys. Acta* **311**, 531–544.

28. Schwartz, A. M. and Perry, J. W. (1978). *Surface Active Agents: Their Chemistry and Technology*, p. 365, R. E. Krieger, Huntington, New York; McBain, J. W. (1942). *Advances in Colloid Science*, Vol. I, pp. 99ff, Interscience, New York.

29. Preston, W. C. (1948). *J. Phys. Chem.* **52**, 84–97; Adamson, A. W. (1982). *Physical Chemistry of Surfaces*, 4th ed., pp. 446–456, Interscience, John Wiley, New York.

30. Donbrow, M., Hamburger, R. and Azaz, E. (1975). *J. Pharm. Pharmacol.* **27**, 160–166.

31. Troller, J. A. and Christian, J. H. B. (1978). *Water Activity and Food*, Academic Press, New York.

32. Bouwstra, J. A., Jousma, H., Van der Meulen, M. M., Vijerberg, C. C., Gooris, G. S. and Spies, F. (1989). *Colloid Polym. Sci.* **267**, 521–538.

33. McBain, J. W. and Burnett, A. J. (1922). *J. Chem. Soc.* **121**, 1320–1333.

34. Laughlin, R. G., Munyon, R. L., Fu, Y.-C. and Fehl, A. J. (1990). *J. Phys. Chem.* **94**, 2546–2552.

35. Laughlin, R. G., Munyon, R. L., Burns, J. L., Coffindaffer, T. W. and Talmon, Y. (1992). *J. Phys. Chem.* **96**, 374–383.

36. Small, D. M. (1986). *The Physical Chemistry of Lipids. From Alkanes to Phospholipids*, Vol. 4, pp. 395, Plenum, New York.

37. Small, D. M. (1968). *New England J. Med.* **279**, 588–593.

38. Wiedemann, H. G. and Bayer, G. (1988). *American Laboratory* 54–61.

Chapter 14

A History of Surfactant Phase Science

14.1 The pre-Phase Rule era, and the Phase Rule

A historically significant landmark in phase science is the encyclopedic treatise, *Die Heterogenen Gleichgewichte vom Standpunkte der Phasenlehre* (Heterogeneous Equilibria from the Perspective of the Phase Rule), edited by H.W. Bakhuis Roozeboom [1–7]. While this seven-part series (published during the period 1901 to 1918) is now usually consigned to rare books collections, it remains a worthwhile reference both to phase data and to phase concepts. It is invaluable as a source of information regarding the history of phase science.

During most of the 19th century the analysis of equilibria among phases was not clearly distinguished from the analysis of chemical equilibria. Early workers were concerned with systems such as calcium carbonate and ammonium chloride, whose phase behavior is very complex in comparison to many systems. The reason is that changes of state are accompanied by changes in molecular structure. The melting of calcium carbonate is accompanied by chemical decomposition to calcium oxide and carbon dioxide, and during the vaporization of ammonium chloride hydrogen chloride and ammonia are formed. Merely defining the number of components in such systems is troublesome.

To make matters worse, the concept of mass action (suggested by Wenzel [8] in 1777, and by Berthollet some years later [9]) was applied indiscriminantly to both chemical equilibria and phase equilibria. While this concept is legitimate for chemical equilibria, its improper application to phase equilibria can lead to serious problems. As an example, varying the quantity of a component in a mixture that is invariant with respect to composition has no effect whatsoever on any aspect of the thermodynamic state of the mixture (Section 4.8.2).

The status of phase science changed dramatically in 1875, when J. Willard Gibbs published his treatise *On the Equilibrium of Heterogeneous Substances* [10]. Gibbs virtually created the concept of "phase". He not only defined this concept from a thermodynamic perspective, but may have introduced the word itself. Roozeboom, who was an important Professor at the University of Amsterdam and a member of the powerful Dutch school of physical chemistry, became aware of Gibbs' ideas in 1887 [1]. In 1897 Bancroft (then at Cornell University) published a book on "The Phase Rule" [11], which influenced Roozeboom's analysis. By 1901, phase science was sufficiently advanced for Roozeboom to write the first volume of the above-mentioned series, which is concerned with the Phase Rule itself and the phase behavior of one-component (unary) systems. Roozeboom also wrote Part 1 of Volume 2 (on binary systems [2]) and edited the later volumes written by his Dutch colleagues [3–7].

Not surprisingly, the significance of the Phase Rule was not immediately recognized. Not only was the concept entirely new, but Gibbs' style of writing is complex and difficult to comprehend [12]. To complicate matters further, the work was published in the obscure *Transactions of the Connecticut Academy of Sciences*. This latter problem was overcome by the translation of Gibbs' papers into German (by Ostwald in 1891 [13]), into French (by Le Chatelier in 1899 [14]), and by the publication of Bancroft's book [11]. In due course, Gibbs' ideas became internationally known and highly respected.

During the first several decades of the 20th century an explosion of research in phase science occurred. With the Phase Rule in hand a firm basis existed for distinguishing good data from bad, progress was rapid, and the phase behavior of virtually every conceivable class of chemical system was explored. Both inorganic and organic systems were investigated, and phase science was widely adopted and effectively utilized – particularly in ceramics, geology, and metallurgy.

By around 1900 the stage was set for the development of surfactant phase science – but no one had actually performed a phase study on a surfactant system. The most important event related to surfactant phase science that had occurred during the pre-Phase Rule period was Chevreul's report, in 1823, of the isolation and characterization of the fatty acids and their salts (the soaps) [15]. Before a surfactant phase diagram was actually determined investigations of the solution physical chemistry of surfactants would be performed, and Friedrich Krafft made important contributions to this area.

14.2   Friedrich Krafft (1852–1923)

The name "Krafft" (whose picture in on the frontispiece) has become one of the most familiar in surfactant science (Section 5.4). Wilhelm Ludwig Friedrich Emil Krafft was a professor of organic and physical chemistry at Heidelburg University, who made two important discoveries related to soap chemistry during the period 1894–1899. He was the first to document (using boiling point elevation data) the anomalous colligative properties of dilute soap solutions in water [16], and their normal behavior in organic solvents such as ethanol [17]. Krafft also discovered that the crystal solubility of soaps displayed a sudden and dramatic increase over a narrow temperature range [18]. He suggested that a relationship existed between the melting point of the fatty acid and the temperature of what is here termed the "Krafft plateau" of soaps (Section 5.4.2) [18], but it is now clear that this was fortuitous [19].

While several decades passed before Krafft's observations were properly interpreted, he must be recognized as being one of the pioneers of surfactant phase science. Perhaps even more important than his specific contributions was the fact that his work attracted the attention of McBain.

14.3 James William McBain (1882–1953)

McBain (Fig. 14.1) was born in Chatham, New Brunswick, Canada, and received his early education in Canada and the United States [20]. He received the BS and MA degrees from the University of Toronto, and the PhD from Heidelberg University (with Bredig), where he learned of the work of Krafft. In 1906 McBain accepted a post as

Fig. 14.1. The young McBain while at Bristol (upper left), the older McBain while at Stanford (upper right), and the Woodland Road Laboratories at Bristol University.

Reader in Physical Chemistry at Bristol University. Three years later he was formally awarded the PhD from Heidelberg. He initiated experimental studies of soap solutions around 1907, and 15 years later (1922) published the first phase diagram of a surfactant system [21].

McBain was elected a Fellow of the Royal Society in 1923. He left Bristol in 1927 for Stanford University, where he continued research on soap chemistry and other aspects of colloid science for another 20 years. He married Mary Evelyn Laing while at Stanford, and remained there until he retired. He was the Davy Medallist in 1939. McBain wrote a book on general surface and colloid science that was published in 1950 [22].

In 1919 McBain was named the first incumbent of the Leverhulme Professorship of Physical Chemistry at Bristol University. This is a prestigious endowed chair which was established by Lord Leverhulme, the founder of an English soap company that later evolved into the modern Unilever Corporation. (This Chair has been held by four distinguished surface chemists. The successors to McBain were W. E. Garner (catalysis chemistry), in 1954 D. H. Everett (surface thermodynamics), and in 1982 R. H. Ottewill (colloid science of surfactants and concentrated latex dispersions).) McBain collaborated with about 20 students during his 20-year tenure at Bristol – three of them women. It is interesting that several of his students (including Millicent Taylor and Mary Evelyn Laing) published, from time to time, a piece of work under their own name [23,24]. Only an acknowledgement, inserted at the end, indicated that McBain had suggested doing the work. McBain's influence was acknowledged by other professors at Bristol [25], and he in turn adopted and used their methods in his own work [25].

McBain contributed to the British war effort during World War I, and to the Allied war effort during World War II. In 1949 he was asked by Prime Minister Nehru to assist in the development of the Indian National Chemical Laboratory at Poona. He returned to Stanford a few months before he died, in 1953.

The characteristics of McBain's work. The hallmarks of McBain's work – upon which rest its significance – were:

(1) the astute selection of experimental methods,
(2) the use of samples of high and well-documented chemical purity, and
(3) the execution of careful quantitative physical studies.

McBain probably benefitted in the analysis and purification of fatty acids from extensive earlier work in this area by other professors at Bristol. He relied principally on measurements of conductivity, freezing points, boiling points, and reaction kinetics in his early research. During his later studies of soap–water phase equilibria he also utilized qualitative observations of the birefringence intensity of mixtures in sealed tubes of known composition as they were heated or cooled, on powder X-ray data, and on dilatometry (the precise measurement of volume as temperature is varied). He recognized the value of X-ray data, and developed the concepts of "expanding" and "non-expanding" liquid crystal phases (Section 8.4.6). His work predates the modern era of chromatographic and spectroscopic analysis.

The motivation for McBain's investigations appears to have been a combination of intellectual curiosity and the practical relevance of the work. His interest in the application of his studies to detergency, and to the contemporary industrial soap manufacturing process ("soap-boiling"), is evident from his participation on the

Council of the Launderer's Research Association [26]. It is also suggested by the fact that his laboratory studies were often performed at 90°C. This is a very inconvenient temperature for quantitative laboratory studies, but is close to the temperature of soap-boiling. By combining his basic data on well-defined systems with French and German technical data on commercial operations, McBain was able (in 1926) to place the art of soap-boiling on a solid scientific basis. His analysis is presented in a chapter in Alexander's *Treatise of Colloid Chemistry* [27].

The course of events. In 1907 McBain was not concerned with the phase behavior of soaps, but rather with their solution physical chemistry and association properties. His work provided early support for the concept that soaps reversibly form electrically charged micellar aggregates in solution [28], but he failed to account properly for the binding of counterions to these aggregates. It was left to Hartley to later show (for example, by use of transport number data) that a considerable fraction of the counterions in ionic surfactant solutions migrate in an electric field together with the oppositely charged micellar aggregates [19]. Extensive data in support of Hartley's revised concept of micellar structure have since been provided by numerous physical studies [29].

McBain was keenly aware of the fact that chemical hydrolysis plays an important role in the physical chemistry of soaps [30], and he discovered that acid soap crystals (rather than fatty acids *per se*) are the hydrolysis product of dilute soap solutions under most conditions (Section 2.2.1) [31]. (However, Rosevear has found that, under extreme conditions of dilution, the fatty acid is formed.) He determined that hexanoate salts are the shortest chain carboxylate salts to display surfactant behavior [32], and that this behavior extends at least to the C_{21} chain length (in behenates). McBain also recognized that crystal phases become progressively more prominent as the chain length is increased [33], and that unsaturation (as in oleates) reduces crystallinity without otherwise strongly affecting solution chemistry, surface chemistry, or those aspects of phase behavior that do not involve the crystal state [34]. He was intrigued by the tendency of soap crystals to assume a ribbon-like morphology, and was thus concerned with the colloidal structure of biphasic mixtures as well as with the phase structure of solutions and liquid crystals (Section 2.7) [35,21,36].

In summary, McBain went about as far in broadly defining the physical science of soaps (and of soap technology) as the available methods could take him.

Strategy and tactics. The strategy that McBain adopted constitutes a beautiful application of the scientific method as outlined 300 years earlier by Francis Bacon. (It is not known whether or not McBain was familiar with Bacon's writings.) The gist of Bacon's philosophy is contained in the following quotation, published in 1620 in *Novum Organum* [37]:

> There are and can exist but two ways of investigating and discovering truth. The one hurries rapidly from the senses and particulars to the most general axioms, and from them, as principles and their supposed indisputable truth, derives and discovers the intermediate axioms. This is the way now in use.
>
> The other constructs its axioms from the senses and particulars, by ascending continually and gradually, till it arrives at the most general axioms, which is the true but unattempted way.

In applying his philosophy, Bacon was as much concerned with destroying incorrect ideas as he was in creating correct ones. Similarly, McBain was at the outset confronted with an array of strongly held ideas regarding soap physical chemistry put forth by others, some of which were right and some of which were wrong. His approach was to isolate a clearly defined and relatively narrow hypothesis, and execute definitive experimental work which tested that hypothesis. An example was the test of the hypothesis (of Krafft) that soaps are not associated in ethanol. This concept was reevaluated, found to be correct, and this result was published [24].

An example of a hypothesis that failed to explain the experimental results was the then-popular hypothesis that the conductivity of soap solutions was due entirely to their hydrolysis reaction. To test this idea he measured the concentration of hydroxide ion that actually existed in soap solutions, and compared the observed conductivity with that calculated using the hydrolysis measurements. In measuring hydroxide concentration electrochemically [38], McBain was forced to utilize a bubbling hydrogen/ platinum electrode – in concentrated soap solutions! (The glass electrode was not yet an established technology.) He was sufficiently unsure of the electrochemical result that he independently determined the hydroxide concentration by measuring the rate of the base-catalyzed decomposition of N-nitrosotriacetoneamine to dinitrogen [25]. The hydroxide ion concentrations measured by the two methods were in agreement, and he was able to refute convincingly the notion that hydrolysis was entirely responsible for the conductivity of soap solutions.

By such processes McBain moved inexorably from an initial narrowly defined concern with the fine points of the solution chemistry of soaps, to a concern with their physical science as a whole, and finally to analysis of their technology. His studies eventually spanned the entire composition range of both binary soap–water and ternary soap–salt–water systems. They also spanned the total range of temperatures within which liquid crystal phases exist in binary systems. He lamented on the high viscosities, the nearly permanent incorporation of air bubbles within viscous phases, and other properties of liquid and liquid crystal phases that still plague research workers in this field [39].

By 1920, McBain's group had compiled a substantial body of information on both homogeneous and heterogeneous soap–water and soap–water–salt mixtures. His publications directly following World War I reflect the highly varied dimensions of this information, a lack of focus on any one aspect of soap science, and a concern with other areas of physical chemistry. He had repeatedly encountered soap "curd", which is formed from soap solutions to which large amounts of salt are added, and a solid phase which persistently separates from soap solutions which he regarded as a 1 : 1 acid soap [30]. In describing the crystal morphology of soap curd, he was strongly influenced by the high quality polarized light microscopy of Zsigmondy *et al.* [36].

It was not until 1921 that the Phase Rule was first mentioned in a publication (with Salmon [40]). In 1922 the first surfactant phase diagram was published (with Burnett [21]). This was a partial isothermal ternary diagram of the sodium laurate (dodecanoate)–sodium chloride–water system at 90°C – hardly the simplest place to start! Similar studies of other soaps and of selected systems over a wide range of temperatures were pursued for several years, so that by 1926 an extensive compilation of ternary soap–water–salt phase diagrams could be published [41].

McBain did not assume *a priori* that the Phase Rule would apply to aqueous soap systems [40]. His writings clearly show that he was aware of the boundary conditions to its application (which had been clearly stated by Gibbs), and that one of these was that surface energies must be sufficiently small (in comparison to bulk phase energies) that they could legitimately be neglected. Since surfactants were by then recognized to be particularly surface active (but see Section 11.2.1), his caution in this regard is understandable.

It was only after many years of research in surfactant phase science that the applicability of the Phase Rule to surfactant systems was firmly established. Perhaps the rigorous test of this hypothesis during studies of the $C_{12}E_6-H_2O$ system, which were reported by Clunie et al. in 1969 [42], may be regarded as the end of this period of uncertainty (Section 7.5). If so, 47 years were required after McBain's first diagram was published to firmly establish that the Phase Rule does indeed apply to surfactant systems. This may seem a long time, but it is actually not excessively long in phase science, as more than 47 years were required to firmly establish the phase diagram of the much simpler sodium chloride–water system [43].

McBain's first diagram was an isothermal ternary diagram; the first binary soap–water diagram (of the potassium oleate–water system) was not published until 1926 [44]. By 1975 when Ekwall published his encyclopedic review of surfactant–water phase equilibria [45], most of the common sodium and potassium soaps, and many other surfactants as well, had been investigated by McBain and others.

Problems in soap phase science. The general form of the phase diagrams of soap–water and of soap–salt–water systems were first suggested by McBain, but he never did get around to describing unary soap systems. His early diagram of the sodium palmitate–sodium chloride–water system was further refined in 1938 by Vold and Ferguson [46]. Dry sodium palmitate was investigated in 1948 by Nordsieck et al., using powder X-ray methods [47], and a dazzling array of structured phase states was found to exist between room temperature and the melting point (Figs 3.2 and 5.9). Using all the available information, Rosevear compiled in 1951 (and published in 1969) a version of the sodium palmitate–water diagram [48] that persisted until the mid-1970s. At that time several soap–water systems were reinvestigated by Madelmont and Perron using improved calorimetric methods [49], which revealed that qualitative errors (largely errors of omission) existed in all the earlier soap–water diagrams. The "intermediate phases" that exist between the hexagonal and lamellar liquid crystal regions were discovered, and their existence subsequently confirmed using other methods [50].

An axiom in phase science is that the phase behavior of a ternary system can never be fully known until the behavior of the three relevant binary systems is defined (Appendix 4.4). Similarly, full knowledge of binary systems hinges upon knowing as well the two relevant unary systems. Nevertheless, the progressive determination of first the simpler systems, then of the more complex, has never been followed during the history of surfactant phase science. The instincts of most people (in both universities and industry) is to start with investigations of those complex biological or industrial systems of immediate interest, and then (perhaps, and often briefly) consider analysis of the simpler ones. The major gaps that presently exist in our knowledge of surfactant phase science can be attributed in large part to this practice – which is precisely the same problem that Bacon decried in 1620 [37].

Viewed in retrospect, it is evident that McBain tackled a problem that was far beyond the capacity of the methods available to him to handle adequately. That he made considerable progress is a tribute to the quality of his intellect, the soundness of his methods, and the considerable energy invested in this work by him and by a dedicated group of coworkers.

14.4 The soap-boiling process

A review of soap-boiling is highly relevant to the history of surfactant phase science. A schematic summary of this process, constructed from the information in McBain's analysis [27], is presented in Fig. 14.2. The scientific disciplines that occupy center stage are phase science and colloid science. The chemical aspect of soap-boiling involves the saponification of aliphatic ester groups, and is superficially straight-forward (except that fats are insoluble in aqueous caustic solutions). The physical aspects of the process, and its engineering on a large scale in an economical manner, are, in striking contrast, very complex. Most of the complexity stems from the phase behavior of the system.

Phase behavior during soap-boiling is manipulated by controlling temperature and soap composition, and by the judicious addition of salt and water. The final steps were controlled in the early days by an artisan (the "soap-boiler") in secrecy, often literally behind a screen [51]. These remarkable individuals utilized subjective appearance, feel, taste, and other information to guide the process.

Early in the soap-boiling process, a heavy liquid phase settles which contains glycerol, salt, unreacted sodium hydroxide or sodium carbonate, and very little soap. After saponification is largely complete, the soap is kept in the lamellar liquid crystal phase. The conditions of the process are dictated by the necessity to maintain a sufficiently fluid state that the coalescence and settling of phases occurs within a reasonable time, and this hinges on staying in this particular phase. Since the process is performed in open kettles, the pressure is one atmosphere.

The details of soap-boiling vary widely depending on the quality of soap desired. Sometimes the fat is simply saponified and the entire mixture processed, but the color is poor, the valuable glycerol is not recovered, and poor quality soap results. During production of the highest quality soap, the contents of the kettle finally settle into three distinct layers (Fig. 14.3). Neglecting the foam at the top, the large upper phase is the neat soap – the lamellar liquid crystal phase. Below the neat soap is the "nigre phase" (a liquid concentrated soap solution), and at the bottom is the nearly soap-free "lye phase" (which is a concentrated electrolyte solution). Cooling the neat soap layer causes the soap to crystallize, and bars or flakes may be formed by suitable processing. Unlike soap products made by simpler processes, those made by soap-boiling are nearly colorless and comparatively free of nonsoap impurities.

Of particular interest with respect to the phase science of soap-boiling is the three-phase state that exists at the end of the process. A biphasic mixture of liquid phase and neat soap exists prior to the final steps, and the third phase is created by the alternate addition of aliquots of salt and water. The diagram of the sodium palmitate–salt–water system at 90°C (Fig. 14.4) is regarded as a realistic model system for commercial mixtures [46]. These three phases coexist within an extremely narrow three-phase triangle in this diagram. It is remarkable that such a tiny region could have been found (using empirical methods) during the earlier period. On the other hand, it *had* to be encountered along the process trajectories utilized; its existence is dictated by the Phase Rule (Section 12.1.2).

The ease with which soap-boilers ended up within this triangle no doubt hinged in part on the scale of the operation. Making an adjustment of a few hundredths of a per cent in total composition is comparatively difficult in small laboratory samples, but is no problem at all on a manufacturing scale.

The Soap Boiling Process (McBain)

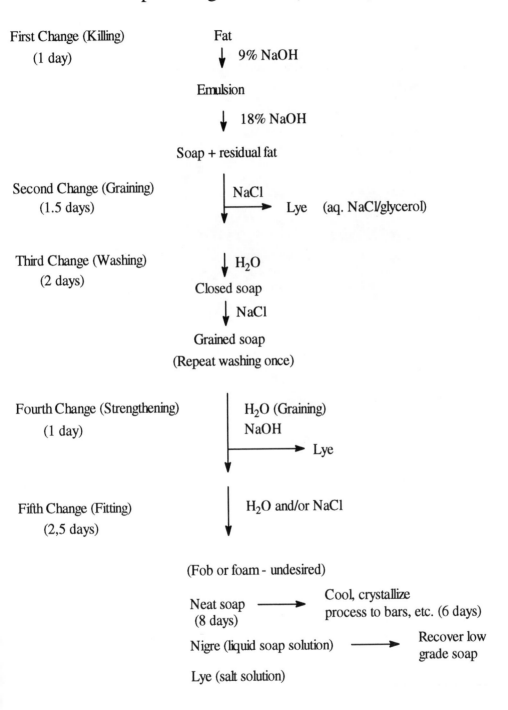

Fig. 14.2. A schematic description of the soap-boiling process [27].

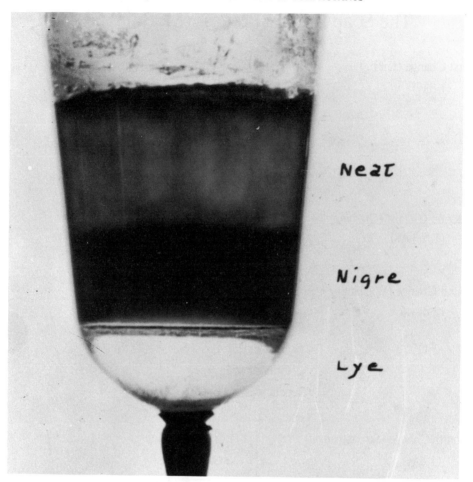

Fig. 14.3. A photograph of a laboratory scale soap "kettle" that has been "pitched", courtesy of F. B. Rosevear. The upper neat, the middle (dark) nigre, and the clear heavy lye phases are readily visible.

Not mentioned in the McBain review were the consequences of failure to finish the process within this three-phase triangle [51]. The neat–nigre–lye triangle must be surrounded by three biphasic regions: the neat–nigre, the nigre–lye, and the neat–lye. By adding water to the neat–nigre region one may enter another triangle within which neat soap, nigre, and "middle phase" coexist. (The middle phase was discovered by McBain [44], and so-named because its composition range fell between the liquid (nigre) and the neat soap. It is now known to have the normal hexagonal liquid crystal phase structure (Chapter 8).) A mistake in the addition of salt or water could result in the formation of substantial quantities of middle phase, which could be very costly because middle phase cannot be pumped or handled like neat soap. In severe cases the middle phase had to be removed from the kettle by hand, and perhaps discarded. An awareness of this problem was likely one of the incentives that McBain and others had for better defining this process.

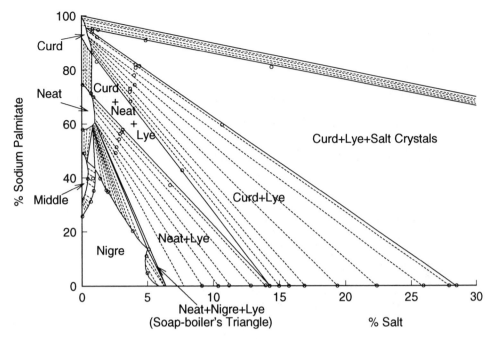

Fig. 14.4. The sodium palmitate–salt–water system at 90°C in the right triangular format (Section 12.1.1). The older terminology is used for the phases that exist. At the end of the process the kettle composition fell within the thin "soap-boiler's triangle".

The micellar nigre phase efficiently extracted colored impurities from the neat soap during soap-boiling (Fig. 14.3). This was the key to the production of soap that was light in color by early manufacturers, but nowadays this separation is addressed differently. In modern soap making the fat is first hydrolyzed by the direct action of water (using a catalyst), the fatty acids produced are purified by vacuum distillation, and the soap is prepared by separately neutralizing them [52]. The capacity to make soap of exceptional color and quality by this process is enormous in comparison with that of the slow (two-week-long) and tedious soap-boiling process. However, the latter is still used in small-scale manufacturing.

It is ironic that soap-boiling began to diminish in importance within a decade after McBain was finally successful in defining the physical science of the process. The phase science is still relevant to fabrication of the final product, of course, and the significance of this pioneering work extends far beyond that of merely improving soap-boiling.

14.5   Jack Henry Schulman (1904–1967)

In the early 1940s, Jack Schulman [53] (then at Cambridge University) published important experimental results – and a model – which were to have a profound (if indirect) influence on surfactant phase science [54]. Schulman discovered that an emulsion of a nonpolar oil with a soap solution could be clarified by titrating into the mixture a short chain alcohol, such as pentanol. Using geometric analysis, and assuming a structure in which spherical droplets of water were "emulsified" within the clear liquid by the soap and pentanol, Schulman estimated that the particles had to have molecular dimensions. The mixtures were regarded as true emulsions, and the small particle size served to explain legitimately the low turbidity of these mixtures (compared to that of ordinary emulsions). In 1959 Schulman, Stoekenius, and Prince suggested the name "microemulsions" for such mixtures – and the world of surfactant science has not since been the same [55]. In 1977 Prince published a book on microemulsions [56].

The word "emulsions" had been (and still is) used for multiphase dispersions of one liquid phase in another long before Schulman's work. Classical "emulsions" are irreversible colloids [57]. Schulman's "microemulsions" are not emulsions at all, but liquid phases that have unusually highly developed phase structures (Sections 8.5.2, 13.4). This unfortunate terminology caused considerable controversy and confusion for about a quarter of a century, but seems presently to have been clarified [58].

Microemulsions caught the fancy of large numbers of chemists and physicists for both academic and practical reasons, and there is no doubt that Schulman was enormously influential in stimulating experimental and theoretical research into surfactant phase science. Unfortunately, he drowned in Italy in 1967 and did not live to experience the full impact of his work and ideas [53].

Although Schulman dealt with phase phenomena, he did not approach either the design of experiments or the analysis of data from the perspective of phase science. He chose instead to view microemulsions from the perspective of molecular models for their structure, and neglected the rigorous thermodynamic analysis of the system. Schulman's contributions to surfactant phase science were, as a result, minimal. It was left to Per Ekwall and his coworkers (initially), and to later workers (particularly in France), to investigate microemulsion-forming systems from a more rigorous perspective.

14.6 Per Ekwall (1895–1990)

Per Ekwall (Fig. 14.5) was born at Pori (Björneborg), Finland, and received his MSc from the University of Helsinki (Helsingfors) in 1920.* He went in 1921 to the newly founded Åbo Akademi Rediviva (The Swedish University of Åbo) in Turku (Åbo), and was awarded the DSc degree from that institution in 1927. He spent 2 years in Germany (on a Rockefeller grant) in Walden's laboratory, and came to know Wolfgang Ostwald. He returned to Åbo, was appointed Associate Professor at the Akademi around 1930, and in 1943 was made Full Professor of Physical Chemistry [59].

Ekwall's phase studies were initiated at Åbo Akademi. In 1963, after retiring as Professor Emeritus from Åbo, he moved to Stockholm and, at the invitation of The Royal Swedish Academy of Engineering Sciences, founded (at age 68) what is now The Swedish Institute for Surface Chemistry. He served as Director of this Institute until 1968.

Ekwall remained scientifically active up to his last hours; he was writing a manuscript on acid soaps when he died. In 1991 a Scandinavian Surface Chemistry Symposium was held to honor his contributions, and a memorial volume has been published [60].

---

*For historical reasons, Finnish cities also usually have Swedish names. In deference to the current political situation, the Finnish name for cities will be used with the corresponding Swedish name indicated in parentheses.

Fig. 14.5.   The young Ekwall while at Åbo Akademi (upper left), the older Ekwall while at the Surface Chemistry Institute in Stockholm (upper right), and the chemistry laboratories at the Åbo Akademi in Turku, Finland.

Ekwall followed the traditional autocratic European style in directing the work of his students. In striking contrast to McBain (Section 14.3), Ekwall sometimes did not even include the names of his students in his early publications. That changed after World War II. Of those dealing with phase science from this later period, Ingvar Danielsson, Leo Mandell, and Krister Fontell were particularly important. Danielsson was influential in orienting the work of the laboratory towards phase science, and was to become Head of the Department of Chemistry at Åbo Akademi. Mandell and Fontell joined Ekwall at the Institute in Stockholm when it was founded and continued the phase studies. When Ekwall retired as Director, Mandell joined a petroleum company and Fontell went to the University of Lund, where he continued research in phase science. Fontell retired in 1987 after having served as a faculty member for 14 years.

Scandinavian surfactant chemists have on the whole been keenly aware of the value of phase science. They have often made good use of phase information in pursuing other aspects of the physical science of surfactants, and routinely include phase studies as an integral aspect of their work. Without question, this tendency is due in large part to Ekwall's influence.

Course of events. There is a striking parallel between the evolution of Ekwall's entry into surfactant phase science and that of McBain's. Like McBain, Ekwall was initially interested in the solution chemistry of surfactants rather than in their phase science. He was an early proponent (along with Jones and Bury) of the notion of a "critical concentration" in the solution chemistry of soaps [61].

The principal focus of McBain's work was the influence of salts on the solution chemistry and phase science of soaps, while Ekwall was more interested in the solubilization of water-insoluble oils by aqueous soap solutions. Starting in 1923, he published a large number of papers in this area [62]. After leaving it briefly, he returned to the subject in 1947 in an important lecture at the 6th Nordic Chemists' Meeting in Lund [63]. A long series of papers followed on the solubilization of carcinogens, fatty alcohols, and hydrocarbons, which led in 1960 to the first report (with Danielsson and Mandell) of surfactant–oil–water phase equilibria. This was the study of the ternary sodium octanoate–decanol–water system at 20°C [64,65]. After Ekwall retired from Åbo Akademi the phase studies were continued in Stockholm for a few years, and resulted in intensive publications of surfactant–oil–water phase diagrams during the late 1960s. Ekwall published in 1975 an extensive survey of both binary and ternary aqueous surfactant phase diagrams [45].

Methods and results. The Ekwall phase studies were performed by preparing numerous mixtures, each containing a few grams, which were mixed, sealed, stored in a constant temperature (20°C) room, and observed over a period of weeks or months [65]. Centrifugation at high gravitational fields was often used to separate the phases in biphasic mixtures. The vials were examined directly between crossed polars, and aliquots were examined using the polarized light microscope. Densities and X-ray powder patterns were obtained. A C_8 soap (sodium octanoate or caprylate) was selected for the initial studies, in preparation for a study of sodium oleate. This was a fortunate choice, for the phases formed by the C_8 soap are far more fluid, phase equilibrium is more quickly attained, and the coalescence and separation of phases is faster in the octanoate system than in the oleate.

Because there were no precedents for this work, the phases observed were initially designated as *A*, *B*, *C*, etc., in the order that they were discovered. Presently, it is generally believed that Ekwall's phases *A*, *B*, and *C* were not single-phase mixtures.

Possibly they were biphasic mixtures of the D and L_1 phases having characteristic optical textures [66]. The D phase is now known to be the lamellar, the E phase the normal hexagonal, and the F phase the inverted hexagonal liquid crystal phase (Chapter 8). The "G" phases included ill-defined crystal phases, but details of the crystal phase chemistry of this system were not clearly defined. Curiously, the existence of acid-soap crystals was not always indicated in his diagrams even though Ekwall was well aware of their existence [67]. The soap–fatty acid–water phase story is far from finished.

Both Ekwall and McBain were keenly interested in acid soap chemistry. The formation of acid soaps by the hydrolysis of soap solutions constantly plagued McBain's investigations [31], and the study of acid soaps also preoccupied Ekwall and his students. Ekwall believed that stoichiometric acid soaps existed in both fluid liquid and liquid crystal phases, as well as in crystalline phase compounds [68].

It was shown that two widely separated and (usually) disconnected liquid phase regions often exist, which Ekwall termed simply the L_1 and L_2 phases. The former has since been termed by others the "water-continuous" or "oil-in-water" (O/W) microemulsion, since it is a water-rich phase within which oil is solubilized. The L_2 phase is an oil-rich phase that contains relatively small amounts of water, which has been termed an "oil-continuous" or "water-in-oil" (W/O) microemulsion" [58]. Ekwall's work came after Schulman's discovery of "microemulsions", and he did not like this term – preferring instead to regard microemulsions as being simply liquid solutions in thermodynamic equilibrium and not emulsions [69]. Schulman visited Ekwall at the Institute in 1965 and the matter was discussed, but the differences in the views of the two men regarding microemulsions were not fully resolved at that time. Schulman's premature death precluded their reaching a mutually agreeable resolution.

It is now clear that the model originally proposed by Schulman for the structure of liquid surfactant–oil–water phases was too simple. Strong support for this model has indeed been developed within the low-water regions of the L_2 phase [70], but it may not apply in other parts of this region.

14.7 Beyond soaps

Soaps are the oldest surfactants known and are still important industrially, so it is hardly surprising that the initial phase studies were performed on these materials. McBain was aware of the existence of "hydrogen soaps", which had been prepared in 1913 by Reychler [71]. (Reychler also prepared the first monoalkyl quaternary ammonium salt surfactants.) These are alkanesulfonic acids which, in the presence of at least one mole of water, exist as a monohydrate that is structurally a hydronium sulfonate salt, RSO_3^-,H_3O^+ (Section 10.5.1).* McBain's interest in these surfactants stemmed from their structural similarity to soaps, and in the fact that the hydrolysis chemistry which plagued his investigations of soaps would not exist. In 1941 the phase diagram of the $C_{12}H_{25}SO_3^-,H_3O^+-H_2O$ system was determined by Marjorie J. Vold

---

*It is evident from its crystal structure that structurally the "monohydrate" of toluenesulfonic acid is hydronium toluenesulfonate. Long-chain aliphatic and aromatic sulfonic acids were to become widely known and commercially important as the catalyst in the Twitchell process for the hydrolysis of triglycerides [102], but aqueous phase studies of the Twitchell catalyst have not been performed.

[72], and found to strongly resemble that of soap–water diagrams. Vold and her husband (Robert D.) were engaged in phase studies of soaps at Procter and Gamble until around 1937, when they left to join McBain at Stanford University. Both Volds had distinguished academic careers in surfactant and colloid chemistry, mostly at the University of Southern California.

Starting with Reychler's work [71], synthetic chemists have explored numerous other molecules which might have the desirable surface properties of soaps but lack their shortcomings. One such shortcoming is the impairment of utility by the calcium and magnesium salts present in "hard" water. Calcium and magnesium soaps are less soluble than sodium soaps, and are constituents of the scum that forms on lavatory or bathtub surfaces when soaps are used in hard water [73].

An early (if naive) effort to overcome this shortcoming was development of the sodium alkyl sulfates $(ROSO_3^-, M^+)$. These surfactants were initially prepared in Germany (by I.G. Farben) and later in the US (by DuPont), mainly for use in the textile industry [74]. Samples were brought from Germany to the US by R. A. Duncan of Procter and Gamble in 1931 [75], and sodium coconut chain-length alkyl sulfates were incorporated into the first commercial synthetic laundry detergent ("Dreft") in 1933. Ester salts of phosphoric acid $(ROPO_3^=, 2M^+)$ constitute a logical extension of the sodium alkyl sulfate concept, and were also investigated [76].

During the late 1940s, Winsor at Shell (in England) contributed important phase (and other) information on the Aerosol sulfosuccinate surfactants [77]. Winsor also developed concepts regarding the influence of molecular geometry on phase behavior [78] that form the basis for more recent developments in this area [79]. At about the same time sodium alkylbenzene sulfonates (ABS) having highly branched alkyl substituents (such as sodium tetrapropylenebenzenesulfonate) became commercially available [80]. Tide, introduced in 1947, was based (in part) on this surfactant. Tide was a revolutionary home laundry product in composition, in that it contained far more builder (50%) than it did surfactant (20%) [81]. (A "builder" is the component of a laundry product that sequesters calcium and magnesium, buffers pH in the mildly basic range, and "peptizes" (stabilizes) soil that has been removed from the fabric as dispersions within the wash water.) The builder at the time was sodium tripolyphosphate. The detergency performance of tide was exceptional.

Around 1965, sodium ABS surfactants were replaced by sodium LAS (*L*inear *A*lkylate *S*ulfonate) surfactants because LAS surfactants are more rapidly biodegradable [82]. Sodium LAS is largely a mixture of positional isomers of sodium straight-chain alkylbenzene sulfonates (Section 11.3.2.1). Phase studies of the individual sodium *x*-dodecylbenzene sulfonate–water systems were performed by the California Research Corporation in the 1940s, but were never published.

It has long been apparent that if anionic functional groups were useful, cationic groups might also be of value [71]. The antimicrobial properties of these surfactants were first recognized in 1935 [83]. They are important today as antimicrobial compounds, but are also widely used in fabric softeners and hair conditioners [84,85]. Phase studies of soluble (mono-long chain) ammonium chloride surfactants were performed at the Armour Company (Chicago) during the 1940s [86], but the phase behavior of a cationic fabric softener was not correctly documented until 1990 [87].

In 1925 polyoxyethylene derivatives of fatty alcohols and alkylphenols were prepared at I. G. Farben in Germany, and in 1928 mono-fatty acid esters of polyoxyethylene glycols were reported [88]. In 1930 it was discovered that monoethers of polyoxy-ethylene glycols were surface active and behaved like soaps [88]. Sugar monoesters were also investigated [89], so that by the early 1950s a fairly large variety of soap-like molecules had been prepared and studied.

In 1957, E. C. Taylor (Princeton University) brought to the attention of Procter and Gamble the fact that alkyldimethylamine oxides were water soluble and surface active. A patent on these compounds as textile surfactants had been filed by the Swiss firm CIBA in 1935 [90], but their potential value as components of detergent products had gone unrecognized. Largely as a result of this discovery, it was realized that the structural requirements of surfactant hydrophilic groups were entirely unobvious. A synthetic and physical research program which addressed this question was therefore initiated around 1958, and sustained for more than a decade.

Two major issues had to be confronted in pursuing this program. One was how to distinguish an amphiphilic compound that is a surfactant from one that is not, and the other was how to rank the relative hydrophilicity of the different polar groups (Chapter 9). Preliminary answers to both questions were found to reside in their aqueous phase behavior. Rigorous phase studies were not usually performed during this program; instead, a simple step-wise dilution test tube method was developed which proved to be adequate for scanning purposes (Appendix 4.4.3). During this period the research group in Newcastle-upon-Tyne (UK) made numerous contributions to both the phase science of surfactants and especially to their solution physical chemistry.

During this period synthetic surfactants such as $C_{12}E_6$, $C_{12}APS$, and $C_{12}AO$ (which are model compounds for commercially important surfactants) were intensively inves-tigated [42,91,92] (see Appendix 1 for definitions of acronyms). A variety of semipolar and zwitterionic surfactants – which a short time before were not even recognized as being surfactants – were investigated. Previously unknown functional groups, such as the sulfone diimines ($-S(NH)_2-$), were synthesized and examined [93]. In addition, a few in-depth studies of binary and ternary systems, such as those by Lang of $C_{10}E_4–H_2O$ [94] and $C_{10}E_4–C_{16}H_{34}–H_2O$ [95], were also performed.

By the early 1970s, amphiphilic compounds containing most of the monopolar and dipolar functional groups that are sufficiently stable (thermally and hydrolytically) to be examined had been synthesized, and their aqueous phase behavior scanned. These data permitted a comprehensive analysis to be performed of those structural features that characterize operative and nonoperative functional groups, which was published in 1978 [96]. This information constitutes the basis for Part IV of this book.

14.8 The microemulsion studies

An enormous number of workers around the world have investigated microemulsions during the past 30 years. This massive effort was stimulated by the pioneering experi-ments and models of Schulman, by the more rigorous phase studies of Ekwall, and by the possibility that microemulsion-based technology was applicable to tertiary oil recovery and other problems. Several books have been written on microemulsions [56,97,99], their physics has been extensively analyzed [99], and considerable know-ledge of these phases exists.

Most phase studies of microemulsion-forming systems have been limited to defining the extent of the microemulsion region itself. The emphasis during this research has been on the structure and physics of microemulsion phases, rather than on the phase science of the system as a whole. Often the work was performed using quaternary or higher systems, which precludes rigorous analysis of phase behavior using currently available methods.

Such work is a useful first step towards defining the physical science of microemulsion-forming systems, but it represents only a preliminary study from the perspective of phase science. Relatively few studies of ternary surfactant–oil–water systems have gone beyond merely determining the liquid phase boundary. The perception that a large amount of work is required for such studies, coupled with the uncertainty that a diagram could be defined after having done the work, is no doubt responsible for this situation. The basic problem is that this perception is valid. Dramatically improved methodology will be required before real progress is made in defining ternary and higher order surfactant systems.

14.9 Summary

The history of surfactant phase science closely parallels the development of fundamental concepts, on the one hand, and of experimental methods, on the other. Aside from Gibbs himself, McBain has without question had the most profound influence on this area. This man had the philosophy, the intellect, and the energy to explore successfully uncharted fields of science. His interactions with colleagues were often productive, and he encouraged both men and women to do physical research of high quality.

McBain, and those that followed in his path, laid the foundation for binary surfactant–water and ternary surfactant–salt–water phase science, while Ekwall and his coworkers later played a similar leading role in clarifying the phase behavior of surfactant–oil–water systems.

In moving beyond soaps, the most important roles have so far been played by the (mostly industrial) synthetic chemists and engineers who synthesize and manufacture surfactants. The surfactants used today are not used because they are the best possible choices. They are used because they are commercially available, and they perform best among those that are available. The conscious design of the optimal surfactant for a particular purpose is a goal which is yet to be attained.

The reason for this has no doubt been, in part, the primitive level of our knowledge of this area. This is illustrated by the fact that a comprehensive view as to which polar functional groups are hydrophilic groups, and which are not, has existed only since 1978 [97]. It is intriguing that a significant fraction of this information was provided by basic research within industrial laboratories. The behavior of soaps was defined in large part by McBain while at Bristol and Stanford Universities, but research at Shell (by Winsor) provided important data on the Aerosol surfactants and work at the Armour Company (Chicago) constituted the pioneering phase research on cationic surfactant salts. The U.S. and European research laboratories of Procter and Gamble later contributed phase studies particularly of various nonionic surfactants. Excepting Tiddy's work of the past decade or so, the published Unilever surfactant research has historically played a more important role in defining the solution thermodynamics and surface physics of surfactant solutions than in describing their phase behavior.

While many unanswered questions remain in surfactant phase science, the level of knowledge that presently exists is impressive in comparison with that which existed just a century ago, when Krafft explored for the first time the solubilities of soaps. With continued effort, the day in which surfactant molecules can in fact be tailored to satisfy a particular need may not be so far off. When that has been accomplished the first large step in surfactant science will finally have been taken, and surfactant scientists can move on to the next.

References

1. Roozeboom, H. W. B. (1901). *Die Heterogenen Gleichgewichte vom Standpunkte der Phasen-lehre. Die Phasenlehre – Systeme aus Einer Komponente*, Vol. 1, F. Vieweg und Sohn, Braunschweig.
2. Roozeboom, H. W. B. (1904). *Die Heterogenen Gleichgewichte vom Standpunkte der Phasen-lehre. Systeme aus zwei Komponenten*, Vol. 2, Part 1, (H. W. B. Roozeboom ed.), F. Vieweg und Sohn, Braunschweig.
3. Buchner, E. H. (1918). *Die Heterogenen Gleichgewichte vom Standpunkte der Phasenlehre. Systeme mit zwei fluessigen Phasen*, Vol. 2, Part 2 (H. W. B. Roozeboom ed.), F. Vieweg und Sohn, Braunschweig.
4. Aten, A. H. W. (1918). *Die Heterogenen Gleichgewichte vom Standpunkte der Phasenlehre. Pseudobinaere Systeme*, Vol. 2, Part 3 (H. W. B. Roozeboom ed.), F. Vieweg und Sohn, Braunschweig.
5. Schreinemaker, F. A. H. (1911). *Die Heterogene Gleichgewichte vom standpunkte der Phasenlehre. Die Taernare Gleichgewichte.*, Vol. 3, Part 1 (H. W. B. Roozeboom ed.), F. Vieweg und Sohn, Braunschweig.
6. Schreinemaker, F. A. H. (1913). *Die Heterogenen Gleichgewichte vom Standpunkte der Pha-senlehre. Systemen mit zwei und mehr flssigen Phasen*, Vol. 3, Part 2 (H. W. B. Roozeboom ed.), F. Vieweg und Sohn, Braunschweig.
7. Schreinemaker, F. A. H. (1913). *Die Heterogenen Gleichgewichte vom Standpunkte der Phasenlehre. Systemen mit zwei und mehr flssigkeiten, ohne mischkristall*, Vol. 3, Part 3 (H. W. B. Roozeboom ed.), F. Vieweg und Sohn, Braunschweig.
8. Wenzel, Carl Friedrich (1777). *Lehre von der Verwandtschaft der Koerper*, G. A. Gerlach, Dresden.
9. Berthollet, C. L. (1801). *Recherches sur les lois de l'affinité*, Memoires d l'Institut national des sciences et arts, Paris; (1803). *Essai de statique chimique*, Demonville et Soers, Paris.
10. Gibbs, J. W. (1928). *The Collected Works of J. Willard Gibbs*, Vol. I (W. R. Longley and R. G. Van Name eds), pp. 55–353, Longmans, Green, New York; Gibbs, J. W. (1961). *The Scientific Papers of J. Willard Gibbs. Thermodynamics*, Vol. 1 (W. R. Longley and R. G. Van Name eds), pp. 55–353, Dover, New York.
11. Bancroft, W. D. (1897). *The Phase Rule*, The Journal of Physical Chemistry, Ithaca, New York.
12. Gerasimov, Y. I. (1976). *Russian J. Phys. Chem. (Eng. ed.)* **50**, 1797–1798.
13. Gibbs, J. W. (1892). *Thermodynamische Studien (Translated by Wo. Ostwald)*, W. Engel-mann, Leipzig.
14. Gibbs, J. W. (1899). *Equilibre des syst'emes chimiques (Translated by H. LeChatelier)*, G. Carre & C. Naud, Paris.
15. Chevreul, M. E. (1823). *Recherches chimiques sur les corps gras d'origines animals*, Paris.
16. Krafft, F. and Wiglow, H. (1895). *Ber. deutsche Chem. Ges.* **28**, 2573–2582.
17. Krafft, F. (1899). *Ber. deutsche Chem. Ges.* **32**, 1584–1596.
18. Krafft, F. (1899). *Ber. deutsche Chem. Ges.* **32**, 1596–1608.
19. Hartley, G. S. (1936). *Aqueous Solutions of Paraffin Chain Salts*, p. 38, Hermann, Paris.
20. Taylor, H. S. (1956). *J. Chem. Soc.* 1918–1920.
21. McBain, J. W. and Burnett, A. J. (1922). *J. Chem. Soc.* **121**, 1320–1333.
22. McBain, J. W. (1950). *Colloid Science*, D.C. Heath, Boston.
23. McBain, J. W. and Taylor, M. (1910). *Ber. deutsche Chem. Ges.* **43**, 321–322.
24. Laing, M. E. (1918). *Trans. Chem. Soc.* **113**, 435–444.

25. Clibbens, D. A. and Francis, F. (1911). *J. Chem. Soc.* **101**, 2358–2371.
26. Anonymous (1921). *The Nonesuch (University of Bristol, Department of Chemistry house organ).*
27. McBain, J. W. (1926). *Colloid Chemistry*, Vol. I (J. Alexander ed.), pp. 137–164, The Chemical Catalog Co., New York.
28. McBain, J. W. and Martin, H. E. (1914). *J. Chem. Soc.* **105**, 957–977; McBain, J. W. and Salmon, C. S. (1920). *J. Am. Chem. Soc.* **42**, 426–460.
29. Wennerstrom, H. and Lindman, B. (1979). *Physics Reports* **52**, 1–86.
30. McBain, J. W. and Taylor, M. (1911). *Zeit. Physik. Chem.* **126**, 179–209.
31. McBain, J. W. (1918). *J. Soc. Chem. Ind.* **37**, 249–252.
32. McBain, J. W., Bunbury, H. M. and Martin, H. E. (1914). *Trans. Chem. Soc.* **105**, 419–433.
33. McBain, J. W. (1920). *Third Rept on Colloid Chemistry, British Assoc.*
34. Krafft, F. and Wiglow, H. (1895). *Ber. deutsche Chem. Ges.* **28**, 2566–2573; McBain, J. W. and Elford, W.J. (1926). *J. Chem. Soc.* 421–438; Laing, M. E. and McBain, J. W. (1920). *Trans. Chem. Soc.* **117**, 1506–1528.
35. McBain, J. W. and Martin, H. E. (1921). *J. Chem. Soc.* **119**, 1769–1774.
36. Zsigmondy, R. and Bachmann, W. (1912). *Kolloid Z.* **11**, 145–157.
37. Francis Bacon (1620). *Novum Organum (Translated from the Latin by Wm. Pickering, 1850)*, pp. 12–13, C. Whittingham, Chiswick.
38. McBain, J. W. and Martin, H. E. (1914). *Trans. Chem. Soc.* **105**, 953–977.
39. McBain, J. M. and Burnett, A. J. (1922). *J. Chem. Soc.* **121**, 1320–1333.
40. McBain, J. W. and Salmon, C. S. (1921). *Trans. Chem. Soc.* **119**, 1374–1383.
41. McBain, J. W., Lazarus, L. H. and Pitter, A. V. (1926). *Z. physik. Chem.* **A147**, 87–117.
42. Clunie, J. S., Goodman, J. F. and Symons, P. C. (1969). *Trans. Faraday Soc.* **65**, 287–296.
43. (1965). *Solubilities*, 4th ed., Vol. II, pp. 851–852, American Chemical Society, Washington, DC.
44. McBain, J. W. and Elford, W. J. (1926). *J. Chem. Soc.* 421–438.
45. Ekwall, P. (1975). *Advances in Liquid Crystals*, Vol. 1 (G. H. Brown ed.), pp. 1–142, Academic Press, New York.
46. Vold, R. D. and Ferguson, R. H. (1938). *J. Am. Chem. Soc.* **60**, 2066–2076.
47. Nordsieck, H., Rosevear, F. B. and Ferguson, R. H. (1948). *J. Chem. Phys.* **16**, 175–180.
48. Rosevear, F. B. (1968). *J. Soc. Cosmetic Chemists* **19**, 581–594.
49. Madelmont, C. and Perron, R. (1974). *Bull. Soc Chim. France* 3430–3434.
50. Rendall, K., Tiddy, G. J. T. and Trevethan, M. A. (1983). *J. Chem. Soc., Faraday Trans. I* **79**, 637–649.
51. F. B. Rosevear, personal communication.
52. Wollatt, E. (1985). *Manufacture of Soaps, Other Detergents and Glycerine. Ellis Horwod Series, Applied Science and Industrial Technol*, pp. 195–236, Ellis Horwood, Chichester.
53. Rosano, H. L. (1969). *J. Coll. Interface Sci.* **29**, i–xii.
54. Hoar, T. P. and Schulman, J. H. (1943). *Nature* **152**, 102–103.
55. Schulman, J. H., Stoekenius, W. and Prince, L. (1959). *J. Phys. Chem.* **63**, 1677–1680.
56. Prince, L. M. (1977). *Microemulsions*, Academic Press, New York.
57. (1952). *Colloid Science, Irreversible Systems*, Vol. I (H. R. Kruyt ed.), pp. 58,335, Elsevier, New York; Becher, P. (1965). *Emulsions: Theory and Practice*, 2nd ed., Reinhold, New York.
58. Danielsson, I. and Lindman, B. (1981). *Colloids and Surfaces* **3**, 391–392.
59. Fontell, K. (1992). *Adv. Coll. Interface Sci.* **41**, pp. vi–xix; K. Fontell, personal communication.
60. (1992). *Adv. Coll. Interface Sci.* **41**, 1–271.
61. Jones, E. R. and Bury, C. R. (1927). *Phil. Mag.* **4**, 841–848.
62. Ekwall, P., Mandell, L. and Fontell, K. (1969). *Mol. Cryst. Liq. Cryst.* **8**, 157–213.
63. Ekwall, P. (1948). Nord. Kemistmötet, 6th Meeting, Lund **1947**, 179–215.
64. Mandell, L. and Ekwall, P. (1968). *Acta Polytech. Scand., Chem. Incl. Met. Ser. (Pt.1)* **74**, 1–116.
65. Ekwall, P., Danielsson, I. and Mandell, L. (1960). *Kolloid Z.* **169**, 113–124.
66. Friman, R., Danielsson, I. and Stenius, P. (1982). *J. Coll. Interface Sci.* **86**, 501–514.
67. Ekwall, P. (1937). *Kolloid Z.* **80**, 177–200.

68. Ekwall, P. (1988). *Coll. Polym. Sci.* **266**, 279–282; Ekwall, P. and Fontell, K. (1988). *Coll. Polym. Sci.* **266**, 184–191.
69. K. Fontell, personal communication.
70. Cebula, D. J., Myers, D. Y. and Ottewill, R. H. (1982). *Colloid Polym. Sci.* **260**, 96–107.
71. Reychler, A. (1913). *Kolloid Z.* **12**, 277–283; *ibid.* **13**, 252–254.
72. Vold, M. J. (1941). *J. Am. Chem. Soc.* **63**, 1427–1432.
73. Linfield, W. M. (1976). *Anionic Surfactants*, Vol. 7 (W. M. Linfield ed.), pp. 1–10, Marcel Dekker, New York; Powe, W. C. (1972). *Detergency. Theory and Test Methods*, Vol. 5 (W. G. Cutler and R. C. Davis eds, M. J. Schick series ed.), pp. 31–64, Marcel Dekker, New York.
74. Hatcher, D. B. (1957). *J. Am. Oil Chemists' Soc.* **34**, 170–172.
75. Lief, A. (1958). *It Floats*, pp. 187–197, Rinehart, New York.
76. Harris, B. R. (1933). *US Patent* 1 917 251, July 17, 1933.
77. Rogers, J. and Winsor, P. A. (1969). *J. Coll. Interface Sci.* **30**, 247–257.
78. Winsor, P. A. (1968). *Chem. Rev.* **68**, 1–40.
79. Israelachvili, J., Mitchell, D. J. and Ninham, B. W. (1976). *J. Chem. Soc., Faraday Trans. 2* **72**, 1525–1568.
80. Feighner, G. C. (1976). *Anionic Surfactants*, Vol. 7 (W. M. Linfield ed.), pp. 288–314, Marcel Dekker, New York; Miller, E. L. and Geiser, P. E. (1957). *J. Am. Oil Chemists' Soc.* **34**, 170–172.
81. Byerly, D. R. (1949). *US Patent* 2 486 921, Nov 1, 1949.
82. Swisher, R. D. (1987). *Surfactant Biodegradation*, 2nd ed., p. 25, Marcel Dekker, New York.
83. Jungermann, E. (1970). *Surfactant Science Series, Cationic Surfactants*, Vol. 4 (Eric Jungermann ed.), pp. 203–418, Marcel Dekker, New York.
84. Laughlin, R. G. (1990). *Cationic Surfactants Physical Chemistry*, 2nd ed., Vol. 37, pp. 449–468, Marcel Dekker, New York.
85. (1980). *Chemical Technology Review. Household and Industrial Fabric Conditioners*, Vol. 152, Noyes Data Corporation, Park Ridge, NJ.
86. Ralston, A. W., Hoerr, C. W. and Hoffman, E. J. (1942). *J. Am. Chem. Soc.* **64**, 1516–1523; Broome, F. K., Hoerr, C. W. and Harwood, H. J. (1951). *J. Am. Chem. Soc.* **73**, 3350–3354; Ralston, A. W., Hoffman, E. J., Hoerr, C. W. and Selby, W. M. (1941). *J. Am. Chem. Soc.* **63**, 1598–1601.
87. Laughlin, R. G., Munyon, R. L., Fu, Y.-C. and Fehl, A. J. (1990). *J. Phys. Chem.* **94**, 2546–2552.
88. Schoenfeldt, N. (1969). *Surface Active Ethylene Oxide Adducts*, pp. 20–24, Pergamon, Oxford.
89. Osipow, L., Snell, F. D. and Finchler, A. (1957). *J. Am. Oil Chemists' Soc.* **34**, 170–172; Benson, F. R. (1967). *Nonionic Surfactants*, Vol. 1 (M. J. Schick ed.), pp. 247–299, Marcel Dekker, New York.
90. British Patent 437 566, 31 Oct. 1935, to Society of Chemical Industry in Basel (CIBA).
91. Nilsson, P-G., Lindman, B. and Laughlin, R. G. (1984). *J. Phys. Chem.* **88**, 6357–6362.
92. Lutton, E. S. (1966). *J. Am. Oil Chemists' Soc.* **43**, 28–30.
93. Laughlin, R. G. and Yellin, W. (1967). *J. Am. Chem. Soc.* **89**, 2435–2443.
94. Lang, J. C. and Morgan, R. D. (1980). *J. Chem. Phys.* **73**, 5849–5861.
95. Lang, J. C. (1984). *Surfactants in Solution*, Vol. 1 (Mittal, Lindman ed.), pp. 35–58, Plenum, New York; (1985). *Proc. Intl. Sch. Phys. "Enrico Fermi", (Phys. Amphiphiles)*, Vol. 90, pp. 336–375.
96. Laughlin, R. G. (1978). *Advances in Liquid Crystals*, Vol. 3 (G. H. Brown ed.), pp. 41–148, Academic Press, New York.
97. Bourrel, M. and Schechter, R. S. (1988). *Microemulsions and Related Systems*, Marcel Dekker, New York.
98. Friberg, S. E. and Bothorel, P. (1987). *Microemulsions: Structure and Dynamics*. CRC Press, Boca Raton, Florida.
99. (1987). *Physics of Amphiphiles. Springer Proceedings in Physics*, Vol. 21 (D. Langevin, J. Meunier and N. Boccara eds), Springer, Heidelberg.

APPENDICES

Glossary and Definitions of Acronyms and Thermodynamic Symbols

A1.1 A glossary of surfactant phase science terms

ABS – an acronym for a salt of an "Alkyl **B**enzene **S**ulfonate" surfactant. Reference is usually made to the sodium salt of the product obtained by sulfonating a mixture of alkyl benzenes such as tetrapropylenebenzene, which is prepared by reacting the tetramer of propylene with benzene via Friedel–Crafts reaction chemistry. See *LAS*.

Acidic molecule – a molecule that is chemically reactive towards a base, via either proton transfer or addition reactions. Two different classes of acidic molecules exist: (1) Brønsted acids (AH), which bear "active" protons that are transferred to the base (B) during the reaction ($AH + B \rightarrow A^- + BH^+$), and (2) Lewis acids (e.g. BF_3), which have an incomplete outer electron shell and accept the nonbonding electron pair of Lewis bases (e.g. $(CH_3)_3N:$) to form a Lewis acid–base adduct ($(CH_3)_3N \rightarrow BF_3$). In aqueous surfactant chemistry one is concerned primarily with Brønsted acids and bases.

Alternation rule – the rule (derived from the Phase Rule) which states that an alternation between even and odd numbers of phases must occur along isothermal or isoplethal process paths within a phase diagram (Section 4.12).

Amidate – the anionic functional group that results from removal of an N–H proton from an amide group ($-CONH_2 + B \rightarrow -CONH^- + BH^+$).

Ammonio – (nomenclature) a substituent group that is a positively charged nitrogen radical (R_3N^+-).

Ammonium ion – the NH_4^+ ion molecule and its substitution derivatives.

Ammonium salt – a salt containing as the cationic partner an ammonium ion.

Amphiphilic molecule – a molecule whose structure includes both polar and nonpolar structural fragments.

Ampholytic molecule – a molecule that may react chemically with either acids or bases to form products that differ in their net charge from the original molecule. Ampholytic molecules may be either acidic only, basic only, or amphoteric (both acidic and basic).

Amphoteric molecule – a molecule that is capable of reacting both with bases (as an acid) and with acids (as a base). Examples are glycine and aluminum hydroxide.

Anionic – possessing a negative charge.

Azeotropic – not "zeotropic". An "azeotropic" phase transition is a potentially reversible phase transformation during which the compositions of the phases involved do not change. The term is ordinarily restricted to mixtures containing two or more components. Numerous examples of constant boiling azeotropic mixtures are encountered in mixed solvent systems. The azeotrope may boil either above the boiling point of the higher boiling solvent, or below the boiling point of the lower boiling solvent. The phase transition at the maximum temperature at which many liquid crystal phase regions exist may be regarded as an azeotropic transition (Section 5.6.1).

Back-bonding – conjugation of the filled nonbonding orbitals on one atom with the empty d orbitals on the adjacent bonded atom.

Basic molecule – a molecule that is capable of reacting with an acid, either by proton transfer or addition reactions (see acidic molecule). As in the case of acidic molecules, basic molecules may react either as Brønsted bases (in which case a proton is transferred to the base from the acid during the reaction) or as Lewis bases (in which case an electron pair on the base forms a covalent bond with an unsaturated atom in the Lewis acid).

Bicontinuous – continuous on both sides and separating two compartments from one another, as in "bicontinuous cubic phase" structures (Section 8.4.8).

Binary system – a system that contains two components.

Binodal – the boundary of the liquid phase along the perimeter of a liquid/liquid miscibility gap.

Birefringent – an optical term which signifies that the refractive index varies with respect to the direction that a light ray takes in passing through a sample (Section 8.2). Birefringent materials appear bright when viewed between crossed polars.

Capping group – a functional group which replaces the terminal −OH group in polyoxyethylene surfactants.

Cationic – having a net positive charge.

Chain length – the length of the longest continuous chain of carbons that exists within the lipophilic group of a surfactant. The term is ambiguous in polyfunctional surfactant molecules.

Chain-melting transition – the eutectic discontinuity in aqueous polar lipid systems at which the lamellar liquid crystal (L_α) phase is formed isothermally by reaction of a more highly ordered phase with a water-rich liquid phase the addition of heat.

Cis double bond – a 1,2-disubstituted ethylenic compound in which the two substitutents lie to the same side of the ethylenic bond.

Cloud point – the temperature at which a clear solution of a surfactant becomes cloudy, usually (but not always) on increasing the temperature.

Cmc – an acronym for "critical micelle concentration".

Coconut chain length – a mixture of straight-chain alkyl lipophilic groups that has the same distribution of homologs as found in the fatty acids of the triglycerides of coconut oil. The carbonyl carbon is usually included in counting the number of carbons – regardless of whether or not it remains in the carbonyl oxidation state. The statistically averaged chain-length is 12.5 carbons.

Coexisting phases – two or more phases which are in direct contact with one another. Coexisting phases either may (or may not) be in equilibrium, but usually the term refers to phases in equilibrium.

Cohesive energy density – a thermodynamic parameter that provides a numerical measure of the attractive forces between molecules, usually within the liquid phase.

Colloidal structure (irreversible) – that level of structure which defines the manner in which coexisting phases (within mixtures containing more than one phase) are arranged in space, relative to one another. See also colloidal structure (reversible or micellar), sol, gel.

Colloidal structure (reversible or micellar) – the structure of the reversibly formed micellar aggregates that exist in liquid solutions of soluble surfactants.

Component – any chemical compound within a mixture whose composition may be varied independently of the other components of the mixture.

Composition – the numerical values of the fractions of each of the components that are present in a mixture. Composition may be expressed either as pure fractions (quantity per unit quantity of total mixture), or as per cent fractions (quantity per 100 units of total mixture). The term "composition" is general and may be used irrespective of the phase behavior of the mixture. See also "concentration".

Composition–temperature space – a surface (or solid) whose dimensions are composition(s) and temperature.

Concentrated phase – the phase of relatively higher solute composition within a mixture containing two coexisting phases. Since water is ordinarily regarded as the solvent in aqueous mixtures, the concentrated phase is usually the phase of lower water composition.

Concentration – the composition of a mixture that is a single phase. It is preferable not to use the term "concentration" for mixtures that consist of two or more phases. The units of concentration may be mass fraction, mole fraction, mass per unit volume, moles per liter (molarity), moles per kilogram of solvent (molality), etc.

Congruent process or transition – a phase reaction in which the reactant and the product phases have the same composition.

Consolute – a term used to describe the boundary of liquid/liquid miscibility gaps. Both "lower consolute boundaries" (having a lower critical point) and "upper consolute boundaries" (having an upper critical point) are known.

Cosurfactant – an amphiphilic compound (usually not a surfactant) whose presence facilitates the formation of microemulsion phases from surfactant–oil–water mixtures.

Counterion – the small ion that is paired with the amphiphilic ion within ionic surfactant salts. The sign of the counterion is positive in anionic surfactants and negative in cationic surfactants. The counterion is usually hydrophilic, but may also be amphiphilic (as in hydrotrope ions).

Critical end point – the numerical values of composition, temperature, and pressure at which two coexisting phases become three as temperature (or pressure) is varied. At upper critical end points the transition from two to three phases occurs as the temperature is lowered, while at lower critical end points the transition from two to three phases occurs as the temperature is raised.

Critical micelle concentration – the narrow range of concentrations in liquid solutions of surfactants within which micellar aggregates first appear as the concentration is increased.

Critical opalescence – the characteristic opalescence (turbidity) that exists within phases near the critical point. Critical opalescence exists within one-phase mixtures; it is *not* the turbidity that exists when one phase is dispersed within another, as occurs beyond the temperature of cloud points.

Critical point – the numerical values of temperature, pressure, and composition at which the differences between two (or more) coexisting phases vanish and the mixture becomes one phase.

Critical pressure – the pressure at the critical point.

Critical temperature – the temperature at the critical point.

Crystal – a class of phase structures characterized by the existence of well-defined periodic structure which extends along all three dimensions in space.

Cubic liquid crystal – a class of liquid crystal structures which are structurally and optically isotropic. Several cubic phases exist which have fundamentally different structures.

Curd – a term used to describe the dense, usually fibrous, soap phases of very low water content that may separate from soap mixtures on cooling. It is now generally agreed that the phase structure within the curd phases is crystalline or at least crystal-like.

Defined – a variable whose numerical value is specified in one manner or another.

Degrees of freedom – the parameter (F) in the Phase Rule which refers to the number of thermodynamic parameters that must be specified in order to fully define the state of a mixture that is held at particular values of the system variables.

Density variables – a defined function of the field variables (Section 3.5). The density variables resemble the extensive thermodynamic variables in that their numerical value is a function of mass.

Determined – a variable whose numerical value is obtained as the result of experimental investigation.

Diglyceride – a diester of fatty acids and glycerol. Neglecting stereoisomerism, two isomeric diglycerides exist which differ in the position of the ester groups (1,2- or 1,3-). 1,2-Diglycerides, and 1,3-diglycerides that have two different fatty acid moieties, may be optically active and exist either as a racemic mixture or as an optical isomer.

Dilong chain – having two long chains (lipophilic groups).

Dilute phase – the phase of relatively lower solute composition in a mixture of two coexisting phases that differ in composition. Typically, the dilute phase is the water-rich phase.

Dispersion – a mixture of two phases in which one phase is subdivided into small, discrete particles that are distributed more or less uniformly through the other continuous phase. "Dispersion" is used as a general term; the dispersed phase and the continuous phase in dispersions may consist of any phase state (gas, liquid, liquid crystal, crystal). See also "emulsions" and "aerosols".

Dissociation constant – the constant for a reaction that occurs in solution which defines the ratio at equilibrium of the activity product of the reaction products to the activity product of the reactants. For the dissociation of an acid AH, $AH \rightleftarrows A^- + H^+$, $K_{\text{dissoc}} = [A^-] \cdot [H^+]/[AH]$, where $[i]$ represents the thermodynamic activity in solution of species i.

Divalent – having two valencies (charges, ligand bonding sites, etc.)

Elaidyl – the 1-(*trans*-9-octadecenyl) radical.

Emulsion – a dispersion of one liquid phase within another.

Equation of state – a relationship among those variables which serve to define the thermodynamic state of a mixture.

Equilibrium – the state of lowest free energy. The term "equilibrium" may refer to a wide range of phenomena, and needs to be defined with reference to a particular phenomenon (Section 4.17.1).

Equilibrium of state – the condition in which the phases present in a mixture are in equilibrium with one another at the phase level of structure.

Ethoxylog – a compound that differs in molecular structure from another compound by the insertion or removal of an oxyethylene ($-OCH_2CH_2-$) group.

Eutectic – an isothermal phase discontinuity at which three phases that differ in composition may coexist, and at which the phase of intermediate composition exists only at temperatures above that of the discontinuity.

Extensive variables – those thermodynamic variables whose numerical value is dependent on mass. Examples are free energy, enthalpy, heat capacity, etc. (Section 3.4).

Field space – a surface (or solid) whose dimensions are those of any of the field variables (temperature, pressure, chemical potential, etc.).

Field variables – a class of thermodynamic variables whose numerical value is independent of mass. Examples are temperature, pressure, chemical potentials, etc. The field variables are essentially the same as the intensive variables. The density variables, however, are defined functions of the field variables (Section 3.5), and so may differ from the extensive variables.

Fluorocarbon – an analog of a hydrocarbon molecule (or structural fragment) in which the hydrogen atoms are replaced by fluorine.

Gel – (1) a structural class of irreversible colloidal states in which the dispersed phase exists as a network that occludes the continuous phase. (2) Any mixture whose appearance and rheology is characteristic of that of gels. (3) One of several phase states often encountered within polar lipid–water systems, within which the chains are highly ordered relative to liquid crystal states as indicated by X-ray, NMR, or other data.

Hardened – a term that refers to naturally occurring fatty compounds which have been hydrogenated so as to transform the olefinic (unsaturated) long-chain compounds originally present into saturated long-chain compounds.

Hexagonal liquid crystal – a liquid crystal phase structure constructed from cylindrical elements having an indefinitely long length. The diameter of the cylinders is similar to that of the spherical micelle (roughly twice the extended length of the molecule), and they are arranged in a hexagonally close-packed array. The core of these cylindrical elements may contain either the lipophilic groups of the surfactants or oils present, or the water and hydrophilic groups. See "normal hexagonal" and "inverted hexagonal".

HLB – hydrophilic lipophilic balance. Two molecules have the same HLB when the hydophilicity of the hydrophilic group and the lipophilicity of the lipophilic group are such that they display similar surface behavior, e.g. as emulsifying agents (Section 11.4).

Homolog – a compound whose molecular structure differs from another compound by the insertion or removal of a methylene ($-CH_2-$) group.

Hydrolysis – as used herein, a chemical reaction of a compound with water that results in the transfer of a proton between water and the reacting molecule. See also "hydrolytic cleavage".

Hydrolytic cleavage – as used herein, a chemical reaction of water with a compound that results in the cleavage of a covalent bond within the compound and the formation of reaction products which incorporate the H and the OH groups of water. Examples are the hydrolytic cleavage of carboxylic or sulfuric acid esters to the corresponding alcohols and acids.

Hydronium ion – the H_3O^+ ion.

Hydrophilic – possessing an affinity for water.

Hydrophilic group – (1) a polar functional group that confers surfactant phase behavior onto a molecule that is operative as a lipophilic group (Section 9.5.1). (2) More generally, a polar functional group that has an affinity for water.

Hydrophilicity – (1) the capacity of a functional group to operate as a hydrophilic group (Section 9.3). (2) More generally, possessing an affinity for water.

Hydrotrope – a subclass of amphiphilic molecules characterized by the existence of lipophilic groups that are too small for aqueous mixtures of the molecule to display surfactant behavior, but too large for the molecule to be regarded as hydrophilic.

Immiscible components – components whose mixtures do not exist as a single phase. This term is preferably used with respect to specific values or ranges of values of the system variables (composition, temperature, and pressure), since in many systems components are miscible under some conditions but not under others.

Incongruent – a phase reaction in which the reactant and the product phases do not have the same composition(s), such as occurs at eutectic and peritectic discontinuities.

Infinitely miscible – a (less preferred) term which signifies that two (or more) components are miscible in all proportions.

Intensive variables – a class of thermodynamic variables whose numerical value is independent of mass. See also extensive variables, field variables, and density variables.

Interface – the surface at which two coexisting phases come together, and which separates one phase region from the other (Section 4.18).

Inverted hexagonal liquid crystal – the hexagonal liquid crystal structure in which the core of the cylindrical structural elements are occupied by the hydrophilic groups of the surfactant molecules, together with the associated water molecules.

Ion – a molecule that bears one or more positive or negative charges.

Ion exchange – the exchange of partners among two or more ionic compounds. Ion exchange may occur either as a physical process within a phase, or as a chemical process if it is accompanied by phase separation and results in the formation of new ionic compounds (Section 2.2).

Ion exchange reaction – an ion exchange process that results in the formation of new chemical compounds. Ion exchange reactions are invariably accompanied by the precipitation of a reaction product as a separate phase.

Ionic – having a nonzero net positive or negative charge or charges.

Ionic compound – a chemical compound which consists of an electrically neutral set of two or more ionic molecules. Synonymous with the generic words "salt" and "electrolyte".

Isoelectronic molecules – molecules that possess exactly the same number of electrons (and protons).

Isopleth – a line on a phase diagram (binary or ternary) along which composition remains constant as temperature is varied.

Isoplethal – an adjective describing those process paths which follow an isopleth.

Isotherm – in binary diagrams, a line on the diagram along which composition remains constant as temperature is varied. In ternary diagrams, a plane surface of constant temperature but variable composition. In quaternary diagrams, a three-dimensional figure of constant temperature but variable composition.

Isothermal – an adjective describing those process paths which fall within an isotherm.

Isotropic – (1) an optical term which signifies that the value of the refractive index is independent of the direction a light ray takes in passing through a material (see also "nonbirefringent"). Isotropic materials appear dark when viewed between crossed polars. (2) A structural term signifying that a structure is identical along any three orthogonal directions in space.

Krafft boundary – the crystal solubility phase boundary in surfactant phase diagrams.

Krafft point – one of several arbitrarily defined points along the Krafft boundary (Section 5.4).

Lamellar liquid crystal – a class of liquid crystal phase structures within which the bilayer is the repeating structural element (Section 8.4.6).

LAS – "linear alkylate sulfonate". A salt (often sodium) of the mixture of alkyl-benzene sulfonate surfactants that results from the sulfonation of "linear alkylates". Linear alkylates result from the alkylation of benzene with a mixture of straight chain olefins or paraffin derivatives. LAS is therefore primarily a mixture of isomers in which the position of the benzene ring on the straight side chain is more or less random (except that no 1-isomer exists).

Lecithin – a polar lipid having a trimethylammonioethylphosphate zwitterionic hydrophilic group and a diglyceride lipophilic group.

Leveling effect – the property of a solvent of restricting the range of acidities or basicities that can be measured in it. Leveling is the consequence of reactions of the solvent itself as an acid or a base. The term is also relevant to the upper limit of hydrophilicity (Section 10.7.3).

Lever rule – the equation used to calculate the ratio of phases that coexist in a mixture that lies within a multiphase region. In ternary systems the lever rule applies equally well to two-phase and three-phase mixtures, but has a somewhat different form in these two cases (Section 12.1.1.6).

Lipophilic – having an affinity for nonpolar or oily phases.

Lipophilic group – the nonpolar or oil-loving structural fragment of an amphiphilic molecule.

Liquid – a class of phase structures which is characterized by the absence of long range periodic order along any three orthogonal directions in space.

Liquid crystal – a class of phase structures which is characterized by the existence of a significant degree of long-range order along at least one direction in space, while at the same time being disordered relative to crystal phases.

Lower consolute boundary – a liquid/liquid miscibility gap having a lower critical point.

Lyothermotropic liquid crystal – a liquid crystal phase which depends for its existence not only upon the presence of a solvent, but also exists only within a particular range of temperatures.

Lyotropic liquid crystal – a liquid crystal phase which depends for its existence upon the presence of a solvent.

Lysopolar lipid – the monolong chain surfactant that results from the hydrolytic cleavage of one long chain group from a polar lipid molecule.

Mass fraction – the fraction of a component in a mixture expressed in units of mass.

Mesogenic solvent – a solvent that is capable of inducing the formation of meso-morphic or liquid crystal phases by interaction with a particular solute.

Mesomorphic phase – a synonym for liquid crystal phase.

Metathesis – a synonym for the "ion exchange" chemical reaction. The term may be used more broadly as well, as during the "metathesis" of olefins.

Microemulsion – "a system (mixture) of water, oil, and amphiphile which is a single optically isotropic and thermodynamically stable liquid solution" (Danielsson, I. and Lindman, B. (1981). *Colloids and Surfaces* 3, 391–392).

Middle – a synonym for the normal hexagonal liquid crystal phase.

Miscible components – components whose mixtures may exist as a single phase. The property of being "miscible or immiscible" usually refers to specific values (or a range of values) of the system variables (composition, temperature, and pressure).

Miscibility gap – an area in a phase diagram within which no equilibrium phase exists.

Mixture – a general term for a sample of defined composition, irrespective of the conditions under which the mixture is held or its phase behavior under these conditions.

Mole – Avogadro's number (6.022×10^{23}) of molecules of a compound. For ionic compounds, one mole is Avogadro's number of all of the ionic molecules that make up the compound ($NaCl$, $MgCl_2$, etc.).

Mole fraction – the fraction of a component in a mixture expressed in units of moles.

Molecular structure – that level of structure which is defined by the manner in which the atoms within a molecule are connected to one another (the "connectivity").

Neat liquid crystal – a synonym for the lamellar liquid crystal phase.

Newtonian – a mixture that follows Newton's law of ideal rheological behavior. In Newtonian fluids the shear stress is proportional to the shear strain; the constant of proportionality is the viscosity.

Nonionic molecule – a molecule that is electrically neutral (bears no net formal charge).

Nonpolar molecule – a molecule that is lacking (or relatively weak) in polarity, as reflected by the absence of substantial charge separation within the bonds or between groups within the molecule.

Normal – (1) (mathematics) a line that is perpendicular to a surface. (2) (nomenclature) An alkyl group having a straight and unbranched chain that is substituted at the terminal position.

Normal hexagonal phase – a liquid crystal having the hexagonal structure in which the lipophilic groups are found within the core of the cylindrical structural elements.

Oil – a compound that is nonpolar or contains only weakly polar functional groups, such as hydrocarbons, fluorocarbons, silicones, triglycerides, waxes, etc. While the term usually refers to such compounds in the liquid phase, it is also used on occasion irrespective of the phase state.

Oil-rich – the phase of larger oil composition among coexisting phases.

Oleoyl – the *cis*-9-octadecenoyl acyl group.

Oleyl – the 1-(*cis*-9-octadecenyl) alkyl group.

Onio – a generic suffix that is substituted for "ium" in naming a positively charge substituent radical. Examples are "phosphonio-" and "ammonio-" (which see).

Orthogonal – as used herein, two directions are orthogonal to one another if the vector component in one direction is zero in the direction of the other. Two lines at right angles, or the three axes of Cartesian coordinates, are orthogonal to one another. The boundaries of triangular ternary diagrams may *not* be orthogonal.

Peritectic – an isothermal phase discontinuity in binary systems at which three phases that differ in composition may coexist, and at which the phase of intermediate composition exists only at temperatures below that of the discontinuity.

Phase – a volume element of a mixture within which a smooth variation in the thermodynamic density variables exists (Section 4.18). The spatial gradient in the density variables is zero within a phase in equilibrium.

Phase behavior – the various aspects of the phase science of a mixture. Phase behavior is defined by stating the number of phases present, their compositions, and their structures.

Phase boundary – (1) in a phase diagram, a line or surface which defines the limits of composition, temperature, or pressure within which a particular phase exists. (2) In a mixture, the physical surface at which two different phase regions come together, and which separates one phase from the other.

Phase prism – the triangular cross-section prism used to portray the phase behavior of ternary mixtures. Typically temperature is scaled along the axis of the prism, while compositions are scaled within the triangular plane cross-sections. Phase prisms may be either equilateral or right-triangular in cross-section.

Phase reaction – a physical process during which one phase reacts to form another phase, either by itself or by interaction with still another phase. Phase structure changes during phase reactions, while molecular structure does not.

Phase Rule – the law of chemistry which defines the relationship at equilibrium between the number of components, C, the number of phases, P, and the degrees of freedom, F, of a particular mixture under specified conditions. The Phase Rule equation is $P + F = C + 2$.

Phase structure – the manner in which molecules are arranged in space within a phase.

Phase transition – a process during which the phase state of a mixture undergoes a change. The term usually refers to discontinuous changes.

Phosphate – a phosphinyl (P → O) compound having three oxygen radicals attached to the phosphinyl group. The parent acid is phosphoric acid. The term applies to both esters and salts having these structural features.

Phosphinate – a phosphinyl (P → O) compound having two carbon or hydrogen and one oxygen radicals attached to the phosphinyl group. The parent acid is a phosphinic acid. The term applies to both esters and salts having these structural features.

Phosphine oxide – a phosphinyl (P → O) compound having three carbon or hydrogen radicals attached to the phosphinyl group.

Phosphinyl – a generic name for the trivalent ≡P → O radical.

Phosphonate – a phosphinyl (P → O) compound having one carbon or hydrogen and two oxygen radicals attached to the phosphinyl group. The parent acid is a phosphonic acid. The term applies to both esters and salts having these structural features.

Pivotal phase behavior – a characteristic kind of phase behavior that may exist when interference between the liquid/liquid miscibility gap and liquid crystal regions occurs (Section 5.10.7).

pKa – the negative logarithm (to the base 10) of the dissociation constant of an acid, K_a. $pKa = -\log(K_a)$.

pKb – the negative logarithm of the hydrolysis constant of a base. The sum of the pKa plus the pKb equals the negative logarithm of the autolysis constant of water, which at 25°C is $c.\,14$. Therefore, $pKb = 14 - pKa$ at this temperature.

Plait point – the coordinates of composition along a biphasic miscibility gap boundary (binodal), within composition–composition space in a ternary diagram, at which the differences between the two coexisting phases vanish. The plait point is analogous to the critical point in binary systems.

Plastic crystals – a subclass of crystal-like phases that is characterized by the existence of substantial disorder within the structure. A difference between plastic and liquid crystals is that the disorder extends along all three dimensions in space in the former, whereas a high level of order typically exists along one or more dimensions within liquid crystals.

POE – polyoxyethylene $-(CH_2CH_2O)_n-$.

Polar lipids – (1) originally, those lipids found in living tissue that were not extractable using the lower alkanes but were extractable using more polar solvents such as chloroform or chloroform-methanol mixtures. (2) Presently, a large and diverse group of (usually) water-insoluble dilong chain surfactant found in the membranes of living cells.

Polyfunctional – containing more than one functional group.

Polytectic – an isothermal phase discontinuity in binary systems at which two of three phases that may coexist have the same composition but differ in structure.

Poor – as used in phase science, that phase within a multiphase mixture in which the composition of a particular component is low relative to its composition in the coexisting phase; e.g. "water-poor".

POP – polyoxypropylene $-(CH_2CH(CH_3)O)_n-$.

Proximate substituent – a substituent in a surfactant molecule that exists on or near the principal (strongest) hydrophilic group.

Remote substituent – a subsituent in a surfactant molecule that exists within the lipophilic group at a site that is far from the hydrophilic group.

Rheology – the science of the flow of materials that occurs in response to application of a shearing force.

Rich – as used in phase science, that phase within a multiphase mixture in which the composition of a particular component is high relative to composition in the coexisting phase, e.g. "water-rich".

Salt – (1) a common name for the compound sodium chloride. (2) A generic term for ionic compounds.

Saturated – incapable of forming additional chemical bonds, either via chemical reactions or by overlap with electron pairs on adjacent atoms.

Separating phase – that phase which separates from solution when the solubility is exceeded.

Silicone – a generic name for a class of (usually polymeric) molecules having the $-O-Si(R_2)-$ repeating structural unit.

Snell's Law – the law of optics which defines the relationship between the angles (relative to the normal to the interface) of rays that pass from one medium through an interface into another of differing refractive index,

Soft tallow – a mixture of fatty chains that has the same distribution of chain lengths found in the triglycerides of natural animal tallow fat, and which has not been hydrogenated (hardened). Soft tallow lipophilic groups contain significant amounts of unsaturated lipophilic groups (principally, oleyl or 18:1). The partial hydrogenation of soft tallow mixtures eliminates most of the linoleyl (18:2) and linolenyl (18:3) polyunsaturated chains; such mixtures are said to have been "touch-hardened".

Sol – a class of irreversible colloidal structures within which the dispersed phase exists as discrete particles. See "*gel*".

Solubility – the composition of a solute in the solvent phase at the limit of the capacity of this phase to dissolve the solute.

Solubility boundary – the dilute boundary of the first miscibility gap encountered as solute is added to solvent.

State – the quantitative thermodynamic dimensions of a mixture. Defining state includes consideration of the number of phases present, their composition and structure, and the numerical values of their thermodynamic variables.

Stearoyl – the octadecanoyl group $(C_{17}H_{35}C(O)-)$.

Stearyl – the 1-*n*-octadecyl group $(C_{18}H_{37}-)$.

Steric effects – those effects that a substituent group has on the physical or chemical properties of a molecule which result from the space that the group occupies (as contrasted with its polarity).

Steric stabilization – the non-electrolytic mechanism of stabilizing colloidal structure that depends upon the existence of (usually large) groups that are attached to the surface of the dispersed phase and extend into the surrounding medium.

Structural surface (plane) – a recognizable surface of atoms or molecules that exists within a structured phase (Section 4.19).

Surface active – capable of modifying the mechanical properties of an interface.

Surfactant – an amphiphilic compound that displays a characteristic aqueous phase behavior. The term is a contraction of the phrase "surface active agent". All surfactants are surface active (as defined above), but so are many nonsurfactants (Section 11.2.1).

Surfactant-rich – the phase of larger surfactant composition among coexisting phases.

System – the number and identities (molecular structures) of the components that are under consideration with respect to an analysis of phase behavior.

System variables – those variables which define the conditions under which a mixture exists. The system variables must be specified in order to define the state of a mixture. Typically, they include temperature, pressure, and composition(s). They may be defined by the investigator, and are therefore independent variables.

Tactile – an adjective referring to the sense of feel and touch.

Tallow – a mixture of homologs in which the distribution of chain-lengths is the same as that found in the fatty acids in the triglycerides in animal tallow fat. The statistically averaged chain-length is 17.3 carbon atoms.

Ternary system – a system containing three components.

Tether – the divalent (usually saturated) radical that connects the formally charged substituent groups within zwitterionic hydrophilic groups. Often (but not always) the tether is a chain of polymethylene groups $(-(CH_2)_n-)$.

Thermotropic liquid crystal – a liquid crystal state that is formed solely by the action of thermal energy (or of electrical or magnetic fields) on a compound. The presence of a solvent is not required in order for thermotropic liquid crystal phases to exist.

Three-phase body – a solid figure, such as may be found within ternary surfactant–oil–water phase prisms, within which three coexisting phases exist.

Tie-line crossing – a phenomenon that occurs during a mixing process which does not follow a tie-line within a two-phase miscibility gap. Complex disproportionation processes, and considerable mass transport within phases and across interfaces, are required to maintain equilibrium when tie-line crossing occurs.

Tie-lines – the lines in a phase diagram which define the compositions of coexisting phases.

Tie-triangles – the (usually irregular) triangles found within ternary phase diagrams which define the compositions of three coexisting phases.

***Trans* double bond** – a 1,2-disubstituted ethylenic compound in which the two substituents are on opposite sides of the ethylenic bond.

Transport – the movement of a quantity (mass, heat, etc.) from one point in space to another.

Triglyceride – a *tris*(fatty acid) ester of glycerol.

Ultralong chain – a distinctive group of surfactants that have lipophilic chain lengths of C_{20} or higher and are water-soluble near room temperature (Section 10.5.3.1).

Unary system – a system containing one component.

Upper consolute boundary – the boundary of a liquid/liquid miscibility gap that displays an upper critical point.

Variance – a synonym for degrees of freedom.

Viscosity – see Newtonian.

Viscous isotropic phase – a synonym for cubic liquid crystal phases (rarely used at present).

Water-rich – the phase of larger water composition among coexisting phases.

Wax – (1) An ester of a fatty alcohol and a fatty acid. (2) A material which has properties that resemble those of waxes.

Zeotropic transition – a phase transition in which the reactant and the product phases do not have the same composition. Normal distillation processes, during which the more volatile component is separated from the less volatile, constitute examples of zeotropic processes. See also "*azeotropic*".

Zwitterionic – a molecule that is electrically neutral (and therefore nonionic) but contains formally charged substituent groups.

A1.2 Acronyms used in the text

For a listing and definition of the various symbols used to designate phase structures, see Table 8.1.

AE_x – an alkylpolyoxyethylene oxide surfactant having x oxyethylene groups. See also C_nE_x.

Aerosol – a trade name for a class of surfactants which are diesters of sodium sulfosuccinate (see AOT).

AES – the ammonioethanesulfonate zwitterionic group, e.g. $-N^+(CH_3)_2CH_2CH_2SO_3^-$.

$AESO_4$ – the ammonioethyl sulfate group, e.g. $-N^+(CH_3)_2CH_2CH_2OSO_3^-$

AOT – "Aerosol OT", a trade name for sodium bis(2-ethylhexyl) sulfosuccinate, Na^+, $(C_4H_9CH(C_2H_5)CH_2O_2CCH_2CH(SO_3^-)CO_2CH_2CH(C_2H_5)C_4H_9)$.

$APSO_3$ – the ammoniopropanesulfonate group, e.g. $-N^+(CH_3)_2CH_2CH_2CH_2SO_3^-$.

$APSO_4$ – the ammoniopropyl sulfate zwitterionic group, e.g. $-N^+(CH_3)_2CH_2CH_2CH_2OSO_3^-$.

BaAOT – the barium salt of AOT.

C – a symbol for the cubic liquid crystal phases.

CaAOT – the calcium salt of AOT.

C_nAA – an alkyldimethylammonioacetate having n carbons in the alkyl group $C_nH_{2n+1}N^+(CH_3)_2CH_2CO_2^-$.

C_nAH – an alkyldimethylammoniohexanoate having n carbons in the alkyl group $C_nH_{2n+1}N^+(CH_3)_2(CH_2)_5CO_2^-$.

C_nAO – an alkyldimethylamine oxide having n carbons in the alkyl group $C_nH_{2n+1}N(\rightarrow O)(CH_3)_2$.

C_nAPS – an alkyldimethylammoniopropane sulfonate having n carbons in the alkyl group $C_nH_{2n+1}N^+(CH_3)_2(CH_2)_3SO_3^-$.

C_nE_x – a polyoxyethylene glycol monoalkyl ether having n carbon atoms in the lipophilic group and x oxyethylene groups, $C_nH_{(2n+1)}(OCH_2CH_2)_xOH$. The acronym C_iE_j is often used as well.

CO_2K – a potassium carboxylate.

CO_2Na – a sodium carboxylate.

C_nPO – an alkyldimethylphosphine oxide having n carbons in the alkyl group, $C_nH_{2n+1}P(\rightarrow O)(CH_3)_2$.

CTAB – cetyl(hexadecyl)trimethylammonium bromide, $C_{16}H_{33}N(CH_3)_3^+, Br^-$.

D – (1) a symbol for the lamellar liquid crystal phase. (2) The Debye unit of bond polarity.

DIT – the "diffusive interfacial transport" phase studies method.

ΔT – a phase criterion for the existence of hydrophilicity (Section 9.5.1).

DOAC – dioctadecylammonium chloride $(C_{18}H_{37})_2NH_2^+, Cl^-$.

DOACS – dioctadecylammonium cumenesulfonate, $(C_{18}H_{37})_2NH_2^+$, $p-(i-C_3H_7)C_6H_4SO_3^-$.

DODMAC – dioctadecyldimethylammonium chloride, $(C_{18}H_{37})_2N(CH_3)_2^+$, Cl^-.

DOMAC – dioctadecylmethylammonium chloride, $(C_{18}H_{37})_2N(CH_3)H^+$, Cl^-.

DPPC – dipalmitoylphosphatidyl choline, $C_{15}H_{31}CO_2CH_2(C_{15}H_{31}CO_2) \cdot CHCH_2OPO_2^- CH_2CH_2N^+(CH_3)_3$.

DSPC – distearoylphosphatidyl choline, $C_{17}H_{35}CO_2CH_2(C_{17}H_{35}CO_2)CHCH_2OPO_2^- \cdot CH_2CH_2N^+(CH_3)_3$.

E – a symbol for the hexagonal liquid crystal phase.

esu – the electrostatic unit of charge.

Et – ethyl.

EtPO – ethyl phosphine oxide.

F – a symbol for the inverted hexagonal liquid crystal phase.

HLB – hydrophilic lipophilic balance.

HOc – octanoic acid, $C_7H_{15}CO_2H$.

L – a general symbol for liquid phases.

LX – a general symbol for liquid crystal phases.

Me – methyl.

MePO – methylphosphine oxide.

MgAOT – the magnesium salt of AOT.

μ_r – the reduced bond moment (Section 9.4).

NaL – sodium laurate Na^+, $C_{11}H_{23}CO_2^-$.

NaM – sodium myristate Na^+, $C_{13}H_{27}CO_2^-$.

NaOc – sodium octanoate (caprylate) Na^+, $C_7H_{15}CO_2^-$.

NaP – sodium palmitate Na^+, $C_{15}H_{31}CO_2^-$.

NaS – sodium stearate Na^+, $C_{17}H_{35}CO_2^-$.

NMR – nuclear magnetic resonance.

NNC=O – the ammonioamidate group, $RC(=O)N^-N^+R_3'$.

Ph – a symbol for "phenyl".

PO – phosphine oxide.

PPSO$_3$ – a phosphoniopropanesulfonate, $C_nH_{2n+1}P^+(CH_3)_2(CH_2)_3SO_3^-$.

R – a generic symbol for a lipophilic (alkyl, alkylaryl, etc.) substituent group.

Rf – the retention fraction of a spot during chromatography. The Rf is defined as the ratio of the distance that the sample moves divided by the distance that the solvent front moves.

SDI – the sulfone diimine $(-S(\rightarrow NH)_2-)$ functional group.

SDS – sodium n-dodecyl sulfate, $C_{12}H_{25}OSO_3^-, Na^+$.

SMI – a sulfone monoimine, or sulfoximine.

SO – a sulfoxide.

SO$_3$Na – a sodium sulfonate.

$v/a_o l$ – the packing parameter. v is the molecular volume, a_o is the area per molecule, and l is the extended length of the molecule (Section 8.8.1).

X – (1) a generic term for the dry crystal phase of a compound. (2) The polar functional group in the generic formula for amphiphilic compounds, RX. R is the lipophilic group.

$X \cdot W_n$ – a crystal hydrate of the compound, X, having n water molecules.

A1.3 Thermodynamic symbols used in the text

For symbols used in the context of describing phase behavior, as a general rule subscripts are numbers and designate components, while superscripts are letters (Greek or Latin) and designate phases.

A – Helmholtz free energy.

a_i – the thermodynamic activity of component i.

a_W – the thermodynamic activity of water (the partial pressure of water divided by the vapor pressure of pure water).

β – compressibility (Section 3.4).

C – number of components, in the Phase Rule.

cal – calory.

C_p – heat capacity at constant pressure.

C_V – heat capacity at constant volume.

$°C$ – degrees Celsius.

Δ – the final numerical value of a variable minus the initial value.

E – energy.

F – variance (or degrees of freedom), in the Phase Rule.

H – enthalpy.

G – Gibbs free energy.

G^0 – standard state Gibbs free energy.

G_r – reduced free energy of mixing (Section 3.11).

G_t – total Gibbs free energy.

h_i – field variable of component i in the Griffiths and Wheeler analysis (Section 3.5).

H_0 – the Hammett acidity function for strong acid media.

J – Joule.

k – the Boltzman constant (per molecule).

K – degrees Kelvin ($°C + 273.13$).

K_H – the Henry's Law constant, a_i/x_i.

kcal – kilocalory.

kJ – kilojoule.

m – as a superscript, signifying metastable.

μ_i – the actual chemical potential of component i.

μ_i^0 – standard state chemical potential of component i.

n – (1) number of moles present. (2) Refractive index.

$n_S \mu_S$ – the "surfactant free energy" contribution to the total free energy.

$n_W \mu_W$ – the "water free energy" contribution to the total free energy.

p – absolute pressure.

p_i – partial vapor pressure of component i in a mixture.

p_i^0 – vapor pressure of pure component i.

P – number of phases, in the Phase Rule.

Pa – Pascal.

% – per cent fraction of component i (mass of i per 100 unit masses of total mixture). Also used to designate the per cent fraction of a particular phase in a mixture.

pKa – the negative logarithm (base 10) of the dissociation constant of an acid.

q – heat.

R – the gas constant (per Avogadro's number of molecules).

S – entropy.

T – absolute temperature.

Torr – unit of pressure, approximately equal to 1 mm mercury.

V – volume.

w – work.

w^j – weight fraction of phase j in a mixture.

w_i – weight fraction of component i in a mixture.

w_i^j – weight fraction of component i (mass of i per unit mass of total mixture) in phase j.

x_i – mole fraction of component i in a mixture.

Literature References to Binary and Ternary Phase Studies

A2.1 Introduction

A listing of the surfactant phase diagram literature is presented in this appendix. A few relevant nonsurfactant diagrams are also listed, including all those found in the text plus several diagrams of nonsurfactant amphiphilic compounds deemed to be of possible interest. The figures of those systems for which diagrams are found in the text may be located by the figure number. Binary diagrams are listed first (A2.2) and then ternary (A2.3). The list of nonaqueous surfactant diagrams does *not* reflect a comprehensive literature search.

Compounds within the lists of surfactants are arranged according to the structural classification scheme for hydrophilic groups (Section 9.7). The classes are arranged in order of decreasing hydrophilicity (anionic, cationic, zwitterionic, semipolar, and single bond). Within each subclass the compounds are arranged in order of *increasing* hydrophilicity. For example, the low oxyethylene nonionic surfactants are listed before the high. Simple hydrophilic groups are listed before complex ones, and where the hydrophilic group is polyfunctional the compound is classified and listed under the strongest hydrophilic group. Zwitterionic compounds are regarded as a monofunctional hydrophilic group, rather than being listed under the anionic subclass.

When more than one homolog has been studied, the homologs are arranged in order of increasing lipophilic group chain length. Lipophilic groups in branched compounds are listed in order of the longest uninterrupted chain of carbon atoms. In the case of alkylbenzene derivatives and related compounds, the benzene ring carbons are included in counting the number of carbons in the lipophilic group. Nonpolar substituents on the lipophilic chain are not counted, but treated as substituents. Saturated compounds are listed before unsaturated. Difficult decisions had to be made in the case of polyfunctional compounds having polar substituents, where the effective lipophilic group chain length is truncated by the substituent group. For purposes of indexing, the simple approach of ignoring the substituent group has been taken. Carbons found between strong hydrophilic groups (including hydroxy and ether groups) or within the tether of zwitterionic groups are not included in the lipophilic group. Double chain compounds are listed according to the length of the longest individual chain (didodecyl will be found among dodecyl compounds). Fluorinated compounds are listed after the hydrocarbon analogs having the same chain length.

In the binary listing only the surfactant component is listed if the diagram is that of the aqueous system, but both components are listed for nonaqueous systems. Each listing spans two (or more) lines. In the first line the number, Figure number in the text, common abbreviations, chemical name, compilation reference, and original reference are listed. The second line contains the chemical formula in line format.

A2.2   Binary systems

No.	Fig.	Abbrev.	Chemical name and formula	Compilation reference	Original reference

Ionic class

Anionic subclass

C_6

| 1 | — | — | Sodium 2-ethylhexyl sulfate | [1] | [11] |

Na^+, $C_4H_9CH(C_2H_5)CH_2OSO_3^-$

| 2 | 10.7 | AOT | Sodium bis(2-ethylhexyl) sulfosuccinate | [1] | [110] |

Na^+, $C_6H_{13}CH(C_2H_5)CH_2OC(O)CH_2CH(SO_3^-)CO_2CH_2CH(C_2H_5)C_6H_{13}$

| 3 | 10.7 | MgAOT | Magnesium bis(2-ethylhexyl)-sulfosuccinate-D_2O | | [67] |

Mg^{++}, $(C_6H_{13}CH(C_2H_5)CH_2OC(O)CH_2CH(SO_3^-)CO_2CH_2CH(C_2H_5)C_6H_{13})_2$

| 4 | 10.7 | CaAOT | Calcium bis(2-ethylhexyl)-sulfosuccinate-D_2O | | [67] |

Ca^{++}, $(C_6H_{13}CH(C_2H_5)CH_2OC(O)CH_2CH(SO_3^-)CO_2CH_2CH(C_2H_5)C_6H_{13})_2$

| 5 | 10.7 | BaAOT | Barium bis(2-ethylhexyl)-sulfosuccinate-D_2O | | [66] |

Ba^{++}, $(C_6H_{13}CH(C_2H_5)CH_2OC(O)CH_2CH(SO_3^-)CO_2CH_2CH(C_2H_5)C_6H_{13})_2$

C_7

| 6 | — | EA C_8 | Ethanolammonium octanoate | | [125] |

$HOCH_2CH_2NH_3^+$, $C_7H_{15}CO_2^-$

| 7 | — | TEA C_8 | Triethanolammonium octanoate | | [63,125] |

$(HOCH_2CH_2)_3NH^+$, $C_7H_{15}CO_2^-$

| 8 | — | — | Octylammonium octanoate | | [62] |

$C_8H_{17}NH_3^+$, $C_7H_{15}CO_2^-$

| 9 | — | — | Lithium perfluorooctanoate | | [72] |

Li^+, $C_7F_{15}CO_2^-$

C_8

| 10 | — | — | Perfluorononanoic acid | | [38] |

H^+, $C_8F_{17}CO_2^-$

| 11 | — | — | Lithium perfluorononanoate | | [38] |

Li^+, $C_8F_{17}CO_2^-$

| 12 | — | — | Sodium perfluorononanoate | | [38] |

Na^+, $C_8F_{17}CO_2^-$

13	—	—	Cesium perfluorononanoate Cs^+, $C_8F_{17}CO_2^-$	[38,12]
14	—	—	Ammonium perfluorononanoate NH_4^+, $C_8F_{17}CO_2^-$	[38,108]
15	—	—	Ethylammonium perfluorononanoate $C_2H_5NH_3^+$, $C_8F_{17}CO_2^-$	[38]
16	—	—	Dimethylammonium perfluorononanoate $(CH_3)_2NH_2^+$, $C_8F_{17}CO_2^-$	[38]
17	—	—	Tetramethylammonium perfluorononanoate $(CH_3)_4N^+$, $C_8F_{17}CO_2^-$	[38,51,108]

C_{10}

18	5.6		Sodium decanesulfonate Na^+, $C_{10}H_{21}SO_3^-$	[130]

C_{11}

19	—	NaL	Sodium dodecanoate (laurate) Na^+, $C_{11}H_{23}CO_2^-$	[1] [94,90,89,91]
20	—	KL	Potassium dodecanoate (laurate) K^+, $C_{11}H_{23}CO_2^-$	[1] [99]
21	—	—	Ethanolammonium dodecanoate $HOCH_2CH_2NH_3^+$, $C_{11}H_{23}CO_2^-$	[125]
22	—	—	Triethanolammonium dodecanoate $(HOCH_2CH_2)_3NH^+$, $C_{11}H_{23}CO_2^-$	[63,125]
23	—	—	*Tris*(2-hydroxypropyl)ammonium dodecanoate $(CH_3CH(OH)CH_2)_3NH^+$, $C_{11}H_{23}CO_2^-$	[125]
24	10.8	—	Tetramethylammonium dodecanoate $(CH_3)_4N^+$, $C_{11}H_{23}CO_2^-$	[59]
25	10.8	—	Tetrabutylammonium dodecanoate $(C_4H_9)_4N^+$, $C_{11}H_{23}CO_2^-$	[59]
26	—	—	Sodium-2,2-dibutyldodecanoate Na^+, $C_{10}H_{21}C(C_4H_9)_2CO_2^-$	[65]
27	—	—	Dodecylammonium dodecanoate $C_{12}H_{25}NH_3^+$, $C_{11}H_{23}CO_2^-$	[61]

C_{12}

28	5.4	SDS	Sodium dodecyl sulfate Na^+, $C_{12}H_{25}OSO_3^-$	[73]
29	—	—	Dodecylammonium dodecylsulfate $C_{12}H_{25}NH_3^+$, $C_{12}H_{25}OSO_3^-$	[60]
30	—	—	Dodecyltrimethylammonium dodecylsulfate $C_{12}H_{25}N(CH_3)_3^+$, $C_{12}H_{25}OSO_3^-$	[60]
31	—	—	Dodecylethyldimethylammonium dodecylsulfate $C_{12}H_{25}N(CH_3)_2(C_2H_5)^+$, $C_{12}H_{25}OSO_3^-$	[61]

32	10.6	—	Dodecanesulfonic acid H^+, $C_{12}H_{25}SO_3^-$	[1]	[116]
C_{13}					
33	—	NaM	Sodium tetradecanoate (myristate) Na^+, $C_{13}H_{27}CO_2^-$	[1]	[92,93,94]
34	—	—	Potassium tetradecanoate (myristate) K^+, $C_{13}H_{27}CO_2^-$	[1]	[99]
35	—	—	Triethanolammonium tetradecanoate $(HOCH_2CH_2)_3NH^+$, $C_{13}H_{27}CO_2^-$		[122]
C_{15}					
36	3.2	NaP	Sodium hexadecanoate (palmitate) Na^+, $C_{15}H_{31}CO_2^-$	[1,2,3]	[117]
37	—	—	Potassium hexadecanoate (palmitate) K^+, $C_{15}H_{31}CO_2^-$	[1]	[99]
C_{17}					
38	—	NaS	Sodium octadecanoate (stearate) Na^+, $C_{17}H_{35}CO_2^-$	[1]	[100]
39	—	—	Potassium octadecanoate (stearate) K^+, $C_{17}H_{35}CO_2^-$	[1]	[99]
40	—	NaOl	Sodium oleate–water Na^+, $c\text{-}C_8H_{17}CH=CHC_7H_{14}CO_2^-$	[1]	[118]
41	—	KOl	Potassium oleate–water K^+, $c\text{-}C_8H_{17}CH=CHC_7H_{14}CO_2^-$	[1]	[99]
42	—	—	Ethanolammonium oleate–water $HOCH_2CH_2NH_3^+$, $c\text{-}C_8H_{17}CH=CHC_7H_{14}CO_2^-$		[125]
43	—	—	Triethanolammonium oleate $(HOCH_2CH_2)_3NH^+$, $c\text{-}C_8H_{17}CH=CHC_7H_{14}CO_2^-$		[122]
44	—	—	4-Hydroxy-2-methyl-2- pentylammonium oleate $CH_3CH(OH)CH_2C(CH_3)_2NH_3^+$, $c\text{-}C_8H_{17}CH=CHC_7H_{14}CO_2^-$	[1]	[46]
C_{19}					
45	—	—	Triethanolammonium eicosanoate–water $(HOCH_2CH_2)_3NH^+$, $C_{19}H_{39}CO_2^-$		[125]
C_{21}					
46	—	—	Triethanolammonium docosanoate–water $(HOCH_2CH_2)_3NH^+$, $C_{21}H_{43}CO_2^-$		[63]
47	—	—	Triethanolammonium erucate (*cis*-13-docosenoate) $(HOCH_2CH_2)_3NH^+$, $c\text{-}C_8H_{17}CH=CHC_{11}H_{22}CO_2^-$		[125]

Miscellaneous

48 — — Disodium chromoglycate [20]
1,3-bis(2-carboxychromon-5-yloxy)-
2-hydroxypropane, disodium salt

Cationic subclass

C_{10}

49 — — Decylammonium chloride [5] [109]
$C_{10}H_{21}NH_3^+$, Cl^-

50 — — Didecyldimethylammonium [126]
bromide
$(C_{10}H_{21})_2N(CH_3)_2^+$, Br^-

C_{12}

51 5.11 — Dodecylammonium chloride [1,5] [13]
$C_{12}H_{25}NH_3^+$, Cl^-

52 — — Dodecylmethylammonium [1,5] [13]
chloride
$C_{12}H_{25}N(CH_3)H_2^+$, Cl^-

53 — — Dodecyldimethylammonium [1,5] [13]
chloride
$C_{12}H_{25}N(CH_3)_2H^+$, Cl^-

54 — — Dodecyltrimethylammonium [1,5]
chloride
$C_{12}H_{25}N(CH_3)_3^+$, Cl^-

55 — — Didodecyldimethylammonium [126]
bromide
$(C_{12}H_{25})_2N(CH_3)_2^+$, Br^-

56 10.9 — 1,26-Bis(dodecyldimethylammonio)- [5]
3,6,9,12,15,18,21,24-octaoxahexa-
cosane dichloride
$C_{12}H_{25}(CH_3)_2N^+(CH_2CH_2O)_8CH_2CH_2N^+(CH_3)_2C_{12}H_{25}$, $2Cl^-$

57 10.9 — 1,26-Bis(dodecyldimethylammonio)-
3,6,9,12,15,18,21,24-octaoxahexa-
cosane bis(p-toluenesulfonate)
$C_{12}H_{25}(CH_3)_2N^+(CH_2CH_2O)_8CH_2CH_2N^+(CH_3)_2C_{12}H_{25}$, $2CH_3C_6H_4SO_3^-$

C_{16}

58 — CTAB Hexadecyltrimethylammonium [5] [55]
bromide
$C_{16}H_{33}N(CH_3)_3^+$, Br^-

59 — — Hexadecyldiethylmethylammonium [55]
bromide
$C_{16}H_{33}N(C_2H_5)_2CH_3^+$, Br^-

60 — — Hexadecylbutyldimethylammonium [55]
bromide
$C_{16}H_{33}N(C_4H_9)(CH_3)_2^+$, Br^-

61	—	CPC	Hexadecylpyridinium bromide	[5]	
			$C_{16}H_{33}NC_5H_5^+$, Br^-		
62	—	—	Hexadecyloctyldimethylammonium bromide	[55,126]	
			$C_{16}H_{33}N(C_8H_{17})(CH_3)_2^+$, Br^-		

C_{18}

| 63 | 2.2 | DODMAC | Dioctadecyldimethylammonium chloride | [5] | [81] |
| | | | $(C_{18}H_{37})_2N(CH_3)_2^+$, Cl^- | | |

Nonionic class

Zwitterionic subclass

C_{10}

| 64 | 5.20 | $C_{10}APSO_4$ | 3-(Decyldimethylammonio)propyl-1-sulfate | [2] | [103] |
| | | | $C_{10}H_{21}N^+(CH_3)_2CH_2CH_2CH_2OSO_3^-$ | | |

C_{12}

| 65 | 10.10 | $C_{12}APS$ | 3-Dodecyldimethylammonio-propane-1-sulfonate | | [103] |
| | | | $C_{12}H_{25}N^+(CH_3)_2CH_2CH_2CH_2SO_3^-$ | | |

C_{14}

| 66 | 10.10 | $C_{14}APS$ | 3-Tetradecyldimethylammonio-propane-1-sulfonate | | [103] |
| | | | $C_{14}H_{29}N^+(CH_3)_2CH_2CH_2CH_2SO_3^-$ | | |

C_{15}

| 67 | — | lysoDPPC | 3-Hexadecanoylglyceryl-1-phosphatidyl choline | | [9] |

$C_{15}H_{31}CO_2CH_2CH(OH)CH_2OPO_2^-OCH_2CH_2N^+(CH_3)_3$

| 68 | 8.11 | DPPC | Dipalmitoylphosphatidyl choline | | [47,115] |

$C_{15}H_{31}CO_2CH_2(C_{15}H_{31}CO_2)CHCH_2OPO_2^-OCH_2CH_2N^+(CH_3)_3$

C_{16}

| 69 | 10.10 | $C_{16}APS$ | 3-Hexadecyldimethylammonio-propane-1-sulfonate | [2] | [103] |
| | | | $C_{16}H_{33}N^+(CH_3)_2CH_2CH_2CH_2SO_3^-$ | | |

C_{17}

| 70 | — | lysoDSPC | 3-Octadecanoylglyceryl-1-phosphatidyl choline | | [9] |

$C_{17}H_{35}CO_2CH_2CH(OH)CH_2OPO_2^-OCH_2CH_2N^+(CH_3)_3$

| 71 | — | — | 3-Oleoylglyceryl-1-phosphatidyl choline | | [9] |

$c\text{-}C_8H_{17}CH=CHC_7H_{14}CO_2CH_2CH(OH)CH_2OPO_2^-OCH_2CH_2N^+(CH_3)_3$

72 —   —   3-Linoleoylglyceryl-1-phosphatidyl   [9]
            choline

c,c-$C_5H_{11}CH{=}CHCH_2CH{=}CHC_7H_{14}CO_2CH_2CH(OH)CH_2OPO_2^-OCH_2CH_2N^+(CH_3)_3$

C_{22}

73 10.11 $C_{22}AE_9SO_4$ 26-Docosyldimethylammonio-
            3,6,9,12,15,18,21,24-octaoxahexa-
            cosyl sulfate   [128]
            $C_{22}H_{45}N^+(CH_3)_2(CH_2CH_2O)_9SO_3^-$

74 10.11 $C_{22}AE_9SO_3$ 26-Docosyldimethylammonio-
            3,6,9,12,15,18,21,24-octaoxahexacosan-
            1-sulfonate   [128]
            $C_{22}H_{45}N^+(CH_3)_2(CH_2CH_2O)_8CH_2CH_2SO_3^-$

75 5.5 $C_{22}AU$ 11-Docosyldimethylammonioundecanoate   [14]
            $C_{22}H_{45}N^+(CH_3)_2(CH_2)_{10}CO_2^-$

Semipolar subclass

C_8

76 — C_8PO Octyldimethylphosphine oxide   [1,4]   [16,53]
            $C_8H_{17}(CH_3)_2PO$

77 10.15 —   Octylmethylsulfoximine   [4]
            $C_8H_{17}(CH_3)S(O)NH$

C_{10}

78 — $C_{10}PO$ Decyldimethylphosphine oxide   [1,4]   [15,53]
            $C_{10}H_{21}(CH_3)_2PO$

79 10.25 $C_{10}E_3PO$ Trioxyethylene glycol decyl   [45]
            2-dimethylphosphinylethyl ether
            $C_{10}H_{21}(OCH_2CH_2)_4(CH_3)_2PO$

80 10.25 $C_{10}E_3AO$ Trioxyethylene glycol decyl   [45]
            2-dimethylaminoethyl ether
            N-oxide
            $C_{10}H_{21}(OCH_2CH_2)_4(CH_3)_2NO$

81         $C_{10}SDI$ Decyl methyl sulfone diimine   [3,4]
            $C_{10}H_{21}(CH_3)S(NH)_2$

C_{11}

82 —   — N-Trimethylammoniododecan-   [1]   [19]
            amidate
            $C_{11}H_{23}C(O)N^-N^+(CH_3)_3$

83 10.3   —   2-Hydroxyundecylmethylsulfoxide   [104]
            (high melting)
            $C_9H_{19}CH(OH)CH_2S(O)CH_3$

84 10.3   —   2-Hydroxyundecyl methyl sulfoxide   [104]
            (low melting)
            $C_9H_{19}CH(OH)CH_2S(O)CH_3$

C_{12}

85 10.5, 10.12 $C_{12}PO$ Dodecyldimethylphosphine oxide [1,4]   [53,21]
            $C_{12}H_{25}(CH_3)_2PO$

86	10.5, 10.13	$C_{12}Et_2PO$	Dodecyldiethylphosphine oxide $C_{12}H_{25}(CH_3CH_2)_2PO$	[2,4]	
87	10.12	$C_{12}AO$	Dodecyldimethylamine oxide $C_{12}H_{25}(CH_3)_2NO$	[1,4]	[88]
88	10.2	γ-OHC$_{12}$PO	γ-Hydroxydodecyldimethyl-phosphine oxide $C_9H_{19}CH(OH)CH_2CH_2(CH_3)_2PO$		[80]
89	5.21	—	Dodecyl-bis(2-cyanoethyl)-phosphine oxide $C_{12}H_{25}P(O)(CH_2CH_2CN)_2$		[80]
90	10.13	—	Dimethyl dodecylphosphonate $C_{12}H_{25}P(O)(OCH_3)_2$		[80]
91	10.14	—	Methyl dodecyl methyl-phosphonate $C_{12}H_{25}OP(O)(CH_3)(OCH_3)$	[2,4]	
92	10.13	—	Dimethyl dodecyl phosphate $C_{12}H_{25}OP(O)(OCH_3)_2$	[2,4]	
93	10.12	$C_{12}SO$	Dodecyl methyl sulfoxide $C_{12}H_{25}(CH_3)SO$	[2,4]	
94	10.15	—	Dodecyl methyl sulfoximine $C_{12}H_{25}(CH_3)S(O)NH$	[2,4]	
95	10.16	—	Dodecyl methyl N-methyl-sulfoximine $C_{12}H_{25}(CH_3)S(O)NCH_3$	[2]	
96	5.18	—	Dodecyl methyl sulfone diimine $C_{12}H_{25}(CH_3)S(NH)_2$	[2,4]	
97	5.2	—	Dodecyldimethylphosphine oxide–octane $C_{12}H_{25}(CH_3)_2PO/C_8H_{18}$	[2]	
98	5.2	—	Dodecyldimethylphosphine oxide–dimethyl sulfoxide $C_{12}H_{25}(CH_3)_2PO/(CH_3)_2SO$	[2]	
99	5.2	—	Dodecyldimethylphosphine oxide–glycerol $C_{12}H_{25}(CH_3)_2PO/CH_2(OH)CH(OH)CH_2OH$		[80]
100	5.2	—	Dodecyldimethylphosphine oxide–aniline $C_{12}H_{25}(CH_3)_2PO/C_6H_5NH_2$		[80]
101	5.2	—	Dodecyldimethylphosphine oxide–tetrachloroethylene $C_{12}H_{25}(CH_3)_2PO/CCl_2{=}CCl_2$		[80]
102	5.2	—	Dodecyldimethylphosphine oxide-1,2-ethylenediamine $C_{12}H_{25}(CH_3)_2PO/CH_2(NH_2)CH_2NH_2$		[80]
103	5.2	—	Dodecyldimethylphosphine oxide-N,N-dimethylformamide $C_{12}H_{25}(CH_3)_2PO/HC(O)N(CH_3)_2$		[80]

104	5.2	—	Dodecyldimethylphosphine oxide–formamide $C_{12}H_{25}(CH_3)_2PO/HC(O)NH_2$		[80]
105	5.2	—	Dodecyldimethylphosphine oxide–methanol $C_{12}H_{25}(CH_3)_2PO/CH_3OH$		[80]
106	5.2	—	Dodecyldimethylphosphine oxide–ethanol $C_{12}H_{25}(CH_3)_2PO/C_2H_5OH$		[80]
107	5.2	—	Dodecyldimethylphosphine oxide–butanol $C_{12}H_{25}(CH_3)_2PO/C_4H_9OH$		[80]
108	5.2	—	Dodecyldimethylphosphine oxide–octanol $C_{12}H_{25}(CH_3)_2PO/C_8H_{17}OH$		[80]
109	5.2	—	Dodecyldimethylphosphine oxide–dodecanol $C_{12}H_{25}(CH_3)_2PO/C_{12}H_{25}OH$		[80]

C_{14}

110	10.5	—	Tetradecyldimethylphosphine oxide [2] $C_{14}H_{29}(CH_3)_2PO$		

C_{16}

111	—	$C_{16}Et_2PO$	Hexadecyldiethylphosphine oxide $C_{16}H_{33}(CH_3CH_2)_2PO$		[21]
112	5.7	$C_{16}AO$	Hexadecyldimethylamine oxide $C_{16}H_{33}(CH_3)_2NO$		[80]

Single bond subclass

C_6

113	—	—	N-hexylribonamide $C_6H_{13}NHC(O)(CH(OH))_4H$		[84]

C_7

114	—	—	N-heptylribonamide $C_7H_{15}NHC(O)(CH(OH))_4H$		[84]

C_8

115	10.19	C_8NH_2	Octylamine $C_8H_{17}NH_2$	[4]	[107]
116	—	C_8E_6	Hexaoxyethylene glycol octyl ether $C_8H_{17}(OCH_2CH_2)_6OH$	[1,2,4]	[18]
117	10.20	—	β-Octylfuranoglucoside $C_8H_{17}OC_6H_{11}O_5$ (tetrahydroxy)		[80]
118	—	—	N-Octylribonamide $C_8H_{17}NHC(O)(CH(OH))_4H$		[84]

C_{10}

119	5.22	$C_{10}E_3$	Trioxyethylene glycol decyl ether $C_{10}H_{21}(OCH_2CH_2)_3OH$	[4]	[7]
120	5.19	$C_{10}E_4NMe_2$	Trioxyethylene glycol decyl 2-N,N-dimethylaminoethyl ether $C_{10}H_{21}(OCH_2CH_2)_4N(CH_3)_2$		[45]
121	4.5, 5.17	$C_{10}E_4$	Tetraoxyethylene glycol decyl ether $C_{10}H_{21}(OCH_2CH_2)_4OH$	[4]	[77]
122	10.25	$C_{10}E_4G$	Tetraoxyethylene glycol decyl 2,3-dihydroxy-1-propyl ether $C_{10}H_{21}(OCH_2CH_2)_4OCH_2CH(OH)CH_2OH$		[45]
123	3.3	$C_{10}E_5$	Pentaoxyethylene glycol decyl ether $C_{10}H_{21}(OCH_2CH_2)_5OH$	[3,4]	[77]
124	3.7	$C_{10}E_6$	Hexaoxyethylene glycol decyl ether $C_{10}H_{21}(OCH_2CH_2)_6OH$	[4]	[77]
125	10.23	$C_{10}PhE_9$	Nonaoxyethylene glycol p-decylphenyl ether $C_{10}H_{21}C_6H_4(OCH_2CH_2)_9OH$		[79]
126	10.24	$C_{10}E_6OMe$	Hexaoxyethylene glycol decyl methyl ether $C_{10}H_{21}(OCH_2CH_2)_4OCH_3$		[45]
127	—	Dobanol 91-E_5	C_9–C_{11} mixture of E_5 (polydisperse)) $C_{9-11}(OCH_2CH_2)_{5(avg)}OH$		[57]
128	—	—	N-Decylribonamide $C_{10}H_{21}NHC(O)(CH(OH))_4H$		[84]
129	—	—	N-Decylisosaccharanamide $C_{10}H_{21}NHC(O)C(OH)(CH_2OH)CH(OH)CH_2OH$		[84]
130	—	—	N-Decylgluconamide $C_{10}H_{21}NHC(O)(CH(OH))_5H$		[84]
131	—	$C_{10}G$	Glycerol 1-decyl ether (1-decyloxy-2,3-dihydroxypropane) $C_{10}H_{21}CH_2CH(OH)CH_2OH$	[4]	[80]

C_{11}

132	—	—	Monododecanoin (monolaurin) $C_{11}H_{23}CO_2CH_2CHOHCH_2OH$	[1]	[129]
133	—	—	Sucrose monolaurate $C_{11}H_{23}CO_2C_{12}H_{21}O_9$ (heptahydroxy)	[1]	[52]
134	10.21	$C_{12}GA$	N-Dodecanoyl-N-methylglucamine $C_{11}H_{23}CON(CH_3)CH_2(CH(OH))_5H$		[44]

C_{12}

135	—	$C_{12}E_3$	Trioxyethylene glycol dodecyl ether $C_{12}H_{25}(OCH_2CH_2)_3OH$		[7,80]

136	—	$C_{12}E_4$	Tetraoxyethylene glycol dodecyl ether $C_{12}H_{25}(OCH_2CH_2)_4OH$	[4]	[101]
137	—	$C_{12}E_5$	Pentaoxyethylene glycol dodecyl ether $C_{12}H_{25}(OCH_2CH_2)_5OH$	[4]	[101]
138	5.8	$C_{12}E_6$	Hexaoxyethyleneglycol dodecyl ether $C_{12}H_{25}(OCH_2CH_2)_6OH$	[1,2]	[17]
139	—	$C_{12}E_{5-7}$	$C_{12}E_{5-7}$ (polydisperse) $C_{12}H_{25}(OCH_2CH_2)_{5-7}OH$ (polydisperse)		[57]
140	—	—	N-Dodecylribonamide $C_{12}H_{25}NHC(O)(CH(OH))_4H$		[84]
141	—	$C_{12}E_8$	Octaoxyethylene glycol dodecyl ether $C_{12}H_{25}(OCH_2CH_2)_8OH$	[4]	[101]

C_{13}

| 142 | — | — | Monotetradecanoin (monomyristin) $C_{13}H_{27}CO_2CH_2CH(OH)CH_2OH$ | [1] | [129] |

C_{14}

| 143 | — | $C_{14}E_{5-7}$ | $C_{14}E_{5-7}$ (polydisperse) $C_{14}H_{29}(OCH_2CH_2)_{5-7}OH$ | | [57] |

C_{15}

| 144 | — | — | Monohexadecanoin (monopalmitin) $C_{15}H_{31}CO_2CH_2CH(OH)CH_2OH$ | [1] | [129] |
| 145 | — | C_9PhE_9 | Nonylphenol-E_9, Triton X-100 $C_9H_{19}C_6H_4(OCH_2CH_2)_9OH$ | | [57] |

C_{16}

146	—	$C_{16}E_3$	Trioxyethylene glycol hexadecyl ether $C_{16}H_{33}(OCH_2CH_2)_3OH$		[6]
147	—	$C_{16}E_4$	Tetraoxyethylene glycol hexadecyl ether $C_{16}H_{33}(OCH_2CH_2)_4OH$	[4]	[101]
148	—	$C_{16}E_8$	Octaoxyethylene glycol hexadecyl ether $C_{16}H_{33}(OCH_2CH_2)_8OH$	[4]	[101]
149	—	$C_{16}E_{12}$	Dodecaoxyethylene glycol hexadecyl ether $C_{16}H_{33}(OCH_2CH_2)_{12}OH$	[4]	[101]
150	—	$C_{10}PhE_9$	Nonaoxyethylene glycol p-decylphenyl ether	[4]	[79]

C_{17}

| 151 | — | — | Monooctadecanoin (monostearin) $C_{17}H_{35}CO_2CH_2CH(OH)CH_2OH$ | [1] | [129] |

152	5.15	—	Monoolein	[1]	[129]

$$c\text{-}C_8H_{17}CH=CHC_7H_{14}CO_2CH_2CH(OH)CH_2OH$$

153	—	—	Monoelaidin	[1]	[129]

$$t\text{-}C_8H_{17}CH=CHC_7H_{14}CO_2CH_2CH(OH)CH_2OH$$

154	—	—	Monolinolein	[1]	[129]

$$c,c\text{-}C_5H_{11}CH=CHCH_2CH=CHC_7H_{14}CO_2CH_2CH(OH)CH_2OH$$

155	—	—	Oleic amide-N-E$_7$-OH (polydisperse)	[57]

$$c\text{-}C_8H_{17}CH=CHC_7H_{14}CONH(CH_2CH_2O)_7H$$

C$_{19}$

156	—	—	Monoeicosanoin (monoarachidin)	[1]	[129]

$$C_{19}H_{39}CO_2CH_2CH(OH)CH_2OH$$

C$_{21}$

157	—	—	Monodocosanoin (monobehenin)	[1]	[129]

$$C_{21}H_{43}CO_2CH_2CH(OH)CH_2OH$$

158	—	—	Monoerucin	[1]	[129]

$$c\text{-}C_8H_{17}CH=CHC_{11}H_{22}CO_2CH_2CH(OH)CH_2OH$$

C$_{22}$

159	—	C$_{22}$E$_7$	Heptaoxyethylene glycol docosyl ether	[63]

$$C_{22}H_{45}(OCH_2CH_2)_7OH$$

160	—	C$_{22}$E$_{14}$	Tetradecaoxyethylene glycol docosyl ether	[63]

$$C_{22}H_{45}(OCH_2CH_2)_{14}OH$$

C$_{27}$

161	—	CholE$_{13}$	Cholesterol E$_{13}$ (polydisperse)	[112]

$$C_{27}H_{45}(OCH_2CH_2)_{13}OH$$

162	—	CholE$_{35}$	Cholesterol E$_{35}$ (polydisperse)	[112]

$$C_{27}H_{45}(OCH_2CH_2)_{35}OH$$

163	—	CholE$_{50}$	Cholesterol E$_{50}$ (polydisperse)	[112]

$$C_{27}H_{45}(OCH_2CH_2)_{50}OH$$

Miscellaneous

164	—	—	N-(12-acryloyloxydodecanoyl)-N-methyl-glucamine	[35]

$$CH_2=CHCO_2(CH_2)_{11}C(=O)N(CH_3)CH_2(CH(OH))_5H$$

165	—	—	Poly(N-(12-acryloyloxydodecanoyl)-N-methyl-glucamine)	[35]

$$-(CH_2CH)_nCO_2(CH_2)_{11}C(=O)N(CH_3)CH_2(CH(OH))_5H$$

166	—	—	Allyloxybiphenyl heptaoxyethylene compound	[85]

$$CH_2=CHCH_2OC_6H_4-C_6H_4-OCH_2CO_2(CH_2CH_2O)_7CH_3$$

167	—	—	Allyloxybiphenyl nonaoxyethylene compound	[86]

$$CH_2=CHCH_2OC_6H_4-C_6H_4-O(CH_2CH_2O)_9CH_3$$

168 —   —   Silylated polymer of No. 166   [86]
$-(OSi(-)(CH_3)CH_2CH_2CH_2OC_6H_4-C_6H_4-O(CH_2CH_2O)_9CH_3)_n$

169 —   —   5-Hexenyl-1-oxybiphenyl nonaoxy-
                           ethylene compound   [86]
$CH_2=CH(CH_2)_4OC_6H_4-C_6H_4-O(CH_2CH_2O)_9CH_3$

Binary nonsurfactant systems

Ionic compounds

170 1.1.   NaCl   Sodium chloride   [127]
                          Na^+, Cl^-

171 11.6   —   Sodium formate   [127]
                          Na^+, HCO_2^-

Nonionic compounds

172 5.16   —   2,2-Dideuteroheneicosane/nonadecane   [97]
                          $C_{19}H_{40}/CH_3CD_2C_{19}H_{39}$

173 11.12   Heptane–perfluoroheptane   [58]
                          C_7H_{16}/C_7F_{16}

174 11.7 $C_{10}Et_2PO$ Decyldiethylphosphine oxide   [80]
                          $C_{10}H_{21}(C_2H_5)PO$

175 11.7 C_8Pr_2PO Octyldipropylphosphine oxide   [80]
                          $C_8H_{17}(C_3H_7)PO$

176 11.7 C_6Me_2PO Dihexylmethylphosphine oxide   [80]
                          $(C_6H_{13})_2(CH_3)PO$

177 11.7 $(C_5)_3PO$ Tripentylphosphine oxide   [80]
                          $(C_5H_{11})_3PO$

178 9.3 $C_{14}OH$ Tetradecanol   [80]
                          $C_{14}H_{29}OH$

179 — E_{23200} Polyoxyethylene glycol   [4]   [111]
                          (dp = 23,200)
                          $H(OCH_2CH_2)_{23200}OH$

180 — E_{480} Polyoxyethylene glycol   [4]   [111]
                          (dp = 480)
                          $H(OCH_2CH_2)_{480}OH$

181 — E_{330} Polyoxyethylene glycol   [4]   [111]
                          (dp = 330)
                          $H(OCH_2CH_2)_{330}OH$

182 — $E_{51.6}$ Polyoxyethylene glycol   [4]   [111]
                          (dp = 51.6)
                          $H(OCH_2CH_2)_{51.6}OH$

183 — E_{51} Polyoxyethylene glycol   [4]   [111]
                          (dp = 51)
                          $H(OCH_2CH_2)_{51}OH$

184 — E_{49} Polyoxyethylene glycol   [4]   [111]
                          (dp = 49)
                          $H(OCH_2CH_2)_{49}OH$

| 185 | 9.2 | — | S-Dodecyl-N-methylsulfonamide-bis(N-methylimine) $C_{12}H_{25}S(NCH_3)_2NHCH_3$ | [2] | |

A2.3 Ternary systems

No.	Fig.	Abbrev.	Chemical name and formula	Compilation reference	Original reference

Ionic class

Dominant surfactant–anionic

C_6

| 186 | — | AOT | Aerosol OT–p-xylene–water @ 20°C Na+, | [1] | [30] |

$C_6H_{13}CH(C_2H_5)CH_2OC(O)CH_2CH(SO_3^-)CO_2CH_2CH(C_2H_5)C_6H_{13}/p\text{-}(CH_3)_2C_6H_4/H_2O$

| 187 | — | AOT | Aerosol OT–decanol–water @ 20°C Na+, | [1] | [30,39] |

$C_6H_{13}CH(C_2H_5)CH_2OC(O)CH_2CH(SO_3^-)CO_2CH_2CH(C_2H_5)C_6H_{13}/C_{10}H_{21}OH/H_2O$

| 188 | — | AOT | Aerosol OT–octanoic acid–water @ 20°C Na+, | [1] | [30,39] |

$C_6H_{13}CH(C_2H_5)CH_2OC(O)CH_2CH(SO_3^-)CO_2CH_2CH(C_2H_5)C_6H_{13}/C_7H_{15}CO_2H/H_2O$

C_7

189	—	—	Sodium octanoate–carbon tetrachloride–water @ 20°C Na+, $C_7H_{15}CO_2^-/CCl_4/H_2O$	[1]	[24,31]
190	12.7	—	Sodium octanoate–n–octane–water @ 20°C Na+, $C_7H_{15}CO_2^-/C_8H_{18}/H_2O$	[1]	[24,31]
191	—	—	Sodium octanoate–1-chloro-octane–water @ 20°C Na+, $C_7H_{15}CO_2^-/C_8H_{17}Cl/H_2O$	[1]	[24,31]
192	—	—	Sodium octanoate–p–xylene–water @ 20°C Na+, $C_7H_{15}CO_2^-/p\text{-}(CH_3)_2C_6H_4/H_2O$	[1]	[24,31]
193	—	—	Sodium octanoate–octane nitrile–water @ 20°C Na+, $C_7H_{15}CO_2^-/C_7H_{15}CN/H_2O$	[1]	[24,31]
194	—	—	Sodium octanoate–methyl octanoate–water @ 20°C Na+, $C_7H_{15}CO_2^-/C_7H_{15}CO_2CH_3/H_2O$	[1]	[24,31]

195	—	—	Sodium octanoate–octanal–water @ 20°C Na^+, $C_7H_{15}CO_2^-/C_7H_{15}CH=O/H_2O$	[1]	[24,31]
196	—	—	Sodium octanoate–methanol–water @ 20°C Na^+, $C_7H_{15}CO_2^-/CH_3OH/H_2O$	[1]	[95]
197	—	—	Sodium octanoate–ethanol–water @ 20°C Na^+, $C_7H_{15}CO_2^-/C_2H_5OH/H_2O$	[1]	[95]
198	—	—	Sodium octanoate–propanol–water @ 20°C Na^+, $C_7H_{15}CO_2^-/C_3H_7OH/H_2O$	[1]	[95]
199	12.17	—	Sodium octanoate–butanol–water @ 20°C Na^+, $C_7H_{15}CO_2^-/C_4H_9OH/H_2O$	[1]	[95]
200	—	—	Sodium octanoate–pentanol–water @ 20°C Na^+, $C_7H_{15}CO_2^-/C_5H_{11}OH/H_2O$	[1]	[95,124]
201	—	—	Sodium octanoate–2-pentanol– water @ 25°C Na^+, $C_7H_{15}CO_2^-/2\text{-}C_5H_{11}OH/H_2O$	[1]	[95,124]
202	—	—	Sodium octanoate–3-pentanol– water @ 25°C Na^+, $C_7H_{15}CO_2^-/3\text{-}C_5H_{11}OH/H_2O$	[1]	[95,124]
203	—	—	Sodium octanoate–hexanol–water @ 20°C Na^+, $C_7H_{15}CO_2^-/C_6H_{13}OH/H_2O$	[1]	[95]
204	—	—	Sodium octanoate–heptanol–water @ 20°C Na^+, $C_7H_{15}CO_2^-/C_7H_{15}OH/H_2O$	[1]	[95]
205	—	—	Sodium octanoate–octanol–water @ 20°C Na^+, $C_7H_{15}CO_2^-/C_8H_{17}OH/H_2O$	[1]	[24,95]
206	—	—	Sodium octanoate–nonanol–water @ 20°C Na^+, $C_7H_{15}CO_2^-/C_9H_{19}OH/H_2O$	[1]	[24,95]
207	12.15	—	Sodium octanoate–decanol–water @ 20°C Na^+, $C_7H_{15}CO_2^-/C_{10}H_{21}OH/H_2O$	[1,3]	[22,23,36,96,43]
208	—	—	Sodium octanoate–ethylene glycol–water @ 20°C Na^+, $C_7H_{15}CO_2^-/CH_2OHCH_2OH/H_2O$	[1]	[27]
209	—	—	Sodium octanoate–glycerol–water @ 20°C Na^+, $C_7H_{15}CO_2^-/CH_2OHCH(OH)CH_2OH/H_2O$	[1]	[27]
210	—	—	Sodium octanoate–1,8-octanediol– water @ 20°C Na^+, $C_7H_{15}CO_2^-/HO(CH_2)_8OH/H_2O$	[1]	[25,26]

211 — — Sodium octanoate–tetraoxyethylene [1] [27]
glycol–water @ 20°C
Na$^+$, C$_7$H$_{15}$CO$_2^-$/H(OCH$_2$CH$_2$)$_4$OH/H$_2$O

212 — — Sodium octanoate–2,5-dimethyl- [1] [31]
phenol–water @ 20°C
Na$^+$, C$_7$H$_{15}$CO$_2^-$/2,5-(CH$_3$)$_2$C$_6$H$_3$OH/H$_2$O

213 — — Sodium octanoate–2-naphthol– [1] [31]
water @ 20°C
Na$^+$, C$_7$H$_{15}$CO$_2^-$/2-C$_{10}$H$_7$OH/H$_2$O

214 — — Sodium octanoate–cholesterol– [1] [33]
water @ 20°C
Na$^+$, C$_7$H$_{15}$CO$_2^-$/C$_{27}$H$_{45}$OH/H$_2$O

215 — — Sodium octanoate–hexanoic [1] [32]
acid–water @ 20°C
Na$^+$, C$_7$H$_{15}$CO$_2^-$/C$_5$H$_{11}$CO$_2$H/H$_2$O

216 — — Sodium octanoate–heptanoic [1] [32]
acid–water @ 20°C
Na$^+$, C$_7$H$_{15}$CO$_2^-$/C$_6$H$_{13}$CO$_2$H/H$_2$O

217 12.22 — Sodium octanoate–octanoic [1] [32]
acid–water @ 20°C
Na$^+$, C$_7$H$_{15}$CO$_2^-$/C$_7$H$_{15}$CO$_2$H/H$_2$O

218 — — Sodium octanoate–decanoic [1] [32]
acid–water @ 20°C
Na$^+$, C$_7$H$_{15}$CO$_2^-$/C$_9$H$_{19}$CO$_2$H/H$_2$O

219 — — Potassium octanoate–octanol– [1] [29]
water @ 20°C
K$^+$, C$_7$H$_{15}$CO$_2^-$/C$_8$H$_{17}$OH/H$_2$O

220 — — Potassium octanoate–octanol–
water @ 25°C [1] [29]
K$^+$, C$_7$H$_{15}$CO$_2^-$/C$_8$H$_{17}$OH/H$_2$O

221 — — Potassium octanoate–decanol–
water @ 20°C [1] [29]
K$^+$, C$_7$H$_{15}$CO$_2^-$/C$_{10}$H$_{21}$OH/H$_2$O

222 — — Octylammonium octanoate–sodium [62]
octanoate–water @ 25°C
C$_8$H$_{17}$NH$_3^+$, C$_7$H$_{15}$CO$_2^-$/Na$^+$, C$_7$H$_{15}$CO$_2^-$/H$_2$O

C$_8$
223 12.19 — Sodium octanesulfonate–decanol– [1] [24,31]
water @ 20°C
Na$^+$, C$_8$H$_{17}$SO$_3^-$/C$_{10}$H$_{21}$OH/H$_2$O

224 — — Sodium octyl sulfate–decanol–water
@ 20°C [1] [31]
Na$^+$, C$_8$H$_{17}$OSO$_3^-$/C$_{10}$H$_{21}$OH/H$_2$O

225 — — Sodium octyl sulfate–decanol–water
@ 27°C [69]
Na$^+$, C$_8$H$_{17}$OSO$_3^-$/C$_{10}$H$_{21}$OH/H$_2$O

226 — —   Magnesium octyl sulfate–decanol–   [68]
        water @ 27°C
        Mg^{++}, $(C_8H_{17}OSO_3^-)_2/C_{10}H_{21}OH/H_2O$

227 — —   Calcium octyl sulfate–decanol–water
        @ 27°C   [69,68]
        Ca^{++}, $(C_8H_{17}OSO_3^-)_2/C_{10}H_{21}OH/H_2O$

228 — C_8AOT Sodium dioctyl sulfosuccinate–   [106]
        butanol–water, T undefined
   Na^+, $C_8H_{17}OC(O)CH_2CH(SO_3^-)CO_2C_8H_{17}/C_4H_9OH/H_2O$

C_9
229 — —   Potassium decanoate–decanoic   [1]   [32]
        acid–water @ 20°C
        K^+, $C_9H_{19}CO_2^-/C_9H_{19}CO_2H/H_2O$

C_{11}
230 — —   Potassium dodecanoate–dodecanoic [1] [29]
        acid–water @ 175°C
        K^+, $C_{11}H_{23}CO_2^-/C_{11}H_{23}CO_2H/H_2O$

231 — —   Sodium-2,2-dibutyldodecanoate–   [65]
        decanol–water @ 25°C
   K^+, $C_{10}H_{21}C(C_4H_9)_2CO_2^-/C_{10}H_{21}OH/H_2O$

C_{12}
232 12.20 SDS   Sodium dodecyl sulfate–pentanol–   [49]
        water @ 18 and 20°C
        Na^+, $C_{12}H_{25}OSO_3^-/C_5H_{11}OH/H_2O$

233 — SDS   Sodium dodecyl sulfate–decanol– [1] [42]
        water @ 25°C
        Na^+, $C_{12}H_{25}OSO_3^-/C_{10}H_{21}OH/H_2O$

234 — SDS   Sodium dodecyl sulfate–decanol–   [42]
        glycerol @ 25°C
   Na^+, $C_{12}H_{25}OSO_3^-/C_{10}H_{21}OH/CH_2OHCH(OH)CH_2OH$

235 — SDS   Sodium dodecyl sulfate–hexanoic [1] [83]
        acid–water @ 25°C
     Na^+, $C_{12}H_{25}OSO_3^-/C_5H_{11}CO_2H/H_2O$

236 — $C_{12}E_1S$ Sodium ethylene glycol dodecyl   [70]
        ether sulfate–decanol–water @ 27°C
    Na^+, $C_{12}H_{25}OCH_2CH_2OSO_3^-/C_{10}H_{21}OH/H_2O$

237 — —   Calcium ethylene glycol dodecyl   [70]
        ether sulfate–decanol–water at 20°C
   Ca^{++}, $(C_{12}H_{25}OCH_2CH_2OSO_3^-)_2/C_{10}H_{21}OH/H_2O$

238 — —   Sodium ethylene glycol dodecyl   [70]
        ether sulfate–calcium ethylene
        glycol dodecyl ether sulfate–water (section)
 Na^+, $C_{12}H_{25}OCH_2CH_2OSO_3^-/Ca^{++}$, $(C_{12}H_{25}OCH_2CH_2OSO_3^-)_2/H_2O$

239 — —   Sodium monooxyethylene glycol   [50]
        dodecyl ether sulfate–calcium
        dodecyl sulfate–water (section)
 Na^+, $C_{12}H_{25}OCH_2CH_2OSO_3^-/Ca^{++}$, $(C_{12}H_{25}OSO_3^-)_2/H_2O$

240   —   —   Sodium dioxyethylene glycol   [50]
dodecyl ether sulfate–calcium
dodecyl sulfate–water (section)
Na^+, $C_{12}H_{25}(OCH_2CH_2)_2OSO_3^-/Ca^{++}$, $(C_{12}H_{25}OSO_3^-)_2/H_2O$

241   —   —   Sodium trioxyethylene glycol   [50]
dodecyl ether sulfate–calcium
dodecyl sulfate–water (section)
Na^+, $C_{12}H_{25}(OCH_2CH_2)_3OSO_3^-/Ca^{++}$, $(C_{12}H_{25}OSO_3^-)_2/H_2O$

242   —   —   Dodecyltrimethylammonium   [60]
dodecylsulfate–dodecanol–water
@ 60°C
$C_{12}H_{25}N(CH_3)_3^+$, $C_{12}H_{25}OSO_3^-/C_{10}H_{21}OH/H_2O$

C_{13}
243   —   NaM   Sodium myristate–ethylbenzene–   [1]   [113]
water @ 82°C
Na^+, $C_{13}H_{27}CO_2^-/C_2H_5C_6H_5/H_2O$

C_{14}
244   — $C_{14}AS$ Sodium tetradecyl sulfate–calcium   [50]
tetradecyl sulfate–water (section)
Na^+, $C_{14}H_{29}OSO_3^-/Ca^{++}$, $(C_{14}H_{29}OSO_3^-)_2/H_2O$

245   —   SDS   Sodium dodecyl sulfate–calcium   [50]
dodecyl sulfate–water (section)
Na^+, $C_{12}H_{25}OSO_3^-/Ca^{++}$, $(C_{12}H_{25}OSO_3^-)_2/H_2O$

246   —   —   Sodium octylbenzenesulfonate–   [78]
pentanol–water @ 25°C
Na^+, $C_8H_{17}C_6H_4SO_3^-/C_5H_{11}OH/H_2O$

C_{15}
247   12.5, 14.4   Sodium hexadecanoate (palmitate)–   [117]
sodium chloride–water at 90°C
Na^+, $C_{15}H_{31}CO_2^-/NaCl/H_2O$

C_{17}
248   —   KOl   Potassium oleate–p-xylene–water   [1]   [31]
@ 20°C
K^+, $c-C_8H_{17}CH=CHC_7H_{14}CO_2^-/p-(CH_3)_2C_6H_4/H_2O$

249   12.16   KOl   Potassium oleate–decanol–water   [1]   [29]
@ 20°C
K^+, $c-C_8H_{17}CH=CHC_7H_{14}CO_2^-/C_{10}H_{21}OH/H_2O$

C_{18}
250   —   CTAB   Hexadecyltrimethylammonium   [71]
bromide–sodium dodecylbenzene-
sulfonate–water @ 25°C
$C_{16}H_{33}N(CH_3)_3^+$, Br^-/Na^+, $C_{12}H_{25}C_6H_4SO_3^-/H_2O$

C_{21}

| 251 | — | — | Ethanolammonium docosanoate–ethanolamine–water @ 25°C $HOCH_2CH_2NH_3^+$, $C_{21}H_{43}CO_2^-/HOCH_2CH_2NH_2/H_2O$ | | [63] |
| 252 | — | — | Triethanolammonium docosanoate–triethanolamine–water @ 25°C $(HOCH_2CH_2)_3NH^+$, $C_{21}H_{43}CO_2^-/(HOCH_2CH_2)_3NH_2/H_2O$ | | [63] |

C_{23}

| 253 | — | — | Sodium desoxycholate–decanol–water @ 20°C Na^+, $C_{24}H_{39}O_4^-/C_{10}H_{21}OH/H_2O$ | [1] | [40] |
| 254 | — | — | Sodium cholate–decanol–water @ 20°C Na^+, $C_{24}H_{39}O_5^-/C_{10}H_{21}OH/H_2O$ | [1] | [40] |

Dominant surfactant–cationic

C_7

| 255 | — | — | Heptyltrimethylammonium bromide–hexanol–water @ 25.5°C $C_7H_{15}N(CH_3)_3^+$, $Br^-/C_6H_{13}OH/H_2O$ | [1] | [82] |

C_8

256	—	—	Octylammonium chloride–p-xylene–water @ 20°C $C_8H_{17}NH_3^+$, $Cl^-/p\text{-}(CH_3)_2C_6H_4/H_2O$	[1]	[31]
257	12.18	—	Octylammonium chloride–decanol–water @ 20°C $C_8H_{17}NH_3^+$, $Cl^-/C_{10}H_{21}OH/H_2O$	[1]	[24,123]
258	—	—	Octylmethylammonium chloride–decanol–water @ 20°C $C_8H_{17}N(CH_3)H_2^+$, $Cl^-/C_{10}H_{21}OH/H_2O$	[1]	[31]
259	—	—	Octyldimethylammonium chloride–decanol–water @ 20°C $C_8H_{17}N(CH_3)_2H^+$, $Cl^-/C_{10}H_{21}OH/H_2O$	[1]	[31]
260	—	—	Octyltrimethylammonium chloride–decanol–water @ 20°C $C_8H_{17}N(CH_3)_3^+$, $Cl^-/C_{10}H_{21}OH/H_2O$	[1]	[31]
261	12.24	—	Octylammonium chloride–octylamine–water @ 20°C $C_8H_{17}NH_3^+$, $Cl^-/C_8H_{17}NH_2/H_2O$		[123]

C_{10}

| 262 | — | $C_{10}C_{10}DMAB$ | Didecyldimethylammonium bromide–hexane–water @ 20°C $(C_{10}H_{21})_2N(CH_3)_2^+$, $Br^-/C_6H_{14}/H_2O$ | | [126] |
| 263 | — | $C_{10}C_{10}DMAB$ | Didecyldimethylammonium bromide–cyclohexane–water @ 20°C $(C_{10}H_{21})_2N(CH_3)_2^+$, $Br^-/c\text{-}C_6H_{12}/H_2O$ | | [126] |

| 264 | — | C$_{10}$C$_{10}$DMAB | Didecyldimethylammonium bromide–octane–water @ 20°C (C$_{10}$H$_{21}$)$_2$N(CH$_3$)$_2^+$, Br$^-$/C$_8$H$_{18}$/H$_2$O | [126] |
| 265 | — | C$_{10}$C$_{10}$DMAB | Didecyldimethylammonium bromide–decane–water @ 20°C (C$_{10}$H$_{21}$)$_2$N(CH$_3$)$_2^+$, Br$^-$/C$_{10}$H$_{22}$/H$_2$O | [126] |

C$_{12}$

266	—	C$_{12}$C$_{10}$DMAB	Dodecyldecyldimethylammonium bromide–cyclohexane–water @ 20°C C$_{12}$H$_{25}$(C$_{10}$H$_{21}$)N(CH$_3$)$_2^+$, Br$^-$/c-C$_6$H$_{12}$/H$_2$O	[126]
267	—	C$_{12}$C$_{10}$DMAB	Dodecyldecyldimethylammonium bromide–hexane–water @ 20°C C$_{12}$H$_{25}$(C$_{10}$H$_{21}$)N(CH$_3$)$_2^+$, Br$^-$/C$_6$H$_{14}$/H$_2$O	[126]
268	—	C$_{12}$C$_{10}$DMAB	Dodecyldecyldimethylammonium bromide–octane–water @ 20°C C$_{12}$H$_{25}$(C$_{10}$H$_{21}$)N(CH$_3$)$_2^+$, Br$^-$/C$_8$H$_{18}$/H$_2$O	[126]
269	—	C$_{12}$C$_{10}$DMAB	Dodecyldecyldimethylammonium bromide–decane–water @ 20°C C$_{12}$H$_{25}$(C$_{10}$H$_{21}$)N(CH$_3$)$_2^+$, Br$^-$/C$_{10}$H$_{22}$/H$_2$O	[126]
270	—	C$_{12}$C$_{10}$DMAB	Dodecyldecyldimethylammonium bromide–dodecane–water @ 20°C C$_{12}$H$_{25}$(C$_{10}$H$_{21}$)N(CH$_3$)$_2^+$, Br$^-$/C$_{12}$H$_{26}$/H$_2$O	[126]
271	12.8	C$_{12}$C$_{12}$DMAB	Didodecyldimethylammonium bromide–hexane–water @ 20°C (C$_{12}$H$_{25}$)$_2$N(CH$_3$)$_2^+$, Br$^-$/C$_6$H$_{14}$/H$_2$O	[37]
272	—	C$_{12}$C$_{12}$DMAB	Didodecyldimethylammonium bromide–cyclohexane–water @ 20°C (C$_{12}$H$_{25}$)$_2$N(CH$_3$)$_2^+$, Br$^-$/c-C$_6$H$_{12}$/H$_2$O	[37]
273	—	C$_{12}$C$_{12}$DMAB	Didodecyldimethylammonium bromide–1-hexene–water @ 20°C (C$_{12}$H$_{25}$)$_2$N(CH$_3$)$_2^+$, Br$^-$/1-C$_6$H$_{12}$/H$_2$O	[37]
274	—	C$_{12}$C$_{12}$DMAB	Didodecyldimethylammonium bromide–octane–water @ 20°C (C$_{12}$H$_{25}$)$_2$N(CH$_3$)$_2^+$,Br$^-$/C$_8$H$_{18}$/H$_2$O	[37]
275	—	C$_{12}$C$_{12}$DMAB	Didodecyldimethylammonium bromide–styrene–water @ 20°C (C$_{12}$H$_{25}$)$_2$N(CH$_3$)$_2^+$, Br$^-$/CH$_2$=CHC$_6$H$_5$/H$_2$O	[8]
276	—	C$_{12}$C$_{12}$DMAB	Didodecyldimethylammonium bromide–decane–water @ 20°C (C$_{12}$H$_{25}$)$_2$N(CH$_3$)$_2^+$, Br$^-$/C$_{10}$H$_{22}$/H$_2$O	[37]
277	—	C$_{12}$C$_{12}$DMAB	Didodecyldimethylammonium bromide–1-decene–water @ 20°C (C$_{12}$H$_{25}$)$_2$N(CH$_3$)$_2^+$, Br$^-$/1-C$_{10}$H$_{20}$/H$_2$O	[37]
278	12.8	C$_{12}$C$_{12}$DMAB	Didodecyldimethylammonium bromide–dodecane–water @ 20°C (C$_{12}$H$_{25}$)$_2$N(CH$_3$)$_2^+$, Br$^-$/C$_{12}$H$_{26}$/H$_2$O	[37]

279 — $C_{12}C_{12}$DMAB   Didodecyldimethylammonium   [37]
                                bromide–tetradecane–water @ 20°C
                                $(C_{12}H_{25})_2N(CH_3)_2^+$, $Br^-/C_{14}H_{30}/H_2O$

C_{14}
280 — $C_{14}C_{14}$DMAB   Ditetradecyldimethylammonium   [126]
                                bromide–hexane–water @ 37°C
                                $(C_{14}H_{29})_2N(CH_3)_2^+$, $Br^-/C_6H_{14}/H_2O$
281 — $C_{14}C_{14}$DMAB   Ditetradecyldimethylammonium   [126]
                                bromide–dodecane–water @ 37°C
                                $(C_{14}H_{29})_2N(CH_3)_2^+$, $Br^-/C_{12}H_{26}/H_2O$
282 — $C_{14}C_{14}$DMAB   Ditetradecyldimethylammonium   [126]
                                bromide–hexadecane–water @ 37°C
                                $(C_{14}H_{29})_2N(CH_3)_2^+$, $Br^-/C_{16}H_{34}/H_2O$

C_{16}
283 —   CTAB   Hexadecyltrimethylammonium   [1]   [28]
                                bromide–hexanol–water @ 25°C
                            $C_{16}H_{33}N(CH_3)_3^+$, $Br^-/C_6H_{13}OH/H_2O$
284 — $C_{16}C_8$DMAB   Hexadecyloctyldimethylammonium   [126]
                                bromide–hexane–water @ 20°C
                            $C_{16}H_{33}(C_8H_{17})N(CH_3)_2^+$, $Br^-/C_6H_{14}/H_2O$
285 — $C_{16}C_8$DMAB   Hexadecyloctyldimethylammonium   [126]
                                bromide–decane–water @ 20°C
                            $C_{16}H_{33}(C_8H_{17})N(CH_3)_2^+$, $Br^-/C_{10}H_{22}/H_2O$
286 — $C_{16}C_8$DMAB   Hexadecyloctyldimethylammonium   [126]
                                bromide–hexadecane–water
                            $C_{16}H_{33}(C_8H_{17})N(CH_3)_2^+$, $Br^-/C_{16}H_{34}/H_2O$

Nonionic class

Dominant surfactant-semipolar

287 — C_{14}AO   Tetradecyldimethylamine   [48]
                                oxide–decane–water.
                                Temperature undefined
                                $C_{14}H_{29}(CH_3)_2NO/C_{10}H_{22}/H_2O$

Dominant surfactant-single bond

C_6
288 — C_6E_6 Hexaoxyethylene glycol hexyl   [1]   [28]
                                ether–benzene–water @ 20°C
                                $C_6H_{13}(OCH_2CH_2)_6OH/C_6H_6/H_2O$
289 — C_6E_6 Hexaoxyethylene glycol hexyl   [1]   [28]
                                ether–cyclohexane–water @ 20°C
                            $C_6H_{13}(OCH_2CH_2)_6OH/c\text{-}C_6H_{12}/H_2O$

290	—	C_6E_6	Hexaoxyethylene glycol hexyl ether–octanol–water @ 20°C $C_6H_{13}(OCH_2CH_2)_6OH/C_8H_{17}OH/H_2O$	[1]	[28]
291	—	C_6E_6	Hexaoxyethylene glycol hexyl ether–4-chloro-3,5-dimethyl-phenol–water @ 20°C $C_6H_{13}(OCH_2CH_2)_6OH/4\text{-}Cl\text{-}3,5\text{-}(CH_3)_2C_6H_2OH/H_2O$	[1]	[28]
292	—	C_6NH_2	Hexylamine–p-xylene–water @ 25°C $C_6H_{13}NH_2/p\text{-}(CH_3)_2C_6H_4/H_2O$		[122]

C_7

293	—	—	Monooctanoin–octanoic acid–water @ 20°C $C_7H_{15}CO_2CH_2CHOHCH_2OH/C_7H_{15}CO_2H/H_2O$	[1]	[24,31]
294	12.21	—	Monooctanoin–decanol–water @ 20°C $C_7H_{15}CO_2CH_2CHOHCH_2OH/C_{10}H_{21}OH/H_2O$	[1]	[24,31]
295	—	C_7NH_2	Heptylamine–p-xylene–water @ 25°C $C_7H_{15}NH_2/p\text{-}(CH_3)_2C_6H_4/H_2O$		[122]

C_8

296	—	C_8E_3	Trioxyethylene glycol octyl ether–decane–water @ 13.6°C $C_8H_{17}(OCH_2CH_2)_3OH/C_{10}H_{22}/H_2O$		[75]
297	—	C_8E_3	Trioxyethylene glycol octyl ether–decane–water @ 15.8°C $C_8H_{17}(OCH_2CH_2)_3OH/C_{10}H_{22}/H_2O$		[75]
298	—	C_8E_3	Trioxyethylene glycol octyl ether–decane–water @ 21.5°C $C_8H_{17}(OCH_2CH_2)_3OH/C_{10}H_{22}/H_2O$		[75]
299	—	C_8E_3	Trioxyethylene glycol octyl ether–decane–water @ 26°C $C_8H_{17}(OCH_2CH_2)_3OH/C_{10}H_{22}/H_2O$		[75]
300	—	C_8E_3	Trioxyethylene glycol octyl ether–decane–water @ 30°C $C_8H_{17}(OCH_2CH_2)_3OH/C_{10}H_{22}/H_2O$		[75]
301	12.12	C_8E_4	Tetraoxyethylene glycol octyl ether–heptane–water @ 25°C $C_8H_{17}(OCH_2CH_2)_4OH/C_7H_{16}/H_2O$		[34]
302	12.14	C_8NH_2	Octylamine–p-xylene–water @ 25°C $C_8H_{17}NH_2/p\text{-}(CH_3)_2C_6H_4/H_2O$		[122]

C_9

| 303 | — | C_9NH_2 | Nonylamine–p-xylene–water @ 25°C $C_9H_{19}NH_2/p\text{-}(CH_3)_2C_6H_4/H_2O$ | | [122] |

C_{10}

| 304 | 12.9 | $C_{10}E_4$ | Tetraoxyethylene glycol decyl ether–hexadecane–water @ 19 and 35°C $C_{10}H_{21}(OCH_2CH_2)_4OH/C_{16}H_{34}/H_2O$ | | [76] |

305 12.10 $C_{10}E_4$ Tetraoxyethylene glycol decyl ether– [76]
hexadecane–water @ 40 and 45°C
$C_{10}H_{21}(OCH_2CH_2)_4OH/C_{16}H_{34}/H_2O$

306 12.11 $C_{10}E_4$ Tetraoxyethylene glycol decyl ether– [76]
hexadecane–water @ 50 and 60°C
$C_{10}H_{21}(OCH_2CH_2)_4OH/C_{16}H_{34}/H_2O$

307 — $C_{10}E_6$ Hexaoxyethylene glycol decyl [1] [28]
ether–4-chloro-3,5-dimethylphenol–
water @ 20°C
$C_{10}H_{21}(OCH_2CH_2)_6OH/4\text{-}Cl\text{-}3,5\text{-}(CH_3)_2C_6H_2OH/H_2O$

308 — $C_{10}NH_2$ Decylamine–p-xylene–water @ 25°C [122]
$C_{10}H_{21}NH_2/p\text{-}(CH_3)_2C_6H_4/H_2O$

C_{12}
309 — $C_{12}E_5$ Pentaoxyethylene glycol dodecyl [75]
ether–decane–water @ 35.4–38.6°C
$C_{12}H_{25}(OCH_2CH_2)_5OH/C_{10}H_{22}/H_2O$

310 — $C_{12}E_{10}$ Decaoxyethylene glycol dodecyl [1] [24,31]
ether–octanoic acid–water @ 20°C
$C_{12}H_{25}(OCH_2CH_2)_{10}OH/C_7H_{15}CO_2H/H_2O$

311 — $C_{12}E_{10}$ Decaoxyethylene glycol dodecyl [1] [24,31]
ether–oleic acid–water @ 20°C
$C_{12}H_{25}(OCH_2CH_2)_{10}OH/c\text{-}C_8H_{17}CH=CHC_7H_{14}CO_2H/H_2O$

C_{15}
312 — C_9PhE_6 Emu-02–p-xylene–water @ 20°C [1] [41]
br-$C_9H_{19}C_6H_4(OCH_2CH_2)_{6(avg)}OH/p\text{-}(CH_3)_2C_6H_4/H_2O$

313 — C_9PhE_9 Triton X-100–benzene–water
@ 29°C [1] [98]
br-$C_9H_{19}C_6H_4(OCH_2CH_2)_{9(avg)}OH/C_6H_6/H_2O$

314 — C_9PhE_9 Triton X-100–decanol–water
@ 20°C [1] [24,31]
br-$C_9H_{19}C_6H_4(OCH_2CH_2)_{9(avg)}OH/C_{10}H_{21}OH/H_2O$

315 — C_9PhE_9 Triton X-100–oleic acid–water
@ 20°C [1] [24,31]
br-$C_9H_{19}C_6H_4(OCH_2CH_2)_{9(avg)}OH/c\text{-}C_8H_{17}CH=CHC_7H_{14}CO_2H/H_2O$

316 — C_9PhE_{12} Emu-09–p-xylene–water @ 20°C [1] [41]
br-$C_9H_{19}C_6H_4(OCH_2CH_2)_{12(avg)}OH/p\text{-}(CH_3)_2C_6H_4/H_2O$

317 — C_9PhE_{12} Emu-09–p-hexadecane–water
@ 20°C [1] [41]
br-$C_9H_{19}C_6H_4(OCH_2CH_2)_{12(avg)}OH/C_{16}H_{34}/H_2O$

C_{22}
318 — $C_{22}E_7$ Heptaoxyethylene glycol docosyl [63]
ether–hexadecane–water @ 50°C
$C_{22}H_{45}(OCH_2CH_2)_7OH/C_{16}H_{34}/H_2O$

319 — $C_{22}E_7$ Heptaoxyethylene glycol docosyl [63]
ether–hexadecane–water @ 55°C
$C_{22}H_{45}(OCH_2CH_2)_7OH/C_{16}H_{34}/H_2O$

| 320 | — | $C_{22}E_7$ | Heptaoxyethylene glycol docosyl ether–hexadecane–water @ 60°C $C_{22}H_{45}(OCH_2CH_2)_7OH/C_{16}H_{34}/H_2O$ | [63] |

Nonsurfactant ternary systems

Salts

321	—	—	Potassium butanoate–octanol–water @ 25°C [1] $K^+, C_3H_7CO_2^-/C_8H_{17}OH/H_2O$	[87]
322	—	—	Butyltrimethylammonium bromide– [1] butanol–water @ 27.5°C $C_4H_9N(CH_3)_3^+, Br^-/C_4H_9OH/H_2O$	[82]
323	—	—	Butyltrimethylammonium bromide– [1] hexanol–water @ 27.5°C $C_4H_9N(CH_3)_3^+, Br^-/C_6H_{13}OH/H_2O$	[82]
324	—	—	Butyltrimethylammonium bromide– [1] octanol–water @ 27.5°C $C_4H_9N(CH_3)_3^+, Br^-/C_8H_{17}OH/H_2O$	[82]
325	—	—	Butyltrimethylammonium bromide– [1] dodecanol–water @ 27.5°C $C_4H_9N(CH_3)_3^+, Br^-/C_{12}H_{25}OH/H_2O$	[82]

Nonionic compounds

326	—	—	Ethanol–hexane–water at 25°C $C_2H_5OH/C_6H_{14}/H_2O$	[126]
327	—	—	Ethanol–heptane–water at 25°C $C_2H_5OH/C_7H_{16}/H_2O$	[126]
328	—	—	Ethanol–octane–water at 25°C $C_2H_5OH/C_8H_{18}/H_2O$	[126]
329	—	—	Ethanol–nonane–water at 25°C $C_2H_5OH/C_9H_{20}/H_2O$	[126]
330	—	—	n-Propanol–hexane–water at 25°C $C_3H_7OH/C_6H_{14}/H_2O$	[127]
331	—	—	n-Propanol–cyclohexane–water @ 25°C $C_3H_7OH/c\text{-}C_6H_{12}/H_2O$	[64]
332	12.3	—	n-Propanol–heptane–water at 25°C $C_3H_7OH/C_7H_{16}/H_2O$	[127]
333	—	—	n-Propanol–toluene–water @ 25°C $2\text{-}C_3H_7OH/CH_3C_6H_5/H_2O$	[64]
334	—	—	n-Propanol–nonane–water at 25°C $C_3H_7OH/C_9H_{20}/H_2O$	[127]
335	—	—	n-Propanol–dodecane–water @ 25°C $C_3H_7OH/C_{12}H_{26}/H_2O$	[64]
336	—	—	Isopropanol–hexane–water at 25°C $2\text{-}C_3H_7OH/C_6H_{14}/H_2O$	[127]

337	—	—	Isopropanol–heptane–water at 25°C [127] 2-C_3H_7OH/C_7H_{16}/H_2O
338	—	—	Isopropanol–octane–water at 25°C [127] 2-C_3H_7OH/C_8H_{18}/H_2O
339	—	—	Isopropanol–nonane–water at 25°C [127] 2-C_3H_7OH/C_9H_{20}/H_2O
340	—	C_3E_2	Dioxyethylene glycol propyl [64] ether–cyclohexane–water @ 25°C $C_3H_7(OCH_2CH_2)_2OH$/c-C_6H_{12}/H_2O
341	—	C_3E_2	Dioxyethylene glycol propyl ether– [64] toluene–water @ 25°C $C_3H_7(OCH_2CH_2)_2OH$/$CH_3C_6H_5$/H_2O
342	—	C_3E_2	Dioxyethylene glycol propyl ether– [64] octane–water @ 25°C $C_3H_7(OCH_2CH_2)_2OH$/C_8H_{18}/H_2O
343	—	C_3E_2	Dioxyethylene glycol propyl ether– [64] dodecane–water @ 25°C $C_3H_7(OCH_2CH_2)_2OH$/$C_{12}H_{26}$/H_2O
344	—	C_4E_1	Ethylene glycol butyl ether–cyclo- [64] hexane–water @ 25°C $C_4H_9OCH_2CH_2OH$/c-C_6H_{12}/H_2O
345	—	C_4E_1	Ethylene glycol butyl ether–toluene– [64] water @ 25°C $C_4H_9OCH_2CH_2OH$/$CH_3C_6H_5$/H_2O
346	—	C_4E_1	Ethylene glycol butyl ether–octane– [64] water @ 25°C $C_4H_9OCH_2CH_2OH$/C_8H_{18}/H_2O
347	—	C_4E_1	Ethylene glycol butyl ether– [64] dodecane–water @ 25°C $C_4H_9OCH_2CH_2OH$/$C_{12}H_{26}$/H_2O
348	—	C_6E_3	Triethylene glycol hexyl ether–cyclo- [64] hexane–water @ 25°C $C_6H_{13}(OCH_2CH_2)_3OH$/c-C_6H_{12}/H_2O
349	—	C_6E_3	Triethylene glycol hexyl ether– [64] toluene–water @ 25°C $C_6H_{13}(OCH_2CH_2)_3OH$/$CH_3C_6H_5$/H_2O
350	—	C_6E_3	Triethylene glycol hexyl ether– [64] octane–water @ 25°C $C_6H_{13}(OCH_2CH_2)_3OH$/C_8H_{18}/H_2O
351	—	C_6E_3	Triethylene glycol hexyl ether– [64] dodecane–water @ 25°C $C_6H_{13}(OCH_2CH_2)_3OH$/$C_{12}H_{26}$/H_2O
352	—	C_6E_4	Tetraethylene glycol hexyl ether– [64] cyclohexane–water @ 25°C $C_6H_{13}(OCH_2CH_2)_4OH$/c-C_6H_{12}/H_2O
353	—	C_6E_4	Tetraethylene glycol hexyl ether– [64] toluene–water @ 25°C $C_6H_{13}(OCH_2CH_2)_4OH$/$CH_3C_6H_5$/H_2O

| 354 | — | C_6E_4 | Tetraethylene glycol hexyl ether–octane–water @ 25°C $C_6H_{13}(OCH_2CH_2)_4OH/C_8H_{18}/H_2O$ | [64] |
| 355 | — | C_6E_4 | Tetraethylene glycol hexyl ether–dodecane–water @ 25°C $C_6H_{13}(OCH_2CH_2)_4OH/C_{12}H_{26}/H_2O$ | [64] |

References

1. Ekwall, P. (1975). *Advances in Liquid Crystals*, Vol. 1 (G. H. Brown ed.), pp. 1–142, Academic Press, New York.
2. Laughlin, R. G. (1978). *Advances in Liquid Crystals*, Vol. 3 (G. H. Brown ed.), pp. 41–148, Academic Press, New York.
3. Laughlin, R. G. (1984). *Surfactants*, (Th. F. Tadros ed.), pp. 53–82, Academic Press, New York.
4. Sjöblom, J., Stenius, P. and Danielsson, I. (1987). *Nonionic Surfactants, Physical Chemistry*, Vol. 23 (Martin J. Schick ed.), pp. 369–434, Marcel Dekker, New York.
5. Laughlin, R. G. (1990). *Cationic Surfactants Physical Chemistry*, 2nd ed., Vol. 37, pp. 1–40, Marcel Dekker, New York.
6. Adam, C. D., Durrant, J. A., Lowry, M. R. and Tiddy, G. J. T. (1984). *J. Chem. Soc., Faraday Trans. I* **80**, 789–801.
7. Ali, A. A. and Mulley, B. A. (1978). *J. Pharm. Pharmacol.* **30**, 205–213.
8. Ström, P. and Anderson, D. M. (1992). *Langmuir* **8**, 691–709.
9. Arvidson, G., Brentel, I., Kahn, A., Lindblom, G. and Fontell, K. (1985). *Eur. J. Biochem.* **152**, 753–759.
10. Balmbra, R. R., Clunie, J. S. and Goodman, J. F. (1969). *Nature* **222**, 1159–1160.
11. Balmbra, R. R., Clunie, J. S. and Goodman, J. F. (1965). *Proceedings Royal Soc., A* **285**, 534–541.
12. Boden, N., Jackson, P. H., McMullen, K. and Holmes, M. C. (1979). *Chem. Phys. Lett.* **65**, 476–479.
13. Broome, F. K., Hoerr, C. W. and Harwood, H. J. (1951). *J. Am. Chem. Soc.* **73**, 3350–3354.
14. Brown, J. P., unreported work.
15. Chernik, G. G. and Sokolova, E. P. (1991). *J. Coll. Interface Sci.* **141**, 409–414.
16. Chernik, G. G., Sokolova, E. and Morachevsky, A. (1987). *Mol. Cryst. Liq. Cryst.* **152**, 143–152.
17. Clunie, J. S., Goodman, J. F. and Symons, P. C. (1969). *Trans. Faraday Soc.* **65**, 287–296.
18. Clunie, J. S., Corkill, J. M., Goodman, J. F., Symons, P. C. and Tate, J. R. (1967). *Trans. Faraday Soc.*, 2839–2845.
19. Clunie, J. S., Corkill, J. M. and Goodman, J. F. (1965). *Proc. Royal Soc., A* **285**, 520–533.
20. Cox, J. S. G., Woodward, G. D. and McCrone, W. C. (1971). *J. Pharm. Sci.* **60**, 1458–1465.
21. Dörfler, H.-D. (1992). *Tenside Surf. Det.* **29**, 351–358.
22. Ekwall, P. (1963). *Finska Kemistsamfundets Medd.* **72**, 59–89.
23. Ekwall, P., Danielsson, I. and Mandell, L. (1960). *Kolloid Z.* **169**, 113–124.
24. Ekwall, P., Mandell, L. and Fontell, K. (1969). *Mol. Cryst. Liq. Cryst.* **8**, 157–213.
25. Ekwall, P., Mandell, L. and Fontell, K. (1968). *Acta Chem. Scand.* **22**, 365–367.
26. Ekwall, P., Mandell, L. and Fontell, K. (1969). *J. Coll. Interface Sci.* **29**, 542–551.
27. Ekwall, P., Mandell, L. and Fontell, L. (1968). *J. Coll. Interface Sci.* **28**, 219–226.
28. Ekwall, P., Mandell, L. and Fontell, K. (1969). *J. Coll. Interface Sci.* **29**, 639–646.
29. Ekwall, P., Mandell, L. and Fontell, K. (1969). *J. Coll. Interface Sci.* **31**, 508–529.
30. Ekwall, P., Mandell, L. and Fontell, K. (1970). *J. Coll. Interface Sci.* **33**, 215–235.

31. Ekwall, P., Mandell, L. and Fontell, K., unpublished work.
32. Ekwall, P. and Mandell, L. (1969). *Kolloid Z.* **233**, 938–944.
33. Ekwall, P. and Mandell, L. (1961). *Acta Chem. Scand.* **15**, 1407–1410.
34. Findenegg, G. H., Hirtz, A., Rasch, R. and Sowa, F. (1989). *J. Phys. Chem.* **93**, 4580–4587.
35. Finkelmann, H. and Schafheutle, M. A. (1986). *Coll. Polym. Sci.* **264**, 786–790.
36. Fontell, K., Mandell, L., Lehtinen, H. and Ekwall, P. (1968). *Acta Polytech. Scand., Chem. Incl. Met. Ser.* **74**, 1–116.
37. Fontell, K. and Jansson, M. (1988). *Colloid Polym. Sci.* **76**, 169–175.
38. Fontell, K. and Lindman, B. (1983). *J. Phys. Chem.* **87**, 3289–3297.
39. Fontell, K. (1973). *J. Coll. Interface Sci.* **43**, 156–164.
40. Fontell, K. (1969). *Proc. Int. Cong. Surface Active Substances, 5th, 1968*, Vol. II, p. 1033.
41. Friberg, S., Mandell, L. and Ekwall, P. (1969). *Acta Chem. Scand.* **23**, 1055–1057.
42. Friberg, S. E. and Liang, P. (1986). *Colloid Polym. Sci.* **264**, 449–453.
43. Friman, R., Danielsson, I. and Stenius, P. (1982). *J. Coll. Interface Sci.* **86**, 501–514.
44. Fu, Y.C., Glardon, A. S. and Laughlin, R. G. (1993). Presented at the American Oil Chemists' Society Meeting, Anaheim, California, 25–29 April.
45. Gibson, T. W., unreported work.
46. Gonick, E. and McBain, J. W. (1946). *J. Am. Chem. Soc.* **68**, 683–685.
47. Grabielle-Madelmont, C. and Perron, R. (1983). *J. Coll. Interface Sci.* **95**, 471–482.
48. Gradzielski, M., Hoffmann, H. and Oetter, G. (1990). *Colloid Polym. Sci.* **268**, 167–178.
49. Guerin, G. and Bellocq, A.M. (1988). *J. Phys. Chem.* **92**, 2550–2557.
50. Hato, M. and Shinoda, K. (1973). *J. Phys. Chem.* **77**, 378–381.
51. Herbst, L., Hoffmann, H., Kalus, J., Reizlein, K., Schmelzer, U. and Ibel, K. (1985). *Ber. Bunsenges. Phys. Chem.* **89**, 1050–1064.
52. Herrington, T. M. and Sarabjit, S. S. (1988). *J. Am. Oil Chem. Soc.* **65**, 1677–1681.
53. Herrmann, K. W., Brushmiller, J. G. and Courchene, W. L. (1966). *J. Phys. Chem.* **70**, 2909–2918.
54. Herrmann, K. W. and Benjamin, L. (1967). *J. Colloid Interface Sci.* **23**, 478–486.
55. Hertel, G. and Hoffmann, H. (1988). *Prog. Coll. Poly. Sci.* **76**, 123–131.
56. Heusch, R. and Kopp, F. (1988). *Prog. Coll. Polymer Sci.* **77**, 77–85.
57. Heusch, R. (1991). *Tenside Surf. Det.* **28**, 38–46.
58. Hildebrand, J. H., Fisher, B. B. and Benisi, H. A. (1950). *J. Am. Chem. Soc.* **72**, 4348–4351.
59. Jansson, M., Jönsson, A., Li, P. and Stilbs, P. (1991). *Coll. and Surf.* **59**, 387–397.
60. Jokela, P. and Jönsson, B. (1988). *J. Phys. Chem.* **92**, 1923–1927.
61. Jokela, P., Jönsson, B. and Kahn, A (1987). *J. Phys. Chem.* **91**, 3291–3298.
62. Jokela, P., Jönsson, B., Eichmuller, B. and Fontell, K (1988). *Langmuir* **4**, 187–192.
63. Jönsson, A., Bokström, J., Malmvik, A-C. and Wärnheim, T. (1990). *J. Am. Oil Chemists Assoc.* **67**, 733–738.
64. Kahlweit, M., Lessner, E. and Strey, R. (1983). *J. Phys. Chem.* **87**, 5032–5040.
65. Kahn, A., Das, K. P., Eberson, L. and Lindman, B. (1988). *J. Coll. Interface Sci.* **125**, 129–138.
66. Kahn, A., Fontell, K. and Lindman, B. (1985). *Prog. Coll. Poly. Sci.* **70**, 30–33.
67. Kahn, A., Fontell, K. and Lindman, B. (1984). *J. Coll. Interface Sci.* **101**, 193–200.
68. Kahn, A., Fontell, K. and Lindman, B. (1984). *Coll. Surfaces* **11**, 401–408.
69. Kahn, A., Fontell, K., Lindblom, G. and Lindman, B. (1982). *J. Phys. Chem.* **86**, 4266–4271.
70. Kahn, A., Lindman, B. and Shinoda, K. (1989). *J. Coll. Interface Sci.* **128**, 396–406.
71. Kaler, E. W., Herrington, K. L., Murthy, A. K. and Zasadzinski, J. A. N. (1992). *J. Phys. Chem.* **96**, 6698–6707.
72. Kekicheff, P. and Tiddy, G. J. T. (1989). *J. Phys. Chem.* **93**, 2520–2526.
73. Kekicheff, P., Grabielle-Madelmont, C. and Ollivon, M. (1989). *J. Coll. Int. Sci.* **131**, 112–132.
74. Krog, N. and Larsson, K. (1968). *Chem. Phys. Lipids* **2**, 129–143.
75. Kunieda, H. and Friberg, S. E. (1981). *Bull. Chem. Soc. Jpn.* **54**, 1010–1014.
76. Lang, J. C. (1984). *Proc. Int. Sch. Phys. "Enrico Fermi" (Phys. Amphiphiles)*, Vol. 90, pp. 336–375.
77. Lang, J. C. and Morgan, R. D. (1980). *J. Chem. Phys.* **73**, 5849–5861.

78. Larché, F. C., Marignan, J., Dussossoy, J. L. and Rouvière, J. (1983). *J. Colloid Interface Sci.* **94**, 564–569.
79. Laughlin, R. G. (1976). *J. Coll. Interface Sci.* **55**, 239–241.
80. R. G. Laughlin, unreported work.
81. Laughlin, R. G., Munyon, R. L., Fu, Y.-C. and Fehl, A. J. (1990). *J. Phys. Chem.* **94**, 2546–2552.
82. Lawrence, A. S. C., Boffey, B., Gingham, A. and Talbot, K. (1967). *Proc. Int. Cong. Surface Active Substances, 4th, 1964*, Vol. II, p. 673.
83. Lawrence, A. S. C. (1959). *Nature* **183**, 1491–1494.
84. Loos, M., Baeyens-Volant, D., David, C., Sigaud, G. and Achard, M. F. (1990). *J. Coll. Interface Sci.* **138**, 128–133.
85. Lühmann, B. and Finkelmann, H. (1966). *Colloid Polym. Sci.* **264**, 189–192.
86. Lühmann, B., Finkelmann, H. and Rehage, G. (1985). *Makromol. Chem.* **186**, 1059–1073.
87. Lumb, E. C. (1951). *Trans. Faraday Soc.* **47**, 1049–1055.
88. Lutton, E. S. (1966). *J. Am. Oil Chem. Soc.* **43**, 28–30.
89. Madelmont, C. and Perron, R. (1973). *Bull. Soc. Chim. France*, 3259–3263.
90. Madelmont, C. and Perron, R. (1974). *Bull. Soc. Chim. France*, 130–134.
91. Madelmont, C. and Perron, R. (1974). *Bull. Soc. Chim. France*, 425–429.
92. Madelmont, C. and Perron, R. (1974). *Bull. Soc. Chim. France*, 1795–1798.
93. Madelmont, C. and Perron, R. (1974). *Bull. Soc. Chim. France*, 1799–1805.
94. Madelmont, C. and Perron, R. (1976). *Colloid and Polymer Sci.* **254**, 6581–6595.
95. Mandell, L., unpublished work.
96. Mandell, L and Ekwall, P (1968). *Acta Polytech. Scand., Chem. Incl. Met. Ser. (Pt.1)* **74**, 1–116.
97. Maroncelli, M., Strauss, H. L. and Snyder, R. G. (1985). *J. Phys. Chem.* **89**, 5260–5267.
98. Marsden, S. S. and McBain, J. W. (1948). *J. Phys. Coll. Chem.* **52**, 110–130.
99. McBain, J. W. and Sierichs, W. C. (1948). *J. Am. Oil Chemists' Soc.* **25**, 221–225.
100. McBain, J. W., Vold, R. D. and Frick, M. (1940). *J. Phys. Chem.* **44**, 1013–1024.
101. Mitchell, D. J., Tiddy, G. J. T., Waring, L., Bostock, T. A. and McDonald, M. P. (1983). *J. Chem. Soc., Faraday Trans. I* **79**, 975–1000.
102. Mulley, B. A. and Metcalf, A. D. (1964). *J. Colloid Sci.* **19**, 501–515.
103. Nilsson, P-G., Lindman, B. and Laughlin, R. G. (1984). *J. Phys. Chem.* **88**, 6357–6362.
104. D. E. O'Connor, unreported work.
105. Persson, P. and Stenius, P. (1984). *J. Colloid Interf. Sci.* **102**, 527–532.
106. Peshek, C. V., O'Neill, K. J. and Osborne, D. W. (1987). *J. Coll. Inter. Sci.* **119**, 289–290.
107. Ralston, A. W., Hoerr, C. W. and Hoffman, E. J. (1942). *J. Am. Chem. Soc.* **64**, 1516–1523.
108. Reizlein, K. and Hoffmann, H. (1984). *Prog. Coll. Poly. Sci.* **69**, 83–93.
109. Rizzatti, M. R. and Gault, J. D. (1986). *J. Coll. Interface Sci.* **110**, 258–262.
110. Rogers, J and Winsor, P. A. (1969). *J. Coll. Interface Sci.* **30**, 247–257.
111. Saeki, S., Kuwahara, N., Makata, M. and Kaneko, M. (1976). *Polymer* **17**, 685–689.
112. Söderlund, H., Sjöblom, J. and Wärnheim, T. (1989). *J. Dispersion Sci. Tech.* **10**, 131–142.
113. Spegt, P. A., Skoulios, A. E. and Luzzati, V. (1961). *Acta Cryst.* **14**, 866–872.
114. Strey, R., Schomaeker, R., Roux, D., Nallet, F. and Olsson, U. (1990). *J. Chem. Soc. Faraday Trans.* **86**, 2253–2261.
115. Ulmius, J., Wennerström, H., Lindblom, G. and Arvidson, G. (1977). *Biochemistry* **16**, 5742–5745.
116. Vold, M. J. (1941). *J. Am. Chem. Soc.* **63**, 1427–1432.
117. Vold, R. D. and Ferguson, R. H. (1938). *J. Am. Chem. Soc.* **60**, 2066–2076.
118. Vold, R. D. (1939). *J. Phys. Chem.* **43**, 1213–1231.
119. Vorob'eva, A. I. and Karapet'yants, M. Kh. (1967). *Russ. J. Phys. Chem.* **41**, 602–605.
120. Vorob'eva, A. I. and Karapet'yants, M. Kh. (1966). *Russ. J. Phys. Chem.* **40**, 1619–1622.
121. Vorob'eva, A. I. and Karapet'yants, M. Kh. (1967). *Russ. J. Phys. Chem.* **41**, 1061–1063.
122. Wärnheim, T. (1986). *Coll. Poly. Sci.* **264**, 1051–1059.
123. Wärnheim, T., Bergenstahl, B., Henriksson, U., Malmvik, A. C. and Nilsson, P. (1987). *J. Coll. Inter. Sci.* **118**, 233–242.
124. Wärnheim, T. and Henriksson, U. (1987). *J. Coll. Inter. Sci.* **115**, 583–586.
125. Wärnheim, T. and Jonsson, A. (1992). *Prog. Coll. Poly. Sci.* **88**, 18–22.

126. Warr, G. G., Sen, R., Evans, D. F. and Trend, J. E. (1988). *J. Phys. Chem.* **92**, 774–783.
127. (1965). *Solubilities*, 4th ed., Vol. II, pp. 851–852, American Chemical Society, Washington, DC.
128. Gosselink, E. P., unreported work.
129. Lutton, E. S. (1965). *J. Am. Oil Chemists Soc.* **42**, 1068–1070.
130. Tartar, H. V. and Wright, K. A. (1939). *J. Am. Chem. Soc.* **61**, 539–544.

Appendix 3

Extracting, Graphing, and Using Data From Published Diagrams

A3.1 Binary diagrams

The phase diagram summarizes the phase behavior of a system in a highly condensed graphical form, and is the focal point of phase science. Phase diagrams are often published in journals or books with meager (or no) documentation. This practice poses genuine difficulties to the prospective user, insofar as the extraction of information from them is concerned (especially quantitative information). Practical methods for dealing with some of these problems have been developed in writing this book, and will be described briefly.

Unfortunately, the reader of this book will have the same problem with the diagrams herein as the author encountered with the original literature. At some point in the future, it may be hoped that access to surfactant phase data in a useful electronic file format will be possible. Such is presently the case with the phase diagrams of metallurgical [1] and ceramic systems [2], but surfactants will have to wait their turn.

Published phase diagrams are not usually accompanied by the numerical experimental data from which they were constructed. In some diagrams these data points may be depicted on the diagram (which is a highly recommended practice), but their numerical values are not reported. On rare occasions, both the numerical values and the graphical figure are reported [3]. Since the diagram is usually reduced in size (often to one column width, or about 70×45 mm), the absence of numerical data poses serious problems for the prospective user who is interested in more than just a pretty picture of phase behavior, or in qualitative information.

The advent of modern computer science offers a means of rectifying this situation. In order to produce the phase diagrams (and other figures) depicted in this book (excepting photographs and those produced using ChemWindows), three software packages (Sigmascan, Sigmaplot, and TableCurve) were used. (These are all available from the Jandel Scientific, 2591 Kerner Blvd, San Rafael CA 94901.) Several programs (macros) were composed to handle various kinds of recurring calculation needs.

Serious problems presently exist in the numerical evaluation of ternary, as well as binary, phase data. The production using computer methods of graphic figures of phase diagrams, and the numerical evaluation of these data, are inextricably linked. Programs (written in either Sigmaplot or QBASIC) are therefore included that permit calculation of the x, y-coordinates of points and lines in ternary phase diagrams from composition data, the inverse conversion of x, y-coordinate data to composition values, limited lever rule calculations in biphasic regions, and the rigorous calculation of phase ratios in ternary tie-triangles.

A3.1.1   Digitizing the diagram

The first step in extracting useful information from a published diagram is to digitize the data points and curves in the diagram. The data from such processes are usually stored in a spreadsheet (*.PRN) type of file, which is a two-dimensional array of (composition, temperature) coordinates. If the source of the diagram is a literature article, the diagram may first have to be enlarged (either photographically or using the enlargement feature of photocopiers). The enlargement is then pasted on a digitizing tablet (such as Sigmascan) and digitized. The puck, rather than the pen attachment, was used. During this process the coordinates of three points are defined by the user, and the cross-hairs of the puck are centered and clicked over each of these three points on the graph. This process calibrates the entire tablet, and one may then click on as many points on the graph (within its active area) as are required to document the diagram. The coordinates at which the puck rests, scaled in a manner determined by the calibration data, are stored on file at each click.* While distortion of the figure may have occurred during its enlargement, the calibration process serves to attenuate (in some degree) the magnitude of such errors. The reproducibility of commercial tablets is about 0.1% of full scale, which is as good as (or better than) the accuracy of all but the highest quality phase data.

It is useful to store in separate arrays within the same file (using delimiter data such as the origin) both the coordinates of the actual data points themselves, and also the coordinates of points taken at frequent intervals along the curves.

A3.1.2   Graphing the data

The *.PRN file produced using the digitizing tablet may be imported into a scientific graphics package (such as Sigmaplot), and a preliminary phase diagram constructed. A number of plots must usually be created within such a graph. For example, one plot may supply the individual data points, another the straight horizontal and vertical lines that exist, and still others the curved lines that describe each phase boundary. Since the boundaries in most phase diagrams are discrete curves described by different mathematical functions, the capacity to create more than one plot on the same graph is essential.

Because of the manual digitizing process, the data describing the curves will typically not be smooth. Selected arrays may be imported from the primary file into TableCurve for further analysis. This software functions as the computer equivalent of a french curve; it mechanically fits a set of data to literally thousands of algorithms of various kinds, and ranks the various fits according to a user-selected statistical parameter (such as the square of residuals, r^2). The user may then select a suitable fit by viewing graphs in which the data points are superimposed on the calculated curve. The desired algorithm is selected, and the data used to produce the displayed graph are stored in a separate Sigmaplot file. This process allows judgement to be exercised as to the form of the curve; those with the highest values of r^2 are often not the best curves. The array necessary for constructing the curve may then be copied from the new Sigmaplot file,

---

*Scanners were considered for this purpose, but rejected because the essential data are easily and accurately digitized by hand, and often the figures are "noisy" so that extracting the desired data from the noise is expected to be time-consuming. The time required to manually digitize most phase diagrams is about 10 minutes, with the more complex diagrams (such as SDS–water, Fig. 5.4) requiring less than 60 minutes.

pasted into the parent file, and a new plot for the curve added to the figure. In simple cases, this entire procedure for creating a smooth curve from a data array requires only a few minutes time – once the user is familiar with the programs.

Phase boundaries are notoriously difficult to fit to mathematical models [4,5], and boundaries having complex curves may have to be subdivided into two or more segments. TableCurve allows editing of the curve file (obviously, *not* the experimental data) so that outlier points that may seriously distort the model may be deleted, artificial points that may be necessary to provide a sensible curve added, etc. The program also allows x- and y-coordinates to be exchanged, so that a curve which bends back on itself may be laid on its side and two values of y do not exist for the same values of x. This interconversion may easily be reversed within Sigmaplot.

This software provides purely empirical models for curves that will likely have no fundamental significance whatsoever. They are, nevertheless, extremely useful for purely numerical analysis and graphics purposes.

Tie-lines. If tie-lines are desired, TableCurve can also be used to provide data for the coordinates of the ends of tie-lines in binary diagrams. The arrays required to define these lines can then easily be constructed from these data within Sigmaplot. The boundary of interest is first modeled, and the selected model is evaluated for x at preselected values of y (for example, at 5°C intervals). If the tie-lines connect two curves, the other boundary must also be similarly modeled and its model evaluated at the same temperatures. As each x, y point is calculated, it too is stored within a *.PRN file that may be imported into the parent phase diagram file. Dashes or text are inserted in the columns to construct a series of discontinuous horizontal lines. The tie-lines in the sodium chloride–water diagram (Fig. 1.1) were constructed in just this manner; the drawing produced is far superior to those in which tie-lines are drawn by hand.

The numerical values for the solubility of sodium chloride that were utilized in the text were similarly determined. Such methods, while purely empirical, utilize the entire data set to provide a model from which boundary coordinates may be evaluated. The resulting estimates should be superior in accuracy to any of the individual data points.

A3.1.3   Adding inner boundaries

A common need in constructing surfactant phase diagrams is designating the existence and approximate position of the liquid crystal boundaries that exist next to the liquid boundary. If the azeotropic point is visible (the liquid crystal region is not squeezed between other phases, Section 5.8), then this inner boundary must be asymptotic to the liquid boundary precisely at its maximum temperature. Constructing such boundaries poses a nontrivial problem, since the algorithm describing the liquid boundary is often very complex and analytically defining the inner boundary would be excessively laborious.

This problem was resolved by utilizing a parabolic correction function to define the liquid crystal boundary. If (as in Fig. A3.1) one has a liquid boundary (curve a–b–c), and wishes to add the liquid crystal boundary to the dilute side of this liquid boundary (b–d), then a parabola may be constructed which straddles the y-axis by defining the values of y_{min}, y_{max}, and the width of the miscibility gap at y_{min}, $d(x)$. This parabolic function may then be evaluated for x for many values of y along the liquid boundary.

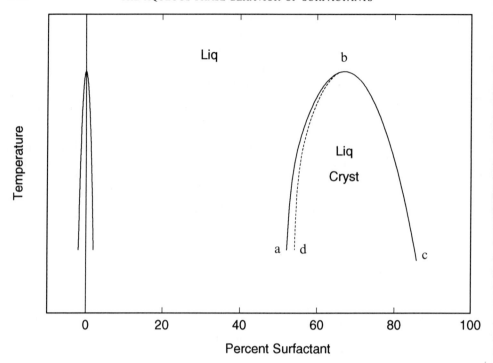

Fig. A3.1. The creation of inner liquid crystal boundaries using a parabolic correction for the abscissa (composition) values. A small parabola is created (left) whose maximum temperature, minimum temperature, and width at the minimum temperature are defined. The addition of the *x*-coordinates of this parabola, at particular values for the *y*-coordinates of the liquid boundary, produces an inner boundary that is asymptotic to the liquid boundary at its maximum temperature.

Adding the *x*-coordinates of the right (positive) half of this parabola to the corresponding *x*-coordinates of the dilute liquid boundary (*a–b*) will produce a new curve (*d–b*). This curve lies at higher compositions, and is necessarily asymptotic to curve *a–b* at point *b* because the slope of the parabola at its vertex is zero.

The values of y_{min}, y_{max}, and $d(x)$ may again be defined for the concentrated half of the liquid boundary, a similar parabola defined, and the *x*-coordinates of the left (negative) half added to the *x*-coordinates of the concentrated liquid boundary. This provides a similar curve which lies just inside the concentrated liquid boundary, and is also asymptotic to it at *b*. The math transforms (programs) by means of which these curves are constructed are listed below, as "DILINBDY.XFM" and "CONINBDY.XFM". The program language used is that of Sigmaplot, which is virtually self-explanatory. Specific values of the coordinates of cells (column,row) or the number of columns exist in these programs, but these may easily be edited during use to fit a particular file.

It is not clear how well this approach to defining the boundary of liquid crystal regions corresponds to reality. The parabolic adjustment will rigorously conform to the relationship required of the liquid and liquid crystal boundaries at the azeotropic point, but may be misleading at lower temperatures. It has never been possible to evaluate the form of these narrow miscibility gaps, however, because the data with which to do so have not existed.

A3.1.4 Miscellaneous curves

In sketching phase diagrams, smooth curves are often required whose general form may be anticipated but for which data do not exist, and would be troublesome to create. Segments of a parabola serve admirably to provide reasonable curves for such purposes. Transforms have been constructed ("PARSECX0.XFM" and "PARSECY0.XFM") that will construct a parabolic curve whose form may be adjusted easily and at will. By specifying the coordinates of two points (either numerically or as their cell coordinates), and also either the x-coordinate of the parabolic axis or the y-coordinate of its vertex, the form of the curve between the points may be varied until the visual appearance is satisfactory. If the axis is placed between the points, then a maximum (or minimum) will exist. If it is placed outside the points (to one side or another), then a section of either one arm or the other of the parabola is produced. By displaying, inspecting, erasing, adjusting, and recalculating these sections (which sounds laborious, but is not), the form of the figure may be adjusted to conform to the intuition of the chemist with reasonable effort.

```
; DILINBDY.XFM
; Parabolic correction to calculate inner bdys
; for dilute half of liquid crystal regions
ymax = cell(26,64) ; define upper limit of y
ymin = cell(26,1)      ; define lower limit of y
dx = 2                ; define delta(x)
b = ymax
a = (ymin − b)/dx^2 ; calc constants of parabola
y = col(26)              ; specify y-column to use
xcorr = sqrt((y − b)/a) ; calc x-correction
x = col(25)
newx = x + xcorr       ; calc new x-data
put newx into col(33) ; add x-data to file
put y into col(34)     ; add y-data to file
```

```
; CONINBDY.XFM
; Parabolic correction to calculate inner bdys
; for conc half of liquid crystal regions
ymax = cell(24,1)   ; define upper limit of y
ymin = cell(24,32) ; define lower limit of y
dx = 2                ; define delta(x)
b = ymax
a = (ymin − b)/dx^2 ; calc constants of parab
y = col(25)
xcorr = sqrt((y − b)/a) ; calc x-correction
x = col(23)
newx = x − xcorr
put newx into col(27) ; add data to file
put y into col(28)
```

```
; Filename PARSECX0.XFM
; Program to calc sec of parabola, specify x0
```

```
n = 20              ; specify no. of points
x1 = cell(47,1)     ; input coordinates of
y1 = cell(48,1)     ; cells containing data
x2 = cell(47,2)
y2 = cell(48,2)
x0 = cell(47,1)     ; specify axis of parabola
a = (y2 − y1)/((x2 − x0)^2 − (x1 − x0)^2)
b = y1 − a * (x1 − x0)^2 ; calc constants of parabola
c = data(0,n)
x = c * (x2 − x1)/n + x1 ; calc x values
y = a*(x − x0)^2 + b ; calc y values
put x into col(47) ; store x values
put y into col(48) ; store y values

; Filename PARSECY0.XFM
; Program to calc sec of parabola, specify y0
n = 10              ; specify no. of points
x1 = cell(1,1)      ; input data
y1 = cell(2,1)
x2 = cell(1,2)
y2 = cell(2,2)
y0 = 0              ; specify vertex of parabola
a = (y2 − y1)/((x2)^2 − (x1)^2)
put a into cell(1,3)
b = (y1 − y0) − a*(x1)^2
put b into cell(2,3)
c = data(0,n)
put c into col(3)
x = c * (x2 − x1)/(n) + x1   ; calc x values
y = a*(x − x0)^2 + b         ; calc y values
put x into col(4)            ; store x values
put y into col(5)            ; store y values
```

A3.2 Ternary diagrams

A3.2.1 Interchanging composition and x,y-coordinate data

Exactly the same procedure may be utilized to digitize and plot ternary diagrams that is used for binary diagrams, but it is necessary (if composition data are used) to first define the outline of the triangular figure and to transform composition values into x,y-coordinates before the diagram can be plotted. In all the triangular figures depicted herein, a file containing the triangle itself was the point of departure. This was constructed using trigonometric transforms that defined the outline of the triangle, as well as each of the desired tick-marks having the desired orientation and length. Data (such as a *.PRN file) were then imported into this file and the diagram constructed.

Rectangular graph coordinates are at right angles to one another, but the axes of triangular phase diagrams are not. It is necessary, therefore, to be able to both express composition coordinates as x, y-coordinates, and also to extract composition data from the values of x, y-coordinates. If the convention of Fig. 12.2 is used in which component 1 is water, component 2 the surfactant, and component 3 the third component (oil or salt), then the x, y-coordinates may be calculated from composition data using the equations

$$y = w_3 \times \text{sqrt}(3)/2, \text{ and}$$

$$x = w_2 + w_3/2$$

If the values of w are expressed as pure fractions, the length of the base of the triangle is taken as one. The calculation may also be performed using per cent values, in which case the base of the triangle is taken as 100. This process enables one to convert composition data from the literature into coordinates that are useful for constructing the figure. Fig. 12.3 was constructed in such a manner.

If the values of the x, y-coordinates are available (e.g. from the digitization of a published triangular diagram using Sigmascan) and the compositions at particular x, y-coordinates are required, these compositions may be calculated using the equations

$$w_3 = y \times 2/\text{sqrt}(3)$$

$$w_2 = x - w_3/2$$

If compositions are expressed in pure fractions

$$w_1 = 1 - (w_2 + w_3)$$

while if compositions are expressed in per cent

$$w_1 = 100 - (w_2 + w_3)$$

The following Sigmaplot transforms perform these calculations:

```
; TERNXY_C.XFM
; Program to convert x, y coordinates to compns of
; components 2 and 3 in ternary diagrams
x = cell(11,1)        ; input x-coordinate data
y = cell(12,1)        ; input y-coordinate data
w3 = y*2/sqrt(3)      ; calc compositions
w2 = x - w3/2
put w2 into cell(13,1) ; add data to file
put w3 into cell(14,1)


; TERNC_XY.XFM
; Program to calculate Cartesian coordinates of
; points within triangular diagrams, defined in
; terms of the fractions of components 2 and 3
; Length of side of triangle taken as unity.
w2 = col(5) ;          input data on component 3
w3 = col(6) ;          input data on component 2
```

$y = w3*\text{sqrt}(3)/2$;   calc coordinates
$x = w2 + w3/2$
put x into col(7) ; add x, y data to files
put y into col(8)

A3.2.2   Tie-line calculations in biphasic mixtures

If ternary data are available, or have been produced by the above process, one may wish to calculate the values of phase ratios in multiphase mixtures. These are trivial calculations using the lever rule in binary systems (Section 4.9), but are more complex in ternary systems. Some of the basic problems that exist in doing such calculations in biphasic regions have been noted (Section 12.1.2), but we will assume for the moment that the compositions of the tie-line of interest are known. It is then possible to specify one composition coordinate (taken to be w_3 in the example below), calculate the coordinates of the mixture for which this is true on the defined tie-line, and calculate the ratios of the two coexisting phases by using the following program. This program is written in QBASIC.

```
REM****************************************************************
REM
REM 2PLEVRUL.BAS
REM LEVER RULE CALCULATIONS OF PHASE RATIOS IN
REM BIPHASIC REGIONS OF 3-COMPONENT SYSTEMS
REM LAUGHLIN/RG
REM 3 DEC 82
REM****************************************************************
PRINT "CALCULATE FRACTION OF EACH PHASE KNOWING TIE-LINE &
ONE COMPOSITION"
REM COMPONENT 1 = SOLVENT
REM FRACTION COMPONENT 2 IN PHASE A IS W2A
REM FRACTION COMPONENT 2 IN PHASE B IS W2B
REM FRACTION COMPONENT 3 IN PHASE A IS W3A
REM FRACTION COMPONENT 3 IN PHASE B IS W3B
REM FRACTION OF PHASE A IS FA
REM FRACTION OF PHASE B IS FB
REM RECTANGULAR COORDINATES OF PHASE A ARE XA AND YA
REM RECTANGULAR COORDINATES OF PHASE B ARE XB AND YB
REM CALCULATE EQUATION OF TIE-LINE
PRINT "INPUT COMPOSITIONS OF COEXISTING PHASES"
INPUT "FRACTION OF 2 IN PHASE A IS"; W2A
INPUT "FRACTION OF 3 IN PHASE A IS"; W3A
INPUT "FRACTION OF 2 IN PHASE B IS"; W2B
INPUT "FRACTION OF 3 IN PHASE B IS"; W3B
D2 = W2B - W2A
D3 = W3B - W3A
R3 = SQR(3)
M = R3 * D2 / (2 * D3 + D2)
REM CALCULATE Y-INTERCEPT OF TIE-LINE EQUATION (B)
```

```
B = W2A - M * (2 * W3A + W2A) / R3
PRINT "SLOPE IS"; M, "Y-INTERCEPT IS"; B
INPUT "DO YOU WISH TO SPECIFY W3 (YES/NO)"; ED$
IF ED$ = "YES" THEN GOTO 100
INPUT "VALUE OF W2 IS"; W2
REM CALCULATE VALUE OF W3 AT INTERSECTION OF Y = W2 WITH TIE-
LINE
W3 = R3 / (2 * M) * (W2 - M * W2 / R3 - B)
PRINT "AT W2 ="; W2, "W3 ="; W3
GOTO 200
100 INPUT "VALUE OF W3 IS"; W3
REM CALCULATE VALUE OF W2 AT INTERSECTION OF CONSTANT W3
LINE WITH TIELINE
W2 = (R3 * B + 2 * M * W3) / (R3 - M)
PRINT "AT W3 ="; W3, "W2 ="; W2
200 REM CALCULATE FRACTIONS OF PHASES A AND B
REM CALCULATE DISTANCE- PHASE B TO INTERSECTION [W2B,W3B -
W2,W3],DIB
DIB = 2 / R3 * ((W2B - W2) ^ 2 + (W3B - W3) * (W2B - W2) + (W3B - W3) ^ 2) ^ .5
REM CALCULATE TIE-LINE LENGTH, DT
DT = 2 / R3 * (D2 ^ 2 + D3 * D2 + D3 ^ 2) ^ .5
FA = DIB / DT
FB = 1 - FA
PRINT "FB ="; FB, "FA ="; FA
INPUT "DO YOU WISH A LINE-PRINT (YES/NO)"; ED$
IF ED$ = "NO" THEN GOTO 300
LPRINT "LEVER RULE CALCULATION FOR 3-COMPONENT SYSTEM"
LPRINT LPRINT "FOR TWO COEXISTENT PHASES A[W2 ="; W2A, "W3 =";
W3A; "]"
LPRINT "AND B[W2 ="; W2B, "W3 ="; W3B; "],THE EQUATION OF THE
TIE-LINE IS"
LPRINT
LPRINT "Y ="; M; "X + "; B
LPRINT
LPRINT "AT W2 ="; W2; ", W3 ="; W3; "AND FA ="; FA; ", FB ="; FB
FOR I = 1 TO 5
LPRINT
NEXT I
300 END
```

A3.2.3 Phase ratios in tie-triangles

A particularly important application of the lever rule to data on ternary systems is calculation of the ratio of the three phases that coexist within a tie-triangle. This calculation can be executed directly with an accuracy that depends only upon the quality of the experimental data; it does not suffer from the uncertainties that exist in dealing with biphasic regions (above). The principle of such calculations is exactly the same as during application of the lever rule to biphasic mixtures, except that lever "arms" are lever "triangles", and the added constraint must be imposed that the sum of the fractions of the three phases present equals one (Section 12.1.2). This calculation is relatively complex, but purely mechanical. It may be performed using the following program (also written in QBASIC).

A test of whether or not the input composition falls within the triangle is built into this program, so as to preclude nonsensical calculations of phase ratios for mixtures that lie outside the triangle. Ascertaining whether or not this is true is altogether nontrivial if the mixture lies near a boundary of the triangle, or in the case of thin tie-triangles (Section 14.4).

```
REM ****************************************************************
REM *** ***
REM *** TIE_TRIA.BAS ***
REM *** LEVER RULE CALCULATIONS IN 3-PHASE REGIONS ***
REM *** OF 3-COMPONENT SYSTEMS ***
REM *** LAUGHLIN/RG ***
REM *** 8 MAR 83 ***
REM *** ***
REM ****************************************************************
PRINT "INPUT COMPOSITIONS OF COEXISTING PHASES AND 3-PHASE
SYSTEM"
PRINT "COMPOSITION UNITS MAY BE EITHER PERCENT OR FRAC-
TIONS"
INPUT "W2A ="; W2(1)
INPUT "W3A ="; W3(1)
INPUT "W2B ="; W2(2)
INPUT "W3B ="; W3(2)
INPUT "W2C ="; W2(3)
INPUT "W3C ="; W3(3)
INPUT "W2M ="; W2(4)
INPUT "W3M ="; W3(4)
REM CALCULATE X,Y-COORDINATES OF POINTS
FOR N = 1 TO 4
GOSUB 100
IF N = 1 THEN YA = Y
IF N = 1 THEN XA = X
IF N = 2 THEN YB = Y
IF N = 2 THEN XB = X
IF N = 3 THEN YC = Y
IF N = 3 THEN XC = X
IF N = 4 THEN YM = Y
IF N = 4 THEN XM = X
NEXT N
REM CALCULATE EQUATIONS AND LENGTHS OF SIDES
REM SIDE A-B
X1 = XA: Y1 = YA: X2 = XB: Y2 = YB
GOSUB 200
AAB = A: BAB = B: CAB = C: DAB = D
```

```
REM SIDE B-C
X1 = XB: Y1 = YB: X2 = XC: Y2 = YC
GOSUB 200
ABC = A: BBC = B: CBC = C: DBC = D
REM SIDE C-A
X1 = XC: Y1 = YC: X2 = XA: Y2 = YA
GOSUB 200 ACA = A: BCA = B: CCA = C: DCA = D
REM PHASE RATIO CALCULATION
REM CALCULATE AREAS OF INDIVIDUAL TRIANGLES
REM POINT M (MIXTURE COMPOSITION) TO SIDE A-B
X = XM: Y = YM: A = AAB: B = BAB: C = CAB: TBASE = DAB
GOSUB 300
ARMAB = AREA
REM POINT M TO SIDE B-C
X = XM: Y = YM: A = ABC: B = BBC: C = CBC: TBASE = DBC
GOSUB 300
ARMBC = AREA
REM POINT M TO SIDE C-A
X = XM: Y = YM: A = ACA: B = BCA: C = CCA: TBASE = DCA
GOSUB 300
ARMCA = AREA
ARTOTAL = ARMAB + ARMBC + ARMCA
PRINT "SUM OF LEVER TRIANGLES = "; ARTOTAL
REM CALCULATE FRACTIONS OF PHASES A, B, AND C
REM FIRST DETERMINE WHETHER OR NOT POINT LIES INSIDE
TRIANGLE
REM CALCULATE DISTANCE OF PERPENDICULAR FROM POINT A TO
SIDE B-C
X = XA: Y = YA: A = ABC: B = BBC: C = CBC: TBASE = DBC
GOSUB 300
TTAREA = AREA
PRINT "TIE-TRIANGLE AREA = "; TTAREA
PRINT : PRINT "IF SUM OF LEVER TRIANGLES IS GREATER THAN
TIE-TRIANGLE AREA THEN",
INPUT "DO NOT CALCULATE. INDICATE WHETHER OR NOT TO
CALCULATE, YES OR NO (Y/N)"; D$
IF D$ = "N" THEN GOSUB 400
FPA = ARMBC / ARTOTAL: FPB = ARMCA / ARTOTAL: FPC = ARMAB /
ARTOTAL
PRINT : PRINT "FOR A MIXTURE OF COMPOSITION W(2) = "; W2(4), "W(3)
= "; W3(4),
PRINT "W(1) = "; (100 - W2(4) - W3(4))
PRINT "IN A TIE-TRIANGLE CONTAINING PHASE A (W(2) = "; W2(1), "W(3)
= "; W3(1), "W(1) = ", (100 - W2(1) - W3(1)), ")"
PRINT "PHASE B (W(2) = "; W2(2), "W(3) = "; W3(2), "W(1) = "; 100 - W2(2) -
W3(2), "), AND"
PRINT "PHASE C (W(2) = "; W2(3), " W(3) = "; W3(3), "W(1) = "; 100 - W2(3) -
W3(3), ")"
```

```
PRINT
PRINT "THE FRACTION OF PHASE A = ", FPA
PRINT "THE FRACTION OF PHASE B = ", FPB
PRINT "THE FRACTION OF PHASE C = ", FPC
PRINT
PRINT "END OF CALCULATION" END
100 REM SUBROUTINE TO CONVERT COMPOSITIONS TO X,Y-COORDI-
NATES
Y = W3(N) * SQR(3) / 2: X = W2(N) + W3(N) / 2
RETURN
200 REM SUBROUTINE TO CALCULATE EQUATIONS, LENGTHS OF LINES,
AND TRIANGULAR
REM AREAS.
M = (Y2 - Y1) / (X2 - X1): YINT = Y1 - M * X1
A = M: B = -1: C = YINT: D = SQR((Y2 - Y1) ^ 2 + (X2 - X1) ^ 2)
RETURN
300 REM SUBROUTINE TO CALCULATE TRIANGULAR AREAS
D = ABS(A * X + B * Y + C) / SQR(A ^ 2 + B ^ 2)
AREA = D * TBASE / 2
RETURN
400 REM SUBROUTINE TO END PROGRAM
PRINT "POINT LIES OUTSIDE TIE-TRIANGLE"
END
```

Happy phase science!

References

1. *Binary Alloy Phase Diagrams*, 2nd ed. Available from ASM International, Metals Park, Ohio. Phone 216/338-5151.
2. NIST Standard Reference Database 31, *Phase Diagrams for Ceramists*, available from The American Ceramic Society, 65 Ceramic Drive, Columbus, Ohio 43214. Phone 614/268-8645.
3. Lang, J. C. and Morgan, R. D. (1980). *J. Chem. Phys. 73*, 5849–5861.
4. Oonk, H. A. J. (1981). *Phase Theory: The Thermodynamics of Heterogeneous Equilibria*, pp. 165–188, Elsevier Scientific, New York.
5. Gorman, J. W. and Cornell, J. A. (1985). *Technometrics*, pp. 3229–3239.

Appendix 4

The Determination of Phase Diagrams

A4.1 An approach to the determination of phase diagrams

The level of development of a field of science is strongly influenced by the scope and limitations of its methodologies, and nowhere is this more clearly evident than in surfactant phase science. In this appendix an approach to the determination of phase diagrams will be suggested, the existing methods will be briefly described and evaluated, and references will be provided to more detailed descriptions.

In the last part, means by which the quality of a published phase diagram can be evaluated will be suggested. Each of the diagrams included in this book have been critically scrutinized using these criteria. Some have been edited by the insertion of dashed lines or boundaries, where such modifications seemed justified.

The first steps in performing a phase study are to define the system in terms of the number of components of interest and their molecular structures, and the range of system variables of interest. The objective of the study is then to define the phase behavior of the selected system within these boundary conditions. It will be recalled (Section 4.1) that the dimensions of phase behavior are:

(1) the number of phases present, when the system is at equilibrium,
(2) the compositions of the phases present, when the system is at equilibrium, and
(3) the structures of the phases present, when the system is at equilibrium.

Attaining equilibrium, and documenting the initial hypothesis that equilibrium has been attained (preferably by using more than one kind of information), must be the paramount concern during a phase study. One may encounter kinetically slow phase reactions, the formation of colloidal states, or other nonequilibrium phenomena during the study. It will be found that information regarding these non-equilibrium aspects of the physical science of the system is most clearly revealed by focusing attention during the process on defining the equilibrium state. Preliminary data typically allow a hypothesis as to the phase behavior to be stated; this hypothesis can then be independently tested by performing qualitatively different experiments.

This cycle of (1) experimental work → (2) phase hypothesis → (3) verification experiments may have to be repeated several times. For simple diagrams of a familiar type, one cycle may suffice. During determination of the DODMAC–water phase diagram [1], about eight different hypothetical diagrams were sketched before one that withstood further testing resulted. During the $C_{10}E_4$–water study [2] an even larger number of hypotheses were created and discarded before one was accepted and published, and even that was still incorrect with respect to a few small details. The soap–water diagrams remain incomplete after 70 years.

This principle may be illustrated using a critical analysis of the process of determining melting points from the perspective of phase chemistry (Section 4.17.1), but first the matter of determining and evaluating the purity of the sample used will be considered.

A4.2   The purity issue

A vexing issue of considerable practical significance to phase science is "how pure must a sample be for phase studies to be meaningful?". If definitive studies are to be performed, acquiring a sample of sufficient purity is just as important as is producing high quality physical data. Modern gas chromatographic analysis of all fatty compounds (either the surfactant *per se*, or the long-chain intermediate used in its synthesis) invariably documents the fact that no samples are 100% pure.

This issue may be resolved by defining the goal of the investigation to be "the determination of physical data that are characteristic of the molecule of interest". If the value of a physical parameter for a particular sample is altered to a perceptible degree by measurably refining the sample, then the value for the refined sample is closer to the true value. If a series of such cycles:

physical study → repurification → redetermination of physical parameter

is carried out until further refinement of the sample no longer alters the value of the parameter, then it may be assumed that a value characteristic of the material has in fact been found.

This approach provides a practical means of deciding "how pure is pure enough?" that does not depend upon arbitrary assumptions as to acceptable numerical values of analytical assays. Such an approach was taken during the $C_{10}E_4$–water study [2], during which it was found that reducing the level of impurities below a certain value (about 0.1%) no longer altered the temperature of the discontinuities at the lower limit of the anomalous phase corridor (Fig. 4.5).

The minimal assay beyond which further refinement does not alter the data will depend upon the precision of the determination. If very precise methods are used, then the minimal level of purity that must be attained to achieve a stable value will be higher than if an impressive method is used.

It is useful to express purity as the per cent of impurities present, rather than as the per cent of the major component. A sample containing 0.1% impurities may be regarded as being twice as pure as one containing 0.2% in terms of the magnitude of the effect of the impurities on colligative properties, even though 99.9% and 99.8% assays look rather similar.

It is also interesting to express the level of impurities as the number of molecules present per mole of major component. Assuming that an impurity has a molecular weight similar to that of the major component, a purity level of 99.8% means that about 10^{21} molecules of impurities exist per mole of major component (or 6×10^{23} molecules)! A level of impurities of 10^{21} molecules/mole would generally be regarded as a very high level of purity, but is far from zero.

A4.3   Unary systems

The usual process for determining melting points [3] involves placing a few milligrams of a sample in a melting point capillary, varying the temperature of the sample, and observing the lowest temperature at which the solid is completely transformed into a liquid. A "melting point range" is invariably reported. The lower limit of this range is the temperature at which a visible change or motion occurs in the sample ("sintering"), and the upper limit is the lowest temperature at which the sample becomes a single liquid phase. Data obtained using this approach may be highly reproducible and have been very useful historically, but this is not the best way to determine this important phase transition.

The melting point of a pure compound is defined thermodynamically as the temperature at which the crystal and the liquid phases coexist in equilibrium with one another at specified pressure (Section 4.14). One notes, first, that the melting point does *not*, in fact, span a range of temperatures. Temperature is not a degree of freedom in a unary system when two coexisting phases exist and pressure is defined; the melting point is a unique temperature under these conditions. Why, then, is a range of temperatures over which the sample "melts" typically observed?

There are many, many possible explanations for this observation. The simplest is that the sample is actually a mixture containing one or more impurities, and the impurity component depresses the temperature at which crystals finally "melt". Quite often this depression occurs as a result of the presence of a eutectic in the binary (major component + impurity) system. If this is the case the eutectic liquid will first appear at the eutectic temperature – but sintering will not usually be observed to occur at this temperature. The sintering temperature probably has no fundamental significance at all, but reflects the temperature at which the fraction of liquid phase present is sufficiently large to be visible. For relatively pure samples, this is probably far above the eutectic temperature.

Lawrence has noted that the "penetration" temperature provides a superior means of defining the temperature of a eutectic in many solvent–solute systems, short of determining the phase diagram [4]. Also, some impurities may not depress the melting point; the presence of water in fatty alcohol mixtures raises the temperature at which crystals "melt", for example (Fig. 10.3). In crystalline hydrocarbons, an impurity that is higher melting than the major component and forms solid solutions near the melting point will raise the "melting point" relative to that of the pure component.

Another possible explanation is that the crystal being studied is, in fact, a phase compound (such as a crystal hydrate) and what is observed is a peritectic phase reaction, perhaps followed by recrystallization of another crystal and then melting. Another possible explanation is the existence of polymorphism in the dry crystal state. In the days when melting points were routinely determined as a means of ascertaining purity, it was observed rather often that compounds melted, recrystallized, and then melted again at a higher temperature. Such observations suggest the existence of polymorphic crystal phases. Still another possible explanation is chemical decomposition.

A phase science approach to the determination of melting points. The determination of melting points in the author's laboratory has for several years been approached as follows. The compound (c. 10 mg) is carefully placed at the bottom of a sealed 3-mm od glass tube about 8 cm long to which a short section of 8-mm od tubing (for attaching a vacuum hose) has been sealed at the open end. A $0.1 \times 1.0\,mm \times 5\,cm$ long rectangular capillary is placed in the 3-mm tube (on top of the sample), and the 3-mm tube is evacuated and sealed above the capillary. The sealed tube is placed in an upright Mettler hot stage and heated to above the melting point. The liquid sample wicks into the capillary, which, being flat, allows textures to be observed without the severe optical distortion that exists in melting point capillaries. After freezing the sample, a rough determination of the melting point is made with the hot stage on a polarized light microscope having long-working-distance universal objectives, the sample is again frozen, and the temperature is adjusted (in step-wise increments) until the melting point condition is determined.

The "melting point condition" has the following characteristics:

(1) the sample is mostly (>95%) liquid phase (so as to most nearly approximate the transition temperature of the strictly pure compound), but contains crystals of the major component;
(2) a small lowering in temperature (0.1°C) causes rapid and extensive (usually incomplete) crystal growth;
(3) a small increase in temperature (0.1°C) causes the crystals of the major component to disappear (but not those of high-melting impurities);
(4) if the melting point temperature is exceeded, the mixture must be cooled several degrees below the true melting point before crystals of the major component reappear. If this happens crystallization must again be induced (by severe cooling) and the process repeated.

The highest temperature at which crystals and liquids coexist (the liquid being the phase present in greatest amount) is taken as the melting point temperature.

Reversibility is established by cooling until the bulk of the sample is frozen, repeating the determination on the same sample, and finding again the same temperature. A progressive decrease in the observed transition temperature on repeating the determination, and/or a change in appearance (such as discoloration or bubble formation), suggests chemical decomposition. The existence of chemical decomposition can be directly evaluated, if required, by breaking open the tube and subjecting the sample to microscope slide thin-layer chromatographic analysis [5].

Many compounds are sufficiently stable in air to melt reversibly during a conventional determination, but others are not. Dioctadecylammonium chloride (DOAC) is as an example of a compound that is thermally stable at its melting point and melts reversibly at 178.9°C, but undergoes oxidative reactions in an open capillary (or between slide and cover slip) that frustrate determination of the melting point by the usual methods.

The above procedure not only provides the melting point temperature, but eliminates the possibility that atmospheric components (principally oxygen and water, sometimes carbon dioxide) influence the determination. An example of the influence of water occurred during a study of the N-dodecanoyl-N-methylglucamine, where it was observed that the temperature of the Lam → Liq transition progressively *increased* (using a slide/cover slip sample) on repeating the determination. This increase was accompanied by changes in the texture of the lamellar phase from a complex undecipherable texture to a recognizable oily streak texture [6]. The form of the phase diagram (after it was finally determined) suggested that these changes were due to absorption of water from the atmosphere. They do not occur in a sealed sample, and a reproducible melting point is readily determined.

The above procedure also provides a permanently sealed sample that (if protected from light) can be reexamined many years later (if required) and expected to have the same composition and properties. By sealing a sample that is thermally stable *in vacuo* and protecting it from light, one has eliminated most of the sources of chemical change in a system. There is one source of chemical instability that cannot be eliminated, however; that is the presence of radioactive isotopes. Small quantities of ^{14}C exist in all organic compounds, and the nuclear chemistry of this and other radioactive nuclei slowly induce chemical change. This change is normally imperceptible unless the abundance of an isotope is enriched – in which case this source of instability is often readily apparent even in the most chemically stable molecules.

This procedure could, in principle, be utilized to determine any transition between condensed phases of non-volatile compounds that is visible by direct observation. Polymorphic phase reactions among crystals may be optically invisible, however, in which case these may only be perceived using calorimetric or structural (X-ray) data.

A4.4 Multicomponent aqueous systems

It was recognized by Hill [7] that all the various methods for determining phase diagrams could be classified as being either isothermal (constant temperature), isoplethal (constant composition), or isobaral (constant pressure). Hill further noted that no isobaral methods existed at that time (1923). That situation has changed with the development of the DIT method, which (at each temperature) produces data that are both isobaral and isothermal [8]. All the phases present are exposed to the same pressure during each experiment, but this pressure varies with temperature.

An excellent axiom to follow in phase science is that one cannot know the behavior of a complex system until one also knows the behavior of the simpler systems from which it is constructed. Unary systems (such as water itself) are the simplest of all, and determining the phase behavior of these systems is essential in order to understand the behavior of binary and ternary systems. The phase behavior of water has been established beyond doubt, but that of surfactants has to be determined.

Occasionally one encounters the point of view that since the primary region of interest is, for example, the dilute concentration region (where a surfactant is actually used), one need not be concerned with the entire phase diagram. This point of view is much too narrow, because phase separation very often leads to concentrated phases. If, for example, a dilute surfactant solution (or liquid crystal phase) is cooled to a temperature such that a crystal separates, then the physical science of the very concentrated crystal state comes into play. This is because the thermodynamics of this state is just as important to the phase behavior within this miscibility gap as is that of the liquid. There is genuine value, from both a fundamental and a practical perspective, to being concerned with the entire composition range. The span of temperatures (or pressures) with which one is concerned may be restricted for various reasons, but it is unwise to exclude from consideration any part of the composition range.

A4.4.1 Crystal hydrates

The vast majority of surfactants form crystal hydrates, and if these phase compounds exist at the temperature of interest they (not the dry crystal) often constitute the coexisting crystal phase (Section 5.7). It is highly desirable during a phase study to determine at an early stage whether or not they exist, to characterize them independently as to composition, structure, and phase behavior, and (ideally) to synthesize them in a pure state.

Clues to the conditions within which crystal hydrates exist are evident from considering the general form of water activity profiles (Section 7.2). These phase compounds will exist within a finite span of water activities at a given temperature, and may be synthesized by isopiestic methods with control of water activity [42]. Such syntheses are not difficult, but may require extended periods of time (months or years). The control of water activity is most easily accomplished by placing a mixture of an inorganic salt and water in the bottom of a small desiccator to control water activity, inserting a filter paper wick against the walls of the desiccator, and placing the sample in a tared weighing bottle on the platform of the desiccator. Provided the empty weighing bottle has been preequilibrated in the desiccator prior to determining its tare weight, the uptake of water may be followed quantitatively and the stoichiometry of the crystal hydrate determined. The sample *must* be dry before starting the experiment if one is to obtain reliable quantitative data. Powder ATR (attenuated total reflectance) FT-IR infrared spectra constitute a superior means of perceiving the presence of even small amounts of water in a crystal, and (along with X-ray powder data) of characterizing a crystal hydrate. Most other kinds of sample preparation methods for obtaining infrared spectra are to be avoided, as they destroy crystal hydrates.

A serious artifact that may be encountered during isopiestic studies at high water activities is the capillary condensation of water in the crevices between crystals, which compromises gravimetric data. The curvature of the interface of such water reduces its vapor pressure (as described by the Kelvin equation [9]), so that condensation will occur if the water activity is too high. However, if the a_w is 0.9 or less it has been found that capillary condensation is eliminated. Exploratory studies at this activity (using a zinc nitrate desiccator) are, for this reason, preferred over studies using pure water.

While only limited experience with these studies exists at this time, it is clear that most surfactants form crystal hydrates (Section 8.3.4).

Whether or not a crystal hydrate coexists with other phases is immediately apparent during DIT-IR studies, since infrared spectra readily distinguish dry crystals from crystal hydrates. Both the O–H fundamental (3300 cm^{-1}) and the bend–stretch combination band of water (5180 cm^{-1}) are useful for detecting and identifying water. The combination band is particularly useful when hydroxy groups also exist in the surfactant (as in the glucamine derivative), as such groups rarely (if ever) interfere with the water combination band.

A4.4.2   Isothermal analytic methods

Isothermal analytic methods were probably the earliest phase study methods used. In the classical application of these methods to the study of partially miscible solvents mixtures are prepared that fall within the miscibility gap, the coexisting phases are separated, and their compositions and structures are determined using appropriate methods. (The approximate composition ranges of miscibility gaps must earlier have been determined by exploratory studies.) When applicable, this is possibly the most rigorous means of determining phase diagrams that exists. Evidence that equilibrium has been attained may be provided by demonstrating that phase compositions are independent of time, and of the path by means of which the final conditions were approached (e.g. by heating or cooling) moreover, each phase is isolated and examined directly.

The isothermal analytic method is restricted to equilibria between fluid phases of low viscosity that separate and coalesce rapidly. The method is not easily applicable even to determination of the phase diagram of the sodium chloride–water system, since crystal phases cannot be cleanly separated and defined. Crystal masses always occlude variable quantities of the liquid phase, and unless this phase can be rinsed out with the solvent (which entails risks), it cannot be characterized. The liquid phase itself may be cleanly separated and characterized, however, so that the liquid solubility boundary in systems such as sodium chloride–water can (and was) determined using this approach. Independent studies were required to determine the composition and structure of the coexisting crystal phase. Elaborate methods for by-passing this problem, when uncertainty existed as to crystal hydrate formation, were developed during the early part of this century: An example is the "wet-residues" method of Schreinemakers [10].

The isothermal analytic method is usually defeated in surfactant systems by the high viscosities of one or both phases, or by the failure of dispersions of one phase in the other to separate cleanly. Both problems are very serious and invariably compromise the method for surfactants of commercial interest. Nevertheless, this method is occasionally useful for binary studies of very short chain-length surfactants such as $C_{10}E_4$. Elegant triple-walled thermostatted sample chambers, excellent temperature control, and well-calibrated high performance liquid chromatography (HPLC) analytical methods were devised which permitted rigorous investigations of the closely spaced pairs of isothermal discontinuities found in this system to be carried out. However, in spite of the care taken later evidence (from swelling studies) suggested that the data were quantitatively uncertain within miscibility gaps involving the lamellar liquid crystal phase, and were qualitatively in error at one point [11]. Otherwise the isothermal and swelling studies were in excellent agreement, and this result strengthens the validity of the diagram.

Ternary surfactant–oil–water studies. Isothermal methods were also the choice of Ekwall and coworkers for the pioneering ternary studies of soap–water systems. Mixtures of a few grams, prepared by direct weighing of analytically well-defined standard samples and water, were held within tightly capped vials and mixed by shaking and diffusion. These studies were conducted within a room held at 20°C. This facilitated the direct observation of numerous mixtures during long periods of time (weeks, months, years), sampling for powder X-ray and density studies (using pycnometry) without disturbing the equilibrium state, and the examination of small representative aliquots using a polarized light microscope kept in the room. The range of temperatures over which this approach is practical is very limited.

Even with the advantages of the lengthy time of study, evidence has mounted during recent years that some of the "phases" initially thought to exist (the "*A*", "*B*", and "*C*" phases) were not, in fact, single phases at all. Instead, they may have been mixtures of phases (often *L* and *D*) that displayed characteristic and reproducible microscopic textures. The position that this is true is taken in this book, but it must be recognized that uncertainties exist in this aspect of ternary equilibria about which legitimate differences of opinion remain. New methodologies will likely be required to resolve these issues unambiguously.

Specialized methods have been described for determining the compositions of three coexisting liquid phases such as are encountered during microemulsion studies [12]. It has been suggested that measurements of phase volumes at several compositions within a tie-triangle can be used to define the compositions of coexisting phases, provided algorithms exist by means of which measured volumes can be converted to estimated compositions. The obvious advantage of this approach is that analytical methods need not be developed for each of the components of the system.

NMR phase studies. In 1977 nuclear magnetic resonance (NMR) methods (principally deuterium NMR) were introduced as a means of conducting phase studies [13]. NMR provided a means of recognizing the phases present without requiring prior separation of the phases. The method has provided useful phase information regarding extremely difficult systems, such as DPPC–water (Fig. 8.11) and other poorly water-soluble surfactants [14]. Its success depends upon the existence of differences in the NMR parameters (line widths or splitting) of deuterium among the various phase states that exist, in mixtures of defined composition that have been stored for a sufficient time for equilibrium to have been established and partial coalescence to have occurred. It is not applicable if these differences are averaged within the NMR time scales. The importance of this approach is enhanced by the fact that NMR data have come to be accepted as an important adjunct to X-ray data for the determination of phase structure [15].

If the NMR method could provide both the identification of phases *in situ* and their ratio, it would constitute a powerful extension indeed of the classical isothermal analytic method. It has not presently attained this status, however, because peak intensities are not easily quantitatively related to the fraction of each phase that is present. Should phase ratios ever become quantitatively determinable using NMR spectral data, this method would become still more valuable. The method requires access to and considerable time on an NMR spectrometer.

Microemulsion studies. Isothermal methods have also been used to investigate microemulsion-forming systems. Because of the ultralow tensions often encountered, such mixtures must often be allowed to stand for long periods of time at constant temperature for the phases to separate. The low viscosities of these phases favor this process, fortunately.

In recent times, robotic systems have been devised which dramatically reduce the manual labor required to prepare, mix, and instrumentally observe (e.g. the turbidity) of large numbers of samples of known composition [43]. Robots offer major advantages in searching for and defining the boundaries of fluid liquid phase regions, but do not circumvent the problems inherent to analytic methods.

Isothermal titration methods. Schulman discovered microemulsions by titrating the cosurfactant (pentanol) into an emulsified mixture of an aqueous solution of the surfactant (potassium oleate) and a hydrocarbon oil. The boundary composition is taken as the point of clarification of the initially turbid mixture. This simple and direct approach has since been utilized by large numbers of workers to define the region of existence of microemulsion phases. Its success, too, hinges upon the assumption that equilibrium of state is rapidly attained during mixing.

A4.4.3 Isoplethal methods

During isoplethal phase studies a composition is selected, a mixture having this composition is prepared and mixed, the temperature of the mixture is varied, and the temperatures at which discontinuities occur in the number of phases that exist are determined. Gross composition remains fixed along isoplethal paths, but the compositions of the phases present may vary widely. The most important of the discontinuities observed are changes from one phase to two (especially of a liquid phase to a mixture of the liquid plus another phase), and isothermal discontinuities such as eutectics, peritectics, and polytectics (Section 4.14). It will be recalled that in passing a eutectic two phases first react with one another to form a third, then one of the three vanishes. Such discontinuities may be observed using any subjective or instrumental physical data that are sensitive to such changes. These include turbidity (scattering intensities at zero scattering angle), birefringence intensities, heat effects, structural data, volume, etc.

Isoplethal studies by direct observation. This most basic of all isoplethal methods involves simply observing a mixture of known composition as the temperature is varied. The principle was first described by Alexejew [16], who utilized it to determine the phase diagrams of partially miscible solvent systems. For this purpose (where the type of boundary encountered is typically an upper or lower consolute boundary), direct observation of the sample in ordinary light is a very sensitive method for determining the boundary temperature. The reason is that separation of the second phase typically occurs throughout the mixture, the separated phase typically exists initially as small droplets, and intense scattering results.

Qualitatively similar isoplethal studies were used by McBain during his early soap–water studies, and were widely adopted by later workers. Mixtures were sealed in glass tubes, thermostatted at various temperatures (with occasional shaking) for long periods of time, and observed in various ways. For those boundaries at which a birefringent phase separates from an isotropic phase, the sensitivity of the observations is considerably enhanced by observing the mixture between crossed polars (either in bulk samples or using a polarized light microscope). This method is of particular value for defining the boundary of liquid phases, especially when used in combination with the step-wise dilution procedure for varying composition (described below).

It is assumed that equilibrium is maintained during isoplethal studies. Mixing is required in order to facilitate mass transport in large (g) samples. Mixing samples within sealed tubes poses a very serious problem, if the phases are viscous. Often mixing is accomplished by sealing the samples in a heavy-walled tube that has a constriction near the middle having an orifice of about 1–2 mm diameter. The mixture is centrifuged back and forth through this orifice a particular temperature. This will eventually provide a homogeneous sample (if the mixture is a single phase), or a dispersion of the phases present (if more than one exists). The process is laborious, requires a large thermostatted centrifuge that spans the temperature range of interest, and the mixing is almost certainly incomplete [17].

A piston mixing device has recently been developed that permits air-free samples to be mixed by forcing the sample back and forth through 25 mm long orifices of known dimensions (c. 0.5 mm). The temperature may be controlled, and small superatmospheric pressures may be tolerated without loss of water [1]. Studies using this device have shown that ordinary mixing (to the point that heterogeneity is not perceptible to the naked eye) does *not* provide truly homogeneous mixtures. Perceptible gradients have been observed (using the DIT apparatus, Section A4.4.4) in both liquid and liquid crystal samples even after passage several times through the piston mixing device, but homogeneous mixtures are eventually obtained using this method with a sufficient number of passes. Concentrated liquid surfactant solutions may also display perceptible heterogeneity, even after 30 min stirring using a conventional magnetic stirrer. Such mixing should be followed by sufficient standing time for residual gradients to relax by diffusion, if rigorously homogeneous mixtures are desired. For many studies in which the sample volume is large (as in NMR), such inhomogeneity may not be important. It is critically important to DIT studies because of the small sample volume taken ($0.8 \mu l$).

The step-wise dilution method. A particularly useful method for scanning phase behavior is the "step-wise dilution method" [18]. This involves a series of isoplethal phase studies, spanning the entire composition range, that are performed using the same sample. Composition is varied by diluting the pure compound with measured aliquots of water in a step-wise manner, and determining at each step the phase behavior of that mixture. The aliquots of water added should be progressively increased in size so as to produce evenly spaced data. When about 5 g is reached mixing is poor in a test tube, and a measured aliquot may be transferred to a new test tube and the process continued. The sample requirement is small (1–2 g), the time required only a few hours (with some practice), and the boundaries of liquid phases can be defined to within about $\pm 1\%$ and $\pm 1°C$ within the temperature span of about -10 to $90°C$. The entire composition range is systematically scanned and observed, and subjective information regarding the viscosity and appearance of many mixtures is efficiently acquired.

To perform such a study the sample is weighed into a 15-mm diameter test tube, an inverted septum that fits over the test tube (and has a hole in it for the thermometer) is wired in place, and a thermometer having a long uncalibrated immersion section (such as is used by organic chemists to measure the temperature inside reaction flasks) is inserted. The thermometer serves both to mix the sample and to measure the temperature. With the septum in place, the loss of water is minimal even at steam temperature (*c.* 10 mg in an hour). The method is expedited if a variety of heating and cooling baths are provided including steam, warm and cool water, an ice slush, and Dry Ice. Aside from the thermometer and test tube, the only apparatus required is a polarized light box (Fig. A4.1) [18].

The procedure involves first heating the sample until the melting point is reached (if below $100°C$), and then determining its temperature. This is accomplished by oscillating the temperature (using the baths) about the melting point while observing the sample between crossed polars. For some purposes the sample is heated or cooled, allowed to drift towards room temperature, and the temperature of the transition noted. In other cases the test tube temperature is varied by brief immersion in a hot or cold bath. The temperature of the transition is ascertained during each oscillation, and the procedure repeated until a reproducible value is obtained. From two to ten cycles (2–15 minutes) is typically required for the data to converge to a reproducible value.

After the melting point of the dry sample has been determined an aliquot of water sufficient to lower the composition by a few per cent is added using a syringe, and the process is repeated. More than one boundary temperature may exist at the same composition, so that after each addition the mixture is first rapidly scanned between 0 and $100°C$ to determine the number of accessible boundaries. (It is advisable not to waste time attempting to determine boundaries for which the method is not suited, such as most liquid crystal/liquid crystal transitions.) Liquid/liquid miscibility gaps are most easily determined using observation in ordinary light, crystal or liquid crystal separation from a liquid phase is best observed using polarized light, and cubic phase separation from a liquid phase is most readily detected by listening to the

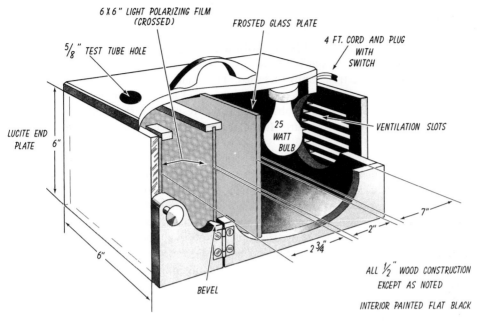

Fig. A4.1.   A construction drawing of a polarized light box suitable for the viewing of test tube samples in polarized light. Reproduced with permission from reference [18].

sample. The separation of a cubic phase from a liquid occurs initially at the surface of the glass, and produces a sharp change in the sound of the thermometer striking the test tube during stirring. *Crystal separation boundaries are determined only while heating the sample.* Establishing the reversibility of a transition by observing the same boundary temperature along both heating and cooling paths is often possible.

Possible artifacts to be aware of include "reflection birefringence" (the existence of spurious brightness that results from the reflection of light at an interface (for example, at the thermometer stem), and critical opalescence near liquid/liquid miscibility gaps. The latter is most pronounced near the critical composition and within 5–10°C of the critical temperature. Critical opalescence is very uniform and, with practice, is readily distinguishable from the more intense, more variable, and characteristic turbidity that results from the separation of a second liquid phase as an emulsion.

The broad features of the phase diagram of the $C_{10}E_3$–water system (Fig. 5.22) were determined in this manner within 4 hours' time, but clarifying the details required a DIT study. The method is also useful as a means of providing phase data that are of considerable value in designing separations based on recrystallization (Figs 5.2 and 5.3).

The step-wise dilution method is limited to defining the boundary of liquid phase regions, although rough data on the location of inner boundaries involving the transformation of a crystal to a liquid crystal are sometimes obtained. Phases are identified after the determination is completed by preparing mixtures centered in the various phase regions, and examining them using microscopic texture or powder X-ray data.

Microscope hot-stage methods. A more sophisticated version of the step-wise dilution method involves use of a polarized light microscope and a hot-stage for the microscope.

In this method a mixture of known composition is prepared, and a representative sample is placed on a microscope slide. The sample is covered with a cover slip, and the cover slip sealed to the slide with a cement. Such a sample may be heated (or cooled, if a cooling stage is available), observed using the microscope, and the temperatures of phase transitions recorded. The small mass of the sample allows the temperature to be rapidly changed, and this method also has the advantage that the optical texture is continuously observed. As a result, information is immediately available regarding phase structure. This approach has been effectively used by several groups [19,20,21].

Microscope studies are laborious; they require long periods of observation through the microscope. Ideally, the temperature measurements should be calibrated against standards, since radiative or convective heat losses from the exposed sample region produce a temperature gradient between the sample and the sensor. (This is also true to a small degree in the Mettler hot stage.) One difficulty is distinguishing between cubic and liquid phases, both of which are isotropic. At room temperature this distinction can be made by shearing the sample (which provides information as to the viscosity of the phase), but the capacity to do this is limited to pressing on the cover slip when the sample is sealed for high-temperature studies.

Microscopic studies are widely used to distinguish isotropic from birefringent phases, but there are times when it is actually more reliable to ascertain whether or not a sample is birefringent in a test tube than in a microscope. This is because the intensity of birefringence is dependent upon sample thickness. The lamellar phase of $C_{10}E_3$, for example, orients between slide and cover slip so easily that no birefringence was observed in this phase using a microscope. The $C_{10}E_3$ phase is grossly birefringent in a test tube, however, because of the thicker sample and more random domain orientation. The lamellar phase of $C_{10}E_4$ also assumes the pseudobirefringent (homeotropic or planar) texture very rapidly and essentially completely within a DIT cell.

Calorimetric methods. Differential scanning calorimetry is widely used for the acquisition of phase information, and numerous attempts have been made to infer phase diagrams using calorimetric data. Calorimetry has been used extensively during physical studies of the polar lipids [22]. Calorimetry provides invaluable information regarding the heat flow that occurs during phase reactions, and thus numerical values for the thermodynamic variables. It is also particularly attractive for studying optically opaque samples, for it readily "sees" the heat effects that occur. Unfortunately, the operator does not also see the sample during calorimetric studies.

During calorimetry a sample of a pure compound, or of a mixture of known composition (1–100 mg, depending on the sensitivity of the calorimeter and the magnitude of the heat effects to be observed), is sealed either in a metal pan (using a metal cover and crimping tool) or a small glass tube. The temperature is programmed up or down, and the difference in heat capacities between the experimental sample and a reference sample (that does not have phase discontinuities within the temperature range of interest) is determined as a function of temperature.

Mixing occurs principally by diffusion after the sample is sealed, so that calorimetrists often cycle the temperature so as to mix the mixture. This practice is also followed in order to obtain "reproducible" thermograms [23]. While cycling will certainly facilitate mixing (if the temperature of a fluid liquid phase is reached), it can be disastrous insofar as attaining equilibrium of state is concerned. This is because the high-temperature phases at phase transitions are also the high-entropy phases; the rate of phase equilibration is in general fast during the heating cycle – but comparatively slow during the cooling cycle. Reheating a cooled sample (that may have partially reached equilibrium) will therefore rapidly destroy any low-entropy phases (such as crystals) that may have started to form. Equilibrium of state exists at the end of the heating cycle, but *not* at the end of the cooling cycle. The attainment of equilibrium from the disordered state during cooling must therefore start anew during each cooling cycle.

If determination of the Krafft boundary is to be attempted, for example, then either a new sample must be used in each scan or conditions such that equilibrium is attained must be maintained after cooling to below the boundary temperature [24]. This is possible, but requires prior knowledge of both the qualitative features of the equilibrium diagram and the rate at which equilibrium is attained.

Changing the temperature of a mixture often dictates that not only heat transport, but also mass transport, must occur if equilibrium of state is to be maintained (Section 5.6.11). Heat flow is relatively fast and predictable, but mass transport is extremely variable and altogether unpredictable (particularly in the absence of the phase diagram). Diffusion coefficients vary from roughly $10^{-4}\,m^2\,s^{-1}$ in the gas phase, to $10^{-10}\,m^2\,s^{-1}$ in the liquid, to $10^{-16}\,m^2\,s^{-1}$ in molecular solids, to perhaps $10^{-22}\,m^2\,s^{-1}$ in covalently cross-linked polymeric solids. Whether or not sufficient time is allowed for mass transport equilibration to occur is determined principally by the scanning rate, and during most calorimetric studies the scanning rates used are far too fast for transport equilibrium to occur. During a recent careful study of the SDS–water system (which is a dynamically fast system in comparison to longer chain-length surfactants [25]), it was established that a scanning rate of $0.2°C\,min^{-1}$ was too fast for mixtures to maintain equilibrium. Qualitative distortion of the form of the thermogram occurred at this scanning rate, as reflected by the absence of heat effects (peaks) that were observed at slower scanning rates. For this reason, this entire study was performed at a scanning rate of $0.08°C\,min^{-1}$.

Some particularly well-executed calorimetric phase studies of surfactant systems are listed below. Reference to the original literature will provide details of the methods used during these studies:

> The soap–water studies [26]
> The study of the DPPC–water system [27].
> The study of the SDS–water system [24].
> The study of the $C_{10}PO$–water system [28,29].

The study of the phosphine oxide–water system is of special interest because recent advances in the treatment of non-isothermal (zeotropic) phase transitions [29] were incorporated into the data analysis. The theoretical form of thermograms for a variety of situations which takes into account the dynamics of heat flow has been suggested by Kessis [30] (and Wunderlich [44]), and appears to have been incorporated into the analysis of thermograms by Madelmont [27]. A particularly important aspect of these theories is that a phase transition which is strictly isothermal (that is, spans a temperature range of precisely zero) inevitably yields a thermogram that spans a finite temperature range.

If a catalog were to be made of the true origin of the heat effects observed during calorimetric studies of surfactant–water mixtures, it would probably be found that less than half are due to equilibrium phase transitions (Section 4.14). The balance will be determined by kinetic factors (either intrinsic or mass transport-related), represent artifacts related to the complex mechanisms by which many phase reactions occur (Section 6.5), etc. Since calorimetric studies are rarely carried out in a manner such that equilibrium of state is approached, characterizing the observed heat effects as the "enthalpy" of a reversibly executed phase reaction is perhaps rarely justified. For all these reasons, exclusive reliance on calorimetry as the basis for determining phase diagrams is a highly questionable practice.

On the positive side, calorimetric studies are invaluable when used in combination with other methods. They often provide a rigorous test of hypotheses as to phase behavior suggested by other methods (and vice versa), disclose the existence of phase transitions that would otherwise go unrecognized (such as polymorphic transitions that are optically invisible), reveal inconsistencies in the composition scales of diagrams produced using other methods [31], expose unanticipated changes of state that may have occurred with the lapse of time [1], etc. Also, calorimetry is a principal method for obtaining numerical thermodynamic data.

Programmed temperature X-ray studies. In 1970 [32] a powder X-ray film camera was modified by inserting a long transverse slit in front of the film, the temperature of the sample was programmed to change, and the film was at the same time moved past the slit at a known rate. With calibration, the position at which the film was exposed could be correlated with the temperature of the sample. The scattering angle varies along the direction of the slit on the exposed film, while temperature varies along the other (Fig. 8.10). So long as the same phase structure exists (approximately) parallel lines result, and phase discontinuities are clearly evident as discontinuities in the pattern of lines. The temperature of the discontinuity may be inferred from such a plot, as well as the influence of temperature on the structural dimensions of the phase. Since the slit must have a finite width, there are really two overlapping temperature scales along this dimension of the film – one for the leading edge of the slit and one for the trailing edge. This must be taken into account in analyzing the temperature of the transition. Further analysis of the impact of the width on the determination of temperature is needed.

In 1970 this technique was applied to the study of triglycerides and aqueous monoglyceride systems, where polymorphism in the crystal state is widely encountered [33,34]. Commercial versions of such instruments now exist which also permit rotation of the sample capillary during the study.

Programmed temperature X-ray studies constitute a powerful tool for surfactant phase studies – especially of unary systems. Since unary surfactant systems are often poorly defined, the widespread use of this method could, practically by itself, rectify this problem. Complications can be imagined to exist during studies of mixtures should spatial gradients in composition develop or the evaporation of water occur, but it should be useful for phase compounds as well as for pure components.

Dilatometry (volume measurements). During the early period of surfactant phase studies dilatometric measurements provided useful information, but are rarely used today. Dilatometry provides data on the change in volume of a mixture that occurs as temperature is varied, and thus information regarding the temperature (and volume dimensions) of phase discontinuities. Typically, the sample is contained within a bulb to which is attached a long capillary tube of uniform dimensions. The space within the bulb contains the sample, plus sufficient mercury to fill the voids within the sample and extend into the capillary. The volume changes that occur as temperature is varied may be quantitatively determined from measurements of the position of the mercury column.

Dilatometry is laborious, time-consuming, and requires careful technique. It is still used as a means of characterizing the liquid/solid ratios in solid shortenings (triglyceride mixtures), although it is being displaced for this use by NMR methods [35].

The water vaporization problem during isoplethal studies. A largely neglected artifact that may exist during isoplethal studies at high temperature is the vaporization of water from condensed phases. While not important at low temperatures and in dilute mixtures, the error in composition introduced at high temperatures becomes serious in mixtures whose composition is >80%. An estimate of the magnitude of the potential problem can be made using data on the vapor pressure of water at the temperature of interest, the volume of the void space that exists, the amount of sample present, and Raoult's Law to estimate the activity of water (Section 7.4). The greater the fraction of void space, the greater is the problem.

Just as important as the total quantity of water that is vaporized is the fact that most of it vaporizes from a thin layer of material at its surface. In preparing a demonstration sample of a cubic phase of a zwitterionic APS surfactant, a sample of about 2 g was heated in a sealed test tube having about 12 ml of void space for long periods of time in a steam bath to provide time for equilibration. After this treatment a thin layer of strongly birefringent (hexagonal) phase existed at the top. This layer persisted for about 6 months at room temperature before it finally vanished.

A4.4.4   Isothermal swelling methods

The penetration experiment. In all the methods for determining phase diagrams described above, it is essential to attain equilibrium of state during the determination. For about 60 years an obviously nonequilibrium experimental technique has been known, however, that quickly and easily provides useful qualitative phase information. This experiment has been variously called the "penetration", or "flooding", experiment [36].

The penetration experiment involves the creation of an interface between a dry sample and water (or other solvent), usually between a microscope slide and cover slip. If the components are partially miscible countercurrent diffusion of the solvent into the solute (and vice versa) produces, within minutes, a series of phase bands along the perimeter of the sample. The pure sample is still found near the center, the pure solvent at the outside, and all the intermediate one-phase compositions that exist are produced within the intervening space. The diffusive transport of each component into the other produces a roughly uniform gradient in composition, but this gradient is interrupted by the miscibility gaps that exist. A discontinuity in compositions exists at these miscibility gaps, and these (usually) result in a discontinuity in refractive index as well.* The only positions at which equilibrium of state exist during this experiment are at the interfaces.

---

*Isooptic conditions such that two coexisting phases have identical refractive indices may exist, in which case the interface is invisible. These conditions are perhaps more common than one might imagine. They have been encountered, occasionally, during DIT studies of the miscibility gap between a concentrated optically negative anisotropic phase (such as the lamellar phase) and a dilute isotropic phase of similar composition (liquid or cubic) as the temperature was varied.

The entire composition range is investigated in every penetration experiment, and often the phase structures of the bands may be inferred from their microscopic textures. Since composition usually varies in a monotonic fashion along radial directions during this experiment, it provides direct information regarding the number of phases that exist and their order of occurrence as a function of composition. It may, or may not, yield information regarding crystal phases.

Routine use of the penetration technique is of considerable value during phase studies, because it is extremely easy to do and quickly provides useful information at ambient temperature. Its use has been extended to higher temperatures by sealing the cover slip to the slide (as during isoplethal microscopic studies) [36].

The penetration experiment may also be executed within a rectangular capillary (such as the 0.1 × 1.0 mm × 5 cm capillaries used for melting point determinations, Section A4.3). Such studies require either that the sample be liquid or melt reversibly, but for such samples penetration experiments in capillaries are usually more clearly defined than are similar studies between slide and cover slip. The reason is that the mass transport which occurs is confined primarily to one dimension, rather than to two. Hydrodynamic forces often distort the bands that exist, especially in systems that form liquid crystal phases of low viscosity.

The qualitative data obtained during penetration studies reflect the same processes that actually occur when a particle is immersed in water. In this case the swelling occurs in three dimensions, rather than two as between slide and cover slip or one as in the capillary experiment. Two- and three-dimensional swelling are more complex hydrodynamically than is one-dimensional swelling, but the phase sequence that ensues must be identical during all three experiments. Restricting swelling to one dimension is highly desirable if one is to utilize swelling methods to gain information regarding phase behavior.

Refined qualitative swelling methods. Important recent developments have occurred in use of the swelling principle to create the phases that exist for further investigations of their structure. An experiment analogous to the capillary penetration experiment described above, termed the "lyotrope gradient" experiment, has been executed within an X-ray capillary. The phases produced were identified and characterized structurally by use of synchrotron X-ray studies. Special apparatus and methods are required to execute this experiment, which must be performed by remote control [37].

The same principle was used during the recent study of the SDS–deuterium oxide system [38]. A demountable cell with mica windows was developed that permitted the loading of a concentrated slurry of SDS crystals. An interface was created between the sample and water, the temperature was adjusted to the desired value, and (after the cell had developed for a period of time) the various phase bands were characterized using synchrotron X-ray studies. With certain knowledge of the qualitative phase behavior in hand, boundary compositions and thermodynamic data were then determined using calorimetry [24]. Additional structural information was provided by NMR and neutron scattering studies [38,39].

It would appear possible, in principle, to extend these elegant synchrotron experiments to more readily available instrumentation (such as laboratory X-ray cameras), but this has not been reported to date. The focus in the above swelling studies (described above) has been on the determination of phase structure. They do not provide information regarding phase compositions, and so do not produce phase diagrams.

Quantitative swelling methods. Isothermal phase reactions are inherently simpler than are isoplethal reactions (Section 4.16) [40], and if the compositions of phases produced by isothermal swelling could be directly ascertained *in situ*, then phase diagrams could (in principle) be constructed directly from a series of swelling studies performed over a range of temperatures.

The Diffusive Interfacial Transport (DIT) method was the first such quantitative phase studies method reported. This experiment is, in essence, a refined penetration experiment. The DIT cell is a long, thin, rectangular cross-section amorphous silica capillary whose sample chamber dimensions are 55 mm × 25 × 450 μm. The shortest dimension is accurately known via direct calibration. The cell is partly filled with surfactant, the balance of the chamber volume is filled with water so as to create an interface between the components, the cell temperature is adjusted to the desired value, and it is then allowed to develop. Well-defined phase bands, separated by transverse interfaces, spontaneously appear (Fig. A4.2). Gradients in composition are at first steep, but with the passage of time (from a few hours to days) diminish and become smooth. In the first version of this method (the DIT–NDX method) quantitative transmission interferometry was utilized to determine compositions via

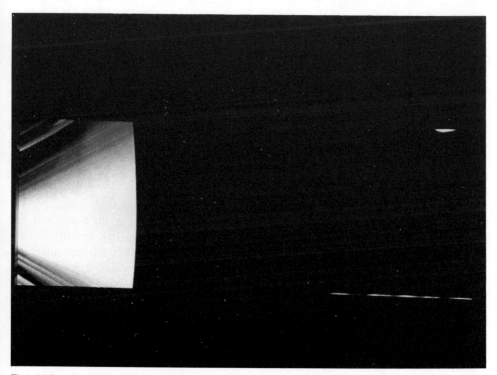

Fig. A4.2. A view in polarized light of the hexagonal–cubic–lamellar phase bands from a DIT study of the $C_{10}E_4G$–water system. The strongly birefringent hexagonal phase and its interface with the cubic phase is readily visible. The lamellar phase is highly oriented into the planar (homeotropic) texture in this mixture, so that it appears isotropic within the center of the chamber. However, the brightness along the edges is characteristic of this texture and serves to identify the phase. The interface between the cubic and the lamellar phase is visible in the actual cell, although invisible in the photograph.

measurements of refractive indices. Although very accurate refractive index data were obtained, the analytical accuracy of these data was unsatisfactory. The acyl methylglucamine diagram (Fig. 10.21) was determined using the DIT–NDX method, but the composition data were adjusted (using as a guide data from calorimetric studies) to produce the diagram shown. The problem is the inherent dependence of the refractive indices of birefringent phases on direction (Section 8.2), coupled with the failure of most such phases to undergo orientation reproducibly within the DIT cell.

A modified version of the DIT method is under development which should provide superior analytical data, and holds promise as a general method for the efficient and reliable determination of phase diagrams. In the DIT–IR method the phases are formed using the same principle and in the same cell as in the original DIT method, but quantitative near-infrared microspectroscopy is used (in place of interferometry) as the basis for analysis. This is not presently a proven method, however.

A4.5   A comparative evaluation of phase study methods

Unary systems. It is always desirable to utilize direct observation of the sample in polarized light, as temperature is varied, as one of the early experiments during a phase study. By use of this method one may both recognize and obtain data on the grossly observable transitions, such as melting points. This method also provides information that is very useful in designing calorimetric or X-ray experiments.

Calorimetric experiments provide temperature data regarding optically detected transitions, may detect as well other transitions that are optically invisible, and provide useful heat effect data. A programmed temperature X-ray scan provides independent confirmation of the results of the direct observation and calorimetric studies, plus definitive information as to phase structure. If programmed temperature facilities are not available, isothermal studies at particular temperatures suggested by the optical and calorimetric studies are useful. If data from these three methods are in full agreement, the unary phase diagram has been rigorously established.

The X-ray scan comes close to being a stand-alone method, but without prior knowledge of the approximate temperatures of transitions is a poor method for conducting exploratory studies. Calorimetry alone indicates that transitions are occurring and (with proper analysis) their temperatures, but does not provide hard information as to phase structure. (The magnitude of the heat effect will distinguish among certain classes of transitions, however. The heat effects observed during the destruction of crystals, for example, are much larger than are those which accompany most transitions involving liquid crystals.) Direct observation alone is unreliable, because some transitions (especially those within the solid state) are likely to be missed.

Binary systems. In studying binary systems it is also highly desirable to first scan the system by fast and simple methods. A combination of penetration experiments and isoplethal studies is attractive. Penetration experiments provide useful qualitative information as to phase behavior near ambient temperature with minimal effort, but are more difficult at elevated temperatures.

Studies that fall outside the temperature range of 0–100°C require the use of mixtures sealed in tubes and long equilibration times. Below 100°C the step-wise dilution method

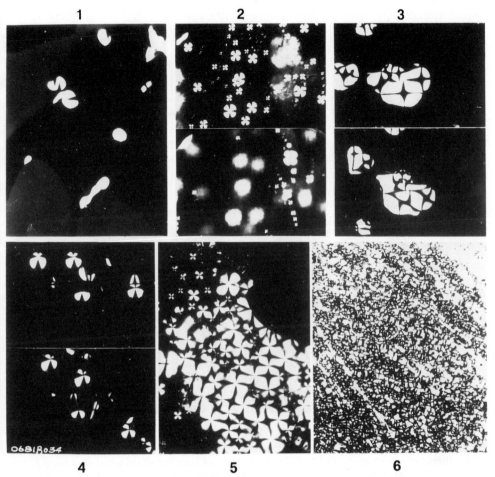

Fig. A4.3.   Common textures of the lamellar (neat) liquid crystal phase. 1. Planar lamellar phase with bright droplets of hexagonal present. 170×. 2. Positive units of 70% potassium oleate lamellar phase. 330×. 3. Negative units of lamellar phase in isotropic liquid matrix. 330×. 4. Fan-like units of lamellar in liquid matrix. 330×. 5. Coarse mosaic texture in a complex network of positive and negative units. 170×. 6. Typical mosaic texture; same geometry as in 5, but finer and less regular (70% potassium oleate). 170×.

is an excellent choice for preliminary investigations because of the small sample and time requirements. The isoplethal microscopic method is a more refined alternative. All three methods provide data on the boundaries of liquid regions and expedite the design of further experiments.

Direct observation does not provide definitive information as to phase structure, so that the further investigation of mixtures at compositions and temperatures near the center of the phase regions that have been observed are required in order to determine phase structure. Often texture in polarized light is sufficient for this purpose, if the texture is easily recognizable. The textures observed by Rosevear [6] are presented in Figs A4.3 to A4.6. Several textures may be observed within the same phase structure. It has been found that these textures are also encountered during DIT studies, but that

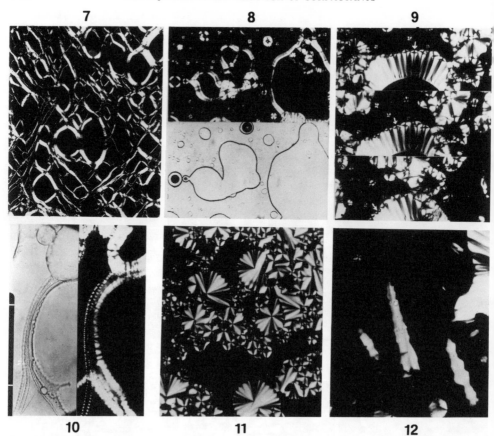

Fig. A4.4.   Common textures of the lamellar (neat) liquid crystal phase. 80×. 7. Oily streaks in a planar (homeotropic) matrix (330×). 8. Large droplets of lamellar phase between crossed polars (top) and in bright-field (below) (330×). Note birefringent borders and planar centers. Circular air bubbles with black centers exist (lower picture). 9. Focal conic detail in birefringent border next to an air bubble. Arrow indicates progression of negative unit from cross to pinwheel to figure 8 as stage is rotated counterclockwise (330×). 10. Terraced drop of lamellar phase in lye (concentrated electrolyte solution) between crossed polars and in bright-field (170×). 11. Fan-like texture of lamellar droplets in liquid phase (330×). 12. Batonnets of lamellar phase in isotropic phase (330×).

others may also be observed since texture depends upon both composition and the system variables. The dry lamellar phase of DOACS, for example, is entirely unrecognizable from these textures; nevertheless, X-ray studies established beyond doubt that this phase was the lamellar liquid crystal.

Phase structure is most reliably determined using powder X-ray studies, ideally in combination with NMR. Liquids typically display a diffuse short spacing at large diffraction angles, crystals display many lines at both long and short spacings that span a wide range of diffraction angles, and liquid crystals display a few sharp lines that are often associated primarily with the long spacings of the structure. A relatively broad liquid-like short spacing that is determined by chain-chain packing is also typically observed [41].

13   **14**   **15**   **16**

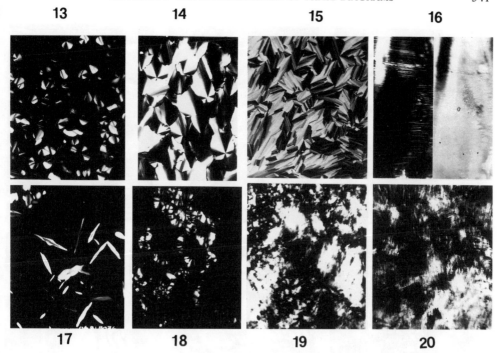

17   **18**   **19**   **20**

Fig. A4.5. Common textures of the hexagonal liquid crystal phase. 13. Fanlike units formed by evaporation of water from a 20% potassium laurate solution (140×). 14. Fanlike texture (70×). 15. Angular texture (70×). Note absence of pinwheels. 16. Zone of uniform extinction (35×). Left – with polars crossed; right – uncrossed 45°. 17. Batonnets of hexagonal in liquid phase (70×). 18. "Oily streaks" of hexagonal in liquid phase (280×). 19. Non-geometric, non-striated hexagonal phase (30% sodium palmitate, 140×). 20. Striations in non-geometric texture (70×).

Isothermal studies. In recent times the examination using NMR of mixtures in which D_2O replaces H_2O as the solvent, that have been equilibrated at a particular temperature, has been widely used during phase studies. The collection and analysis of the spectral data upon which phase identification is based are described elsewhere [13,14]. The combination of powder X-ray data with NMR data constitutes a powerful means of both determining and identifying phase structure. One may (on rare occasion) find that phases separate cleanly under the influence of gravity, and then classical isothermal analytic methods are appropriate. Application of strong gravitational fields to accelerate separation is, however, dangerous (Section 4.17.5).

Ideally, the preliminary studies of mixtures using isoplethal methods would be followed by isothermal swelling studies which provide quantitative data regarding phase boundary compositions. Since such a method does not presently exist, that approach is reserved for the future.

A4.6 Evaluating the quality of phase diagrams

Since 1875, the Phase Rule has provided a solid thermodynamic basis for evaluating the quality of a phase diagram. Phase diagrams that do not conform to the Phase Rule may

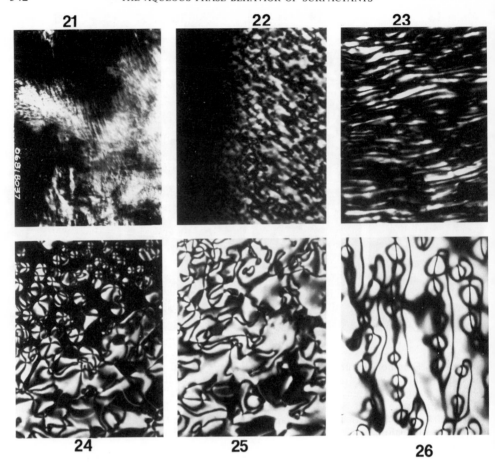

Fig. A4.6. Common textures of the nematic liquid crystal phase. 21. Simple nematic texture. 22. Stippled nematic phase in contact with liquid phase. 23. Striated nematic texture. 24. Sinuous nematic texture in mixture of nematic and liquid phases. Note distorted extinction crosses. 25. Sinuous nematic texture. 26. Chains of extinction crosses (alternating positive and negative) in nematic phase. Lamellar phase in liquid (330×).

be assumed *a priori* to in error, but phase diagrams that do conform may be either correct or incorrect. The fact that a published diagram conforms to the Phase Rule signifies that the suggested behavior is consistent with the basic principles of phase science, but does not guarantee that the diagram is correct. The truth of this is particularly evident to those who determine phase diagrams, which typically undergo evolutionary refinement from one model to another (all of which should conform to the Phase Rule) during the process.

The alternation rule, which is derived from and is based upon the Phase Rule, is of particular value in evaluating diagrams. As noted in Section 4.12 the alternation rule states that *an alternation between odd and even numbers of phases occurs* (*in increments of one and in a logical fashion*) *along isoplethal and isothermal trajectories* in the diagram. It is important to note that not only must the alternation rule apply in a purely numerical sense, but that the sequence of phases that are inserted or deleted along process paths must be sensible. As noted in Sections 5.6.10 and 4.10, changes that occur in the number of phases present must result either from the insertion (creation) of a new phase, or from the deletion (destruction) of an existing phase. The substitution of one phase for another does not occur.

In evaluating a binary diagram, one of the first things to look for is whether or not one-phase regions are adjacent to one another without being separated by a miscibility gap. Unless the phase transition is second order (and no aqueous surfactant transitions have, to date, been found to be second order), this is a clear violation of the alternation rule.

It has not been possible, historically, to determine quantitatively both boundary compositions at small miscibility gaps, but it has often been possible to document the fact that they exist. Therefore, it is advisable in constructing a phase diagram to indicate the existence of the boundaries of liquid crystal phases by sketching in dashed lines at reasonable positions. Then, someone who is not expert in this field will at least be alerted as to their existence.

Another feature of surfactant diagrams that is often ignored is the nature of the phase reactions that occur at the upper or lower temperature limits of a phase. Sometimes the liquid crystal phase boundary is shown to be separated at its maximum temperature from the boundary of the coexisting liquid phase. This is not a violation of the alternation rule, but it *is* a violation of the Phase Rule. It would suggest that the liquid crystal may decompose to two liquid phases without passing through an intervening state in which the liquid crystal coexists with another phase. Such boundaries must either touch at an azeotropic point (which *is* a two-phase state), the liquid crystal must decompose at a peritectic discontinuity, or some other acceptable phase behavior must exist. It is useful to provide a reasonable hypothesis, if not a proven result, to describe the phase behavior at these discontinuities. Such a hypothesis may then, at some later time, be subjected to experimental evaluation.

Very often the crystal phase region of surfactant phase diagrams is not clearly defined simply because it is not known. It is of value to determine the composition, if not the structure, of at least those crystal phases that coexist with fluid phases. If more than one crystal phase exists, then (ideally) that phase should be indicated as well. If polymorphism results in the existence of a polytectic discontinuity, then which end of the polytectic corresponds to the two-phase state should be designated in some way. Otherwise, this discontinuity is ambiguous.

Metastable phase boundaries are often designated in phase diagrams; an example is the crystal heptahydrate of sodium sulfate and the supercooled hexagonal liquid crystal boundary in N-dodecanoyl-N-methylglucamine. These features too must conform to the alternation rule, but their special status should be indicated (as by dashed lines).

Sometimes obviously nonequilibrium phenomena (such as regions of stable emulsions, vesicle dispersions, etc.), are indicated on phase diagrams. While it is useful to superimpose such information on to a phase diagram, it is preferable not to present it in such a way as to suggest that these features constitute equilibrium phenomena.

Violations of the condition of equilibrium may be encountered in phase diagrams that do not constitute a violation of the alternation rule. An example is a non-horizontal tie-line in a binary diagram, which would suggest that two phases in equilibrium have different temperatures.

Ternary diagrams. The alternation rule applies to ternary diagrams in much the same way as it does to binary – providing one takes into account the larger number of composition variables that must be stated to define ternary mixtures (Section 12.1.1.1). The coexistence of two phases in binary systems is signified by a (one-dimensional) line, while such regions in ternary diagrams span a (two-dimensional) area within the isotherm. The boundaries of two-phase regions must either touch the boundaries of one-phase regions (including "regions" of fixed composition), or touch three-phase tie-triangles. Each tie-triangle touches a one-phase region at each vertex (which defines the three phases that coexist within the triangle), and two-phase regions along its sides. Along linear process paths within an isotherm there is, as a result, an alternation of an even and an odd number of phases, in a logical sequence, throughout. The same must be true along isopleths, although to visualize them one must have a three-dimensional view of the ternary prism. If pressure diagrams were available, the same rules would have to be followed as this field variable is changed.

Data points. Adherance to Schreinemaker's Rule is also required (Section 12.2.2) [45]. It is preferable to depict on the diagram the actual data upon which the diagram is based, as well as smooth curves drawn through these points [38]. This combination provides the reader, at a glance, a subjective feeling for the degree of uncertainty in the boundary positions. For a serious study a table of numerical data is advisable [2]. Many diagrams are published in which neither the data points nor the data are reported, unfortunately, and considerable improvement is possible in this aspect of phase science.

References

1. Laughlin, R. G., Munyon, R. L., Fu, Y.-C. and Fehl, A. J. (1990). *J. Phys. Chem.* **94**, 2546–2552.
2. Lang, J. C. and Morgan, R. D. (1980). *J. Chem. Phys.* **73**, 5849–5861.
3. Skau, E. L., Arthur, J. C. Jr and Wakeham, H. (1959). *Technique of Organic Chemistry. Physical Methods of Organic Chemistry*, 3rd ed., Vol. I, pp. 312–338, Interscience, New York.
4. Lawrence, A. S. C. (1969). *Liquid Crystals 2*, Vol. 1, Gordon & Breach, London.
5. Randerath, K. (1963). *Thin-Layer Chromatography*, (translated by D. D. Libman), Academic Press, New York.
6. Rosevear, F. B. (1954). *J. Am. Oil Chemists Soc.* **31**, 628–638.
7. Hill, A. E. (1923). *J. Am. Chem. Soc.* **45**, 1143–1155.
8. Laughlin, R. G. and Munyon, R. L. (1987). *J. Phys. Chem.* **91**, 3299–3305.
9. Adamson, A. W. (1982). *Physical Chemistry of Surfaces*, 4th ed., pp. 584–591, Interscience, John Wiley, New York.
10. Schreinemakers, F. A. H. (1893). *Z. physik. Chem.* **11**, 75–109.
11. Laughlin, R. G. (1990). *Food Emulsion And Foams: Theory and Practice*, Vol. 86 (D. L. Gaden, series ed., P. J. Wan, volume ed.), pp. 7–15, American Institute of Chemical Engineers, New York.
12. Smith, D. H. and Lane, B. J. (1987). *J. Disp. Sci. Tech.* **8**, 217–247.
13. Ulmius, J., Wennerstrom, H., Lindblom, G. and Arvidson, G. (1977). *Biochemistry* **16**, 5742–5745; Stilbs, P. (1987). *Prog. Nucl. Mag. Reson. Spectrosc.* **19**, 1–45.
14. Kahn, A., Fontell, K. and Lindman, B. (1984). *Coll. Surfaces* **11**, 401–408; Kahn, A., Fontell, K., Lindblom, G. and Lindman, B. (1982). *J. Phys. Chem.* **86**, 4266–4271; Kahn, A., Fontell, K. and Lindman, B. (1985). *Prog. Coll. Poly. Sci.* **70**, 30–33; Kahn, A., Lindman, B. and Shinoda, K. (1989). *J. Coll. Interface Sci.* **128**, 396–406; Kahn, A., Das, K. P., Eberson, L. and Lindman, B. (1988). *J. Coll. Interface Sci.* **125**, 129–138;

Kahn, A., Jonsson, B. and Wennerstrom, H. (1985). *J. Phys. Chem.* **89**, 5180–5184; Arvidson, G., Brentel, I., Kahn, A., Lindblom, G. and Fontell, K. (1985). *Eur. J. Biochem.* **152**, 753–759; Kahn, A., Fontell, K., Lindblom, G. and Lindman, B. (1982). *J. Phys. Chem.* **86**, 4266–4271; Kahn, A., Lindman, B. and Shinoda, K. (1989). *J. Coll. Interface Sci.* **128**, 396–406; Kahn, A., Fontell, K. and Lindman, B. (1985). *Prog. Coll. Poly. Sci.* **70**, 30–33; Kahn, A., Fontell, K. and Lindman, B. (1984). *J. Coll. Interface Sci.* **101**, 193–200; Jokela, P, Jonsson, B. and Kahn, A. (1987). *J. Phys. Chem.* **91**, 3291–3298.

15. Lindblom, G. and Rilfors, L. (1989). *Biochim. Biophys. Acta* **988**, 221–256.
16. Alexejew, W. (1882). *J. Prakt. Chem.* **25**, 518–520.
17. Laughlin, R. G., unreported work.
18. Laughlin, R. G. (1976). *J. Coll. Interface Sci.* **55**, 239–241.
19. Balmbra, R. R. , Bucknall, D. A. B. and Clunie, J. S. (1970). *Mol. Cryst. Liq. Cryst.* **11**, 173–186; Clunie, J. S., Corkill, J. M. and Goodman, J. F. (1965). *Proc. Royal Soc.* **A285**, 520–533; Balmbra, R. R., Clunie, J. S. and Goodman, J. F. (1965). *Proc. Royal Soc.* **A285**, 534–541.
20. Mitchell, D. J., Tiddy, G. J. T., Waring, L., Bostock, T. A. and McDonald, M. P. (1983). *J. Chem. Soc., Faraday Trans. I*, 975–1000; Adam, C. D., Durrant, J. A., Lowry, M. R. and Tiddy, G. J. T. (1984). *J. Chem. Soc., Faraday Trans. I* **80**, 789–801.
21. Luehmann, B., Finkelmann, H. and Rehage, G. (1985). *Makromol. Chem.* **186**, 1059–1073; Luehmann, B. and Finkelmann, H. (1966). *Colloid Polym. Sci.* **264**, 189–192; Heusch, R. (1991). *Tenside Surf. Det.* **28**, 38–46.
22. NIST Standard Reference Database 34, *Lipid Thermotropic Phase Transition Database*, available from the National Institute of Standards and Technology, 221/A320, Gaithersburg, MD 20896. Phone 301/975–2208.
23. Hattori, Mineyuki, Fukuda, Shinichi, Nakamura, Daiyu, and Ikeda, Ryuichi (1990). *J. Chem. Soc., Faraday Trans.* **86**, 3777–3783.
24. Kekicheff, P., Grabielle-Madelmont, C. and Ollivon, M. (1989). *J. Coll. Int. Sci.* **131**, 112–132.
25. Gormally, J., Gettins, W. and Wyn-Jones, E. (1981). *Mol. Interact.* **2**, 143–147.
26. Madelmont, C. and Perron, R. (1973). *Bull. Soc. Chim. France* 3263–3268, 3259–3263; Madelmont, C. and Perron, R. (1974). *Bull. Soc. Chim. France* 1795–1798, 1799–1805, 3425–3429, 3430–3434; Madelmont, C. and Perron, R. (1976). *Colloid and Polymer Sci.* **254**, 6581–6595.
27. Grabielle-Madelmont, C. and Perron, R. (1983). *J. Coll. Interface Sci.* **95**, 471–482, 483–493.
28. Chernik, G. G. (1991). *J. Coll. Interface Sci.* **141**, 400–408; Chernik, G. G. and Sokolova, E. P. (1991). *J. Coll. Interface Sci.* **141**, 409–414.
29. Chernik, G. G. and Fillipov, V. K. (1991). *J. Coll. Interface Sci.* **141**, 415–424.
30. Kessis, J.-J. (1970). *C. R. Acad. Sci. Paris, Serie C* **270**, 1–4, 120–122, 265–267.
31. Fu, Y. C., Glardon, A. S. and Laughlin, R. G. (1993). Presented at the American Oil Chemists' Society Meeting, Anaheim, California, April 25–29.
32. Ruener, Ue. (1970). *Lebensm.-Wiss. in Tech.* **3**, 101–106.
33. Hernquist, L. and Larsson, K. (1982). *Fett-Seifen-Anstrichmittel* **84**, 349.
34. Larsson, K., Fontell, K. and Krog, N. (1980). *Chem. Phys. Lipids* **27**, 321–328; Krog, N. and Lauridsen, J. B. (1976). *Food Emulsions*, Vol. 5 (S. Friberg ed.), pp. 67–140, Marcel Dekker, New York; Larsson, K. and Krog, N. (1973). *Chem. Phys. Lipids* **10**, 177–180; Larsson, K., Fontell, K. and Krog, N. (1980). *Chem. Phys. Lipids* **27**, 321–328.
35. Madison, B. L. and Hill, R. C. (1978). *J. Am. Oil Chemists' Soc.* **55**, 328–331.
36. Laughlin, R. G. (1992). *Adv. Coll. Interface Sci.* **41**, 57–79.
37. Caffrey, M. (1987). *Biophys. J.* **51**, 444a.
38. Kekicheff, P. and Cabane, B. (1987). *J. Phys.* **48**, 1571–1583; Kekicheff, P. (1989). *J. Coll. Int. Sci.* **131**, 133–152.
39. Kekicheff, P. and Cabane, B. (1988). *Acta Cryst. B* **44**, 395–406.
40. Laughlin, R. G. (1990). *J. Am. Oil Chem. Soc.* **67**, 705–710.
41. Luzzati, V. (1968). *Biological Membranes, Physical Fact and Function* (D. Chapman and D. F. H. Wallach eds), pp. 71–124, Academic Press, New York.

42. (1992). *CRC Handbook of Chemistry and Physics*, 72nd ed., pp. 15–21. CRC Press, Boca Raton, FL.
43. Peshek, C. V., O'Neill, K. J. and Osborne, D. W. (1987). *J. Coll. Int. Sci.* **119**, 289–290.
44. Wunderlich, B. (1990). *Thermal Analysis*, pp. 158–180. Academic Press, Boston.
45. Francis, A. W. (1961). *Critical Solution Temperatures*, Vol. 31 (R. F. Gould, ed.), pp. 72–76, American Chemical Society, Washington, D.C.

Index

RN CH： ：RARY   ：：-3753

MAY 19 '95